Handbook of Cell Biology

Volume I

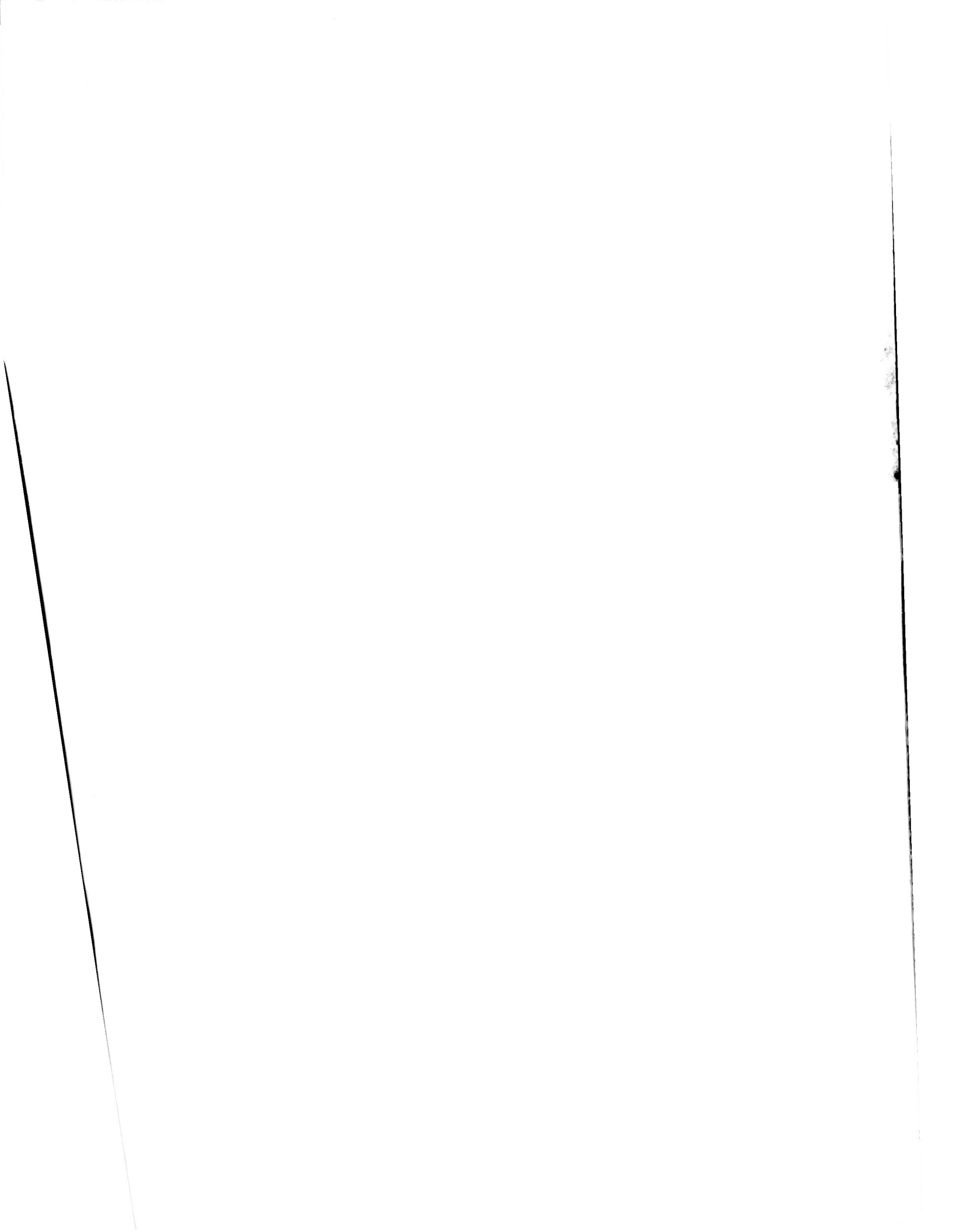

Handbook of Cell Biology
Volume I

Edited by **Samantha Granger**

R CALLISTO REFERENCE

New York

Published by Callisto Reference,
106 Park Avenue, Suite 200,
New York, NY 10016, USA
www.callistoreference.com

Handbook of Cell Biology: Volume I
Edited by Samantha Granger

International Standard Book Number: 978-1-63239-378-4 (Hardback)

Printed in the United States of America.

Contents

Preface

Cell is the fundamental unit of life, the building block of an organism. All the essential activities for living organisms take place in cells. The study of cells, their structures and functions is known as cell biology. To know about any organism, one should start with the cells which together make up tissues, which further come up together to form an organ, a group of organs form an organ system and many organ systems make up an organism. Cell biology focuses on the functions of the cell, the metabolic reactions taking place, and its structure- the components and their respective functions. It has now emerged as the science of identifying and treating diseases at the cellular level. Also, at the nuclear level of the cell, studies are being done, as it is considered the controlling unit of the cell.

The linker histone h1.2 is an intermediate in the apoptotic response to cytokine deprivation in t-effectors, linking peroxiredoxin and vacuolar-atpase functions in calorie restriction-mediated life span extension. Similarly, important subjects of enquiry in the field of cell biology are 2-cys peroxiredoxins: emerging hubs determining redox dependency of mammalian signaling networks, u1 snrnp-dependent suppression of polyadenylation: physiological role and therapeutic opportunities in cancer, quantifying changes in the cellular thiol-disulfide status during differentiation of b cells into antibody-secreting plasma cells, alternative splicing regulation of cancer-related pathways in caenorhabditis elegans: an in vivo model system with a powerful reverse genetics toolbox etc.

This book is the result of synchronised efforts of the scientists, academicians, writers and contributors. I wish to thank them and everyone else involved in the completion of this book.

Editor

The Plasma Membrane Potential and the Organization of the Actin Cytoskeleton of Epithelial Cells

Silvia Chifflet[1] and Julio A. Hernández[2]

[1] Departamento de Bioquímica, Facultad de Medicina, Universidad de la República, Gral. Flores 2125, 11800 Montevideo, Uruguay
[2] Sección Biofísica, Facultad de Ciencias, Universidad de la República, Iguá 4225 esq. Mataojo, 11400 Montevideo, Uruguay

Correspondence should be addressed to Silvia Chifflet, schiffle@mednet.org.uy

Academic Editor: Michael Hortsch

The establishment and maintenance of the polarized epithelial phenotype require a characteristic organization of the cytoskeletal components. There are many cellular effectors involved in the regulation of the cytoskeleton of epithelial cells. Recently, modifications in the plasma membrane potential (PMP) have been suggested to participate in the modulation of the cytoskeletal organization of epithelia. Here, we review evidence showing that changes in the PMP of diverse epithelial cells promote characteristic modifications in the cytoskeletal organization, with a focus on the actin cytoskeleton. The molecular paths mediating these effects may include voltage-sensitive integral membrane proteins and/or peripheral proteins sensitive to surface potentials. The voltage dependence of the cytoskeletal organization seems to have implications in several physiological processes, including epithelial wound healing and apoptosis.

1. Introduction

The transport of water and solutes across epithelial layers represents a major achievement of biological evolution and constitutes the basis for the existence of higher organisms [1]. To accomplish their transport properties, the epithelial cells acquire characteristic structural and functional features. An epithelial layer must constitute a well-defined macroscopic permeability barrier, which results in the selective transport of solutes and water across the overall tissue. For this, transport epithelia must develop a complex set of cell junctions and a polarized distribution of membrane molecules, which localize at distinct apical and basolateral domains of the plasma membrane [2–5].

The establishment and maintenance of the polarized epithelial phenotype require a characteristic organization of the cytoskeletal components. There are many cellular effectors involved in the regulation of the cytoskeleton of epithelial cells [6–11]. Recently, modifications in the plasma membrane potential (PMP) have been shown to participate in the modulation of the cytoskeletal organization. The purpose of this paper is to review evidence relating PMP modifications to changes in the cytoskeletal organization of epithelia, with

an emphasis on the actin cytoskeleton, and discuss possible molecular paths mediating these effects. Prior to this, we briefly review the basic characteristics of the cytoskeleton and the generation of plasma membrane potentials in epithelial cells.

2. General Morphological Aspects and Organization of the Cytoskeleton of Epithelial Cells

Mature epithelia are characterized by two fundamental morphological and functional features: a tight cellular packing, supported by the existence of strong adhesive forces between neighboring cells and a polarized cellular phenotype [5]. These two properties are interdependent, since the establishment of intercellular junctions represents the main positional cue triggering the development of cell polarization. All the anchoring junctions, either between cells or between cells and substrate, are associated to cytoskeletal components that are crucial for the junction stability. The characteristic organization of the cytoskeleton of epithelial cells greatly depends upon these interactions with the cell junctions, as depicted in

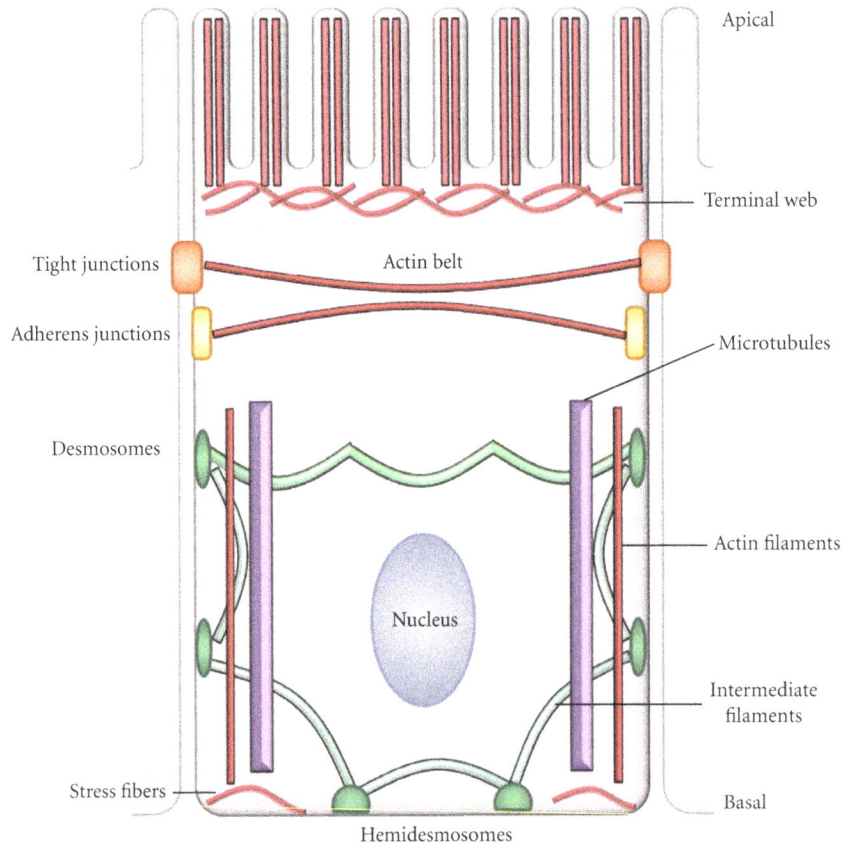

FIGURE 1: Schematic representation of the cytoskeletal organization of a transporting epithelial cell. The scheme shows the cell-cell and cell-substrate junctions connecting the actin, tubulin, and intermediate filament cytoskeletons. See text for details.

Figure 1. In particular, the actin cytoskeleton associates with diverse cellular structures in mature epithelial cells (Figure 1). In these cells, the most conspicuous actin structure is the circumferential actin belt, a bundle of actomyosin fibers located immediately beneath and associated with the tight and adherens junctions. Here, the microfilaments and the cell junctions interact via a complex set of multifunctional proteins (Figure 2(a)). Actin filaments can also be found along the lateral membrane [12–14], where it codistributes with myosin I [15–17]. Short actin filaments are part of the spectrin-based membrane skeleton network. Besides its classical role in membrane domain organization [18], this network is involved in the biogenesis of the lateral membrane of epithelial cells and in the maintenance of their columnar shape [19, 20]. Stress fibers at the basal domain are also actomyosin structures associated to focal contacts and other cell-substrate junctions [21]. At the apical domain, parallel bundles of crosslinked actin filaments, extending from the terminal web, constitute the core of the microvilli.

Polarized epithelial cells exhibit a characteristic pattern of microtubule organization (Figure 1) [6, 22, 23]. Unlike fibroblastic-like cells, where centrosomes are responsible for the nucleation and organization of microtubules, in epithelial cells, most microtubules are acentrosomal. They also present distinctive properties in their behavior, stability, dynamics and regulation [24–27]. Typically, in epithelia microtubules organize in long apicobasal-oriented parallel fibers

that span the whole length of the cell, with their minus end pointing towards the apical surface and in disordered short filament networks underneath the apical and basal membranes [28–31]. This particular arrangement of microtubules is crucial for the targeted traffic of vesicles that sustains the existence of distinctive apical and basolateral domains of epithelial cells [23, 32]. Figure 1 also schematically shows that the other major component of the cytoskeleton, the intermediate filaments, traverses epithelial cells connecting desmosomes and hemidesmosomes [33].

Intercellular anchoring junctions are key structures for epithelial organization and function. Among these, the tight junctions (zonula occludens) selectively seal the intercellular space at the apical side and prevent the exchange of membrane proteins and lipids between the apical and basolateral domains. Underneath the tight junctions, the adherens junctions (zonula adherens) and desmosomes are the main responsible for the mechanical strength of the cell-cell contacts. A schematic description of the structure of anchoring junctions is depicted in Figure 2(a). As emphasized by Matter and Balda [34] for the case of tight junctions, three levels of organization can be recognized. This conception can be generalized to all the anchoring junctions. The first level corresponds to the integral membrane proteins, the second to a set of proteins mediating, among others, the interaction between the cytoskeleton and the adhesion proteins, and the third to the cytoskeletal proteins. The second level is made up

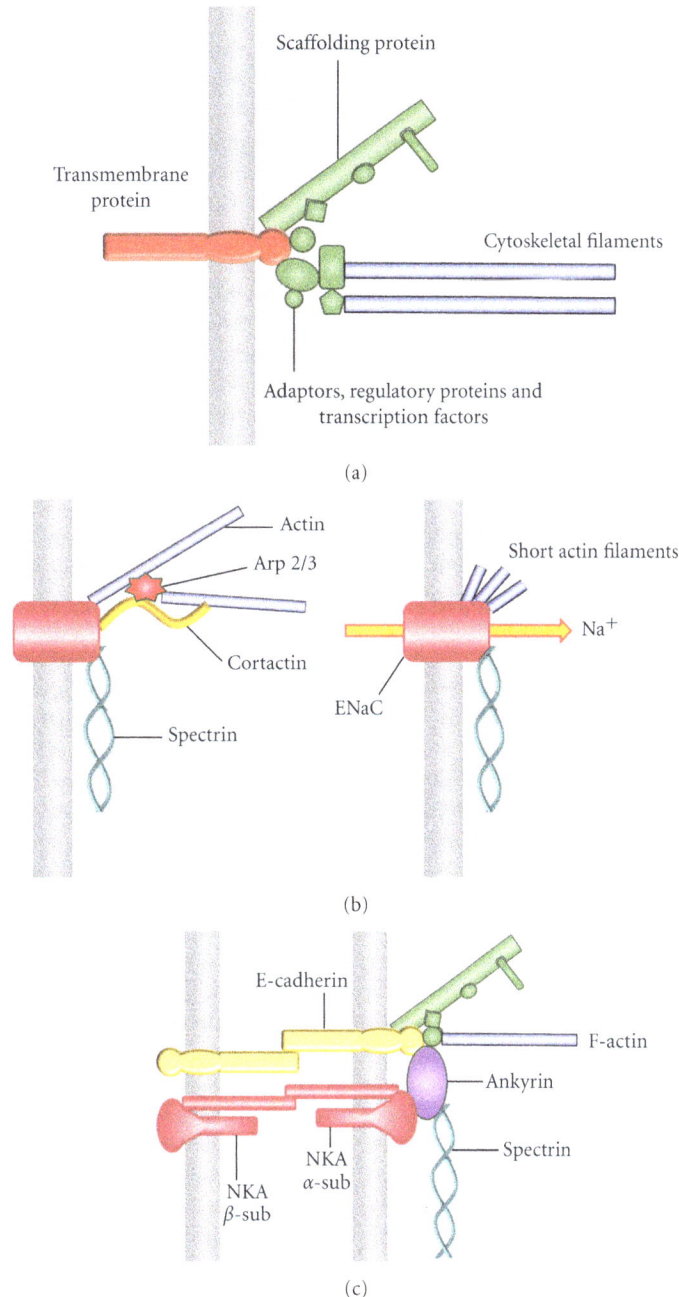

FIGURE 2: (a) General schematic representation of an anchoring junction. The scheme depicts the basic organization of the proteins comprising an anchoring junction. The three organizational levels (see text) are represented by different colors; modified from Matter and Balda [34]. (b) Scheme representing some of the interactions between ENaC and the cytoskeleton. The channel is kept in its membrane location via the spectrin-based cytoskeleton. In the nonstimulated state, ENaC is bound to F-actin directly and/or by cortactin-Arp2/3 interaction (left panel). Diverse stimuli promote formation of short actin filaments that activate ENaC by direct binding to the channel (right panel). See main text for references. (c) Scheme representing some of the interactions between the Na$^+$-K$^+$ ATPase (NKA) and the cytoskeleton; modified from Bennett and Healy [18].

of a complex array of soluble proteins, with many of which being capable to transit between other cellular compartments and their juxtamembrane location at the corresponding cell junction [33]. This complex includes transcription factors, regulatory proteins, and scaffolding proteins and is a critical site for the transduction of diverse types of signals [35]. Among these, the acquisition of the epithelial phenotype is triggered by the recruiting of proteins at the second and third junction levels, initially induced by the establishment of cell-cell and cell-substrate contacts [36]. The composition and interactions of the proteins of the second level are highly complex and still remain matters of active investigation. The general organizational pattern shown in Figure 2(a) for the anchoring junctions can also be extended to include the linking between other types of integral membrane proteins, such as ion channels and pumps and the cytoskeleton (see below).

Adherens junctions (AJs) have a central role in the establishment and maintenance of the epithelial phenotype. In this respect, classical cadherins have been recognized to be involved in the establishment of early intercellular contacts and in the organization of microfilaments and microtubules [8, 11, 36]. The association between AJ and these cytoskeletal components is interdependent, since the actin and tubulin cytoskeletons in turn contribute to AJ formation, stability and strength [37–43]. Thus, when actin cytoskeleton is disrupted, AJ formation is impaired [38, 39, 44]. Moreover, actin cytoskeletal reorganization, as induced for instance by extracellular calcium deprivation, may determine AJ and tight junction disruption with the consequent loss of the epithelial integrity [45, 46]. Likewise, cell-cell contacts stabilize the plus and minusends of microtubules at the adherens junctions, while the blocking of their dynamic turnover provokes brakeage of cell-cell contacts [26, 43, 47].

Even in the quiescent state, the cytoskeleton of epithelial cells is highly dynamical, undergoing constant assembly and disassembly of its structural units. For this reason, it is subject to complex mechanisms of regulation, many of which involve components of cell-cell junctions and other membrane proteins. Since it is not the purpose of this work to make a thorough revision of this rather involved issue, the reader is referred to several specialized reviews [11, 36, 48, 49]. In Section 5 we will only mention some regulatory pathways that are possible candidates to mediate the responses of the cytoskeleton to modifications in the plasma membrane potential.

3. Interactions between Ionic Transport Systems of the Plasma Membrane and the Cytoskeleton

Besides its associations with the cell junctions, the cytoskeleton also interacts, directly or indirectly, with diverse ion transport systems of the plasma membrane [50–53]. Similarly to the case of the cell junctions, many of the associations of the cell membrane channels and transporters with the cytoskeleton are interdependent (for detailed reviews, see [50, 52]). In this way, several transport systems are anchoring sites for the cortical cytoskeleton [54–62]. Conversely, the binding to the cytoskeleton modulates diverse ion transport activities [56, 61, 63–68]. This interrelation between the cytoskeleton and ionic transport systems plays a relevant role in the physiological properties of transport epithelia [69, 70]. For example, in the renal medullary thick ascending limb the Na^+/H^+ exchanger regulates bicarbonate absorption by controlling the organization of the actin cytoskeleton [71]. In human syncitiotrophoblast, gelsolin (an actin regulatory protein) stimulates nonselective cation channels of the TRP family in the presence of calcium [72]. In HEK 293 cells, calcium-activated chloride channels require cytoskeletal interactions to achieve full activation [73]. A most noteworthy example of the modulation of ionic transport by the cytoskeleton is provided by volume-sensitive ion transport systems, that modify transport rates in response to changes in cytoskeletal tension [74–76]. In several epi-

thelial cells, where water and salt transcellular transport determine modifications in the cellular volume, stretch-activated potassium and chloride channels participate in the regulation of salt transport [77–79]. Among other examples of the regulation of ionic conductances by the cytoskeleton, in mammary adenocarcinoma cells a well-organized actin network is necessary for the proper activation of CFTR by cAMP [80].

The interrelationship between ion transport systems and the cytoskeleton is also crucial in cell adhesion and migration, processes requiring significant cytoskeletal remodeling and modifications of ion transport. In this respect, it has been shown that inhibition of the CFTR-dependent conductance impairs lamellipodia formation in bronchial epithelial cells [81]. In T-cells, Kv1.3 channels participate in promoting adhesion by establishing complexes with $\beta 1$ integrin [82]. In neuroblastoma cell lines, TRPM7 channels affect cell adhesion by direct interaction with the actomyosin cytoskeleton [83]. TRPV4 has also been reported to form complexes with the actin cytoskeleton and regulatory kinases, involved in lamellipodial formation [61]. Focal adhesion kinase, an enzyme involved in integrin-mediated focal adhesion, is activated by forming complexes with Kv1.2 channels both in epithelial and nonepithelial cells [84]. The inhibition of either the Na^+/H^+ exchanger transport activity or its actin cytoskeletal anchoring significantly decreases migration of PS-120 fibroblasts [66]. Also, ENaC has been reported to participate in the processes of cell migration and wound healing in epithelia [85, 86] and other tissues [87]. The molecular nature of this role may imply direct interactions between this channel and cytoskeletal components. In this respect, it has been shown that the alpha subunit of ENaC associates with spectrin [54, 88] and short actin filaments [89]. In addition, there is evidence that the direct binding between ENaC and these filaments modifies the channel conductance [90–92]. More recently, Ilatovskaya et al. [68] have shown that, in renal epithelial cells, the actin cytoskeleton regulates ENaC activity via a cortactin-Arp2/3 complex. As an example of the structural relationships between an ion channel and the cytoskeleton, Figure 2(b) resumes diverse evidence in a schematic diagram of the interactions between ENaC and cytoskeletal components during nonactivated and activated conditions [50, 54, 68, 88]. To be noted, the mechanism through which the cytoskeleton promotes modifications in ionic conductance may depend on the particular transport system. For the case of ENaC, Figure 2(b) shows that the channel activation requires direct binding to short actin filaments, produced by direct PKA phosphorylation of actin or via an actin regulatory protein [50]. As another example, in vascular endothelia 4.1 proteins have been proposed to be necessary for the stable expression of TRPC4 in the plasma membrane [93].

The sodium pump has also classically been recognized to establish interactions with the cortical cytoskeleton [94]. More recently, it was shown that this enzyme participates in tight junction assembly in MDCK cells via activation of RhoA and stress fiber formation [95–97]. In caveolar structures the sodium pump is the core of a multiprotein complex, the sodium pump signalosome, that contains several proteins involved in cytoskeletal regulation [98]. Moreover, in vitro experiments reinforce the existence of interrelations between the Na^+-K^+-ATPase and actin [64]. The molecular nature of the interactions between the sodium pump and the cytoskeletal components, and their physiological relevance, are beginning to be known in more detail. In this respect, the critical role of the interaction of ankyrin-G and the spectrin-based actin cytoskeleton in the membrane retention of the sodium pump has been well established [19, 99]. The binding of ankyrin with the α_1-subunit of the Na^+-K^+ ATPase has been shown to be crucial not only for the membrane anchoring of the enzyme, but also for its trafficking in the polarized cell [100]. Based upon the available evidence, the scheme of Figure 2(c) summarizes some of the recognized interactions between the sodium pump and the cytoskeleton in vertebrate cells. In *D. melanogaster*, Dubreuil and coworkers demonstrated the existence of ankyrin-independent interactions between the sodium pump and spectrin [101, 102].

4. The Plasma Membrane Potential of Epithelial Cells

The classical dogma of the ionic transport properties of epithelia was founded by Ussing and coworkers in the 1950s and has ever since become the basic paradigm of epithelial transport [103–105]. According to this model, in its essential terms, the polarized distribution of the sodium pump and sodium channels is the basic process that transforms a homogeneous cell into an epithelial cell, capable of performing net transepithelial transport of salt between two separated compartments (Figure 3). Although the general scheme of Figure 3(b) may apply to different epithelial cell types, it is particularly characteristic of tight epithelia [106]. In this respect, it was recognized that epithelia can be divided into two categories, tight and leaky, according to the electrical resistance of the tight junctions [107]. In addition to this difference, the two types are distinguished by other transport properties. Thus, besides the low electrical resistance, leaky epithelia also exhibit a high water permeability and establish low-transepithelial-potential differences (see below), while the reverse is true for tight epithelia [108].

In general, since the apical and basolateral domains of epithelial cells have different compositions of ionic transport systems, the mechanisms of generation of the membrane potential differ between them. For the case shown in Figure 3(b), the PMP across the basolateral membrane is approximately given by a diffusion potential dominated by potassium. Among other differential contributions, the sodium pump may be responsible for the generation of the basolateral PMP in an electrogenic fashion, particularly if potassium permeability is low [109]. The composition of the apical membranes greatly varies among the different epithelial cell types. Therefore, the generation of the apical PMP depends on the specific epithelia considered. For example, in intestinal epithelial cells the sodium-glucose electrogenic cotransport may be the major contributor to the apical PMP [110]. The selective modifications in the ionic conductances at the apical or basolateral domains may

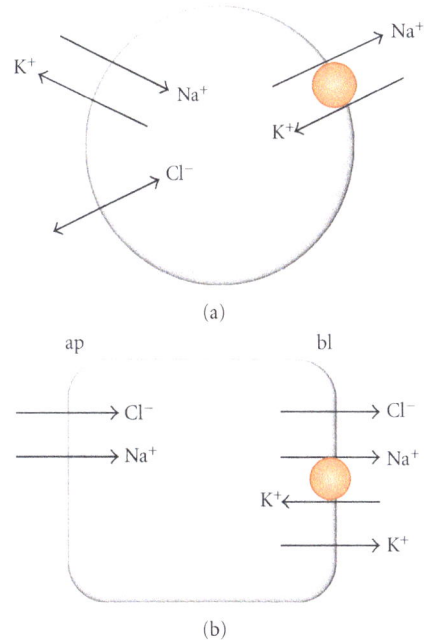

FIGURE 3: Schemes showing some ionic transport systems of the plasma membrane. (a) In symmetric, nonpolarized cells the sodium pump maintains the electrochemical gradients of sodium and potassium across the plasma membrane. The sodium and potassium channels underlie the generation of a diffusion potential across the plasma membrane. Chloride is usually maintained at activities close to equilibrium. (b) In polarized epithelial cells the asymmetric distribution of, mainly, sodium channels and the sodium pump into distinctive apical and basolateral membranes may determine a net transcellular transport of sodium chloride. (ap: apical; bl: basolateral; the orange circles represent the sodium pump, and the other arrows correspond to ionic channels).

play physiological roles in diverse epithelia. Thus, in the acinar cells of parotid salivary glands, modifications in the potassium conductance of the apical membrane produce changes in the fluid flow via changes in the apical PMP [111–113]. In pancreatic acinar ducts, modifications in the apical depolarization mediated by the sodium-glucose cotransporter modulate the amount of basolateral uptake of chloride and bicarbonate [114]. Another example is given by the basolateral membranes of tracheal and intestinal cells, where potassium channels containing the *KCNE* family of β-subunits control the chloride flux via modifications in the basolateral PMP [115].

In epithelial cells, the differential generation of the PMP at the two membrane domains determines the existence of a net transepithelial potential difference (TPD). As mentioned above, tight epithelia are capable to maintain large TPDs as a consequence, among other factors, of the existence of a large electrical resistance of the paracellular pathway [108]. On the contrary, the TPDs across leaky epithelia are generally small. As an example, Figure 4 shows a scheme of some of the ionic transport systems of corneal endothelium, a typical leaky epithelium that pumps salt and water from the corneal stroma to the aqueous compartment of the eye [116, 117]. As can be seen, the presence at the two domains of sodium-bicarbonate cotransporters with different stoichiometric

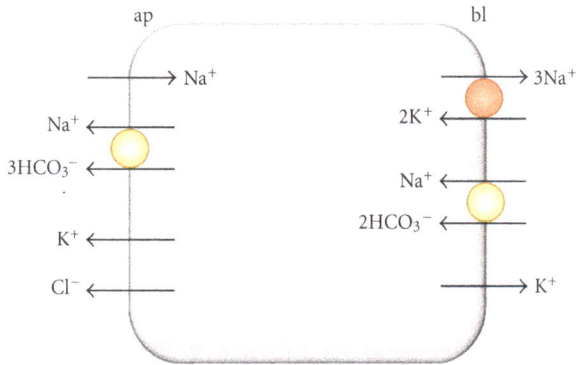

FIGURE 4: Scheme of a corneal endothelial cell showing electrogenic transport systems of the apical (ap) and basolateral (bl) membranes. Orange circle and arrows as in Figure 3, pale orange circles represent sodium-bicarbonate cotransporters; modified from Montalbetti and Fischbarg [117].

ratios is the cause for a small TPD (i.e., less than 1 mV). Among other possible roles, as in paracellular ion movement, this TPD may have a relevant function in the mechanism of solute-solvent coupling across corneal endothelium [118].

Epithelial cells in culture may acquire some structural and physiological characteristics of the in situ epithelial cells, including the generation of a TPD [119–121]. For the case of small TPDs, as is the case of corneal endothelium [116], the employment of voltage-sensitive fluorescent probes provides with global PMP values which are approximately equal to the apical and basolateral PMPs of these cells in culture [117, 122] but does not permit to distinguish between them. Electrophysiological procedures are required to determine the PMP at each one of the plasma membrane domains, both under in situ or culture conditions.

5. Modulation of the Epithelial Actin Cytoskeleton by Modifications in the Plasma Membrane Potential

Several authors have suggested that the plasma membrane potential of nonexcitable cells could play a role in diverse cellular processes [123–125]. In MDCK cells, Vaaraniemi et al. [126] found that activation of protein kinase C determined PMP depolarization and reorganization of the spectrin-based and actin cytoskeletons. Consistently with these findings, we showed that the nonspecific modifications of the PMP (i.e., depolarization or hyperpolarization) promote changes in the organization of the actin and tubulin cytoskeletons in bovine corneal endothelial (BCE) cells in culture [127, 128]. In particular, the changes observed for the actin cytoskeleton consisted, for the PMP depolarization, in a gradual loss of the peripheral ring, an increase in F-actin throughout the cytoplasm, appearance of intercellular gaps and, for sufficiently prolonged treatments, eventual cell detachment [127]. Conversely, it was noteworthy to confirm that PMP hyperpolarization determined the opposite response, that is, an increase in the compactness of actin at the peripheral ring and an augmented resistance

of intercellular adhesion to diverse destabilizing procedures [128]. These cytoskeletal responses to PMP modifications were characteristic of some cultured epithelia in confluence displaying a typical epithelial phenotype exhibiting, among other characteristics, a well-defined circumferential actin ring, whereas nonconfluent or undifferentiated epithelial cell lines did not manifest a recognizable response [128, 129]. Effects of PMP depolarization on microfilaments were also observed in kidney tubular cells, where the authors demonstrated that Rho activation and the consequent increase in phosphorylation of the light myosin chain are involved [130, 131]. The role of the PMP on the cytoskeletal organization was also supported by the finding that, in vascular endothelia, depolarization decreases cell stiffness by affecting the cortical actin cytoskeleton [132, 133].

The activities of diverse signaling intermediates are sensible to modifications of the PMP. Thus, in excitable cells, regulators of G protein signaling (RGS), Rho proteins and PKA are activated by the calcium increase provoked by PMP depolarization [134–140]. In renal epithelia, however, the activation of Rho determined by depolarization is not mediated by cytosolic calcium increase [130, 141]. Among the integral membrane proteins that could mediate cytoskeletal responses to diverse effectors, the phosphatidylinositol phosphatase Ci-VSP [142] and its homologs [143, 144] contain a voltage sensor in the transmembrane domain [145] and produce PIP2, a well-known regulator of the actin cytoskeleton [146, 147]. Interestingly, this enzyme, present in epithelial and nonepithelial cells [143, 144], is activated by PMP depolarization [142]. More recently, in *Xenopus* oocytes Zhang et al. [148] described an alternative voltage-sensitive mechanism to increase PIP2 level in response to PMP depolarization, via activation of a PI4 kinase. The G protein-coupled receptors (GPCR) constitute another family of membrane proteins shown to initiate signaling paths leading to actin remodeling [149, 150]. These receptors, activated by a variety of extracellular effectors, are directly regulated by the PMP [151–156].

At this point, it should be reminded that the "plasma membrane potential," as determined from typical electrophysiological procedures, refers to net electrical potential differences between the intra- and extracellular bulk compartments. This difference comprises a series of intermediate electrical potential changes that include surface potentials at membrane proximities and the transmembrane potential [123, 157, 158]. Changes in the surface potentials could affect peripheral proteins, many of them involved in cytoskeletal regulation [123, 158]. For example, diverse putative peripheral proteins can bind to the inner surface of the plasma membrane by electrostatic interactions and modify their degree of attachment in response to modifications in the surface potential, such as MARCKS [159], PTEN [160], K-Ras [161], c-Src: [162], Rac1 [163], and ERM proteins [164]. For the particular case of K-Ras it has been shown that, apart from the inner surface potential, the transmembrane potential also affects its binding to the plasma membrane [165].

The effects of the PMP on the cytoskeletal organization could also be mediated by membrane ionic transport systems directly or indirectly connected to cytoskeletal elements, such as the ones mentioned in Section 3. In principle, there are two main mechanisms through which the PMP could affect the interrelationship between ion transport systems and the cytoskeleton: (a) by modifying ionic currents and thus the ionic environment near the cytoskeletal binding regions, and (b) by determining electroconformational modifications of the ion transporting proteins that can be propagated to the cytoskeletal components, as could be the case for other non-transporting integral membrane proteins. As an example of this latter possibility, several ion channels have been proposed to transmit signals via conformational coupling with integrins, irrespectively of changes in the ionic fluxes [166]. Whatever the mechanism, it must be noted that the particular ion transport system that would mediate the cytoskeletal response to a certain PMP modification (i.e., a hyperpolarization or a depolarization) may depend on the specific procedure employed for the modification. For instance, in a certain cell type PMP hyperpolarization may be achieved by increasing the potassium conductivity, determining augmented potassium efflux, or by increasing the chloride conductivity, producing increased chloride influx. Correspondingly, the intermediate path and specific organizational response of the cytoskeleton to the particular PMP change provoked may also depend on the specific procedure and ionic path (cf. Section 3).

Another possible mechanism mediating the effects of PMP modifications on cytoskeletal rearrangements could be the direct conduction of electrical signals generated at the plasma membrane by the cytoskeletal components themselves. In support of this idea is the finding that actin filaments can propagate electrical signals per se [167].

6. Possible Physiological, Pathological, and Medical Implications of a Regulation of the Actin Cytoskeleton by the Membrane Potential

In principle, it could be expected that the transition of an epithelial cell from a quiescent to a specific secretion or absorption state occurs with characteristic modifications both in cytoskeletal organization and ionic conductances. As suggested from the evidence reviewed in this work, these concurrent modifications could be mediated by the complex regulatory framework provided by the interactions between cytoskeletal components and diverse membrane transport systems. From the results reviewed in the previous section the modifications in the plasma membrane potential could participate in the regulation of the organization of the cytoskeleton of epithelial cells, possibly via effects mediated by ionic transport systems. The examples shown in Section 3 support this notion by describing diverse examples of epithelial cells where ionic transport is associated to cytoskeletal modifications.

As a further physiological counterpart to the results commented in Section 5, we put into evidence that PMP depolarization occurs during wound healing in bovine corneal endothelial cells, as a consequence of the increased expression of the epithelial sodium channel (ENaC), and that

it may have a role in the healing process [85]. To be noted, the border cells actively participating in the healing response of epithelia experience characteristic reorganizations of the actin cytoskeleton, which are blocked by ENaC inhibition (ibid). ENaC-dependent PMP depolarization was also observed in the course of wound healing by others in a cell line of human trophoblast [86] and by us in other epithelia in culture (unpublished results). Interestingly, in healing corneal endothelium and epithelium Watsky [168] described a late hyperpolarizing potassium current that, in view of the results described above, could have the role of restituting the membrane potential to its basal value. A role for ENaC in the processes of wound healing was also proposed by Grifoni et al. [87] for smooth vascular muscle cells.

Actin has been found to participate in the development of the apoptotic response [169]. Thus, interference with actin dynamics by inhibition of its depolymerization enhances apoptotic activity in HL-60 cells [170, 171]. However, in T-cells disruption of the actin cytoskeleton promotes caspase-3-mediated apoptosis [172]. A concurrent finding of interest within the conceptual framework of this paper is that cells undergo PMP depolarization in the course of apoptosis, a fact that has been speculated to play a role in the cytoskeletal reorganization that takes place during this process [173].

The finding that hyperpolarization of the PMP determines actin compaction along the adherens junctions and increases junction stability [128] may have application in the design of therapeutic strategies. In this respect, the loss of epithelial intercellular adhesions is at the basis of diverse pathologies [174, 175]. Some of these represent major medical challenges, such as cancer progression [176–179] ischemic injuries [180] and bowel inflammatory diseases [181].

7. Concluding Remarks

The modifications in the plasma membrane potential have mostly been classically associated with the physiology of excitable tissues. In both excitable and nonexcitable cells, the PMP is an energetic component of the electrochemical gradients responsible of membrane ionic transport. The findings reviewed in this work contribute to the concept that the PMP may also participate in other cellular processes, including the establishment and maintenance of the morphological and functional features of epithelial cells. In particular, we have emphasized here the possible role of the PMP in the regulation of the actin cytoskeleton. Although some knowledge about signaling pathways involved in the transduction of electrical signals at the plasma membrane to mechanical modifications of the cytoskeleton has been unraveled, the involvement of the cytoskeleton in many relevant physiological cellular phenomena permits to anticipate great progress in this respect in the near future.

Acknowledgments

This work received financial support by grants from CSIC and PEDECIBA (Universidad de la República, Uruguay), ANII (Ministerio de Educación y Cultura, Uruguay), and

Comisión Honoraria de Lucha contra el Cáncer (Uruguay). The authors thank two anonymous reviewers for most fruitful suggestions.

References

[1] T. Zeuthen, "Molecular mechanisms of water transport," in *Molecular Biology Intelligence Unit*, pp. 97–126, R.G. Landes Co., Austin, Tex, USA, 1996.

[2] E. Rodriquez-Boulan and W. J. Nelson, "Morphogenesis of the polarized epithelial cell phenotype," *Science*, vol. 245, no. 4919, pp. 718–725, 1989.

[3] C. Yeaman, K. K. Grindstaff, and W. J. Nelson, "New perspectives on mechanisms involved in generating epithelial cell polarity," *Physiological Reviews*, vol. 79, no. 1, pp. 73–98, 1999.

[4] W. J. Nelson, "Adaptation of core mechanisms to generate cell polarity," *Nature*, vol. 422, no. 6933, pp. 766–774, 2003.

[5] D. M. Bryant and K. E. Mostov, "From cells to organs: building polarized tissue," *Nature Reviews Molecular Cell Biology*, vol. 9, no. 11, pp. 887–901, 2008.

[6] J. Jaworski, C. C. Hoogenraad, and A. Akhmanova, "Microtubule plus-end tracking proteins in differentiated mammalian cells," *International Journal of Biochemistry and Cell Biology*, vol. 40, no. 4, pp. 619–637, 2008.

[7] J. E. Eriksson, T. Dechat, B. Grin et al., "Introducing intermediate filaments: from discovery to disease," *Journal of Clinical Investigation*, vol. 119, no. 7, pp. 1763–1771, 2009.

[8] K. P. Harris and U. Tepass, "Cdc42 and vesicle trafficking in polarized cells," *Traffic*, vol. 11, no. 10, pp. 1272–1279, 2010.

[9] S. Etienne-Manneville, "From signaling pathways to microtubule dynamics: the key players," *Current Opinion in Cell Biology*, vol. 22, no. 1, pp. 104–111, 2010.

[10] C. M. Niessen, D. Leckband, and A. S. Yap, "Tissue organization by cadherin adhesion molecules: dynamic molecular and cellular mechanisms of morphogenetic regulation," *Physiological Reviews*, vol. 91, no. 2, pp. 691–731, 2011.

[11] B. Baum and M. Georgiou, "Dynamics of adherens junctions in epithelial establishment, maintenance, and remodeling," *Journal of Cell Biology*, vol. 192, no. 6, pp. 907–917, 2011.

[12] A. L. Hartman, N. M. Sawtell, and J. L. Lessard, "Expression of actin isoforms in developing rat intestinal epithelium," *Journal of Histochemistry and Cytochemistry*, vol. 37, no. 8, pp. 1225–1233, 1989.

[13] M. S. Balda, L. Gonzalez-Mariscal, K. Matter, M. Cereijido, and J. M. Anderson, "Assembly of the tight junction: the role of diacylglycerol," *Journal of Cell Biology*, vol. 123, no. 2, pp. 293–302, 1993.

[14] C. J. Maples, W. G. Ruiz, and G. Apodaca, "Both microtubules and actin filaments are required for efficient postendocytotic traffic of the polymeric immunoglobulin receptor in polarized Madin-Darby canine kidney cells," *Journal of Biological Chemistry*, vol. 272, no. 10, pp. 6741–6751, 1997.

[15] J. Breckler and B. Burnside, "Myosin-I in retinal pigment epithelial cells," *Investigative Ophthalmology and Visual Science*, vol. 35, no. 5, pp. 2489–2499, 1994.

[16] M. J. Tyska and M. S. Mooseker, "MYO1A (brush border myosin I) dynamics in the brush border of LLC-PK1-CL4 cells," *Biophysical Journal*, vol. 82, no. 4, pp. 1869–1883, 2002.

[17] M. C. Wagner, B. L. Blazer-Yost, J. Boyd-White, A. Srirangam, J. Pennington, and S. Bennett, "Expression of the unconventional myosin Myo1c alters sodium transport in M1 collecting duct cells," *American Journal of Physiology, Cell Physiology*, vol. 289, no. 1, pp. C120–C129, 2005.

[18] V. Bennett and J. Healy, "Organizing the fluid membrane bilayer: diseases linked to spectrin and ankyrin," *Trends in Molecular Medicine*, vol. 14, no. 1, pp. 28–36, 2008.

[19] K. Kizhatil and V. Bennett, "Lateral membrane biogenesis in human bronchial epithelial cells requires 190-kDa ankyrin-G," *Journal of Biological Chemistry*, vol. 279, no. 16, pp. 16706–16714, 2004.

[20] K. Kizhatil, J. Q. Davis, L. Davis, J. Hoffman, B. L. M. Hogan, and V. Bennett, "Ankyrin-G is a molecular partner of E-cadherin in epithelial cells and early embryos," *Journal of Biological Chemistry*, vol. 282, no. 36, pp. 26552–26561, 2007.

[21] B. Geiger, A. Bershadsky, R. Pankov, and K. M. Yamada, "Transmembrane extracellular matrix—cytoskeleton crosstalk," *Nature Reviews Molecular Cell Biology*, vol. 2, no. 11, pp. 793–805, 2001.

[22] T. J. Keating and G. G. Borisy, "Centrosomal and non-centrosomal microtubules," *Biology of the Cell*, vol. 91, no. 4-5, pp. 321–329, 1999.

[23] A. Musch, "Microtubule organization and function in epithelial cells," *Traffic*, vol. 5, no. 1, pp. 1–9, 2004.

[24] R. Pepperkok, M. H. Bre, J. Davoust, and T. E. Kreis, "Microtubules are stabilized in confluent epithelial cells but not in fibroblasts," *Journal of Cell Biology*, vol. 111, no. 6, pp. 3003–3012, 1990.

[25] E. Shelden and P. Wadsworth, "Observation and quantification of individual microtubule behavior in vivo: microtubule dynamics are cell-type specific," *Journal of Cell Biology*, vol. 120, no. 4, pp. 935–945, 1993.

[26] C. M. Waterman-Storer, W. C. Salmon, and E. D. Salmon, "Feedback interactions between cell-cell adherens junctions and cytoskeletal dynamics in newt lung epithelial cells," *Molecular Biology of the Cell*, vol. 11, no. 7, pp. 2471–2483, 2000.

[27] A. Reilein, S. Yamada, and W. J. Nelson, "Self-organization of an acentrosomal microtubule network at the basal cortex of polarized epithelial cells," *Journal of Cell Biology*, vol. 171, no. 5, pp. 845–855, 2005.

[28] R. Bacallao, C. Antony, C. Dotti, E. Karsenti, E. H. K. Stelzer, and K. Simons, "The subcellular organization of Madin-Darby canine kidney cells during the formation of a polarized epithelium," *Journal of Cell Biology*, vol. 109, no. 6 I, pp. 2817–2832, 1989.

[29] M. H. Bre, R. Pepperkok, A. M. Hill et al., "Regulation of microtubule dynamics and nucleation during polarization in MDCK II cells," *Journal of Cell Biology*, vol. 111, no. 6, pp. 3013–3021, 1990.

[30] T. Meads and T. A. Schroer, "Polarity and nucleation of microtubules in polarized epithelial cells," *Cell Motility and the Cytoskeleton*, vol. 32, no. 4, pp. 273–288, 1995.

[31] K. K. Grindstaff, R. L. Bacallao, and W. J. Nelson, "Apiconuclear organization of microtubules does not specify protein delivery from the trans-Golgi Network to different membrane domains in polarized epithelial cells," *Molecular Biology of the Cell*, vol. 9, no. 3, pp. 685–699, 1998.

[32] O. A. Weisz and E. Rodriguez-Boulan, "Apical trafficking in epithelial cells: signals, clusters and motors," *Journal of Cell Science*, vol. 122, no. 23, pp. 4253–4266, 2009.

[33] W. W. Franke, "Discovering the molecular components of intercellular junctions—a historical view," *Cold Spring Harbor Perspectives in Biology*, vol. 1, no. 3, Article ID a003061, 2009.

[34] K. Matter and M. S. Balda, "Signalling to and from tight junctions," *Nature Reviews Molecular Cell Biology*, vol. 4, no. 3, pp. 225–236, 2003.

[35] B. N. G. Giepmans and S. C. D. van IJzendoorn, "Epithelial cell-cell junctions and plasma membrane domains," *Biochimica et Biophysica Acta*, vol. 1788, no. 4, pp. 820–831, 2009.

[36] M. Cavey and T. Lecuit, "Molecular bases of cell-cell junctions stability and dynamics," *Cold Spring Harbor Perspectives in Biology*, vol. 1, no. 5, Article ID a002998, 2009.

[37] S. H. Jaffe, D. R. Friedlander, F. Matsuzaki, K. L. Crossin, B. A. Cunningham, and G. M. Edelman, "Differential effects of the cytoplasmic domains of cell adhesion molecules on cell aggregation and sorting-out," *Proceedings of the National Academy of Sciences of the United States of America*, vol. 87, no. 9, pp. 3589–3593, 1990.

[38] C. L. Adams, W. J. Nelson, and S. J. Smith, "Quantitative analysis of cadherin-catenin-actin reorganization during development of cell-cell adhesion," *Journal of Cell Biology*, vol. 135, no. 6, pp. 1899–1911, 1996.

[39] B. Angres, A. Barth, and W. J. Nelson, "Mechanism for transition from initial to stable cell-cell adhesion: kinetic analysis of E-cadherin-mediated adhesion using a quantitative adhesion assay," *Journal of Cell Biology*, vol. 134, no. 2, pp. 549–557, 1996.

[40] J. Zhang, M. Betson, J. Erasmus et al., "Actin at cell-cell junctions is composed of two dynamic and functional populations," *Journal of Cell Science*, vol. 118, no. 23, pp. 5549–5562, 2005.

[41] F. Pilot, J. M. Philippe, C. Lemmers, and T. Lecuit, "Spatial control of actin organization at adherens junctions by the synaptotagmin-like protein Btsz," *Nature*, vol. 442, no. 7102, pp. 580–584, 2006.

[42] K. Abe and M. Takeichi, "EPLIN mediates linkage of the cadherin-catenin complex to F-actin and stabilizes the circumferential actin belt," *Proceedings of the National Academy of Sciences of the United States of America*, vol. 105, no. 1, pp. 13–19, 2008.

[43] W. Meng, Y. Mushika, T. Ichii, and M. Takeichi, "Anchorage of microtubule minus ends to adherens junctions regulates epithelial cell-cell contacts," *Cell*, vol. 135, no. 5, pp. 948–959, 2008.

[44] V. Vasioukhin, C. Bauer, M. Yin, and E. Fuchs, "Directed actin polymerization is the driving force for epithelial cell-cell adhesion," *Cell*, vol. 100, no. 2, pp. 209–219, 2000.

[45] A. I. Ivanov, I. C. McCall, C. A. Parkos, and A. Nusrat, "Role for actin filament turnover and a myosin II motor in cytoskeleton-driven disassembly of the epithelial apical junctional complex," *Molecular Biology of the Cell*, vol. 15, no. 6, pp. 2639–2651, 2004.

[46] A. I. Ivanov, I. C. McCall, B. Babbin, S. N. Samarin, A. Nusrat, and C. A. Parkos, "Microtubules regulate disassembly of epithelial apical junctions," *BMC Cell Biology*, vol. 7, article no. 12, 2006.

[47] S. J. Stehbens, A. D. Paterson, M. S. Crampton et al., "Dynamic microtubules regulate the local concentration of E-cadherin at cell-cell contacts," *Journal of Cell Science*, vol. 119, no. 9, pp. 1801–1811, 2006.

[48] A. I. Ivanov, C. A. Parkos, and A. Nusrat, "Cytoskeletal regulation of epithelial barrier function during inflammation," *American Journal of Pathology*, vol. 177, no. 2, pp. 512–524, 2010.

[49] J. Nance and J. A. Zallen, "Elaborating polarity: PAR proteins and the cytoskeleton," *Development*, vol. 138, no. 5, pp. 799–809, 2011.

[50] C. Mazzochi, D. J. Benos, and P. R. Smith, "Interaction of epithelial ion channels with the actin-based cytoskeleton," *American Journal of Physiology, Renal Physiology*, vol. 291, no. 6, pp. F1113–F1122, 2006.

[51] P. R. Kiela and F. K. Ghishan, "Ion transport in the intestine," *Current Opinion in Gastroenterology*, vol. 25, no. 2, pp. 87–91, 2009.

[52] A. J. Baines, P. M. Bennett, E. W. Carter, and C. Terracciano, "Protein 4.1 and the control of ion channels," *Blood Cells, Molecules, and Diseases*, vol. 42, no. 3, pp. 211–215, 2009.

[53] H. Yu, Y. Zhang, L. Ye, and W. G. Jiang, "The FERM family proteins in cancer invasion and metastasis," *Frontiers in Bioscience*, vol. 16, no. 4, pp. 1536–1550, 2011.

[54] J. B. Zuckerman, X. Chen, J. D. Jacobs, B. Hu, T. R. Kleyman, and P. R. Smith, "Association of the epithelial sodium channel with Apx and α-spectrin in A6 renal epithelial cells," *Journal of Biological Chemistry*, vol. 274, no. 33, pp. 23286–23295, 1999.

[55] V. Bennett and A. J. Baines, "Spectrin and ankyrin-based pathways: metazoan inventions for integrating cells into tissues," *Physiological Reviews*, vol. 81, no. 3, pp. 1353–1392, 2001.

[56] S. A. Rajasekaran, L. G. Palmer, K. Quan et al., "Na,K-ATPase β-subunit is required for epithelial polarization, suppression of invasion, and cell motility," *Molecular Biology of the Cell*, vol. 12, no. 2, pp. 279–295, 2001.

[57] S. P. Denker and D. L. Barber, "Ion transport proteins anchor and regulate the cytoskeleton," *Current Opinion in Cell Biology*, vol. 14, no. 2, pp. 214–220, 2002.

[58] Y. Noda and S. Sasaki, "Actin-binding channels," *Progress in Brain Research*, vol. 170, pp. 551–557, 2008.

[59] R. Padányi, Y. Xiong, G. Antalffy et al., "Apical scaffolding protein NHERF2 modulates the localization of alternatively spliced plasma membrane Ca2+ pump 2B variants in polarized epithelial cells," *Journal of Biological Chemistry*, vol. 285, no. 41, pp. 31704–31712, 2010.

[60] G. H. Lee, T. Ahn, D. S. Kim et al., "Bax inhibitor 1 increases cell adhesion through actin polymerization: involvement of calcium and actin binding," *Molecular and Cellular Biology*, vol. 30, no. 7, pp. 1800–1813, 2010.

[61] C. Goswami, J. Kuhn, P. A. Heppenstall, and T. Hucho, "Importance of non-selective cation channel TRPV4 interaction with cytoskeleton and their reciprocal regulations in cultured cells," *PLoS ONE*, vol. 5, no. 7, Article ID e11654, 2010.

[62] Z. Wang and K. L. Schey, "Aquaporin-0 interacts with the FERM domain of ezrin/radixin/moesin proteins in the ocular lens," *Investigative Ophthalmology & Visual Science*, vol. 52, no. 8, pp. 5079–5087, 2011.

[63] E. M. Schwiebert, J. W. Mills, and B. A. Stanton, "Actin-based cytoskeleton regulates a chloride channel and cell volume in a renal cortical collecting duct cell line," *Journal of Biological Chemistry*, vol. 269, no. 10, pp. 7081–7089, 1994.

[64] H. F. Cantiello, "Actin filaments stimulate the Na(+)-K(+)-ATPase," *American Journal of Physiology*, vol. 269, no. 5, pp. F637–F643, 1995.

[65] B. K. Berdiev, A. G. Prat, H. F. Cantiello et al., "Regulation of epithelial sodium channels by short actin filaments," *Journal of Biological Chemistry*, vol. 271, no. 30, pp. 17704–17710, 1996.

[66] S. P. Denker and D. L. Barber, "Cell migration requires both ion translocation and cytoskeletal anchoring by the Na-H exchanger NHE1," *Journal of Cell Biology*, vol. 159, no. 6, pp. 1087–1096, 2002.

[67] B. Cha and M. Donowitz, "The epithelial brush border Na+/H+ exchanger NHE3 associates with the actin cytoskeleton by binding to ezrin directly and via PDZ domain-containing Na+/H+ exchanger regulatory factor (NHERF) proteins," *Clinical and Experimental Pharmacology and Physiology*, vol. 35, no. 8, pp. 863–871, 2008.

[68] D. V. Ilatovskaya, T. S. Pavlov, V. Levchenko, Y. A. Negulyaev, and A. Staruschenko, "Cortical actin binding protein cortactin mediates ENaC activity via Arp2/3 complex," *FASEB Journal*, vol. 25, no. 8, pp. 2688–2699, 2011.

[69] S. Khurana, "Role of actin cytoskeleton in regulation of ion transport: examples from epithelial cells," *Journal of Membrane Biology*, vol. 178, no. 2, pp. 73–87, 2000.

[70] E. A. Papakonstanti and C. Stournaras, "Cell responses regulated by early reorganization of actin cytoskeleton," *FEBS Letters*, vol. 582, no. 14, pp. 2120–2127, 2008.

[71] B. A. Watts, T. George, and D. W. Good, "The basolateral NHE1 Na+/H+ exchanger regulates transepithelial HCO3- absorption through actin cytoskeleton remodeling in renal thick ascending limb," *Journal of Biological Chemistry*, vol. 280, no. 12, pp. 11439–11447, 2005.

[72] N. Montalbetti, Q. Li, G. A. Timpanaro et al., "Cytoskeletal regulation of calcium-permeable cation channels in the human syncytiotrophoblast: role of gelsolin," *Journal of Physiology*, vol. 566, no. 2, pp. 309–325, 2005.

[73] Y. Tian, P. Kongsuphol, M. Hug et al., "Calmodulin-dependent activation of the epithelial calcium-dependent chloride channel TMEM16A," *FASEB Journal*, vol. 25, no. 3, pp. 1058–1068, 2011.

[74] E. K. Hoffmann and S. F. Pedersen, "Sensors and signal transduction pathways in vertebrate cell volume regulation," *Contributions to Nephrology*, vol. 152, pp. 54–104, 2006.

[75] B. D. Matthews, D. R. Overby, R. Mannix, and D. E. Ingber, "Cellular adaptation to mechanical stress: role of integrins, Rho, cytoskeletal tension and mechanosensitive ion channels," *Journal of Cell Science*, vol. 119, no. 3, pp. 508–518, 2006.

[76] K. Hayakawa, H. Tatsumi, and M. Sokabe, "Actin stress fibers transmit and focus force to activate mechanosensitive channels," *Journal of Cell Science*, vol. 121, no. 4, pp. 496–503, 2008.

[77] B. C. Tilly, M. J. Edixhoven, L. G. J. Tertoolen et al., "Activation of the osmo-sensitive chloride conductance involves P21(rho) and is accompanied by a transient reorganization of the F-actin cytoskeleton," *Molecular Biology of the Cell*, vol. 7, no. 9, pp. 1419–1427, 1996.

[78] E. K. Hoffmann, T. Schettino, and W. S. Marshall, "The role of volume-sensitive ion transport systems in regulation of epithelial transport," *Comparative Biochemistry and Physiology, A Molecular and Integrative Physiology*, vol. 148, no. 1, pp. 29–43, 2007.

[79] F. Lang, "Mechanisms and significance of cell volume regulation," *Journal of the American College of Nutrition*, vol. 26, no. 5, 2007.

[80] H. Cantiello, "Role of actin filament organization in CFTR activation," *Pflugers Archiv*, vol. 443, no. 1, supplement 1, pp. S75–S80, 2001.

[81] K. R. Schiller, P. J. Maniak, and S. M. O'Grady, "Cystic fibrosis transmembrane conductance regulator is involved in airway epithelial wound repair," *American Journal of Physiology, Cell Physiology*, vol. 299, no. 5, pp. C912–C921, 2010.

[82] M. Levite, L. Cahalon, A. Peretz et al., "Extracellular K+ and opening of voltage-gated potassium channels activate T cell integrin function: physical and functional association between Kv1.3 channels and $\beta 1$ integrins," *Journal of Experimental Medicine*, vol. 191, no. 7, pp. 1167–1176, 2000.

[83] K. Clark, M. Langeslag, B. Van Leeuwen et al., "TRPM7, a novel regulator of actomyosin contractility and cell adhesion," *EMBO Journal*, vol. 25, no. 2, pp. 290–301, 2006.

[84] J. F. Wei, L. Wei, X. Zhou et al., "Formation of Kv2.1-FAK complex as a mechanism of FAK activation, cell polarization and enhanced motility," *Journal of Cellular Physiology*, vol. 217, no. 2, pp. 544–557, 2008.

[85] S. Chifflet, J. A. Hernández, and S. Grasso, "A possible role for membrane depolarization in epithelial wound healing," *American Journal of Physiology, Cell Physiology*, vol. 288, no. 6, pp. C1420–C1430, 2005.

[86] S. del Mónaco, Y. Assef, and B. A. Kotsias, "Epithelial sodium channel in a human trophoblast cell line (BeWo)," *Journal of Membrane Biology*, vol. 223, no. 3, pp. 127–139, 2008.

[87] S. C. Grifoni, K. P. Gannon, D. E. Stec, and H. A. Drummond, "ENaC proteins contribute to VSMC migration," *American Journal of Physiology, Heart and Circulatory Physiology*, vol. 291, no. 6, pp. H3076–H3086, 2006.

[88] D. Rotin, D. Bar-Sagi, H. O'Brodovich et al., "An SH3 binding region in the epithelial Na$^+$ channel (αrENaC) mediates its localization at the apical membrane," *EMBO Journal*, vol. 13, no. 19, pp. 4440–4450, 1994.

[89] C. Mazzochi, J. K. Bubien, P. R. Smith, and D. J. Benos, "The carboxyl terminus of the α-subunit of the amiloride-sensitive epithelial sodium channel binds to F-actin," *Journal of Biological Chemistry*, vol. 281, no. 10, pp. 6528–6538, 2006.

[90] H. F. Cantiello, J. L. Stow, A. G. Prat, and D. A. Ausiello, "Actin filaments regulate epithelial Na+ channel activity," *American Journal of Physiology*, vol. 261, no. 5, pp. C882–C888, 1991.

[91] B. K. Berdiev, R. Latorre, D. J. Benos, and I. I. Ismailov, "Actin modifies Ca2+ block of epithelial Na+ channels in planar lipid bilayers," *Biophysical Journal*, vol. 80, no. 5, pp. 2176–2186, 2001.

[92] S. J. Copeland, B. K. Berdiev, H. L. Ji et al., "Regions in the carboxy terminus of α-bENaC involved in gating and functional effects of actin," *American Journal of Physiology, Cell Physiology*, vol. 281, no. 1, pp. C231–C240, 2001.

[93] D. L. Cioffi, S. Wu, M. Alexeyev, S. R. Goodman, M. X. Zhu, and T. Stevens, "Activation of the endothelial store-operated ISOC Ca2+ channel requires interaction of protein 4.1 with TRPC4," *Circulation Research*, vol. 97, no. 11, pp. 1164–1172, 2005.

[94] W. J. Nelson and P. J. Veshnock, "Ankyrin binding to (Na++K+)ATPase and implications for the organization of membrane domains in polarized cells," *Nature*, vol. 328, no. 6130, pp. 533–536, 1987.

[95] R. G. Contreras, L. Shoshani, C. Flores-Maldonado, A. Lázaro, and M. Cereijido, "Relationship between Na(+),K(+)-ATPase and cell attachment," *Journal of Cell Science*, vol. 112, no. 23, pp. 4223–4232, 1999.

[96] S. A. Rajasekaran, L. G. Palmer, S. Y. Moon et al., "Na,K-ATPase activity is required for formation of tight junctions, desmosomes, and induction of polarity in epithelial cells," *Molecular Biology of the Cell*, vol. 12, no. 12, pp. 3717–3732, 2001.

[97] I. Larre, A. Lazaro, R. G. Contreras et al., "Ouabain modulates epithelial cell tight junction," *Proceedings of the National Academy of Sciences of the United States of America*, vol. 107, no. 25, pp. 11387–11392, 2010.

[98] W. Schoner and G. Scheiner-Bobis, "Endogenous and exogenous cardiac glycosides: their roles in hypertension, salt

metabolism, and cell growth," *American Journal of Physiology, Cell Physiology*, vol. 293, no. 2, pp. C509–C536, 2007.

[99] V. Bennett and J. Healy, "Membrane domains based on an-kyrin and spectrin associated with cell-cell interactions," *Cold Spring Harbor Perspectives in Biology*, vol. 1, no. 6, Article ID a003012, 2009.

[100] P. R. Stabach, P. Devarajan, M. C. Stankewich, S. Bannykh, and J. S. Morrow, "Ankyrin facilitates intracellular trafficking of α1- Na+-K+-ATPase in polarized cells," *American Journal of Physiology, Cell Physiology*, vol. 295, no. 5, pp. C1202–C1214, 2008.

[101] R. R. Dubreuil, P. Wang, S. Dahl, J. Lee, and L. S. B. Goldstein, "Drosophila β spectrin functions independently of α spectrin to polarize the Na,K ATPase in epithelial cells," *Journal of Cell Biology*, vol. 149, no. 3, pp. 647–656, 2000.

[102] A. Das, C. Base, S. Dhulipala, and R. R. Dubreuil, "Spectrin functions upstream of ankyrin in a spectrin cytoskeleton assembly pathway," *Journal of Cell Biology*, vol. 175, no. 2, pp. 325–335, 2006.

[103] V. Koefoed-Johnsen and H. H. Ussing, "The nature of the frog skin potential," *Acta physiologica Scandinavica*, vol. 42, no. 3-4, pp. 298–308, 1958.

[104] S. G. Schultz, "A century of (epithelial) transport physiology: from vitalism to molecular cloning," *American Journal of Physiology*, vol. 274, no. 1, pp. C13–C23, 1998.

[105] L. G. Palmer and O. S. Andersen, "The two-membrane model of epithelial transport: Koefoed-Johnsen and ussing (1958)," *Journal of General Physiology*, vol. 132, no. 6, pp. 607–612, 2008.

[106] L. Reuss, "Ussing's two-membrane hypothesis: the model and half a century of progress," *Journal of Membrane Biology*, vol. 184, no. 3, pp. 211–217, 2001.

[107] E. Frömter and J. Diamond, "Route of passive ion permeation in epithelia," *Nature: New biology*, vol. 235, no. 53, pp. 9–13, 1972.

[108] N. K. Wills, L. Reuss, and S. A. Lewis, *Epithelial Transport : A Guide to Methods and Experimental Analysis*, Chapman & Hall, London, UK, 1996.

[109] J. A. Hernández and S. Chifflet, "Electrogenic properties of the sodium pump in a dynamic model of membrane transport," *Journal of Membrane Biology*, vol. 176, no. 1, pp. 41–52, 2000.

[110] E. M. Wright, D. D.F. LOO, and B. A. Hirayama, "Biology of human sodium glucose transporters," *Physiological Reviews*, vol. 91, no. 2, pp. 733–794, 2011.

[111] D. I. Cook and J. A. Young, "Effect of K+ channels in the apical plasma membrane on epithelial secretion based on secondary active Cl- transport," *Journal of Membrane Biology*, vol. 110, no. 2, pp. 139–146, 1989.

[112] J. E. Melvin, D. Yule, T. Shuttleworth, and T. Begenisich, "Regulation of fluid and electrolyte secretion in salivary gland acinar cells," *Annual Review of Physiology*, vol. 67, pp. 445–469, 2005.

[113] L. Palk, J. Sneyd, T. J. Shuttleworth, D. I. Yule, and E. J. Crampin, "A dynamic model of saliva secretion," *Journal of Theoretical Biology*, vol. 266, no. 4, pp. 625–640, 2010.

[114] S. Futakuchi, H. Ishiguro, S. Naruse et al., "High glucose inhibits HCO3(-) and fluid secretion in rat pancreatic ducts," *Pflugers Archiv*, vol. 459, no. 1, pp. 215–226, 2009.

[115] P. Preston, L. Wartosch, D. Günzel et al., "Disruption of the K+ channel β-subunit KCNE3 reveals an important role in intestinal and tracheal Cl- transport," *Journal of Biological Chemistry*, vol. 285, no. 10, pp. 7165–7175, 2010.

[116] J. Fischbarg and F. P. J. Diecke, "A mathematical model of electrolyte and fluid transport across corneal endothelium," *Journal of Membrane Biology*, vol. 203, no. 1, pp. 41–56, 2005.

[117] N. Montalbetti and J. Fischbarg, "Frequency spectrum of transepithelial potential difference reveals transport-related oscillations," *Biophysical Journal*, vol. 97, no. 6, pp. 1530–1537, 2009.

[118] F. Jorge, "Fluid transport across leaky epithelia: central role of the tight junction and supporting role of aquaporins," *Physiological Reviews*, vol. 90, no. 4, pp. 1271–1290, 2010.

[119] D. Sas, M. Hu, O. W. Moe, and M. Baum, "Effect of claudins 6 and 9 on paracellular permeability in MDCK II cells," *American Journal of Physiology , Regulatory Integrative and Comparative Physiology*, vol. 295, no. 5, pp. R1713–R1719, 2008.

[120] M. Avella, O. Ducoudret, D. F. Pisani, and P. Poujeol, "Swelling-activated transport of taurine in cultured gill cells of sea bass: physiological adaptation and pavement cell plasticity," *American Journal of Physiology, Regulatory Integrative and Comparative Physiology*, vol. 296, no. 4, pp. R1149–R1160, 2009.

[121] R. Montesano, H. Ghzili, F. Carrozzino, B. C. Rossier, and E. Feraille, "cAMP-dependent chloride secretion mediates tubule enlargement and cyst formation by cultured mammalian collecting duct cells," *American Journal of Physiology, Renal Physiology*, vol. 296, no. 2, pp. F446–F457, 2009.

[122] S. Chifflet and J. A. Hernandez, "Use of standard fluorescence microscopy to assess modifications in the plasma membrane potential and in the intracellular concentration of inorganic ions in cultured cells," in *Modern Research and Educational Topics in Microscopy*, A. M. Vilas and J. D. Alvarez, Eds., vol. 1, pp. 229–236, Formatex, 2007.

[123] M. Olivotto, A. Arcangeli, M. Carlà, and E. Wanke, "Electric fields at the plasma membrane level: a neglected element in the mechanisms of cell signalling," *BioEssays*, vol. 18, no. 6, pp. 495–504, 1996.

[124] J. A. Hernández and E. Cristina, "Modeling cell volume regulation in nonexcitable cells: the roles of the Na+ pump and of cotransport systems," *American Journal of Physiology-Cell Physiology*, vol. 275, no. 4, pp. C1067–C1080, 1998.

[125] F. Lang, G. L. Busch, M. Ritter et al., "Functional significance of cell volume regulatory mechanisms," *Physiological Reviews*, vol. 78, no. 1, pp. 247–306, 1998.

[126] J. Vaaraniemi, V. Huotari, V. P. Lehto, and S. Eskelinen, "Effect of PMA on the integrity of the membrane skeleton and morphology of epithelial MDCK cells is dependent on the activity of amiloride-sensitive ion transporters and membrane potential," *European Journal of Cell Biology*, vol. 74, no. 3, pp. 262–272, 1997.

[127] S. Chifflet, J. A. Hernández, S. Grasso, and A. Cirillo, "Nonspecific depolarization of the plasma membrane potential induces cytoskeletal modifications of bovine corneal endothelial cells in culture," *Experimental Cell Research*, vol. 282, no. 1, pp. 1–13, 2003.

[128] V. Nin, J. A. Hernández, and S. Chifflet, "Hyperpolarization of the plasma membrane potential provokes reorganization of the actin cytoskeleton and increases the stability of adherens junctions in bovine corneal endothelial cells in culture," *Cell Motility and the Cytoskeleton*, vol. 66, no. 12, pp. 1087–1099, 2009.

[129] S. Chifflet, V. Correa, V. Nin, C. Justet, and J. A. Hernández, "Effect of membrane potential depolarization on the organization of the actin cytoskeleton of eye epithelia. The role of

adherens junctions," *Experimental Eye Research*, vol. 79, no. 6, pp. 769–777, 2004.

[130] K. Szászi, G. Sirokmány, C. di Ciano-Oliveira, O. D. Rotstein, and A. Kapus, "Depolarization induces Rho-Rho kinase-mediated myosin light chain phosphorylation in kidney tubular cells," *American Journal of Physiology, Cell Physiology*, vol. 289, pp. C673–C685, 2005.

[131] F. Waheed, P. Speight, G. Kawai, Q. Dan, A. Kapus, and K. Szászi, "Extracellular signal-regulated kinase and GEF-H1 mediate depolarization-induced Rho activation and paracellular permeability increase," *American Journal of Physiology, Cell Physiology*, vol. 298, no. 6, pp. C1376–C1387, 2010.

[132] H. Oberleithner, C. Callies, K. Kusche-Vihrog et al., "Potassium softens vascular endothelium and increases nitric oxide release," *Proceedings of the National Academy of Sciences of the United States of America*, vol. 106, no. 8, pp. 2829–2834, 2009.

[133] C. Callies, J. Fels, I. Liashkovich et al., "Membrane potential depolarization decreases the stiffness of vascular endothelial cells," *Journal of Cell Science*, vol. 124, no. 11, pp. 1936–1942, 2011.

[134] S. S. Grewal, A. M. Horgan, R. D. York, G. S. Withers, G. A. Banker, and P. J. S. Stork, "Neuronal calcium activates a Rap1 and B-Raf signaling pathway via the cyclic adenosine monophosphate-dependent protein kinase," *Journal of Biological Chemistry*, vol. 275, no. 5, pp. 3722–3728, 2000.

[135] M. Ishii, A. Inanobe, S. Fujita, Y. Makino, Y. Hosoya, and Y. Kurachi, "Ca2+ elevation evoked by membrane depolarization regulates G protein cycle via RGS proteins in the heart," *Circulation Research*, vol. 89, no. 11, pp. 1045–1050, 2001.

[136] M. Mita, H. Yanagihara, S. Hishinuma, M. Saito, and M. P. Walsh, "Membrane depolarization-induced contraction of rat caudal arterial smooth muscle involves Rho-associated kinase," *Biochemical Journal*, vol. 364, no. 2, pp. 431–440, 2002.

[137] S. Sakurada, N. Takuwa, N. Sugimoto et al., "Ca2+-dependent activation of Rho and Rho kinase in membrane depolarization-induced and receptor stimulation-induced vascular smooth muscle contraction," *Circulation Research*, vol. 93, no. 6, pp. 548–556, 2003.

[138] C. Liu, J. Zuo, E. Pertens, P. B. Helli, and L. J. Janssen, "Regulation of Rho/ROCK signaling in airway smooth muscle by membrane potential and [Ca2+]i," *American Journal of Physiology, Lung Cellular and Molecular Physiology*, vol. 289, no. 4, pp. L574–L582, 2005.

[139] M. H. Roos, W. F. Van Rodijnen, A. A. Van Lambalgen, P. M. Ter Wee, and G. J. Tangelder, "Renal microvascular constriction to membrane depolarization and other stimuli: pivotal role for rho-kinase," *Pflugers Archiv*, vol. 452, no. 4, pp. 471–477, 2006.

[140] M. Fernández-Tenorio, C. Porras-González, A. Castellano, A. Del Valle-Rodríguez, J. López-Barneo, and J. Ureña, "Metabotropic regulation of RhoA/Rho-associated kinase by l-type Ca2+ Channels: new mechanism for depolariza- tion-Evoked mammalian arterial contraction," *Circulation Research*, vol. 108, no. 11, pp. 1348–1357, 2011.

[141] C. X. Bai, S. Kim, A. J. Streets, A. C. M. Ong, and L. Tsiokas, "Activation of TRPP2 through mDia1-dependent voltage gating," *EMBO Journal*, vol. 27, no. 9, pp. 1345–1356, 2008.

[142] Y. Murata and Y. Okamura, "Depolarization activates the phosphoinositide phosphatase Ci-VSP, as detected in Xenopus oocytes coexpressing sensors of PIP2," *Journal of Physiology*, vol. 583, no. 3, pp. 875–889, 2007.

[143] H. Neuhaus and T. Hollemann, "Kidney specific expression of cTPTE during development of the chick embryo," *Gene Expression Patterns*, vol. 9, no. 8, pp. 568–571, 2009.

[144] M. Ogasawara, M. Sasaki, N. Nakazawa, A. Nishino, and Y. Okamura, "Gene expression profile of Ci-VSP in juveniles and adult blood cells of ascidian," *Gene Expression Patterns*, vol. 11, pp. 233–238, 2011.

[145] Y. Okamura and J. E. Dixon, "Voltage-sensing phosphatase: its molecular relationship with PTEN," *Physiology*, vol. 26, no. 1, pp. 6–13, 2011.

[146] R. P. Bhattacharyya, A. Reményi, B. J. Yeh, and W. A. Lim, "Domains, motifs, and scaffolds: the role of modular interactions in the evolution and wiring of cell signaling circuits," *Annual Review of Biochemistry*, vol. 75, pp. 655–680, 2006.

[147] T. Takenawa, "Phosphoinositide-binding interface proteins involved in shaping cell membranes," *Proceedings of the Japan Academy Series B*, vol. 86, no. 5, pp. 509–523, 2010.

[148] X. Zhang, X. Chen, C. Jia, X. Geng, X. Du, and H. Zhang, "Depolarization increases phosphatidylinositol (PI) 4,5-bisphosphate level and KCNQ currents through PI 4-kinase mechanisms," *Journal of Biological Chemistry*, vol. 285, no. 13, pp. 9402–9409, 2010.

[149] M. Cotton and A. Claing, "G protein-coupled receptors stimulation and the control of cell migration," *Cellular Signalling*, vol. 21, no. 7, pp. 1045–1053, 2009.

[150] S. H. Lee and R. Dominguez, "Regulation of actin cytoskeleton dynamics in cells.," *Molecules and cells*, vol. 29, no. 4, pp. 311–325, 2010.

[151] M. Cohen-Armon and M. Sokolovsky, "Evidence for involvement of the voltage-dependent Na+ channel gating in depolarization-induced activation of G-proteins," *Journal of Biological Chemistry*, vol. 268, no. 13, pp. 9824–9838, 1993.

[152] Y. Ben-Chaim, O. Tour, N. Dascal, I. Parnas, and H. Parnas, "The M2 muscarinic G-protein-coupled receptor is voltage-sensitive," *Journal of Biological Chemistry*, vol. 278, no. 25, pp. 22482–22491, 2003.

[153] J. Martinez-Pinna, I. S. Gurung, C. Vial et al., "Direct voltage control of signaling via P2Y1 and other Gαq-coupled receptors," *Journal of Biological Chemistry*, vol. 280, no. 2, pp. 1490–1498, 2005.

[154] Y. Ben-Chaim, B. Chanda, N. Dascal, F. Bezanilla, I. Parnas, and H. Parnas, "Movement of 'gating charge' is coupled to ligand binding in a G-protein-coupled receptor," *Nature*, vol. 444, no. 7115, pp. 106–109, 2006.

[155] L. Ohana, O. Barchad, I. Parnas, and H. Parnas, "The metabotropic glutamate G-protein-coupled receptors mGluR3 and mGluR1a are voltage-sensitive," *Journal of Biological Chemistry*, vol. 281, no. 34, pp. 24204–24215, 2006.

[156] K. Sahlholm, D. Marcellino, J. Nilsson, K. Fuxe, and P. Arhem, "Voltage-sensitivity at the human dopamine D2S receptor is agonist-specific," *Biochemical and Biophysical Research Communications*, vol. 377, no. 4, pp. 1216–1221, 2008.

[157] S. Genet, R. Costalat, and J. Burger, "A few comments on electrostatic interactions in cell physiology," *Acta Biotheoretica*, vol. 48, no. 3-4, pp. 273–287, 2000.

[158] T. Yeung and S. Grinstein, "Lipid signaling and the modulation of surface charge during phagocytosis," *Immunological Reviews*, vol. 219, no. 1, pp. 17–36, 2007.

[159] U. Dietrich, P. Kruger, and J. A. Kas, "Structural investigation on the adsorption of the MARCKS peptide on anionic lipid monolayers—effects beyond electrostatic," *Chemistry and Physics of Lipids*, vol. 164, no. 4, pp. 266–275, 2011.

[160] R. E. Redfern, D. Redfern, M. L. M. Furgason, M. Munson, A. H. Ross, and A. Gericke, "PTEN phosphatase selectively

binds phosphoinositides and undergoes structural changes," *Biochemistry*, vol. 47, no. 7, pp. 2162–2171, 2008.

[161] J. F. Hancock, H. Paterson, and C. J. Marshall, "A polybasic domain or palmitoylation is required in addition to the CAAX motif to localize p21(ras) to the plasma membrane," *Cell*, vol. 63, no. 1, pp. 133–139, 1990.

[162] C. T. Sigal, W. Zhou, C. A. Buser, S. McLaughlin, and M. D. Resh, "Amino-terminal basic residues of Src mediate membrane binding through electrostatic interaction with acidic phospholipids," *Proceedings of the National Academy of Sciences of the United States of America*, vol. 91, no. 25, pp. 12253–12257, 1994.

[163] M. A. O. Magalhaes and M. Glogauer, "Pivotal advance: phospholipids determine net membrane surface charge resulting in differential localization of active Rac1 and Rac2," *Journal of Leukocyte Biology*, vol. 87, no. 4, pp. 545–555, 2010.

[164] K. Hamada, T. Shimizu, T. Matsui, S. Tsukita, S. Tsukita, and T. Hakoshima, "Structural basis of the membrane-targeting and unmasking mechanisms of the radixin FERM domain," *EMBO Journal*, vol. 19, no. 17, pp. 4449–4462, 2000.

[165] G. A. Gomez and J. L. Daniotti, "Electrical properties of plasma membrane modulate subcellular distribution of K-Ras," *FEBS Journal*, vol. 274, no. 9, pp. 2210–2228, 2007.

[166] A. Arcangeli and A. Becchetti, "Complex functional interaction between integrin receptors and ion channels," *Trends in Cell Biology*, vol. 16, no. 12, pp. 631–639, 2006.

[167] J. A. Tuszyński, S. Portet, J. M. Dixon, C. Luxford, and H. F. Cantiello, "Ionic wave propagation along actin filaments," *Biophysical Journal*, vol. 86, no. 4, pp. 1890–1903, 2004.

[168] M. A. Watsky, "Nonselective cation channel activation during wound healing in the corneal endothelium," *American Journal of Physiology, Cell Physiology*, vol. 268, no. 5, pp. C1179–C1185, 1995.

[169] V. E. Franklin-Tong and C. W. Gourlay, "A role for actin in regulating apoptosis/programmed cell death: evidence spanning yeast, plants and animals," *Biochemical Journal*, vol. 413, no. 3, pp. 389–404, 2008.

[170] J. Y. Rao, Y. S. Jin, Q. Zheng, J. Cheng, J. Tai, and G. P. Hemstreet, "Alterations of the actin polymerization status as an apoptotic morphological effector in HL-60 cells," *Journal of Cellular Biochemistry*, vol. 75, no. 4, pp. 686–697, 1999.

[171] D. P. Cioca and K. Kitano, "Induction of apoptosis and CD10/neutral endopeptidase expression by jaspamide in HL-60 line cells," *Cellular and Molecular Life Sciences*, vol. 59, no. 8, pp. 1377–1387, 2002.

[172] H. Suria, L. A. Chau, E. Negrou, D. J. Kelvin, and J. Madrenas, "Cytoskeletal disruption induces T cell apoptosis by a caspase-3 mediated mechanism," *Life Sciences*, vol. 65, no. 25, pp. 2697–2707, 1999.

[173] R. Franco, C. D. Bortner, and J. A. Cidlowski, "Potential roles of electrogenic ion transport and plasma membrane depolarization in apoptosis," *Journal of Membrane Biology*, vol. 209, no. 1, pp. 43–58, 2006.

[174] E. M. Fish and B. A. Molitoris, "Alterations in epithelial polarity and the pathogenesis of disease states," *New England Journal of Medicine*, vol. 330, no. 22, pp. 1580–1588, 1994.

[175] M. Cereijido, R. G. Contreras, D. Flores-Benítez et al., "New diseases derived or associated with the tight junction," *Archives of Medical Research*, vol. 38, no. 5, pp. 465–478, 2007.

[176] P. Cowin, T. M. Rowlands, and S. J. Hatsell, "Cadherins and catenins in breast cancer," *Current Opinion in Cell Biology*, vol. 17, no. 5, pp. 499–508, 2005.

[177] C. Salon, S. Lantuejoul, B. Eymin, S. Gazzeri, C. Brambilla, and E. Brambilla, "The E-cadherin-beta-catenin complex and its implication in lung cancer progression and prognosis," *Future Oncology*, vol. 1, no. 5, pp. 649–660, 2005.

[178] N. Bonitsis, A. Batistatou, S. Karantima, and K. Charalabopoulos, "The role of cadherin/catenin complex in malignant melanoma," *Experimental Oncology*, vol. 28, no. 3, pp. 187–193, 2006.

[179] U. Cavallaro, S. Liebner, and E. Dejana, "Endothelial cadherins and tumor angiogenesis," *Experimental Cell Research*, vol. 312, no. 5, pp. 659–667, 2006.

[180] T. A. Sutton, H. E. Mang, S. B. Campos, R. M. Sandoval, M. C. Yoder, and B. A. Molitoris, "Injury of the renal microvascular endothelium alters barrier function after ischemia," *American Journal of Physiology, Renal Physiology*, vol. 285, no. 2, pp. F191–F198, 2003.

[181] M. G. Laukoetter, P. Nava, and A. Nusrat, "Role of the intestinal barrier in inflammatory bowel disease," *World Journal of Gastroenterology*, vol. 14, no. 3, pp. 401–407, 2008.

Brain Miffed by Macrophage Migration Inhibitory Factor

Nic E. Savaskan,[1] Günter Fingerle-Rowson,[2] Michael Buchfelder,[1] and Ilker Y. Eyüpoglu[1]

[1] *Department of Neurosurgery, University of Erlangen-Nuremberg, Schwabachanlage 6, 91054 Erlangen, Germany*
[2] *Clinic I for Internal Medicine, University Hospital Cologne, Kerpener Straße 62, 50924 Cologne, Germany*

Correspondence should be addressed to Nic E. Savaskan, Nicolai.savaskan@uk-erlangen.de and
Ilker Y. Eyüpoglu, ilker.eyupoglu@uk-erlangen.de

Academic Editor: Pier Giorgio Mastroberardino

Macrophage migration inhibitory factor (MIF) is a cytokine which also exhibits enzymatic properties like oxidoreductase and tautomerase. MIF plays a pivotal role in innate and acquired immunity as well as in the neuroendocrine axis. Since it is involved in the pathogenesis of acute and chronic inflammation, neoangiogenesis, and cancer, MIF and its signaling components are considered suitable targets for therapeutic intervention in several fields of medicine. In neurodegenerative and neurooncological diseases, MIF is a highly relevant, but still a hardly investigated mediator. MIF operates via intracellular protein-protein interaction as well as in CD74/CXCR2/CXCR4 receptor-mediated pathways to regulate essential cellular systems such as redox balance, HIF-1, and p53-mediated senescence and apoptosis as well as multiple signaling pathways. Acting as an endogenous glucocorticoid antagonist, MIF thus represents a relevant resistance gene in brain tumor therapies. Alongside this dual action, a functional homolog-annotated D-dopachrome tautomerase/MIF-2 has been uncovered utilizing the same cell surface receptor signaling cascade as MIF. Here we review MIF actions with respect to redox regulation in apoptosis and in tumor growth as well as its extracellular function with a focus on its potential role in brain diseases. We consider the possibility of MIF targeting in neurodegenerative processes and brain tumors by novel MIF-neutralizing approaches.

1. Introduction

Macrophage migration inhibitory factor was one of the first cytokines identified after interferon [1] and represents a key regulator of the immune system (MIF is historically also known as glycosylation-inhibiting factor, GIF) [2, 3]. MIF was initially described as a proinflammatory soluble factor derived from T cells under various conditions such as delayed-type hypersensitivity responses and inflammation guiding site-specific migration of immunocompetent cells [2, 4]. It soon became apparent that MIF possesses immunoregulatory effects and is even constitutively detectable in various body fluids and cells of the mammalian organism. MIF levels are higher at sites of inflammation, within immune and brain cells and various cancer cells (Figure 1). Later, MIF was shown to contribute to neuroendocrine modulation, as a pituitary gland-derived hormone, inflammation, atherosclerosis, cancer development, and cancer progression [5–11]. MIF was first cloned from T cells in 1989, which revealed not only its primary sequence and conserved domains but also led to the discovery that MIF exhibits two catalytic centers, one for thiol-protein oxidoreductase activity and another one for tautomerase activity [12–14]. These findings fueled speculation that MIF was not only a cytokine, but a possible combination of enzyme and cytokine "cytozyme" [12, 13, 15, 16]. Hence, MIF's conserved gene structure and structural homology with D-dopachrome tautomerase (DDT/MIF-2) aroused further speculation surrounding its proposed enzymatic actions and cytokine properties [17, 18]. This enigmatic property of MIF fostered the development of genetic approaches towards a better understanding of its biology in physiology and disease.

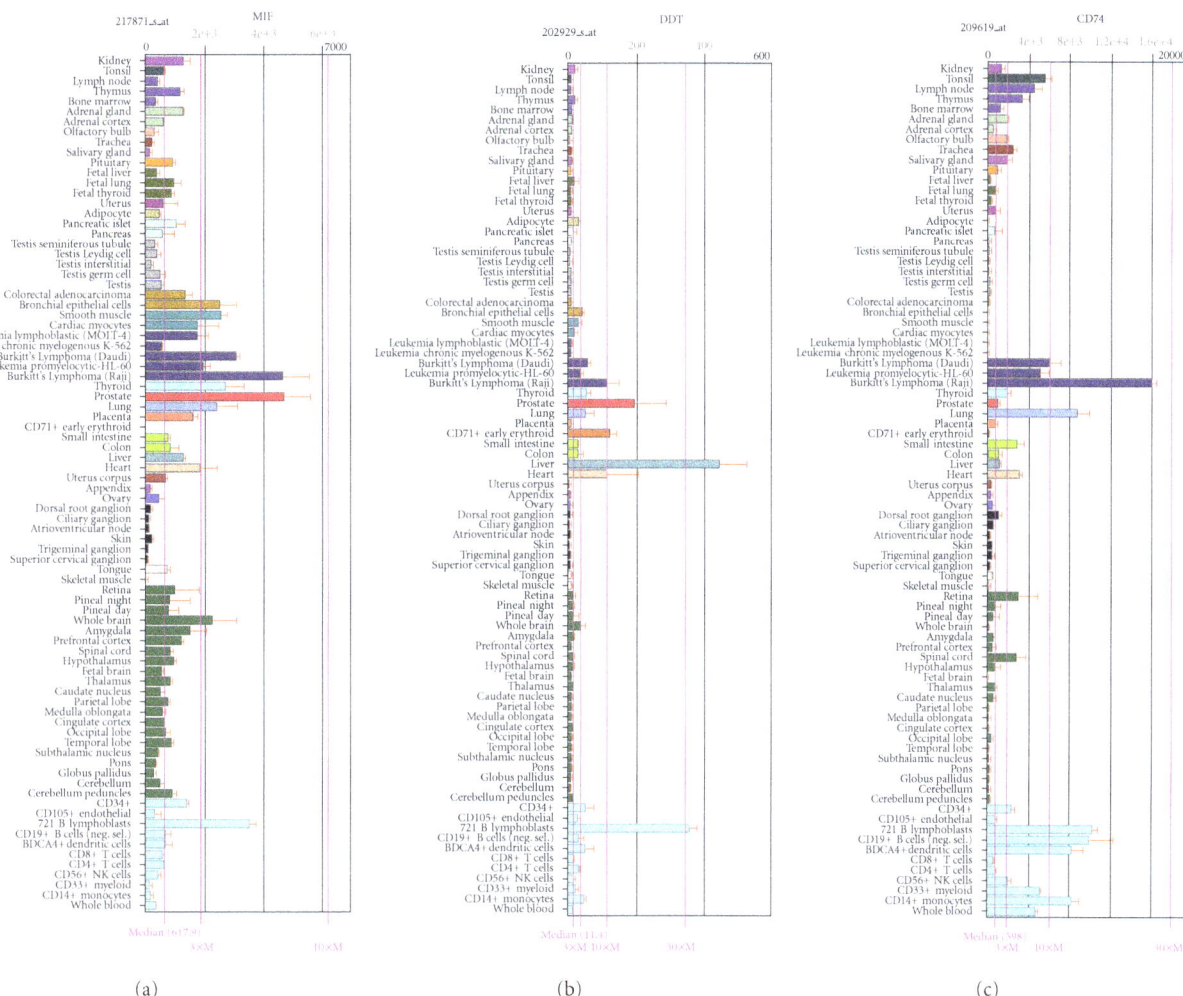

FIGURE 1: MIF, DDT, and CD74 distribution in human tissues. Comparative analysis of MIF, DDT (MIF-2), and its receptor CD74 expression in various human tissues. For human mRNA expression analysis, the BioGPS database (http://biogps.gnf.org profile graph) with the Affymetrix chip Human U133A was acquired. Note in particular the different expression values of MIF and DDT in brain tissue. For details on the Affymetrix chip analysis, see [37, 38].

To date, it is known that MIF induces pleiotropic functions in inflammation, malignant transformation, and endocrine and metabolic processes. In this paper, we focus on MIF-dependent signaling in redox regulation and brain cancer progression and discuss recent findings in MIF neurobiology.

2. MIF Structure and Function

The small and highly conserved protein MIF with an approximate molecular weight of 12.5 kDa (human MIF contains 115 aa) does not exhibit any similarities with known cytokines [12, 19, 20]. MIF protein does not require an N-terminal export-specific leader sequence for secretion as it is secreted via an alternative, nonclassical pathway.

However, MIF contains two conserved domain motifs (Figure 2). The CXXC domain motif (Cys-X-X-Cys at position 56–60) in the center of MIF has been shown to exhibit

catalytic activity [21–23]. It is a consensus sequence of proteins of the thiol-protein oxidoreductase superfamily, other members of which include thioredoxins, glutaredoxins, and peroxiredoxins [24, 25]. Common to this enzyme superfamily is that all members are involved in disulfide-mediated redox reactions and glutathione metabolism in which the CXXC domain takes center stage. In the case of MIF, the CXXC domain is potentially involved in forming MIF homodimers and trimers, the most likely active form of MIF [26–28]. Hence, the CXXC domain of MIF has been shown to exhibit low redox catalytic activity *in vitro* (compared to thioredoxin and glutaredoxins) and modulates cellular redox stress responses by elevating the intracellular glutathione (GSH) pool [14, 29–34]. In particular, reactive oxygen species (ROS) induce elevated MIF mRNA and protein expression in neurons, and MIF represents a negative regulator for angiotensin-II-induced chronotropic action

(a)

(b)

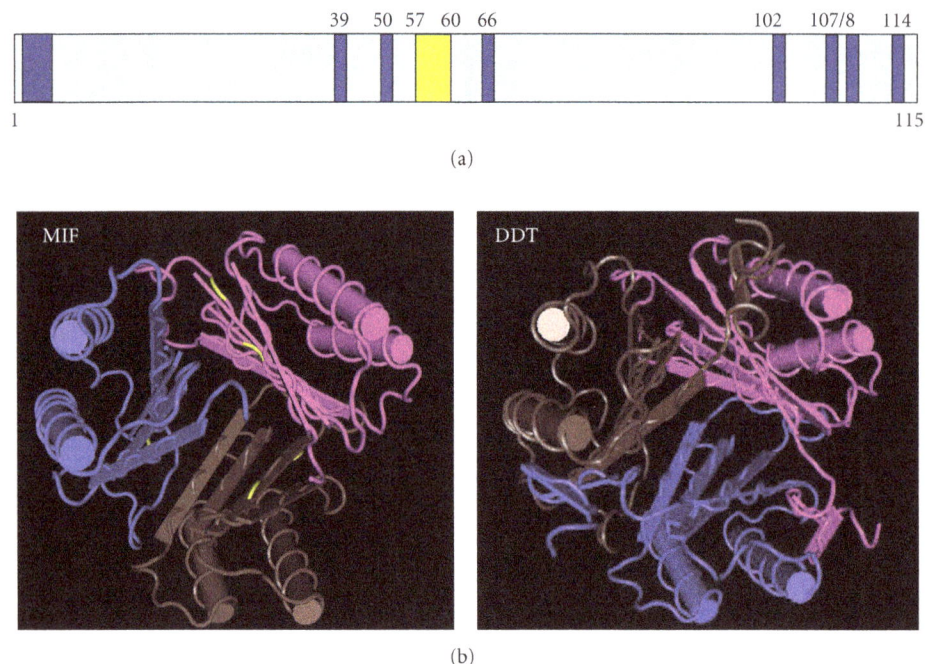

Figure 2: Structural homologies of MIF and DDT. (a) Primary structural scheme of the human MIF gene. Yellow region indicates the CXXC domain, the blue boxed domains indicate the proposed tautomerase/isomerase domains and clustered amino acids (Phe3, Val39, Gly50, Lys66, Asn102, Gly107, Trp108, Phe113, and Ala114) [18]. (b) Structural comparison of human MIF and DDT trimers. The catalytically important CXXC domain is shown in yellow. β-sheets are given as arrows, and α-sheets are shown as columns. Data were obtained from the NCBI database (http://www.ncbi.nlm.nih.gov/Structure/mmdb/mmdbsrv.cgi?uid=89970) based on the study of [28].

and firing in neurons [33, 35, 36]. In addition, MIF has been found to protect from oxidative stress in an ischemia/reperfusion cardiac lesion model [29, 34].

It is worthy to note that the CXXC domain in MIF seems to be essential in facilitating the inhibition of angiotensin II. Evidence for this comes from MIF peptide fragments containing the CXXC domain (ΔMIF$^{50-65}$) which mimic the wild-type MIF action whereas a mutant ΔMIF$^{50-65}$ replacing the second cystine to serine (C60/S60) does not [36]. Redox stress is known to be elevated under conditions of hypoxia or malignant transformation. Hypoxia-inducible MIF elevation has been reported in head and neck cancer cells, pancreas, cervical carcinoma cell lines, and glial tumors [40–43]. Further studies revealed that MIF transcription is induced by hypoxia-inducible factor 1α (HIF1α) and is physically linked to HIF1α through COP9 signalosome subunit 5 (CSN5) interaction [42, 44, 45]. MIF can potentially inhibit apoptosis and p53-mediated growth arrest and its depletion impairs cell proliferation in cancer [6, 7, 46–48]. It was suggested that MIF's action of blocking apoptosis is dependent on its catalytic oxidoreductase activity. However, whether MIF deletion in tumors makes them prone to hypoxia and affects tumor vasculature *in vivo* remains to be thoroughly investigated. First studies already indicate that MIF expression and MIF signalling are associated with tumor angiogenesis [49–51].

The second enzymatic domain of MIF is its enigmatic tautomerase activity which has spurred intensive research on the physiologic substrate and function. In an attempt to identify the enzyme responsible for converting the non-naturally occurring substrate L-dopachrome into dihydroxyindole carboxylic acid (a catalytic step important in biosynthesis of melanin), MIF was purified and subsequently identified by peptide sequencing from bovine lens tissue [13]. Further investigations of the structure of MIF revealed that the tautomerase/isomerase activity is located at the N-terminal portion with a conserved proline residue at position 2 [27, 28, 52–54]. The three-dimensional protein structure of MIF revealed striking similarities with D-dopachrome tautomerase (DDT/PPT2) although MIF shares solely ~30% amino acid sequence homology with DDT [28, 53] (Figure 2). These findings led to various enzymatic and mutational investigations identifying the N-terminal portion of MIF as essential for tautomerase activity [15, 16, 55]. However, since the finding of MIF's *in vitro* tautomerase activity investigations have focused on the identification of its physiological substrate and biological role which is still ongoing. Genetic studies in the meantime revealed that catalytically dead mutants still exert MIF-specific functions. Moreover, tautomerase-null MIF knock-in mice compensate the MIF gene deletion (MIF$^{-/-}$ or MIF null mutant) phenotype which leads to the argument that the tautomerase activity may be possibly dispensable *in vivo* [18, 56, 57]. MIF's highly conserved substrate pocket may represent a vestigial relict reflecting its ancestral origin in innate immunity and be dispensable at least for its function in promoting cellular growth and tumorigenesis *in vivo* [11, 47, 58, 59]. However, the catalytically dead MIF mutant (P1G-MIF) shows reduced binding to some protein interaction partners, such as its cell surface receptor CD74 and the c-jun amino-terminal kinase activator Jab1/CSN5. This indicates that the N-terminal proline and the catalytic pocket may play a role in protein-protein interaction of MIF with its binding partners [57]. Noteworthy were findings reporting more pronounced phenotype and defects in CD74 knock-out mice (MIF receptor) than in solely MIF-deficient mice [59–64]. This led to the hypothesis that more MIF-like ligands acting on CD74 receptor may exist. The group of Bucala and colleagues recently identified D-dopachrome tautomerase (DDT) as a MIF-like cytokine with overlapping functions [65]. Neutralizing antibodies against DDT can protect mice from lethal endotoxic shock to a comparable extent as MIF neutralization, by reducing circulating TNF-α, IFN-γ, IL12, and IL-1β [5, 60, 65–68]. It has subsequently been suggested to redefine DDT as MIF-2 due to their structural homologies and functional resemblance with data on DDT knock-out mice and combined neutralization studies to unravel this renaming.

3. MIF Distribution in the Brain

Distribution and microarray expression profiles (BioGPS analysis) of MIF, DDT, and their joint receptor CD74 already suggest spatial overlapping as well as ancillary functions (Figure 1). MIF is widely expressed in the body and shows high levels in lymphocytes, thyroid, prostate, placenta, and lungs. In the murine brain, MIF transcripts and protein are mainly present in the cortex, hippocampus, and pituitary gland [5, 69] and thus differ in distribution and expression level in comparison to DDT (Figure 3). In particular, MIF immunoreactivity has been found in neurons of the hippocampus within fiber structures and terminals such as the mossy fibers of the dentate gyrus and in dendrites of the hippocampal CA regions [69]. Furthermore, MIF is upregulated in neurons and in macrophages following intracranial LPS stimulation. Interestingly, MIF is also found in microglial cells, the resident macrophages of the brain as well as in cerebrospinal fluid (CSF), and shows elevation after experimental LPS treatment, too. Moreover, MIF pretreatment can reduce the number of invading microglial cells and macrophage into allogeneic fetal mesencephalic grafts in rodents [70]. However, this MIF treatment did not affect the outcome on graft function and survival leaving the potential of MIF as a neuroimmune modulator in Parkinson's disease open. It has recently been shown that MIF can promote the growth of neural progenitor cells *in vitro* [71], indicating already a growth-promoting effect in particular cell populations. Contrary to such growth promoting effect is one report on elevated MIF levels in the cerebrospinal fluid of Alzheimer patients and the beneficial effects of MIF inhibition after amyloid β protein-induced neurotoxicity *in vitro* [72]. As indicated above, MIF may function in a context-dependent manner with various effects on different neural and glial cells. The presence of MIF in hippocampal structures which are prone to glucocorticoid-induced tissue damage has led to speculations of MIF and its association with glucocorticoid action under normal and pathophysiological processes.

4. MIF Signaling, Glucocorticoids, and Metabolism

MIF was one of the first cytokine-mediated activities derived from T cells described. It then became apparent that MIF is also expressed by monocytes/macrophages and signals in both an autocrine and paracrine manner [2, 4]. Gene-targeting experiments and neutralization approaches affirmed its upstream role in the inflammatory cascade promoting proinflammatory mediators such as TNF-α, IL-12, IL-1β, and PGE$_2$ [7, 59, 60]. MIF's role as an autocrine innate immune regulator has been exemplified by its "auto-loop" route through TNF-α, which in turn leads to further MIF secretion in macrophages [73]. Thus, it became apparent that MIF follows two signaling principles. First, MIF executes its biological function as a secreted molecule requiring specific receptor(s) at the cell surface of its target cells, that is, transcellular signaling. Secondly, MIF acts as an intracellular or autocrine signaling molecule with catalytic activity and specific binding partners due to its structural features (intracellular domains and mechanisms; see section above).

The identification of MIF's receptor-mediated signaling gave rise to a hub for the discovery of intracellular and extracellular interaction partners and functions [63, 74–79]. To date receptor-mediated MIF signaling has been identified through the cell surface receptor complexes CD74 (CD74/invariant chain—CD44 signaling complex), CXCR2, CXCR4, and CD74-CXCR2/4 [63, 74, 76, 80] (Figure 4). Especially the structural homology of the canonical CXCL8 ligand, a so called pseudo-(E)LR motif present in MIF and binding to CXCR2 and CXCR4 qualified MIF as a non-cognate chemokine ligand [27, 63, 81]. These receptors bind MIF to the surface of cells and mediate activation of extracellular-regulated mitogen-activated protein (ERK-MAP), phosphatidylinositol 3/protein kinase B (PI3K/AKT), and Src-tyrosine kinases through CD44, already indicating the presence of a link to oncogenic signaling utilized by cancer cells (Figure 4).

In particular, MIF impacts macrophage and lymphocyte functions and thereby regulates innate and acquired immunity [73, 82, 83]. In mice, MIF was cloned as an immunoregulatory peptide from the pituitary gland and was shown to specifically counteract glucocorticoid effects such as suppression of TNF-α, IL-8, and IL-1β secretion [5, 84, 85]. Moreover, MIF's impact on the innate immune system can be fatal in lethal endotoxic shock by counteracting the protective effects of glucocorticoids at various levels [86, 87]. Glucocorticoids and steroid analogues such as dexamethasone are widely used and are most effective anti-inflammatory drugs, acting through various mechanisms and recruiting downstream effectors such as NFκB, histone deacetylase 2 (HDAC2), $\alpha 1\beta 1$ integrin, and phospholipase

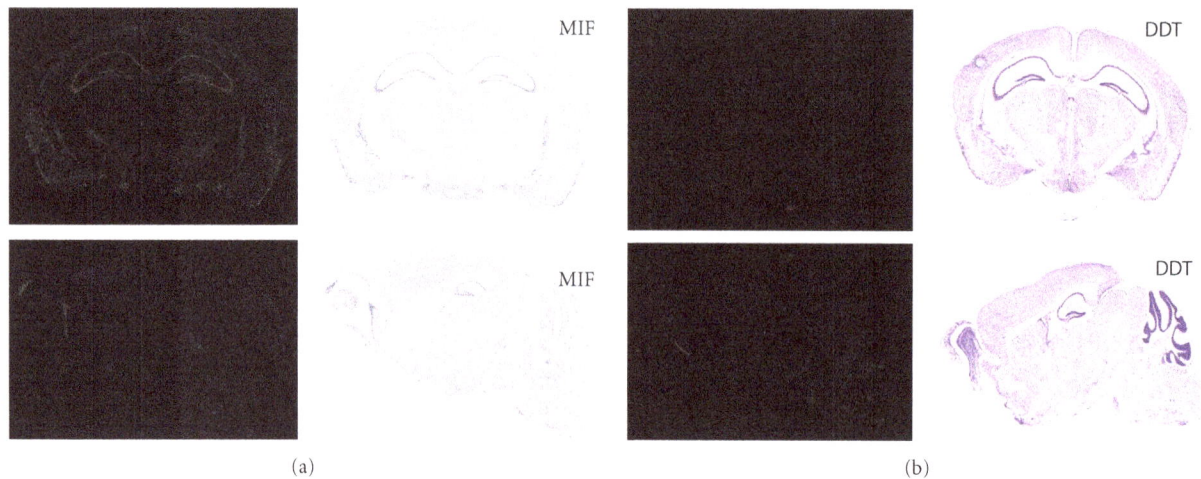

FIGURE 3: MIF and DDT distribution in the brain. Representative in situ hybridization images of MIF mRNA (a) and DDT mRNA distribution (b) in adult mouse brain (left) with consecutive counterstained brain section (Nissl stain, right). Upper panels of (a) and (b) represent coronal plane; lower panels show sagittal plane. Data were provided from the Allan Brain Atlas website (http://www.brain-map.org/), and the Brain Explorer 1.3 software was utilized for the visualization of gene expression [39].

A2 (PLA2) [88, 89]. In particular, glucocorticoids have been used for decades for the treatment of various neuroinflammatory, neurotrauma, and neurooncological disease conditions. One reason lies in that glucocorticoids are one of the most powerful classes of agents in reducing tumor-associated edema and tissue swelling and can thus reduce the incidence of fatal herniation in space occupying lesions to a certain extent. MIF in this pathway is therefore of clinical significance.

MIF counteracts glucocorticoid signalling by decreasing IκB levels leading to NFκB activation, upregulates PLA2, and downregulates MAP kinase phosphatase 1 [86, 87, 90]. The bell-shaped MIF regulation by glucocorticoids is worthy of note with increased MIF release from monocytes/macrophages at low physiological amounts of glucocorticoids and inhibited MIF release at high glucocorticoid concentrations [84, 91]. In this manner, MIF inhibition offers an alternative strategy for anti-inflammatory therapy in neuroinflammation such as multiple sclerosis and Guillain-Barré syndrome, although the effects of MIF on prescribed glucocorticoid analogues in patients require further consideration. Hence, MIF can control glucose catabolism in muscle cells by elevating the level of the key enzyme phosphofructo-2-kinase leading to lactate production [92]. MIF also modulates downstream AMP-activated protein kinase effects in cardiac cells such as the glucose transport function [93]. Whether MIF upregulates phosphofructo-2-kinase and glycolysis in brain tumor cells with subsequently increased lactate release has not yet been tested. Since the Warburg effect is one characteristic feature of malignant gliomas (i.e., primary brain tumors derived from glial and precursor cells), further investigation into the metabolic effects of MIF in brain tumor cells would be highly desirable.

5. MIF Links Inflammation with Cell Cycle Regulation

MIF has a central role as monocytes/macrophages in the global regulator of monocyte/macrophage-derived cytokines. It is an interesting finding that distinct thresholds of MIF affect monocytes/macrophages differentially. At low concentrations MIF induces the release of TNF-α, IL-12, IL-1β, and PGE$_2$ and, in a distinct difference from other "common" cytokines, involves MAPK, Akt, and PI3K activation and regulation of Jab1 and p53 [6, 7, 59, 94, 95]. In particular, the latter is involved in the resolution of inflammation by inducing p53-dependent, activation-induced cell death [6]. High and sustained MIF action, for instance, in chronic inflammation, also promotes the release of macrophage effector cytokines such as TNF-α, IL-12, IL-1β, and PGE$_2$. On the other hand it also prevents cytoplasmatic accumulation of the tumor suppressor gene p53, thus inhibiting apoptosis (Figure 4). This peculiarity of MIF caught the attention of the cancer research field. Bypassing p53-mediated growth arrest is an important feature of cancer cells and of a tumor promoting microenvironment. TP53, the human gene encoding the p53 protein, mutates at a high frequency (approx. 30%) in adult malignant gliomas and glioblastomas. The increased expression of MIF in malignant gliomas is of particular interest since MIF suppresses p53-dependent signaling and thereby enhances susceptibility to further oncogenic mutations. Hence, MIF interacts with Jab1/CSN5 and negatively regulates the cullin-1-containing ubiquitin E3 ligase complex with effects on p27- and E2F1-3-dependent cell cycle control [96, 97]. Conversely, loss of MIF in a p53-deficient background leads to uncoupled DNA damage checkpoint response, thereby aggravating tumorigenesis in p53$^{-/-}$/MIF$^{-/-}$ mice [96]. It has recently

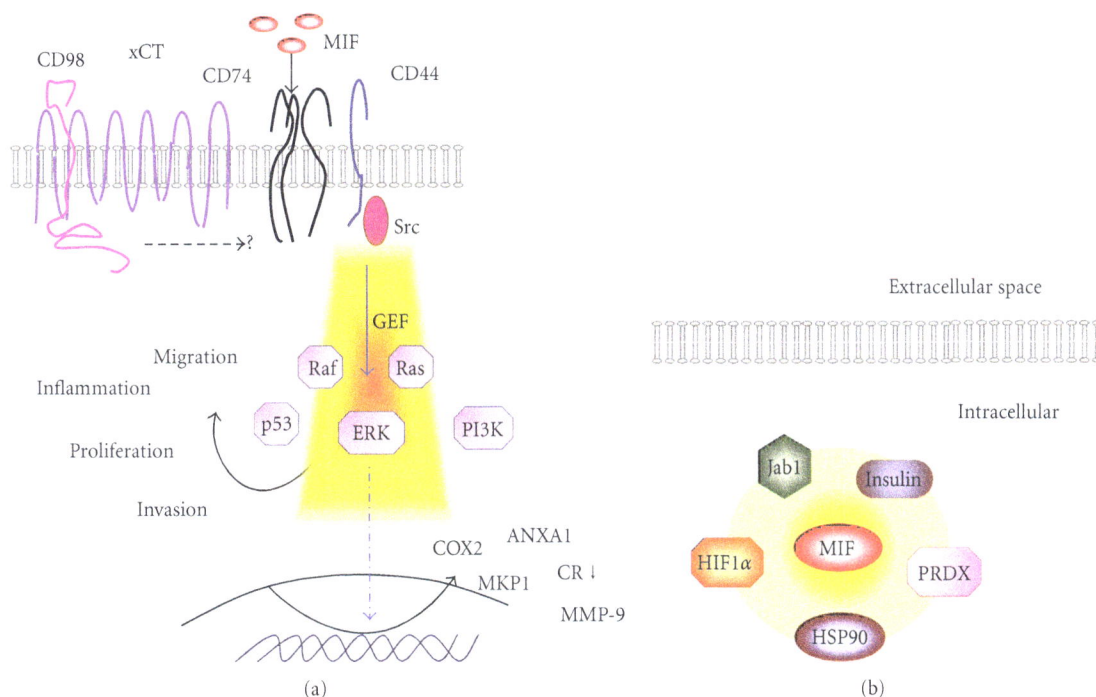

FIGURE 4: MIF receptor signalling and downstream effectors. (a) Schematic model of receptor-mediated MIF signalling involving CD74 and CXCRs. The involvement of the glutamate antiporter xCT (system x_c^-, xCT forms a heterodimer with CD98 as indicated) in CD74/CD44-dependent signalling is proposed, indicated by the dotted arrow. (b) MIF binding partners with link to brain cancer. Note that the indicated MIF-binding partners given in the scheme are far from complete. Abbreviations used: COX2, cyclooxygenase 2; ERK, extracellular signal-regulated kinases; GC, glucocorticoids; GR, glucocorticoid receptor; GEF, guanosine exchange factor; HIF1α, hypoxy Jab1, Jun-activation domain-binding protein-1; MKP1, mitogen-activated protein kinase phosphatases; MMP-9, matrix metallopeptidase or type IV collagenase/gelatinase B; PRDX, peroxiredoxin; Src, sarcoma protooncogene.

been shown that the chaperone HSP90 stabilizes MIF for E3-ubiquitin-ligase-dependent proteasome degradation in various tumor cells, leading to increased MIF levels even under siRNA-mediated transcriptional silencing [98]. This regulatory protein stabilization feature secures persistent MIF action in cancer cells independent of transcriptional and translational levels.

6. MIF, Brain Tumors, Angiogenesis, and Tumor Microenvironment

MIF is produced by neuroendocrine and immune tissues and possesses several features that allow it to be classified as a neuroendocrine mediator [5, 99]. This cytokine has glucocorticoid-antagonist properties within the immune system and participates in the regulation of several endocrine circuits under physiological conditions. Further, initial *in vitro* studies indicate a growth-promoting activity of MIF on neural progenitor cells [71]. In this context, MIF controls the site-specific migration of the immunocompetent cells of the brain, the microglia. These cell entities are considered to be the resident macrophages of the brain and are involved in almost all pathophysiological mechanisms, including trauma, autoimmune and neuroinflammatory disease, and brain tumors. The precise role of these immunocompetent cells of the CNS in tumor progression is subject of much

controversy since its specific role is not yet completely understood. Immunological "escape mechanisms" could play a decisive role in tumor invasion and proliferation.

The association of MIF with the progression of malignant brain tumors places this cytokine in center stage [100, 101]. It is suggested that brain tumors secrete MIF to control the activity of accumulating tumor-promoting cells, which in turn might have inductive tumor-progressive as well as proangiogenic effects [58, 101]. Thus, based on its localization and functional features, MIF would be well in a position to execute important control of the tumor microenvironment. A conceptual framework has been sketched to reflect the metabolic and immune cell complexity of brain tumors in a simplified model classifying the tumor into three distinct zones (Figure 5). Although each border may depict a smooth shift into the next transition zone, Tumor Zone 1 (TZ1) consists of the main tumor—bulk, corresponding to contrast enhancing regions in clinical MRI settings. Here, MIF is mainly produced and secreted into the surrounding tissue. TZ2 represents the area of perifocal edema, which is characterized by its specific proangiogenic microenvironment and transitory glioma cells. Apart from these cells, there is a pronounced accumulation of microglial cells, which also infiltrate the TZ1. The TZ3 is the most challenging and intractable zone for therapeutic intervention, since this zone consists mainly of healthy brain parenchyma. However,

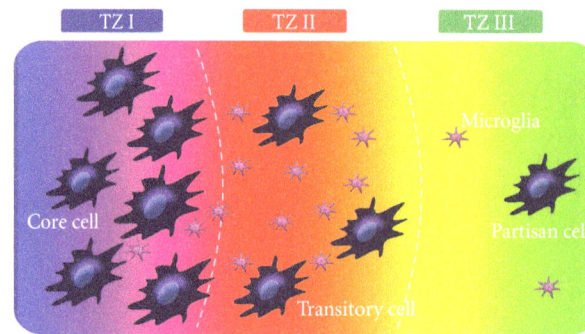

FIGURE 5: The brain tumor microenvironment, heterogeneous tumor zones and MIF actions. Conceptual framework depicting the metabolic and immune cell complexity of malignant brain tumors, (glioblastomas, GBM) is given as a simplified model classifying the tumor into three distinct tumor zones (TZ1–TZ3). Tumor Zone 1 (TZ1) consists of the main tumor—bulk and core glioma cells, corresponding to contrast enhancing regions in MRI images. MIF is mainly produced in TZ1 and secreted into the extracellular space. TZ2 represents the area of perifocal edema, which is characterized by its specific proangiogenic microenvironment and presence of transitory glioma cells. In addition, this tumor zone shows pronounced accumulation of microglial cells, which also infiltrate TZ1. TZ3 is the most awkward zone for therapeutic intervention, since this tumor zone consists mainly of healthy brain parenchyma. However, isolated glioma-initiating cells termed partisan cells colonize TZ3 and are most probably responsible for tumor recurrence following surgery. TZ2 is probably biologically most active, influencing TZ1 and TZ3 by tumor-derived metabolites impacting the immune system, angiogenesis, and cell fate.

isolated glioma-initiating cells termed partisan cells colonize TZ3 and are most probably responsible for tumor recurrence following surgery. The TZ2 is probably biologically most active, influencing TZ1 and TZ3 through tumor-derived metabolites impacting the immune system, angiogenesis, and cell fate. With regard to MIF, however, production and secretion of MIF occur in TZ1, while its receptors are mainly expressed by microglial cells in TZ2 and on glioma cells themselves. MIF could therefore act in a dual fashion both as an autocrine factor as well as a tumor-derived factor which influences the immune micromilieu (Figure 5). Another relevant aspect is that malignant gliomas secrete neurotoxic concentrations of the oncometabolite glutamate as a consequence of their metabolic alterations, and increased glutathione needs [102, 103]. Further, glutamate stimulates the migration and activation of microglial cells [104]. This aspect has not been given much attention from a neurooncological point of view. The metabolic cytokine crosstalk reveals its clinical implication: CD44 as coreceptor of CD74 is also a regulatory component of the glutamate transporter xCT controlling cancer redox state [105]. Additionally, as a specific surface cell receptor in mesenchymal stem cells, CD44 regulates the vascular architecture of highly vascularized tumors such as malignant gliomas through the activation of these stem cells, thereby playing a possible role in their progression. Nevertheless, it needs to be unambiguously demonstrated whether MIF is primarily effective in an autocrine or intracellular manner in malignant gliomas. Thus, further studies on this matter will be decisive for future MIF-neutralizing approaches. Two approaches are available in experimental and clinical studies for the therapeutic targeting of MIF. Firstly, MIF-neutralizing antibodies have been experimentally tested in a murine arthritis model and in rodent glomerulonephritis models with promising efficacy [106–108]. Along the same line, CD74-neutralizing antibodies have been applied to B-cell

malignancies, although comparable data of these approaches are missing. Additionally, soluble CD74 molecules have been isolated in vitro. Secondly, there are now effective, small-molecule MIF antagonists available, with ISO-1 being the most widely accepted one [78, 108]. Based on these findings further small compound library screenings and computational drug design studies are now underway. This approach will probably identify promising small-molecule MIF inhibitors in the future. Due to the lack of immunological responses, low-molecular-weight inhibitors are so far most promising for MIF-neutralizing approaches in humans.

Considering data from clinical studies as well, MIF expression also has predictive values, as patients with malignant gliomas and high MIF expression levels show worse prognosis and earlier tumor recurrence [109]. Interestingly, MIF abundance is associated with increased microvessels and elevated IL-8 expression. Moreover, the MIF receptor CD74 has been shown to contribute to temozolomide resistance [109, 110]. Taking all these facts into account, the underlying molecular mediators and metabolites and immunological crosstalk remain only partially understood despite the central role of dysregulated metabolism in brain tumors. A comprehensive understanding of the dynamics and hierarchy of MIF as a glioma-derived oncometabolite as well as immunological and vascular consequences is therefore critical in identifying effective drug targets in the development of multimodal managements of brain tumors. In order to achieve this target, a detailed analysis of MIF action in this disease with high unmet medical need appears mandatory. Future studies will show whether available MIF and CD74 receptor inhibitors could be efficiently used in our armamentarium against malignant brain tumors.

Conflict of Interests

The authors declare no competing financial interests.

Authors' Contribution

N. E. Savaskan and I. Y. Eyüpoglu conceived and designed the paper and figures with contributions from G. Fingerle-Rowson and M. Buchfelder. All authors shaped the final paper.

Acknowledgments

The authors thank all members of the neurooncology laboratory team and Dr. Schlachetzki for valuable comments on this paper. They are grateful to the editing work of Dr. N. Hore. They gratefully acknowledge the "Tumorzentrum Erlangen" for continuous support and for clinical data exchange. Their work is supported by the German Research Foundation (Deutsche Forschungsgemeinschaft, DFG Grant Ey 94/2-1).

References

[1] A. Isaacs and J. Lindenmann, "Virus interference. I. The interferon," *Proceedings of the Royal Society B*, vol. 147, no. 927, pp. 258–267, 1957.

[2] B. R. Bloom and B. Bennett, "Mechanism of a reaction in vitro associated with delayed-type hypersensitivity," *Science*, vol. 153, no. 3731, pp. 80–82, 1966.

[3] B. R. Bloom, B. Bennett, H. F. Oettgen, E. P. McLean, and L. J. Old, "Demonstration of delayed hypersensitivity to soluble antigens of chemically induced tumors by inhibition of macrophage migration," *Proceedings of the National Academy of Sciences of the United States of America*, vol. 64, no. 4, pp. 1176–1180, 1969.

[4] J. R. David, "Delayed hypersensitivity in vitro: its mediation by cell-free substances formed by lymphoid cell-antigen interaction," *Proceedings of the National Academy of Sciences of the United States of America*, vol. 56, no. 1, pp. 72–77, 1966.

[5] J. Bernhagen, T. Calandra, R. A. Mitchell et al., "MIF is a pituitary-derived cytokine that potentiates lethal endotoxaemia," *Nature*, vol. 365, no. 6448, pp. 756–759, 1993.

[6] J. D. Hudson, M. A. Shoaibi, R. Maestro, A. Carnero, G. J. Hannon, and D. H. Beach, "A proinflammatory cytokine inhibits p53 tumor suppressor activity," *Journal of Experimental Medicine*, vol. 190, no. 10, pp. 1375–1382, 1999.

[7] R. A. Mitchell, H. Liao, J. Chesney et al., "Macrophage migration inhibitory factor (MIF) sustains macrophage proinflammatory function by inhibiting p53: regulatory role in the innate immune response," *Proceedings of the National Academy of Sciences of the United States of America*, vol. 99, no. 1, pp. 345–350, 2002.

[8] E. F. Morand, M. Leech, and J. Bernhagen, "MIF: a new cytokine link between rheumatoid arthritis and atherosclerosis," *Nature Reviews Drug Discovery*, vol. 5, no. 5, pp. 399–410, 2006.

[9] A. Mikulowska, C. N. Metz, R. Bucala, and R. Holmdahl, "Macrophage migration inhibitory factor is involved in the pathogenesis of collagen type II-Induced arthritis in mice," *Journal of Immunology*, vol. 158, no. 11, pp. 5514–5517, 1997.

[10] J. H. Pan, G. K. Sukhova, J. T. Yang et al., "Macrophage migration inhibitory factor deficiency impairs atherosclerosis in low-density lipoprotein receptor-deficient mice," *Circulation*, vol. 109, no. 25, pp. 3149–3153, 2004.

[11] C. Bifulco, K. McDaniel, L. Leng, and R. Bucala, "Tumor growth-promoting properties of macrophage migration inhibitory factor," *Current Pharmaceutical Design*, vol. 14, no. 36, pp. 3790–3801, 2008.

[12] W. Y. Weiser, P. A. Temple, J. S. Witek-Giannotti, H. G. Remold, S. C. Clark, and J. R. David, "Molecular cloning of a cDNA encoding a human macrophage migration inhibitory factor," *Proceedings of the National Academy of Sciences of the United States of America*, vol. 86, no. 19, pp. 7522–7526, 1989.

[13] E. Rosengren, R. Bucala, P. Åman et al., "The immunoregulatory mediator macrophage migration inhibitory factor (MIF) catalyzes a tautomerization reaction," *Molecular Medicine*, vol. 2, no. 1, pp. 143–149, 1996.

[14] R. Kleemann, A. Kapurniotu, R. W. Frank et al., "Disulfide analysis reveals a role for macrophage migration inhibitory factor (MIF) as thiol-protein oxidoreductase," *Journal of Molecular Biology*, vol. 280, no. 1, pp. 85–102, 1998.

[15] K. Bendrat, Y. Al-Abed, D. J. E. Callaway et al., "Biochemical and mutational investigations of the enzymatic activity of macrophage migration inhibitory factor," *Biochemistry*, vol. 36, no. 49, pp. 15356–15362, 1997.

[16] S. L. Stamps, M. C. Fitzgerald, and C. P. Whitman, "Characterization of the role of the amino-terminal proline in the enzymatic activity catalyzed by macrophage migration inhibitory factor," *Biochemistry*, vol. 37, no. 28, pp. 10195–10202, 1998.

[17] N. Esumi, M. Budarf, L. Ciccarelli, B. Sellinger, C. A. Kozak, and G. Wistow, "Conserved gene structure and genomic linkage for D-dopachrome tautomerase (DDT) and MIF," *Mammalian Genome*, vol. 9, no. 9, pp. 753–757, 1998.

[18] M. Swope, H. W. Sun, P. R. Blake, and E. Lolis, "Direct link between cytokine activity and a catalytic site for macrophage migration inhibitory factor," *EMBO Journal*, vol. 17, no. 13, pp. 3534–3541, 1998.

[19] V. Paralkar and G. Wistow, "Cloning the human gene for macrophage migration inhibitory factor (MIF)," *Genomics*, vol. 19, no. 1, pp. 48–51, 1994.

[20] W. R. Pearson, "MIF proteins are not glutathione transferase homologs," *Protein Science*, vol. 3, no. 3, pp. 525–527, 1994.

[21] J. Bernhagen, M. Bacher, T. Calandra et al., "An essential role for macrophage migration inhibitory factor in the tuberculin delayed-type hypersensitivity reaction," *Journal of Experimental Medicine*, vol. 183, no. 1, pp. 277–282, 1996.

[22] R. Kleemann, R. Mischke, A. Kapurniotu, H. Brunner, and J. Bernhagen, "Specific reduction of insulin disulfides by macrophage migration inhibitory factor (MIF) with glutathione and dihydrolipoamide: potential role in cellular redox processes," *FEBS Letters*, vol. 430, no. 3, pp. 191–196, 1998.

[23] M. Thiele and J. Bernhagen, "Link between macrophage migration inhibitory factor and cellular redox regulation," *Antioxidants and Redox Signaling*, vol. 7, no. 9-10, pp. 1234–1248, 2005.

[24] E. Herrero and M. A. De La Torre-Ruiz, "Monothiol glutaredoxins: a common domain for multiple functions," *Cellular and Molecular Life Sciences*, vol. 64, no. 12, pp. 1518–1530, 2007.

[25] E. C. Meng and P. C. Babbitt, "Topological variation in the evolution of new reactions in functionally diverse enzyme superfamilies," *Current Opinion in Structural Biology*, vol. 21, no. 3, pp. 391–397, 2011.

[26] H. Sugimoto, S. I. Oda, T. Otsuki, T. Hino, T. Yoshida, and Y. Shiro, "Crystal structure of human indoleamine 2,3-dioxygenase: catalytic mechanism of O_2 incorporation by a

heme-containing dioxygenase," *Proceedings of the National Academy of Sciences of the United States of America*, vol. 103, no. 8, pp. 2611–2616, 2006.

[27] H. W. Sun, J. Bernhagen, R. Bucala, and E. Lolis, "Crystal structure at 2.6-Å resolution of human macrophage migration inhibitory factor," *Proceedings of the National Academy of Sciences of the United States of America*, vol. 93, no. 11, pp. 5191–5196, 1996.

[28] M. Suzuki, H. Sugimoto, A. Nakagawa, I. Tanaka, J. Nishihira, and M. Sakai, "Crystal structure of the macrophage migration inhibitory factor from rat liver," *Nature Structural Biology*, vol. 3, no. 3, pp. 259–266, 1996.

[29] M. Takahashi, J. Nishihira, M. Shimpo et al., "Macrophage migration inhibitory factor as a redox-sensitive cytokine in cardiac myocytes," *Cardiovascular Research*, vol. 52, no. 3, pp. 438–445, 2001.

[30] M. T. Nguyen, J. Beck, H. Lue et al., "A 16-residue peptide fragment of macrophage migration inhibitory factor, MIF-(50–65), exhibits redox activity and has MIF-like biological functions," *Journal of Biological Chemistry*, vol. 278, no. 36, pp. 33654–33671, 2003.

[31] M. Tuyet Nguyen, H. Lue, R. Kleemann et al., "The cytokine macrophage migration inhibitory factor reduces pro-oxidative stress-induced apoptosis," *Journal of Immunology*, vol. 170, no. 6, pp. 3337–3347, 2003.

[32] J. Ceccarelli, L. Delfino, E. Zappia et al., "The redox state of the lung cancer microenvironment depends on the levels of thioredoxin expressed by tumor cells and affects tumor progression and response to prooxidants," *International Journal of Cancer*, vol. 123, no. 8, pp. 1770–1778, 2008.

[33] R. A. Harrison and C. Sumners, "Redox regulation of macrophage migration inhibitory factor expression in rat neurons," *Biochemical and Biophysical Research Communications*, vol. 390, no. 1, pp. 171–175, 2009.

[34] K. Koga, A. Kenessey, S. R. Powell, C. P. Sison, E. J. Miller, and K. Ojamaa, "Macrophage migration inhibitory factor provides cardioprotection during ischemia/reperfusion by reducing oxidative stress," *Antioxidants and Redox Signaling*, vol. 14, no. 7, pp. 1191–1202, 2011.

[35] S. Busche, S. Gallinat, M. A. Fleegal, M. K. Raizada, and C. Sumners, "Novel role of macrophage migration inhibitory factor in angiotensin II regulation of neuromodulation in rat brain," *Endocrinology*, vol. 142, no. 11, pp. 4623–4630, 2001.

[36] C. Sun, H. Li, L. Leng, M. K. Raizada, R. Bucala, and C. Sumners, "Macrophage migration inhibitory factor: an intracellular inhibitor of angiotensin II-induced increases in neuronal activity," *Journal of Neuroscience*, vol. 24, no. 44, pp. 9944–9952, 2004.

[37] A. I. Su, T. Wiltshire, S. Batalov et al., "A gene atlas of the mouse and human protein-encoding transcriptomes," *Proceedings of the National Academy of Sciences of the United States of America*, vol. 101, no. 16, pp. 6062–6067, 2004.

[38] C. Wu, C. Orozco, J. Boyer et al., "BioGPS: an extensible and customizable portal for querying and organizing gene annotation resources," *Genome Biology*, vol. 10, no. 11, article R130, 2009.

[39] E. S. Lein, M. J. Hawrylycz, N. Ao et al., "Genome-wide atlas of gene expression in the adult mouse brain," *Nature*, vol. 445, no. 7124, pp. 168–176, 2007.

[40] A. C. Koong, N. C. Denko, K. M. Hudson et al., "Candidate genes for the hypoxic tumor phenotype," *Cancer Research*, vol. 60, no. 4, pp. 883–887, 2000.

[41] M. Bacher, J. Schrader, N. Thompson et al., "Up-regulation of macrophage migration inhibitory factor gene and protein expression in glial tumor cells during hypoxic and hypoglycemic stress indicates a critical role for angiogenesis in glioblastoma multiforme," *American Journal of Pathology*, vol. 162, no. 1, pp. 11–17, 2003.

[42] M. Winner, A. C. Koong, B. E. Rendon, W. Zundel, and R. A. Mitchell, "Amplification of tumor hypoxic responses by macrophage migration inhibitory factor-dependent hypoxia-inducible factor stabilization," *Cancer Research*, vol. 67, no. 1, pp. 186–193, 2007.

[43] C. A. Dumitru, H. Gholaman, S. Trellakis et al., "Tumor-derived macrophage migration inhibitory factor modulates the biology of head and neck cancer cells via neutrophil activation," *International Journal of Cancer*, vol. 129, no. 4, pp. 859–869, 2011.

[44] S. M. Welford, B. Bedogni, K. Gradin, L. Poellinger, M. B. Powell, and A. J. Giaccia, "HIF1α delays premature senescence through the activation of MIF," *Genes and Development*, vol. 20, no. 24, pp. 3366–3371, 2006.

[45] S. Oda, T. Oda, K. Nishi et al., "Macrophage migration inhibitory factor activates hypoxia-inducible factor in a p53-dependent manner," *PLoS ONE*, vol. 3, no. 5, Article ID e2215, 2008.

[46] S. Iwata, T. Hori, N. Sato et al., "Adult T cell leukemia (ATL)-derived factor/human thioredoxin prevents apoptosis of lymphoid cells induced by L-cystine and glutathione depletion: possible involvement of thiol-mediated redox regulation in apoptosis caused by pro-oxidant state," *Journal of Immunology*, vol. 158, no. 7, pp. 3108–3117, 1997.

[47] F. Talos, P. Mena, G. Fingerle-Rowson, U. Moll, and O. Petrenko, "MIF loss impairs Myc-induced lymphomagenesis," *Cell Death and Differentiation*, vol. 12, no. 10, pp. 1319–1328, 2005.

[48] D. Z. Xiao, B. Dai, J. Chen et al., "Loss of macrophage migration inhibitory factor impairs the growth properties of human HeLa cervical cancer cells," *Cell Proliferation*, vol. 44, no. 6, pp. 582–590, 2011.

[49] X. Xu, B. Wang, C. Ye et al., "Overexpression of macrophage migration inhibitory factor induces angiogenesis in human breast cancer," *Cancer Letters*, vol. 261, no. 2, pp. 147–157, 2008.

[50] B. Liao, B. L. Zhong, Z. Li, X. Y. Tian, Y. Li, and B. Li, "Macrophage migration inhibitory factor contributes angiogenesis by up-regulating IL-8 and correlates with poor prognosis of patients with primary nasopharyngeal carcinoma," *Journal of Surgical Oncology*, vol. 102, no. 7, pp. 844–851, 2010.

[51] V. Veillat, C. Carli, C. N. Metz, Y. Al-Abed, P. H. Naccache, and A. Akoum, "Macrophage migration inhibitory factor elicits an angiogenic phenotype in human ectopic endometrial cells and triggers the production of major angiogenic factors via CD44, CD74, and MAPK signaling pathways," *Journal of Clinical Endocrinology and Metabolism*, vol. 95, no. 12, pp. E403–E412, 2010.

[52] H. Sugimoto, M. Suzuki, A. Nakagawa, I. Tanaka, and J. Nishihira, "Crystal structure of macrophage migration inhibitory factor from human lymphocyte at 2.1 Å resolution," *FEBS Letters*, vol. 389, no. 2, pp. 145–148, 1996.

[53] H. Sugimoto, M. Taniguchi, A. Nakagawa, I. Tanaka, M. Suzuki, and J. Nishihira, "Crystallization and preliminary X-ray analysis of human D-dopachrome tautomerase," *Journal of Structural Biology*, vol. 120, no. 1, pp. 105–108, 1997.

[54] A. B. Taylor, W. H. Johnson, R. M. Czerwinski, H. S. Li, M. L. Hackert, and C. P. Whitman, "Crystal structure of

macrophage migration inhibitory factor complexed with (E)-2-fluoro-p-hydroxycinnamate at 1.8 Å resolution: implications for enzymatic catalysis and inhibition," *Biochemistry*, vol. 38, no. 23, pp. 7444–7452, 1999.

[55] J. B. Lubetsky, M. Swope, C. Dealwis, P. Blake, and E. Lolis, "Pro-1 of macrophage migration inhibitory factor functions as a catalytic base in the phenylpyruvate tautomerase activity," *Biochemistry*, vol. 38, no. 22, pp. 7346–7354, 1999.

[56] A. Hermanowski-Vosatka, S. S. Mundt, J. M. Ayala et al., "Enzymatically inactive macrophage migration inhibitory factor inhibits monocyte chemotaxis and random migration," *Biochemistry*, vol. 38, no. 39, pp. 12841–12849, 1999.

[57] G. Fingerle-Rowson, D. R. Kaleswarapu, C. Schlander et al., "A tautomerase-null macrophage migration-inhibitory factor (MIF) gene knock-in mouse model reveals that protein interactions and not enzymatic activity mediate MIF-dependent growth regulation," *Molecular and Cellular Biology*, vol. 29, no. 7, pp. 1922–1932, 2009.

[58] B. E. Rendon, S. S. Willer, W. Zundel, and R. A. Mitchell, "Mechanisms of macrophage migration inhibitory factor (MIF)-dependent tumor microenvironmental adaptation," *Experimental and Molecular Pathology*, vol. 86, no. 3, pp. 180–185, 2009.

[59] G. Fingerle-Rowson, O. Petrenko, C. N. Metz et al., "The p53-dependent effects of macrophage migration inhibitory factor revealed by gene targeting," *Proceedings of the National Academy of Sciences of the United States of America*, vol. 100, no. 16, pp. 9354–9359, 2003.

[60] M. Bozza, A. R. Satoskar, G. Lin et al., "Targeted disruption of migration inhibitory factor gene reveals its critical role in sepsis," *Journal of Experimental Medicine*, vol. 189, no. 2, pp. 341–346, 1999.

[61] N. Honma, H. Koseki, T. Akasaka et al., "Deficiency of the macrophage migration inhibitory factor gene has no significant effect on endotoxaemia," *Immunology*, vol. 100, no. 1, pp. 84–90, 2000.

[62] F. Chagnon, C. N. Metz, R. Bucala, and O. Lesur, "Endotoxin-induced myocardial dysfunction: effects of macrophage migration inhibitory factor neutralization," *Circulation Research*, vol. 96, no. 10, pp. 1095–1102, 2005.

[63] J. Bernhagen, R. Krohn, H. Lue et al., "MIF is a noncognate ligand of CXC chemokine receptors in inflammatory and atherogenic cell recruitment," *Nature Medicine*, vol. 13, no. 5, pp. 587–596, 2007.

[64] Y. Gore, D. Starlets, N. Maharshak et al., "Macrophage migration inhibitory factor induces B cell survival by activation of a CD74-CD44 receptor complex," *Journal of Biological Chemistry*, vol. 283, no. 5, pp. 2784–2792, 2008.

[65] M. Merk, S. Zierow, L. Leng et al., "The D-dopachrome tautomerase (DDT) gene product is a cytokine and functional homolog of macrophage migration inhibitory factor (MIF)," *Proceedings of the National Academy of Sciences of the United States of America*, vol. 108, no. 34, pp. 577–585, 2011.

[66] T. Calandra, L. A. Spiegel, C. N. Metz, and R. Bucala, "Macrophage migration inhibitory factor is a critical mediator of the activation of immune cells by exotoxins of Gram-positive bacteria," *Proceedings of the National Academy of Sciences of the United States of America*, vol. 95, no. 19, pp. 11383–11388, 1998.

[67] M. A. McDevitt, J. Xie, G. Shanmugasundaram et al., "A critical role for the host mediator macrophage migration inhibitory factor in the pathogenesis of malarial anemia," *Journal of Experimental Medicine*, vol. 203, no. 5, pp. 1185–1196, 2006.

[68] A. Arjona, H. G. Foellmer, T. Town et al., "Abrogation of macrophage migration inhibitory factor decreases West Nile virus lethality by limiting viral neuroinvasion," *Journal of Clinical Investigation*, vol. 117, no. 10, pp. 3059–3066, 2007.

[69] M. Bacher, A. Meinhardt, H. Y. Lan et al., "MIF expression in the rat brain: implications for neuronal function," *Molecular Medicine*, vol. 4, no. 4, pp. 217–230, 1998.

[70] S. C. Schwarz, J. Schwarz, J. Sautter, and W. H. Oertel, "Effects of macrophage migration inhibitory factor and macrophage migration stimulatory factor on function and survival of foetal dopaminegic grafts in the 6-hydroxydopamine rat model of Parkinson's disease," *Experimental Brain Research*, vol. 120, no. 1, pp. 95–103, 1998.

[71] S. Ohta, A. Misawa, R. Fukaya et al., "Macrophage migration inhibitory factor (MIF) promotes cell survival and proliferation of neural stem/progenitor cells," *Journal of Cell Science*. In press.

[72] M. Bacher, O. Deuster, B. Aljabari et al., "The role of macrophage migration inhibitory factor in alzheimer's disease," *Molecular Medicine*, vol. 16, no. 3-4, pp. 116–121, 2010.

[73] T. Calandra, J. Bernhagen, R. A. Mitchell, and R. Bucala, "The macrophage is an important and previously unrecognized source of macrophage migration inhibitory factor," *Journal of Experimental Medicine*, vol. 179, no. 6, pp. 1895–1902, 1994.

[74] L. Leng, C. N. Metz, Y. Fang et al., "MIF signal transduction initiated by binding to CD74," *Journal of Experimental Medicine*, vol. 197, no. 11, pp. 1467–1476, 2003.

[75] T. Calandra and T. Roger, "Macrophage migration inhibitory factor: a regulator of innate immunity," *Nature Reviews Immunology*, vol. 3, no. 10, pp. 791–800, 2003.

[76] X. Shi, L. Leng, T. Wang et al., "CD44 is the signaling component of the macrophage migration inhibitory factor-CD74 receptor complex," *Immunity*, vol. 25, no. 4, pp. 595–606, 2006.

[77] L. Leng and R. Bucala, "Insight into the biology of Macrophage Migration Inhibitory Factor (MIF) revealed by the cloning of its cell surface receptor," *Cell Research*, vol. 16, no. 2, pp. 162–168, 2006.

[78] S. Balachandran, A. Rodge, P. K. Gadekar et al., "Novel derivatives of ISO-1 as potent inhibitors of MIF biological function," *Bioorganic and Medicinal Chemistry Letters*, vol. 19, no. 16, pp. 4773–4776, 2009.

[79] W. L. Jorgensen, S. Gandavadi, X. Du et al., "Receptor agonists of macrophage migration inhibitory factor," *Bioorganic and Medicinal Chemistry Letters*, vol. 20, no. 23, pp. 7033–7036, 2010.

[80] V. Schwartz, H. Lue, S. Kraemer et al., "A functional heteromeric MIF receptor formed by CD74 and CXCR4," *FEBS Letters*, vol. 583, no. 17, pp. 2749–2757, 2009.

[81] C. Weber, S. Kraemer, M. Drechsler et al., "Structural determinants of MIF functions in CXCR2-mediated inflammatory and atherogenic leukocyte recruitment," *Proceedings of the National Academy of Sciences of the United States of America*, vol. 105, no. 42, pp. 16278–16283, 2008.

[82] S. Onodera, K. Suzuki, T. Matsuno, K. Kaneda, M. Takagi, and J. Nishihira, "Macrophage migration inhibitory factor induces phagocytosis of foreign particles by macrophages in autocrine and paracrine fashion," *Immunology*, vol. 92, no. 1, pp. 131–137, 1997.

[83] M. Bacher, C. N. Metz, T. Calandra et al., "An essential regulatory role for macrophage migration inhibitory factor in T-cell activation," *Proceedings of the National Academy of Sciences of the United States of America*, vol. 93, no. 15, pp. 7849–7854, 1996.

[84] T. Calandra, J. Bernhagen, C. N. Metz et al., "MIF as a glucocorticoid-induced modulator of cytokine production," *Nature*, vol. 377, no. 6544, pp. 68–71, 1995.

[85] S. C. Donnelly, C. Haslett, P. T. Reid et al., "Regulatory role for macrophage migration inhibitory factor in acute respiratory distress syndrome," *Nature Medicine*, vol. 3, no. 3, pp. 320–323, 1997.

[86] T. Roger, A. L. Chanson, M. Knaup-Reymond, and T. Calandra, "Macrophage migration inhibitory factor promotes innate immune responses by suppressing glucocorticoid-induced expression of mitogen-activated protein kinase phosphatase-1," *European Journal of Immunology*, vol. 35, no. 12, pp. 3405–3413, 2005.

[87] J. M. Daun and J. G. Cannon, "Macrophage migration inhibitory factor antagonizes hydrocortisone-induced increases in cytosolic IκBα," *American Journal of Physiology*, vol. 279, no. 3, pp. R1043–R1049, 2000.

[88] M. Löwenberg, C. Stahn, D. W. Hommes, and F. Buttgereit, "Novel insights into mechanisms of glucocorticoid action and the development of new glucocorticoid receptor ligands," *Steroids*, vol. 73, no. 9-10, pp. 1025–1029, 2008.

[89] C. Piette, C. Munaut, J. M. Foidart, and M. Deprez, "Treating gliomas with glucocorticoids: from bedside to bench," *Acta Neuropathologica*, vol. 112, no. 6, pp. 651–664, 2006.

[90] D. Aeberli, Y. Yang, A. Mansell, L. Santos, M. Leech, and E. F. Morand, "Endogenous macrophage migration inhibitory factor modulates glucocorticoid sensitivity in macrophages via effects on MAP kinase phosphatase-1 and p38 MAP kinase," *FEBS Letters*, vol. 580, no. 3, pp. 974–981, 2006.

[91] L. Santos, P. Hall, C. Metz, R. Bucala, and E. F. Morand, "Role of macrophage migration inhibitory factor (MIF) in murine antigen-induced arthritis: interaction with glucocorticoids," *Clinical and Experimental Immunology*, vol. 123, no. 2, pp. 309–314, 2001.

[92] F. Benigni, T. Atsumi, T. Calandra et al., "The proinflammatory mediator macrophage migration inhibitory factor induces glucose catabolism in muscle," *Journal of Clinical Investigation*, vol. 106, no. 10, pp. 1291–1300, 2000.

[93] E. J. Miller, J. Li, L. Leng et al., "Macrophage migration inhibitory factor stimulates AMP-activated protein kinase in the ischaemic heart," *Nature*, vol. 451, no. 7178, pp. 578–582, 2008.

[94] R. A. Mitchell, C. N. Metz, T. Peng, and R. Bucala, "Sustained mitogen-activated protein kinase (MAPK) and cytoplasmic phospholipase A2 activation by macrophage migration inhibitory factor (MIF): regulatory role in cell proliferation and glucocorticoid action," *Journal of Biological Chemistry*, vol. 274, no. 25, pp. 18100–18106, 1999.

[95] R. Kleemann, A. Hausser, G. Geiger et al., "Intracellular action of the cytokine MIF to modulate AP-1 activity and the cell cycle through Jab1," *Nature*, vol. 408, no. 6809, pp. 211–216, 2000.

[96] A. Nemajerova, P. Mena, G. Fingerle-Rowson, U. M. Moll, and O. Petrenko, "Impaired DNA damage checkpoint response in MIF-deficient mice," *EMBO Journal*, vol. 26, no. 4, pp. 987–997, 2007.

[97] A. Nemajerova, U. M. Moll, O. Petrenko, and G. Fingerle-Rowson, "Macrophage migration inhibitory factor coordinates DNA damage response with the proteasomal control of the cell cycle," *Cell Cycle*, vol. 6, no. 9, pp. 1030–1034, 2007.

[98] R. Schulz, N. D. Marchenko, L. Holembowski et al., "Inhibiting the HSP90 chaperone destabilizes macrophage migration inhibitory factor and thereby inhibits breast tumor progression," *The Journal of Experimental Medicine*, vol. 209, no. 2, pp. 275–289, 2012.

[99] G. R. Fingerle-Rowson and R. Bucala, "Neuroendocrine properties of macrophage migration inhibitory factor (MIF)," *Immunology and Cell Biology*, vol. 79, no. 4, pp. 368–375, 2001.

[100] M. Mittelbronn, M. Platten, P. Zeiner et al., "Macrophage migration inhibitory factor (MIF) expression in human malignant gliomas contributes to immune escape and tumour progression," *Acta Neuropathologica*, vol. 122, no. 3, pp. 353–365, 2011.

[101] T. Engelhorn, N. E. Savaskan, M. A. Schwarz et al., "Cellular characterization of the peritumoral edema zone in malignant brain tumors," *Cancer Science*, vol. 100, no. 10, pp. 1856–1862, 2009.

[102] N. E. Savaskan, A. Heckel, E. Hahnen et al., "Small interfering RNA-mediated xCT silencing in gliomas inhibits neurodegeneration and alleviates brain edema," *Nature Medicine*, vol. 14, no. 6, pp. 629–632, 2008.

[103] N. E. Savaskan and I. Y. Eyüpoglu, "XCT modulation in gliomas: relevance to energy metabolism and tumor microenvironment normalization," *Annals of Anatomy*, vol. 192, no. 5, pp. 309–313, 2010.

[104] O. Ullrich, A. Diestel, I. Y. Eyüpoglu, and R. Nitsch, "Regulation of microglial expression of integrins by poly(ADP-ribose) polymerase-1," *Nature Cell Biology*, vol. 3, no. 12, pp. 1035–1042, 2001.

[105] T. Ishimoto, O. Nagano, T. Yae et al., "CD44 variant regulates redox status in cancer cells by stabilizing the xCT subunit of system xc⁻ and thereby promotes tumor growth," *Cancer Cell*, vol. 19, no. 3, pp. 387–400, 2011.

[106] L. L. Santos, A. Dacumos, J. Yamana, L. Sharma, and E. F. Morand, "Reduced arthritis in MIF deficient mice is associated with reduced T cell activation: down-regulation of ERK MAP kinase phosphorylation," *Clinical and Experimental Immunology*, vol. 152, no. 2, pp. 372–380, 2008.

[107] A. Y. Hoi, M. J. Hickey, P. Hall et al., "Macrophage migration inhibitory factor deficiency attenuates macrophage recruitment, glomerulonephritis, and lethality in MRL/lpr mice," *Journal of Immunology*, vol. 177, no. 8, pp. 5687–5696, 2006.

[108] L. Leng, L. Chen, J. Fan et al., "A small-molecule macrophage migration inhibitory factor antagonist protects against glomerulonephritis in Lupus-Prone NZB/NZW F1 and MRL/lpr mice," *Journal of Immunology*, vol. 186, no. 1, pp. 527–538, 2011.

[109] X. B. Wang, X. Y. Tian, Y. Li, B. Li, and Z. Li, "Elevated expression of macrophage migration inhibitory factor correlates with tumor recurrence and poor prognosis of patients with gliomas," *Journal of Neuro-Oncology*, vol. 106, no. 1, pp. 43–51, 2012.

[110] G. J. Kitange, B. L. Carlson, M. A. Schroeder et al., "Expression of CD74 in high grade gliomas: a potential role in temozolomide resistance," *Journal of Neuro-Oncology*, vol. 100, no. 2, pp. 177–186, 2010.

3

The Cytoplasm-to-Vacuole Targeting Pathway: A Historical Perspective

Midori Umekawa and Daniel J. Klionsky

Life Sciences Institute, University of Michigan, Ann Arbor, MI 48109-2216, USA

Correspondence should be addressed to Daniel J. Klionsky, klionsky@umich.edu

Academic Editor: Fulvio Reggiori

From today's perspective, it is obvious that macroautophagy (hereafter autophagy) is an important pathway that is connected to a range of developmental and physiological processes. This viewpoint, however, is relatively recent, coinciding with the molecular identification of autophagy-related (Atg) components that function as the protein machinery that drives the dynamic membrane events of autophagy. It may be difficult, especially for scientists new to this area of research, to appreciate that the field of autophagy long existed as a "backwater" topic that attracted little interest or attention. Paralleling the development of the autophagy field was the identification and analysis of the cytoplasm-to-vacuole targeting (Cvt) pathway, the only characterized biosynthetic route that utilizes the Atg proteins. Here, we relate some of the initial history, including some never-before-revealed facts, of the analysis of the Cvt pathway and the convergence of those studies with autophagy.

1. The Background

To understand the origin of the studies that led to the identification of the Cvt pathway, we need to briefly step back into the early days of yeast molecular genetics. Randy Schekman's group was studying the secretory pathway and isolating mutants defective in various steps including endoplasmic reticulum (ER)-to-Golgi transport as well as secretion to the cell surface. Two former postdocs from the Schekman lab, Scott Emr and Tom Stevens, decided to pursue a similar direction, but to avoid a direct overlap with Randy Schekman by focusing on a pathway that branches off from the secretory pathway, the delivery of proteins to the vacuole. The Emr and Stevens labs isolated a new set of mutants initially named *vpt* (vacuolar protein targeting) [1] and *vpl* (vacuolar protein localization) [2], and subsequently *vps* (vacuolar protein sorting), which are defective in the delivery of resident proteins to the vacuole. Being interested in protein sorting, one of us (D.J.K.) went to Scott Emr's lab to learn about yeast.

While in the Emr lab, I characterized the vacuolar delivery of proteinase A (Pep4) and vacuolar alkaline phosphatase (Pho8). Around that time, the sequence of the gene encoding another vacuolar hydrolase, aminopeptidase I (Ape1) was published [3, 4]. It is important to keep in mind that this was the late 1980s, quite some time before the *Saccharomyces cerevisiae* genome was sequenced in its entirety. In fact, automated sequencing was relatively new, so it was still a major accomplishment when a gene was sequenced. Until then, only the sequences of Pep4 [5, 6], Prc1 (carboxypeptidase Y) [7], Pho8 [8], Prb1 (proteinase B) [9], and Ams1 (α-mannosidase) [10] were known among the vacuolar hydrolases. Thus, it was quite exciting to those of us studying vacuolar protein targeting when a new protein sequence became available. One of my main goals in the Emr lab was to identify the vacuolar-targeting motif and determine a consensus sequence (mapping consensus targeting or retention signals was very popular in those days), a task that was all the more difficult due to the limited number of proteins available for comparison. Hence, I was particularly interested in having a new protein that I could analyze.

Ape1 was known to be a vacuolar hydrolase, and it was characterized as being a glycoprotein [11]. The latter finding fit with the fact that all of the characterized vacuolar hydrolases traffic through the secretory pathway to the Golgi complex and from there are diverted to the vacuole. One interesting feature of the protein sequence for the precursor form of Ape1 (prApe1), however, was that it lacked a standard signal sequence. Accordingly, I assumed that it entered the ER by a unique mechanism. This seemed to add some additional interest to the analysis, as the idea of analyzing yet onc morc vacuolar hydrolasc was gctting somewhat tedious. When I discussed the idea of analyzing the targeting of prApe1 with Scott Emr, however, he was not interested. After all, even if the details of the process were slightly unusual, we were still talking about the characterization of another vacuolar hydrolase that transits through a portion of the secretory pathway. Indeed, at the time, there seemed to be more interesting projects to pursue, so the analysis of prApe1 was left on the "back burner".

Shortly after that time, I started an independent position at the University of California, Davis. To stay clear of the Emr lab (which, for a new assistant professor, loomed like an 800-pound gorilla), I pursued an analysis of the vacuolar H$^+$-translocating ATPase and vacuolar acid trehalase. At that time, Scott forwarded to me a letter (this was just before email became widely used) from a postdoc applicant that he was not able to invite to his lab. That postdoc, Nieves Garcia Alvarez, was from one of the labs, that of Paz Suarez-Rendueles, which was involved in characterizing yeast vacuolar hydrolases, and I agreed to offer her a position. Nieves initially worked on the vacuolar ATPase project. I knew, however, that her lab in Spain was one of two that had essentially simultaneously sequenced the APE1/LAP4 gene encoding prApe1 [4]. During Nieves' time in my lab, I wrote to Beth Jones who had published one paper on Ape1 [12] and asked if she intended to pursue this topic; I did not want to compete with her, but she indicated that she was not going to be working on it, and I was welcome to it. Thus, I obtained the gene from the Suarez-Rendueles lab and a new postdoc from that lab, Rosaria Cueva Noval, along with my postdoc Debbie Yaver and me, began to examine the vacuolar targeting of prApe1.

The initial experiments on prApe1 were confusing, because I could not find any evidence for glycosylation or for the existence of the protein within the compartments of the secretory pathway [13]. (As a side note, our first paper on Ape1 was published back-to-back with the first paper from Yoshinori Ohsumi's lab on the characterization of autophagy in yeast [14]. This was coincidental, and, to be honest, I paid no attention to the Ohsumi paper at that time, because it was on the topic of autophagy; I was studying protein targeting, not some presumed "garbage" pathway that was only used for protein degradation.) Eventually, it dawned on me that the published data were incorrect and that Ape1 was not a glycoprotein. At this time, Fred Dice was making headlines with his analysis of the KFERQ—(KFERQ being the consensus sequence for the recognized substrates) or pentapeptide-dependent pathway for the transport of proteins into the lysosome (the current name for this pathway, "chaperone-mediated autophagy," had not been coined yet) [15]. Considering that Ape1 was not a glycoprotein, and that it did not enter the endoplasmic reticulum, I reasoned that it entered the vacuole by translocating directly across the limiting membrane. Accordingly, I further assumed that

there must be protein machinery, similar to the as yet uncharacterized components involved in the KFERQ pathway, in the vacuolar membrane just waiting for me to come along and identify them.

Therefore, in order to identify the vacuolar membrane translocation components, we generated a chimera of prApe1 fused to the HIS3 gene. Our initial screen was based on the idea that a his3 mutant strain of yeast would not be able to grow in the absence of histidine if the chimera was efficiently delivered to the vacuole. Accordingly, we could isolate mutants that were able to grow without histidine, and they would have defects in the various components of the translocation machinery. It became clear early on that the screen was not working, although we did not know why; we could not easily follow the localization of the chimera because the green fluorescent protein was not yet being used for cell biology studies. Randy Schekman was giving a seminar on campus at that time, and I told him about our project. He suggested that we generate antibodies that only recognized prApe1 and carry out a screen looking for mutants that accumulate the precursor form of the protein. We did attempt that approach, using colony blots after transferring cells to nitrocellulose, but it was very difficult to score positive colonies. However, we also noticed that wild-type cells analyzed by western blot, when grown appropriately, had essentially no prApe1; all of the protein was in the mature form. We also determined (using a pep4Δ mutant as the control) that we could easily detect the precursor that accumulated when one out of ten colonies was defective for prApe1 maturation. Accordingly, even though it was laborious, Tanya Harding, and later Ann Hefner-Gravink, in my lab began to analyze random mutants in batches of ten for the accumulation of prApe1.

We isolated a series of such mutants and placed them into complementation groups [16]. This was quite exciting as we were finally about to identify the long-awaited translocation machinery for the vacuole. To be sure that we were not going to waste our time analyzing mutants that were already known, we began to compare our mutants with all other previously identified mutants that affected vacuolar protein delivery. Of course this included the vps mutants from Tom Stevens and Scott Emr, but also endocytosis mutants and vacuolar morphology (vam) mutants. Even though we did not expect overlaps from the latter, we wanted to be thorough. In fact, we were so careful that we even requested protein extracts from Yoshinori Ohsumi and Michael Thumm, who had isolated apg [17] and aut [18] mutants, respectively, that are defective in autophagy. Obviously (or so we thought at the time), there was not going to be an overlap; autophagy is a degradative pathway, and our mutants (then named cvt) were defective in a biosynthetic pathway. Imagine our surprise, and disappointment, when we found an essentially complete overlap among these three sets of genes [19, 20]. The disappointment was for two reasons. First, instead of having a unique set of mutants that we could study on our own, we knew we immediately had competitors. Second, we were being dragged against our will into the field of autophagy.

Nonetheless, we continued with our studies of prApe1 targeting and began to clone the CVT/APG/AUT genes and

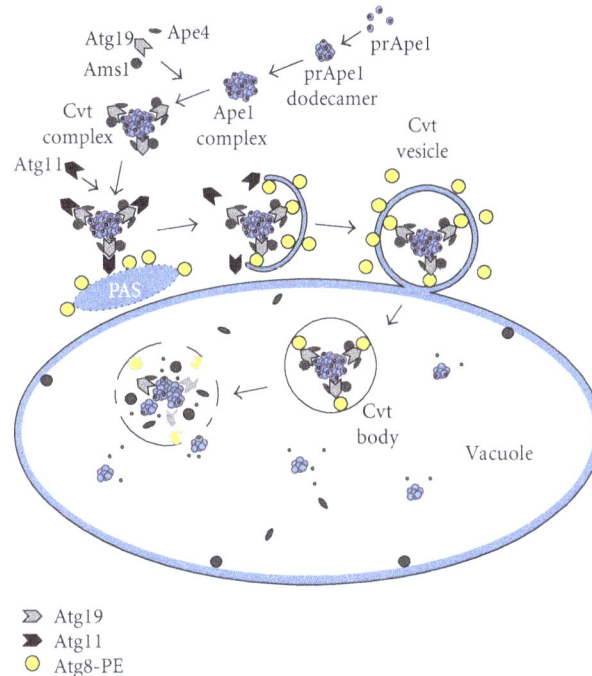

FIGURE 1: Overview of the Cvt pathway. (1) Formation of the Cvt complex: Precursor Ape1 forms a dodecamer. Multiple dodecamers assemble into an Ape1 complex. The Ape1 complex binds Atg19 via the prApe1 propeptide to form the Cvt complex. Other Cvt cargo, including Ams1 and Ape4, bind Atg19 at distinct domains. (2) Movement to the PAS: Atg19 binds the scaffold protein Atg11, and the Cvt complex moves to the PAS. (3) Formation of the Cvt vesicle: Atg19 binds Atg8–PE, which drives the sequestration of the Cvt complex by the double-membrane phagophore. (4) Fusion of the Cvt vesicle with the vacuole: After completion of the Cvt vesicle, the outer membrane fuses with the vacuole, releasing the single membrane Cvt body into the lumen. The Cvt body is broken down by the Atg15 lipase, allowing access to vacuolar hydrolases. Atg19 and Atg8 are degraded. The propeptide of prApe1 is removed and the enzyme becomes active.

analyze the gene products. After discovering the overlap with the *APG* genes, we sent purified antisera against Ape1 to the Ohsumi lab to be used in an electron microscopy analysis by Misuzu Baba. I can still remember Yoshinori Ohsumi cryptically telling me about some striking and exciting results that "could not be described" over the phone, but that had to be seen in person. This resulted in a visit to Japan, and the viewing of images that were indeed striking, revealing that prApe1 import was morphologically similar to autophagy (Figure 1) [21]. Much of the initial work on the characterization of the Atg proteins was done in collaboration with the Ohsumi lab [20, 22–27] and also with the lab of Bill Dunn [22, 28–32], who was studying peroxisome degradation in *Pichia pastoris*. Having established the historical perspective, we now present some of the details of those initial studies of the Cvt pathway, starting with the characterization of aminopeptidase I import by a mechanism that is independent of the secretory pathway, identification of the vacuolar targeting domain, the isolation of mutants defective in prApe1 delivery to the vacuole, and concluding with the genetic and morphological studies that revealed the overlap with autophagy.

2. The Transport of prApe1 to the Vacuole Is Mediated by the Cvt Pathway

Ape1 was initially characterized as a vacuolar enzyme that hydrolyzes leucine peptides (hence the original nomencla-

ture leucine aminopeptidase, or LAP, which is unfortunately confusing because LAPI is encoded by the *LAP4* gene, whereas LAPIV is encoded by *LAP2*, etc.) [33]. The hydrolase is synthesized as an inactive zymogen containing a propeptide that may sterically block its active site; it is processed to its mature form in the vacuole by proteinase B in a *PEP4*-dependent manner [34]. As mentioned above, published data suggested that the Ape1 precursor was transported through part of the secretory pathway, because it was characterized as a glycoprotein [11]. However, a detailed characterization of prApe1 biosynthesis suggested that its delivery to the vacuole was independent of the secretory pathway: (1) prApe1 lacks a signal sequence for transport into the ER, and it is not glycosylated; (2) the half-life of processing (i.e., removal of the propeptide in the vacuole) of prApe1 is substantially longer (~30 min) than that of Prc1 or Pep4 (~6 min), both of which are transported to the vacuole via part of the secretory pathway; (3) vacuolar import of prApe1 is relatively unaffected by *sec* mutants [13].

The obvious question then became, how does prApe1 target to and enter the vacuole? A series of biochemical analyses were performed to address this issue. After it is synthesized as a 61-kDa protein in the cytosol, prApe1 is proteolytically processed to a mature 50-kDa form in the vacuole. The prApe1 propeptide plays an essential role in the transport process [35]. A detailed mutagenesis analysis carried out by Mike Oda revealed that the first amphipathic α-helix in the propeptide is critical for the vacuolar targeting of the enzyme. Deletion of the precursor region or mutations that affect the first α-helical region inhibit its binding to the membrane fraction and prevent subsequent vacuolar delivery and processing. Further analysis by John Kim revealed that prApe1 is assembled as a dodecamer (~669 kDa) in the cytoplasm prior to vacuolar delivery, which argued against direct translocation across the vacuole limiting membrane [36]. The propeptide of prApe1 is not required for its oligomerization. A pulse chase analysis showed that the oligomeric assembly and the subsequent membrane association are very rapid events with a half-life of ~3 min. These results suggested that the long half-life of prApe1 transport may be due to the rate limiting step of the import of the dodecameric enzyme into the vacuole lumen after its binding to membrane.

3. The Cvt and Autophagy Pathways Share the Same Machinery

The oligomerization of prApe1 and the slow kinetics of import into the vacuole argued against transport through the secretory pathway. To understand the mechanism of vacuolar delivery, a detailed biochemical and genetic analysis was carried out in *S. cerevisiae*, which revealed that autophagy and the Cvt pathway largely share the same machinery for double-membrane vesicle formation [16, 19, 20, 27]. A genetic screen to analyze the Cvt pathway was carried out by monitoring the accumulation of prApe1 as described in Section 1. From the initial screen, five *cvt* mutants (*cvt2/atg7, cvt3, cvt5/atg8, cvt6* and *cvt7/atg9*) were isolated, which showed a complete block in prApe1 processing, but were not defective in the maturation of the precursor form of Prc1 or Pep4 [16]. Most of these mutants also showed a defect in nonselective autophagy [19, 20]. Just prior to the isolation of the *cvt* mutants, Michael Thumm in Dieter Wolf's lab isolated a series of *aut* mutants, based on defects in the degradation of the fatty acid synthase. The *aut* mutants including *aut3* (*cvt10/atg1*), *aut5* (*cvt17/atg15*), *aut7* (*cvt5/atg8*), and *aut9* (*cvt7/atg9*) also displayed a significant block in the maturation of prApe1, providing genetic evidence for a role of these proteins in both the Cvt pathway and autophagy [19]. A similar analysis of the *apg* mutants from Yoshinori Ohsumi's lab also revealed an extensive overlap [20]. Subsequently, all of the *ATG* genes, except *ATG11, ATG17, ATG19, ATG22, ATG29,* and *ATG31* were found to be required for both pathways. In 2003, the nomenclature for these *CVT* and *APG/AUT* genes was unified as "*ATG*" for "autophagy related" [37].

4. Precursor Aminopeptidase I Is Imported by a Vesicular Mechanism

The genetic overlap between the *cvt* and *apg/aut* mutants gave rise to the idea of a vesicle-mediated mechanism for prApe1 import. Indeed, electron microscopy analyses performed by Misuzu Baba revealed that the prApe1 dodecamers further assembled into a large complex composed of multiple dodecamers (called an Ape1 complex), and that in the cytoplasm this complex is surrounded by a double

membrane-bound structure, followed by fusion with the vacuolar membrane [21], similar to what was observed in bulk autophagy [38]. This result demonstrated the use of an autophagy-like mechanism for the Cvt pathway. However, the double membrane structure enwrapping the Ape1 complex (termed a Cvt vesicle) is ~150-nm in diameter, in contrast with that of the autophagosome, which is 300–900 nm. In addition, the Cvt vesicle, in contrast to the autophagosome, excludes bulk cytoplasm. Furthermore, while autophagy is induced under starvation conditions, the Cvt pathway occurs constitutively in growing conditions. Finally, as we mentioned above, the Cvt pathway is a selective, biosynthetic pathway, whereas autophagy is generally nonselective and is degradative. How then could we explain the apparent overlap in the import machinery? Importantly, when cells are subjected to starvation, the Cvt complex is sequestered within a larger autophagosome [38], although the kinetics for import are essentially the same as during vegetative growth. Thus, while the biosynthetic Cvt pathway can be distinguished from autophagy, the Ape1 complex can be taken up by autophagosomes under starvation conditions, again suggesting that the Cvt pathway and autophagy utilize much of the same machinery.

In *S. cerevisiae*, the biogenesis and the vacuolar transport of both autophagosomes and Cvt vesicles include the following steps: (1) membrane from various sources generates vesicles containing Atg9 (see below) as a critical integral membrane protein, and these vesicles form into tubulovesicular clusters in a SNARE-dependent manner; (2) one or more clusters contribute to the formation of a perivacuolar phagophore assembly site (PAS), which is considered to be a foundation/nucleation site that (3) leads to formation of the phagophore, the initial sequestering compartment; (4) two ubiquitin-like protein conjugation systems including Atg8 and its conjugation to phosphatidylethanolamine (PE) contribute to the formation and elongation of the phagophore to generate the double-membrane Cvt vesicle and autophagosome; (5) the completed vesicles dock and fuse with the vacuole, releasing the inner vesicle into the lumen where the single-membrane structures are referred to as Cvt or autophagic bodies.

Both autophagosomes and Cvt vesicles are said to be formed *de novo*, to emphasize the fact that their generation occurs by a mechanism that is distinct from that used in the budding of transient transport vesicles in the secretory pathway. Although the details of sequestering vesicle biogenesis are still not clear, almost all of the Atg proteins are localized at least transiently to the PAS [39]. Atg9, which is the sole integral membrane protein in yeast that is essential for Cvt vesicle and autophagosome formation, is relatively unique in that it is localized at multiple sites including the PAS. The population of Atg9 at the non-PAS sites (Atg9 reservoirs) corresponds to the tubulovesicular clusters and is proposed to traffic between these sites and the PAS, providing membrane for phagophore expansion. The function of most of the Atg proteins is still not known. For example, Atg8–PE participates in cargo recognition during selective types of autophagy and is also involved in determining the size of the autophagosome [40], but the details of these mechanisms are not known.

5. Discovery of the Cvt-Specific Genes

As mentioned above, not all of the *cvt* and *apg/aut* mutants displayed an overlap; some mutants were defective only in autophagy or the Cvt pathway, but not both. For example, the *atg11* mutant shows a complete block in the maturation of prApe1, but is essentially normal for autophagy [19, 23]. These results suggested that the Cvt pathway and autophagy share most of the same machinery, but that they also need some molecules that are specific for each pathway. One of the fundamental differences between the Cvt pathway and autophagy concerns their temporal and physiological activity. The Cvt pathway is active during vegetative growth, consistent with its role as a biosynthetic trafficking route. In contrast, autophagy is induced under starvation conditions, where it can break down cellular macromolecules to supply building blocks and energy. A complex of proteins including Atg13, which is required both for the Cvt pathway and autophagy, appears to be partly responsible for switching these pathways in response to changes in the environment. In starvation conditions, Atg13 interacts with the Atg1 complex including Atg17, Atg29, and Atg31 to induce autophagy [22, 41–43]. Under vegetative conditions, Atg13 may have a lower affinity for Atg1, a condition that may promote the Cvt pathway. Atg13 is regulated by its phosphorylation status in a TORC1-dependent manner; Atg13 is highly phosphorylated in growing conditions but dephosphorylated in starvation conditions [41, 44].

Another characteristic of the Cvt pathway is the specificity for its cargo, whereas macroautophagy is a nonselective process, suggesting that the Cvt pathway requires a receptor, which recognizes the substrate. In this case, the substrate corresponds to the cargo of the Cvt vesicles, which is comprised primarily of the Ape1 complex. A systematic yeast two-hybrid screen in *S. cerevisiae* was performed and the gene product of *YOL082W* was found as a potential interacting protein with prApe1 [45]. Biochemical analysis demonstrated that *YOL082W* encodes a protein that functions as a receptor for the targeting of prApe1 by the Cvt pathway, and the gene was renamed *CVT19* [46, 47] and later *ATG19* [37]. In *atg19Δ* cells, the precursor form of Ape1 accumulates in the cytoplasm in both nutrient rich and starvation conditions, suggesting that Atg19 is necessary for the targeting of prApe1 both by the Cvt pathway and autophagy. An important point in this regard is that import of prApe1 by autophagy is still a selective process that utilizes a receptor protein; this explains why the kinetics of import are the same as for the Cvt pathway and are much faster than would be expected for bulk uptake of cytoplasm.

An immunoprecipitation analysis showed that Atg19 physically interacts with the propeptide of prApe1, and the coiled-coil domain of Atg19 mediates this interaction [48]. Atg19 localizes at the PAS with the Ape1 complex [49]; the combination of the Ape1 complex bound to Atg19 is referred to as the Cvt complex. In *atg19Δ* cells, GFP-Ape1 forms a dodecamer, but it does not localize at the PAS. The

kinetics of the maturation of prApe1 and the degradation of Atg19 are quite similar. Together with the localization data, these findings suggest that Atg19 is delivered to the vacuole by the Cvt pathway along with the precursor Ape1 dodecamer. Interestingly, deletion of *APE1* results in a dispersed Atg19 distribution, and Atg19 does not localize to the PAS in *ape1Δ* cells, suggesting that the Ape1 complex itself is required for concentrating its soluble receptor at this site. Further analyses revealed that Atg19-prApe1 movement to the PAS is dependent on Atg11, which we now know acts as an adaptor or scaffold protein for selective autophagy pathways, such as the Cvt pathway, and the selective autophagic degradation of peroxisomes and mitochondria (termed pexophagy and mitophagy, resp.) [22, 50]. Atg11 may mediate the transport of Atg9 to the PAS for selective autophagy during vegetative growth [51], whereas Atg17 may carry out this role for bulk autophagy during starvation. Atg11 has certain characteristics of a scaffold protein in that it interacts with several Atg proteins, including Atg1, Atg9, Atg17, Atg19, Atg20, and itself [51, 52].

In the Cvt pathway, Atg19 binds the prApe1 propeptide independent of any other Atg proteins. Atg11 can then interact with Atg19, allowing movement of the cargo to the PAS. Once at the PAS, Atg19 also interacts with Atg8–PE; it is not known if both Atg8 and Atg11 bind Atg19 at the same time, as their binding sites are distinct, but very close to each other. Thus, Atg19 is a receptor that is responsible for recognizing the prApe1 dodecamer to target it to the PAS due to its interaction with Atg11. Furthermore, Atg19 leads to the incorporation of the Cvt complex into a double-membrane vesicle (i.e., a Cvt vesicle or autophagosome) via its interaction with Atg8 [48]. In the absence of other Atg proteins such as Atg1, Cvt vesicles, and autophagosomes do not form; however, the Cvt complex is still targeted to the PAS, suggesting that Atg19 transport of prApe1 to the PAS occurs independent of the vesicle formation steps. Atg19 is both ubiquitinated and deubiquitinated *in vivo*, and these modifications of Atg19 are required for the efficient trafficking of prApe1 via the Cvt pathway [53]. Atg19 interacts with the deubiquitinating enzyme Ubp3, and the deletion of *UBP3* leads to decreased targeting of prApe1. Furthermore, the mutation on the ubiquitin acceptor site, Lys213 and Lys216 of Atg19, reduces the interaction of Atg19 with prApe1. Thus, the ubiquitination and deubiquitination of Atg19 are likely to play a structural or mechanistic role in the normal progression of the Cvt pathway, instead of serving as a degradation signal for the proteasome.

As described above, many of the yeast Atg proteins responsible for the Cvt pathway and autophagy have been identified, and the general mechanism involved in these processes has been explored through genetic and biochemical approaches. Nevertheless, the molecular mechanism underlying nucleation of the sequestering phagophore remains largely unknown. Many processes involving membrane rearrangement and movement, such as endocytosis or membrane ruffling, require the cytoskeleton. The actin cytoskeleton is required for the selective Cvt pathway, but not for nonselective autophagy in yeast [54]. Actin plays a role in trafficking of Atg9 to the PAS and recruitment of the Cvt cargo in growing conditions. Further studies identified actin-related proteins, including components of the Arp2/3 complex, as playing a role in the transport of Atg9 for specific types of autophagy [55]. The Arp2 protein itself interacts with Atg9 and regulates the dynamics of Atg9 movement. Thus, the Arp2/3 complex may allow Atg9, along with its associated membrane, to move in a directed fashion to the PAS along actin cables. The specific autophagy factors such as Atg19 and Atg11, and perhaps other molecular components, may serve as adaptors between the Cvt cargo and the actin cytoskeleton.

6. Discovery of Other Cvt Cargo, Ams1 and Ape4

Prior to the analysis of the Cvt pathway, Ams1 was shown to enter the vacuole independent of the secretory pathway [56], although the mechanism of import was unclear. We found that Ams1 is another hydrolase targeted to the vacuole by the Cvt pathway [57], as its delivery is blocked in cvt (atg) mutants. Similar to prApe1, Ams1 forms oligomers composed of 4 to 6 of the 122-kDa species in the cytosol, and the oligomeric state is maintained during the import process. Ams1 transport is also mediated by Atg19 [47] and its binding site is distinct from that used by prApe1 [48]. Thus, Ams1 is part of a prApe1-Atg19-Ams1 Cvt complex. In apeΔ cells, the Ams1-Atg19 interaction still occurs, but this complex is dispersed in the cytosol, whereas deletion of AMS1 does not affect the transport of the prApe1-Atg19 complex. These results indicate that Ams1, which is synthesized at a level that is substantially lower than prApe1, might exploit the prApe1-Atg19 import system to achieve its own efficient transport to the vacuole.

Recently, it was shown that Ams1 is delivered to the vacuole in an Atg19-independent manner under starvation conditions [58]. During autophagy, Atg34 (Yol083w), a homolog of Atg19, functions as a receptor for Ams1. In atg19Δ cells, Ams1 targeting is disrupted in nutrient-rich conditions [47, 48]. However, Ams1 is efficiently transported into the vacuole under starvation conditions by autophagy even in atg19Δ cells. A genome-wide yeast two-hybrid screen suggested that Yol083w is an Ams1 interacting protein [45], and Atg34 indeed physically interacts with Ams1 [58]. Similar to Atg19, Atg34 binds Atg8 and Atg11 using distinct domains, and these interactions are essential for its function in targeting Ams1 into an autophagosome; an Atg34 mutant that lacks its Atg8 interacting motif forms a complex with Ams1, but shows a defect in sequestration into autophagosomes. Importantly, the transport of Ams1 mediated by Atg34 in starvation conditions is prApe1 independent, unlike that mediated by Atg19 in growing conditions.

Also recently, aspartyl aminopeptidase (Yhr113w/Ape4) was found to be a third Cvt cargo protein [59]. Yeast two-hybrid analyses suggested that Ape4 can associate with Atg19 and prApe1 [60]. Unlike prApe1, Ape4 does not possess a propeptide region and it does not self-assemble into aggregates [59]; however, it still binds to Atg19. An immunoprecipitation analysis with truncated versions of Atg19 revealed that the three identified Cvt cargo components, prApe1,

Ams1, and Ape4, associate with Atg19 by binding to distinct sites. GFP-Ape4 colocalizes with RFP-Ape1 at the PAS in growing conditions, and this localization is dependent on Atg19. Notably, Ape4 transport to the vacuole by the Cvt pathway is significantly decreased in ape1Δ cells, suggesting that Ape4 relies on the prApe1-Atg19 complex for its targeting, similar to Ams1 in vegetative conditions. In atg11Δ cells, Ape4 can colocalize with prApe1, but it does not localize at the PAS.

7. Conclusions

An intriguing question has been why yeast cells have utilized the Cvt pathway to import a resident vacuolar hydrolase. In higher eukaryotes, there is no evidence for a Cvt pathway, and the ATG genes specifically involved in this pathway are not conserved; in contrast, those genes that are also needed for autophagy are highly conserved [61]. However, selective types of autophagy clearly take place in higher eukaryotes, including mitophagy and pexophagy. The molecular machinery involved in these processes in mammalian cells has not been completely elucidated, but it is likely that the general mechanism is conserved. For example, receptors such as BNIP3L and BNIP3 function as receptors in mammalian mitophagy, whereas Atg32 carries out this function in yeast; BNIP3L and BNIP3 are not homologs of Atg32, but they are functional counterparts, supporting the concept of mechanistic conservation. Furthermore, most of the machinery for the Cvt pathway is also used for pexophagy and mitophagy, which, as noted above, take place in higher eukaryotes. This means that with regard to the Atg proteins, the apparent absence of the Cvt pathway in mammals may be viewed as a deficiency in the specific receptor Atg19, rather than a major difference between yeast and other eukaryotes.

Returning to the initial question regarding the origin of the Cvt pathway, one possibility is that the oligomeric structure of prApe1 or Ams1 is critical for stability and/or function. The size of the oligomeric form of these hydrolases would prevent translocation through the ER translocon, necessitating a vesicle-mediated import process. Also, Ams1 does not appear to be synthesized as a zymogen. Thus, it would be problematic for this hydrolase to traverse the secretory pathway along with other newly synthesized glycosylated proteins. These vacuolar hydrolases are likely required in large amounts when the cell is starved or when aggregated proteins or damaged organelles accumulate, and the synthesis of most vacuolar hydrolases increases substantially during starvation. Under these conditions, the efficient transport of these hydrolases as oligomers by means of a vesicle-mediated mechanism such as autophagy would be extremely efficient. It would seem reasonable for the cell to modify the autophagy pathway very slightly with the addition of a small number of specificity components to take advantage of the existing autophagy machinery and allow it to be used for various types of selective sequestration processes.

Acknowledgment

This work was supported by NIH public health service Grant GM53396 to D. J. Klionsky.

References

[1] V. A. Bankaitis, L. M. Johnson, and S. D. Emr, "Isolation of yeast mutants defective in protein targeting to the vacuole," *Proceedings of the National Academy of Sciences of the United States of America*, vol. 83, no. 23, pp. 9075–9079, 1986.

[2] J. H. Rothman and T. H. Stevens, "Protein sorting in yeast: mutants defective in vacuole biogenesis mislocalize vacuolar proteins into the late secretory pathway," *Cell*, vol. 47, no. 6, pp. 1041–1051, 1986.

[3] Y. H. Chang and J. A. Smith, "Molecular cloning and sequencing of genomic DNA encoding aminopeptidase I from *Saccharomyces cerevisiae*," *The Journal of Biological Chemistry*, vol. 264, no. 12, pp. 6979–6983, 1989.

[4] R. Cueva, N. Garcia-Alvarez, and P. Suarez-Rendueles, "Yeast vacuolar aminopeptidase yscI. Isolation and regulation of the APE1 (LAP4) structural gene," *FEBS Letters*, vol. 259, no. 1, pp. 125–129, 1989.

[5] G. Ammerer, C. P. Hunter, J. H. Rothman, G. C. Saari, L. A. Valls, and T. H. Stevens, "*PEP4* gene of *Saccharomyces cerevisiae* encodes proteinase A, a vacuolar enzyme required for processing of vacuolar precursors," *Molecular and Cellular Biology*, vol. 6, no. 7, pp. 2490–2499, 1986.

[6] C. A. Woolford, L. B. Daniels, F. J. Park, E. W. Jones, J. N. Van Arsdell, and M. A. Innis, "The *PEP4* gene encodes an aspartyl protease implicated in the posttranslational regulation of *Saccharomyces cerevisiae* vacuolar hydrolases," *Molecular and Cellular Biology*, vol. 6, no. 7, pp. 2500–2510, 1986.

[7] L. A. Valls, C. P. Hunter, J. H. Rothman, and T. H. Stevens, "Protein sorting in yeast: the localization determinant of yeast vacuolar carboxypeptidase Y resides in the propeptide," *Cell*, vol. 48, no. 5, pp. 887–897, 1987.

[8] Y. Kaneko. Y., N. Hayashi, A. Toh-e, I. Banno, and Y. Oshima, "Structural characteristics of the *PHO8* gene encoding repressible alkaline phosphatase in *Saccharomyces cerevisiae*," *Gene*, vol. 58, no. 1, pp. 137–148, 1987.

[9] C. M. Moehle, R. Tizard, S. K. Lemmon, J. Smart, and E. W. Jones, "Protease B of the lysosomelike vacuole of the yeast *Saccharomyces cerevisiae* is homologous to the subtilisin family of serine proteases," *Molecular and Cellular Biology*, vol. 7, no. 12, pp. 4390–4399, 1987.

[10] M. J. Kuranda and P. W. Robbins, "Cloning and heterologous expression of glycosidase genes from *Saccharomyces cerevisiae*," *Proceedings of the National Academy of Sciences of the United States of America*, vol. 84, no. 9, pp. 2585–2589, 1987.

[11] G. Metz and K. H. Roehm, "Yeast aminopeptidase I. Chemical composition and catalytic properties," *Biochimica et Biophysica Acta*, vol. 429, no. 3, pp. 933–949, 1976.

[12] B. Distel, E. J. M. Al, H. F. Tabak, and E. W. Jones, "Synthesis and maturation of the yeast vacuolar enzymes carboxypeptidase Y and aminopeptidase I," *Biochimica et Biophysica Acta*, vol. 741, no. 1, pp. 128–135, 1983.

[13] D. J. Klionsky, R. Cueva, and D. S. Yaver, "Aminopeptidase I of *Saccharomyces cerevisiae* is localized to the vacuole independent of the secretory pathway," *The Journal of Cell Biology*, vol. 119, no. 2, pp. 287–300, 1992.

[14] K. Takeshige, M. Baba, S. Tsuboi, T. Noda, and Y. Ohsumi, "Autophagy in yeast demonstrated with proteinase-deficient mutants and conditions for its induction," *The Journal of Cell Biology*, vol. 119, no. 2, pp. 301–312, 1992.

[15] J. F. Dice, H. L. Chiang, E. P. Spencer, and J. M. Backer, "Regulation of catabolism of microinjected ribonuclease A. Identification of residues 7–11 as the essential pentapeptide," *The Journal of Biological Chemistry*, vol. 261, no. 15, pp. 6853–6859, 1986.

[16] T. M. Harding, K. A. Morano, S. V. Scott, and D. J. Klionsky, "Isolation and characterization of yeast mutants in the cytoplasm to vacuole protein targeting pathway," *The Journal of Cell Biology*, vol. 131, no. 3, pp. 591–602, 1995.

[17] M. Tsukada and Y. Ohsumi, "Isolation and characterization of autophagy-defective mutants of *Saccharomyces cerevisiae*," *FEBS Letters*, vol. 333, no. 1-2, pp. 169–174, 1993.

[18] M. Thumm, R. Egner, B. Koch et al., "Isolation of autophagocytosis mutants of *Saccharomyces cerevisiae*," *FEBS Letters*, vol. 349, no. 2, pp. 275–280, 1994.

[19] T. M. Harding, A. Hefner-Gravink, M. Thumm, and D. J. Klionsky, "Genetic and phenotypic overlap between autophagy and the cytoplasm to vacuole protein targeting pathway," *The Journal of Biological Chemistry*, vol. 271, no. 30, pp. 17621–17624, 1996.

[20] S. V. Scott, A. Hefner-Gravink, K. A. Morano, T. Noda, Y. Ohsumi, and D. J. Klionsky, "Cytoplasm-to-vacuole targeting and autophagy employ the same machinery to deliver proteins to the yeast vacuole," *Proceedings of the National Academy of Sciences of the United States of America*, vol. 93, no. 22, pp. 12304–12308, 1996.

[21] M. Baba, M. Osumi, S. V. Scott, D. J. Klionsky, and Y. Ohsumi, "Two distinct pathways for targeting proteins from the cytoplasm to the vacuole/lysosome," *The Journal of Cell Biology*, vol. 139, no. 7, pp. 1687–1695, 1997.

[22] J. Kim, Y. Kamada, P. E. Stromhaug et al., "Cvt9/Gsa9 functions in sequestering selective cytosolic cargo destined for the vacuole," *The Journal of Cell Biology*, vol. 153, no. 2, pp. 381–396, 2001.

[23] S. V. Scott, D. C. Nice, J. J. Nau et al., "Apg13p and Vac8p are part of a complex of phosphoproteins that are required for cytoplasm to vacuole targeting," *The Journal of Biological Chemistry*, vol. 275, no. 33, pp. 25840–25849, 2000.

[24] M. D. George, M. Baba, S. V. Scott et al., "Apg5p functions in the sequestration step in the cytoplasm-to-vacuole targeting and macroautophagy pathways," *Molecular Biology of the Cell*, vol. 11, no. 3, pp. 969–982, 2000.

[25] T. Noda, J. Kim, W.-P. Huang et al., "Apg9p/Cvt7p is an integral membrane protein required for transport vesicle formation in the Cvt and autophagy pathways," *The Journal of Cell Biology*, vol. 148, no. 3, pp. 465–479, 2000.

[26] N. Mizushima, T. Noda, T. Yoshimori et al., "A protein conjugation system essential for autophagy," *Nature*, vol. 395, no. 6700, pp. 395–398, 1998.

[27] S. V. Scott, M. Baba, Y. Ohsumi, and D. J. Klionsky, "Aminopeptidase I is targeted to the vacuole by a nonclassical vesicular mechanism," *The Journal of Cell Biology*, vol. 138, no. 1, pp. 37–44, 1997.

[28] K. A. Tucker, F. Reggiori, W. A. Dunn, and D. J. Klionsky, "Atg23 is essential for the cytoplasm to vacuole targeting pathway and efficient autophagy but not pexophagy," *The Journal of Biological Chemistry*, vol. 278, no. 48, pp. 48445–48452, 2003.

[29] H. Abeliovich, C. Zhang, W. A. Dunn, K. M. Shokat, and D. J. Klionsky, "Chemical genetic analysis of Apg1 reveals a non-kinase role in the induction of autophagy," *Molecular Biology of the Cell*, vol. 14, no. 2, pp. 477–490, 2003.

[30] J. Guan, P. E. Stromhaug, M. D. George et al., "Cvt18/Gsa12 is required for cytoplasm-to-vacuole transport, pexophagy, and autophagy in *Saccharomyces cerevisiae* and *Pichia pastoris*," *Molecular Biology of the Cell*, vol. 12, no. 12, pp. 3821–3838, 2001.

[31] C.-W. Wang, J. Kim, W.-P. Huang et al., "Apg2 is a novel protein required for the cytoplasm to vacuole targeting, autophagy, and pexophagy pathways," *The Journal of Biological Chemistry*, vol. 276, no. 32, pp. 30442–30451, 2001.

[32] H. Abeliovich, W. A. Dunn Jr., J. Kim, and D. J. Klionsky, "Dissection of autophagosome biogenesis into distinct nucleation and expansion steps," *The Journal of Cell Biology*, vol. 151, no. 5, pp. 1025–1034, 2000.

[33] J. Frey and K. H. Roehm, "Subcellular localization and levels of aminopeptidases and dipeptidase in *Saccharomyces cerevisiae*," *Biochimica et Biophysica Acta*, vol. 527, no. 1, pp. 31–41, 1978.

[34] R. J. Trumbly and G. Bradley, "Isolation and characterization of aminopeptidase mutants of *Saccharomyces cerevisiae*," *Journal of Bacteriology*, vol. 156, no. 1, pp. 36–48, 1983.

[35] M. N. Oda, S. V. Scott, A. Hefner-Gravink, A. D. Caffarelli, and D. J. Klionsky, "Identification of a cytoplasm to vacuole targeting determinant in aminopeptidase I," *The Journal of Cell Biology*, vol. 132, no. 6, pp. 999–1010, 1996.

[36] J. Kim, S. V. Scott, M. N. Oda, and D. J. Klionsky, "Transport of a large oligomeric protein by the cytoplasm to vacuole protein targeting pathway," *The Journal of Cell Biology*, vol. 137, no. 3, pp. 609–618, 1997.

[37] D. J. Klionsky, J. M. Cregg, W. A. Dunn et al., "A unified nomenclature for yeast autophagy-related genes," *Developmental Cell*, vol. 5, no. 4, pp. 539–545, 2003.

[38] M. Baba, K. Takeshige, N. Baba, and Y. Ohsumi, "Ultrastructural analysis of the autophagic process in yeast: detection of autophagosomes and their characterization," *The Journal of Cell Biology*, vol. 124, no. 6, pp. 903–913, 1994.

[39] K. Suzuki, T. Kirisako, Y. Kamada, N. Mizushima, T. Noda, and Y. Ohsumi, "The pre-autophagosomal structure organized by concerted functions of *APG* genes is essential for autophagosome formation," *The EMBO Journal*, vol. 20, no. 21, pp. 5971–5981, 2001.

[40] Z. Xie, U. Nair, and D. J. Klionsky, "Atg8 controls phagophore expansion during autophagosome formation," *Molecular Biology of the Cell*, vol. 19, no. 8, pp. 3290–3298, 2008.

[41] Y. Kamada, T. Funakoshi, T. Shintani, K. Nagano, M. Ohsumi, and Y. Ohsumi, "Tor-mediated induction of autophagy via an Apg1 protein kinase complex," *The Journal of Cell Biology*, vol. 150, no. 6, pp. 1507–1513, 2000.

[42] T. Kawamata, Y. Kamada, K. Suzuki et al., "Characterization of a novel autophagy-specific gene, *ATG29*," *Biochemical and Biophysical Research Communications*, vol. 338, no. 4, pp. 1884–1889, 2005.

[43] Y. Kabeya, T. Kawamata, K. Suzuki, and Y. Ohsumi, "Cis1/Atg31 is required for autophagosome formation in *Saccharomyces cerevisiae*," *Biochemical and Biophysical Research Communications*, vol. 356, no. 2, pp. 405–410, 2007.

[44] Y. Kamada, K. I. Yoshino, C. Kondo et al., "Tor directly controls the Atg1 kinase complex to regulate autophagy," *Molecular and Cellular Biology*, vol. 30, no. 4, pp. 1049–1058, 2010.

[45] P. Uetz, L. Glot, G. Cagney et al., "A comprehensive analysis of protein-protein interactions in *Saccharomyces cerevisiae*," *Nature*, vol. 403, no. 6770, pp. 623–627, 2000.

[46] R. Leber, E. Silles, I. V. Sandoval, and M. J. Mazón, "Yol082p, a novel CVT protein involved in the selective targeting of aminopeptidase I to the yeast vacuole," *The Journal of Biological Chemistry*, vol. 276, no. 31, pp. 29210–29217, 2001.

[47] S. V. Scott, J. Guan, M. U. Hutchins, J. Kim, and D. J. Klionsky, "Cvt19 is a receptor for the cytoplasm-to-vacuole targeting pathway," *Molecular Cell*, vol. 7, no. 6, pp. 1131–1141, 2001.

[48] T. Shintani, W.-P. Huang, P. E. Stromhaug, and D. J. Klionsky, "Mechanism of cargo selection in the cytoplasm to vacuole targeting pathway," *Developmental Cell*, vol. 3, no. 6, pp. 825–837, 2002.

[49] J. Kim, W.-P. Huang, P. E. Stromhaug, and D. J. Klionsky, "Convergence of multiple autophagy and cytoplasm to vacuole targeting components to a perivacuolar membrane compartment prior to de Novo vesicle formation," *The Journal of Biological Chemistry*, vol. 277, no. 1, pp. 763–773, 2002.

[50] T. Kanki and D. J. Klionsky, "Mitophagy in yeast occurs through a selective mechanism," *The Journal of Biological Chemistry*, vol. 283, no. 47, pp. 32386–32393, 2008.

[51] C. He, H. Song, T. Yorimitsu et al., "Recruitment of Atg9 to the preautophagosomal structure by Atg11 is essential for selective autophagy in budding yeast," *The Journal of Cell Biology*, vol. 175, no. 6, pp. 925–935, 2006.

[52] T. Yorimitsu and D. J. Klionsky, "Atg11 links cargo to the vesicle-forming machinery in the cytoplasm to vacuole targeting pathway," *Molecular Biology of the Cell*, vol. 16, no. 4, pp. 1593–1605, 2005.

[53] B. K. Baxter, H. Abeliovich, X. Zhang, A. G. Stirling, A. L. Burlingame, and D. S. Goldfarb, "Atg19p ubiquitination and the cytoplasm to vacuole trafficking pathway in yeast," *The Journal of Biological Chemistry*, vol. 280, no. 47, pp. 39067–39076, 2005.

[54] F. Reggiori, I. Monastyrska, T. Shintani, and D. J. Klionsky, "The actin cytoskeleton is required for selective types of autophagy, but not nonspecific autophagy, in the yeast *Saccharomyces cerevisiae*," *Molecular Biology of the Cell*, vol. 16, no. 12, pp. 5843–5856, 2005.

[55] I. Monastyrska, C. He, J. Geng, A. D. Hoppe, Z. Li, and D. J. Klionsky, "Arp2 links autophagic machinery with the actin cytoskeleton," *Molecular Biology of the Cell*, vol. 19, no. 5, pp. 1962–1975, 2008.

[56] T. Yoshihisa and Y. Anraku, "A novel pathway of import of α-mannosidase, a marker enzyme of vacuolar membrane, in *Saccharomyces cerevisiae*," *The Journal of Biological Chemistry*, vol. 265, no. 36, pp. 22418–22425, 1990.

[57] M. U. Hutchins and D. J. Klionsky, "Vacuolar localization of oligomeric α-mannosidase requires the cytoplasm to vacuole targeting and autophagy pathway components in *Saccharomyces cerevisiae*," *The Journal of Biological Chemistry*, vol. 276, no. 23, pp. 20491–20498, 2001.

[58] K. Suzuki, C. Kondo, M. Morimoto, and Y. Ohsumi, "Selective transport of α-mannosidase by autophagic pathways: identification of a novel receptor, Atg34p," *The Journal of Biological Chemistry*, vol. 285, no. 39, pp. 30019–30025, 2010.

[59] M. Yuga, K. Gomi, D. J. Klionsky, and T. Shintani, "Aspartyl aminopeptidase is imported from the cytoplasm to the vacuole by selective autophagy in *Saccharomyces cerevisiae*," *The Journal of Biological Chemistry*, vol. 286, no. 15, pp. 13704–13713, 2011.

[60] T. Ito, T. Chiba, R. Ozawa, M. Yoshida, M. Hattori, and Y. Sakaki, "A comprehensive two-hybrid analysis to explore the yeast protein interactome," *Proceedings of the National Academy of Sciences of the United States of America*, vol. 98, no. 8, pp. 4569–4574, 2001.

[61] W. H. Meijer, I. J. Van Der Klei, M. Veenhuis, and J. A. K. W. Kiel, "*ATG* genes involved in non-selective autophagy are conserved from yeast to man, but the selective Cvt and pexophagy pathways also require organism-specific genes," *Autophagy*, vol. 3, no. 2, pp. 106–116, 2007.

4

Selective Autophagy in *Drosophila*

Ioannis P. Nezis[1, 2, 3]

[1] *Department of Biochemistry, Institute for Cancer Research, Oslo University Hospital, Montebello, 0310 Oslo, Norway*
[2] *Centre for Cancer Biomedicine, Faculty of Medicine, University of Oslo, Montebello, 0310 Oslo, Norway*
[3] *Laboratory of Cell Biology, Department of Biological Applications and Technologies, University of Ioannina, 45110 Ioannina, Greece*

Correspondence should be addressed to Ioannis P. Nezis, ioannis.nezis@rr-research.no

Academic Editor: Anne Simonsen

Autophagy is an evolutionarily conserved process of cellular self-eating and is a major pathway for degradation of cytoplasmic material by the lysosomal machinery. Autophagy functions as a cellular response in nutrient starvation, but it is also associated with the removal of protein aggregates and damaged organelles and therefore plays an important role in the quality control of proteins and organelles. Although it was initially believed that autophagy occurs randomly in the cell, during the last years, there is growing evidence that sequestration and degradation of cytoplasmic material by autophagy can be selective. Given the important role of autophagy and selective autophagy in several disease-related processes such as neurodegeneration, infections, and tumorigenesis, it is important to understand the molecular mechanisms of selective autophagy, especially at the organismal level. *Drosophila* is an excellent genetically modifiable model organism exhibiting high conservation in the autophagic machinery. However, the regulation and mechanisms of selective autophagy in *Drosophila* have been largely unexplored. In this paper, I will present an overview of the current knowledge about selective autophagy in *Drosophila*.

1. Introduction

Macroautophagy (from hereafter referred to as autophagy) is an evolutionarily conserved process by which a portion of the cytosol and organelles are sequestered by isolation membranes called phagophores. The phagophore engulfs portions of the cytoplasm and forms a double-membrane-layered organelle called the autophagosome. The autophagosome then fuses with a lysosome and generate the autolysosome that has a single limiting membrane, where its sequestered components are degraded [1]. Autophagy serves as a cellular response in nutrient starvation, but it is also responsible for the removal of aggregated proteins, damaged organelles, and developmental remodeling and therefore plays an important role in the quality control of proteins and organelles and in cellular homeostasis [1]. Genetic inhibition of autophagy induces degeneration that resembles degeneration observed during ageing, and physiological ageing is associated with reduced autophagic activity [2]. Autophagy is implicated in neurodegeneration, infections, tumorigenesis, heart disease, liver and lung disease, myopathies, and in lysosomal storage disorders [2]. Interestingly, it has been shown that induction of autophagy can increase longevity in multiple animal species [3]. Contrary to the belief that autophagy is a nonselective process, recent evidence suggests that degradation of proteins, protein aggregates, organelles, and bacteria can be selective through adaptor proteins [4]. It is therefore important to elucidate the role of selective autophagy in normal and pathological conditions using model organisms. The fruit fly *Drosophila melanogaster* is a genetically modifiable model organism and is an excellent model for investigating the mechanisms of selective autophagy in the context of the physiology of the cell, the system, and the living organism. This paper will summarize the current knowledge about selective autophagy in *Drosophila*.

2. Selective Autophagy in *Drosophila*

Studies in *Drosophila* so far revealed the presence of highly conserved autophagic machinery compared to yeast and mammals [5]. *atg* (autophagy-related) genes and their regulators in *Drosophila* in many cases, in contrast to mammalian systems, have single orthologs, allowing for nonredundant genetic studies [5]. However, the regulation and mechanisms of selective autophagy have not been described in details, and there is only limited evidence for the presence of selective autophagy and autophagic cargo receptors. Additionally, cellular processes related to selective autophagy like mitophagy (selective autophagy of mitochondria), xenophagy (selective autophagy of bacteria and viruses), nucleophagy (selective autophagy of nucleus), and pexophagy (selective autophagy of peroxisomes) are largely unexplored in *Drosophila*. In the following text, I will describe what is reported so far in the literature about selective autophagy and selective autophagy-related proteins in *Drosophila*.

2.1. Selective Autophagy Receptors in Drosophila

2.1.1. Ref(2)P, the Drosophila Homologue of the Mammalian Selective Autophagy Receptor p62/SQSTM1.
In mammals, six proteins have been identified as selective autophagy receptors so far: p62/SQSTM1, NBR1, NDP52, Nix, optineurin, and Stbd1 [4, 6, 7]. These proteins contain a LIR/LRS (LC3-interacting region/LC3 recognition sequence) motif and have been shown to interact with the autophagosomal membrane protein LC3 (microtubule-associated protein 1 light chain 3) [4]. The phosphatidylinositol-3-phosphate-(PI3P-) binding protein Alfy (autophagy-linked FYVE domain containing protein) was also shown to be required for selective degradation of aggregated proteins such as polyQ [8, 9] although a LIR/LRS motif has not yet been identified in Alfy sequence.

Landmark studies from Johansen's group indicated that mammalian p62/SQSTM1 is degraded selectively by autophagy and introduced the significant role of p62/SQSTM1 in autophagy [10, 11]. p62/SQSTM1 is the first identified and most studied autophagy cargo receptor. It is a multifunctional scaffold protein that serves a large variety of cellular functions [4, 12, 13]. The human p62 protein is 440 amino acids long and contains several structural and functional motifs [4] (Figure 1(a)). A Phox and Bem1p domain (PB1 domain) is located at the N-terminus and is required for di- and multimerization of the protein as well as interaction with the protein kinases MEKK3, MEK5, ERK, PKCζ, and PKCλ/ι and autophagy receptor NBR1 [4]. A zinc-finger-type (ZZ-type) domain follows the PB1 domain and is the binding site of receptor-interacting serine-threonine kinase 1 (RIP1) [12, 14]. Subsequently, there is a TRAF6-binding (TB) domain which contains the binding site of E3 ubiquitin-protein ligase TRAF6 [12, 14]. Nuclear-cytoplasmic shuttling of the protein is mediated by nuclear localization signals (NLSs) and nuclear export signal (NES) which are also present [15]. p62/SQSTM1 contains a LIR/LRS motif and a kelch-like ECH-associated protein 1 (KEAP1) interacting region (KIR) motif responsible for the interaction with LC3 and KEAP1, respectively [11, 16, 17]. The C-terminus of p62 harbors a ubiquitin-associated (UBA) domain required for its binding to mono- and polyubiquitin [4] (Figure 1(a)).

The *Drosophila* single p62 homologue, Ref(2)P (*refractory to Sigma P ref(2)P/CG10360*), has 599 amino acids and contains an N-terminal PB1 domain followed by a ZZ-type zinc finger domain and a C-terminal UBA domain (Figure 1(a)) [13, 18]. Although Ref(2)P has not been shown to be a selective autophagic substrate directly, several lines of evidence support this. First, it has been shown that Ref(2)P is a major component of protein aggregates in flies that are defective in autophagy, in flies that have impaired proteasomal function, in *Drosophila* models of human neurodegenerative diseases, and in protein aggregates formed during normal aging in *Drosophila* adult brain [18] (Figure 2). The abilities of Ref(2)P to oligo- and multimerize (through its PB1 domain) and to bind ubiquitinated proteins (through its UBA domain) were shown to be required during the *in vivo* formation of protein aggregates in the adult brain of *Drosophila* [18].

Second, bioinformatic analysis of the sequence of Ref(2)P reveals the presence of a putative LIR motif. The human p62 LIR motif is a 22 amino acid long sequence which contains an evolutionarily conserved motif of three acidic residues followed by a tryptophan (DDDW in p62) [4]. Johansen and Lamark implemented a sequence logo from 25 different LIR motifs from 21 different proteins that all have been tested for binding to ATG8 family proteins. They showed that the LIR motif seems to be eight amino acids long and proposed that the consensus LIR motif could be written as D/E-D/E-D/E-W/F/Y-X-X-L/I/V. It seems that there is a requirement for aromatic residues in the W-site (W/F/Y) and also a requirement for large, hydrophobic residues in the L-site (L/I/V) [4]. Bioinformatic analysis of Ref(2)P sequence reveals the presence of a putative LIR between amino acids 451–458 with a sequence DPEWQLID, which fits very well with the criteria for aromatic residues at W site (W) and hydrophobic residues at L site (I) (Figure 1(b)). Bioinformatic prediction also reveals the presence of a putative KIR motif spanning between the amino acids residues 484–496 (Figure 1(b)). The functional roles of putative LIR and KIR motifs of Ref(2)P have to be tested experimentally *in vitro* and *in vivo*. Taken together, the above information suggest that Ref(2)P is a selective autophagy cargo receptor in *Drosophila melanogaster*.

Ref(2)P was initially characterized in a screen for modifiers of sigma virus multiplication [19–21]. Sigma virus belongs to the family of rhabdoviruses which have two natural hosts, either insect and vertebrate or insect and plant [22]. Sigma virus is an atypical rhabdovirus, since there are no known plants or vertebrate hosts, and it only infects *Drosophila* [23]. Sigma virus is widespread in natural populations of *Drosophila*, and flies infected with the virus exhibit reduced viability of infected eggs and lower survival over winter [23–25]. ref(2)P is the best characterized locus among five host loci which are involved in the control of Sigma virus infection and multiplication, including ref(1)H, ref(2)P and ref(3)D [19, 26–28]. *Drosophila* flies in nature contain two types of alleles: the permissive alleles of ref(2)P which allow efficient sigma virus multiplication, and the restrictive alleles which reduce the replication of the virus [19, 23]. In flies having the permissive alleles, the probability of infection

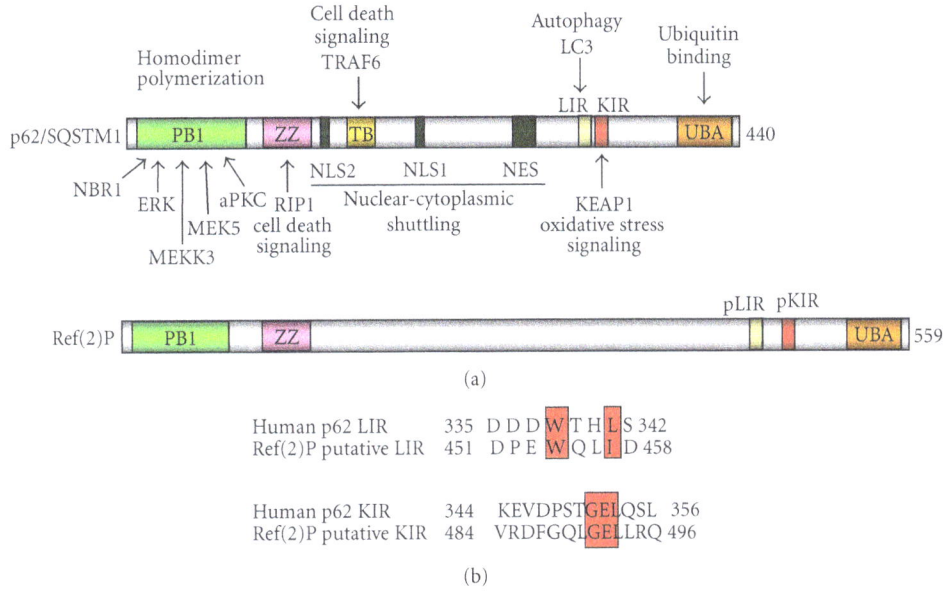

FIGURE 1: Schematic presentation of functional and structural domains of p62 and its *Drosophila* orthologue, Ref(2)P. (a) p62 consists of a PB1 domain (Phox and Bem1p domain) which is responsible for the interaction with the autophagy receptor NBR1 and the protein kinases ERK, MEKK3, MEK5, PKCζ, and PKCλ/ι. The PB1 domain is followed by a ZZ-type zinc finger domain which contains the binding site for RIP1 and a TB domain which harbors the binding site of TRAF6. Nuclear localization signals (NLSs) and nuclear export signal (NES) are also present. p62 contains a LIR (LC3-interacting region) and a KIR (KEAP1-interacting region) motif and a C-terminal UBA (ubiquitin associated) domain responsible for binding to ubiquitin. Ref(2)P has similar structural and functional domains compared to p62. It consists of a PB1 domain which is followed by a ZZ-type zinc finger domain and a C-terminal UBA domain responsible for binding to ubiquitin. Ref(2)P also contains putative LIR and KIR motifs. (b) Bioinformatic prediction of Ref(2)P's putative LIR and KIR motifs and alignment with human p62's motifs. The functional roles of putative LIR and KIR motifs of Ref(2)P have to be tested experimentally.

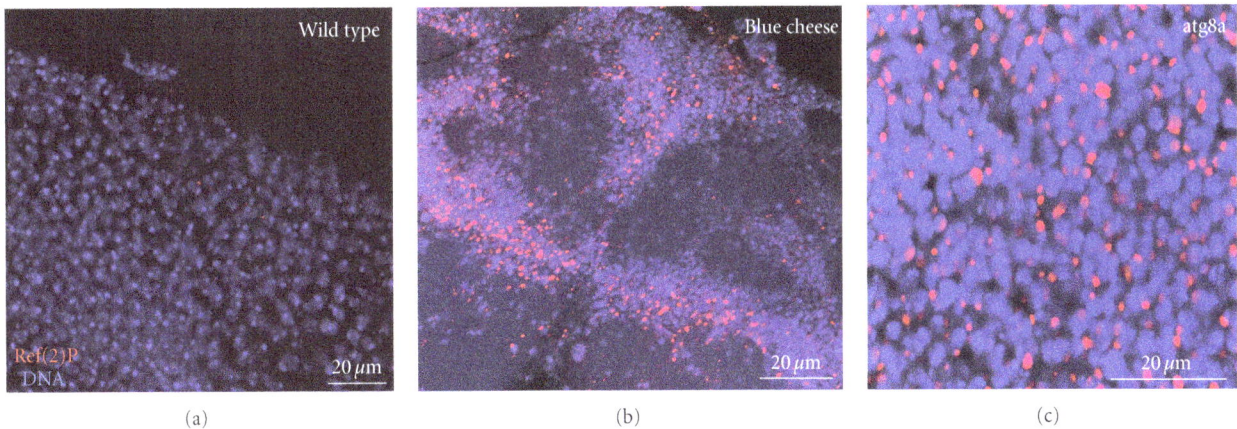

FIGURE 2: Ref(2)P accumulates in the adult brain of atg8a and blue cheese mutant flies. Confocal micrographs of superficial sections of the adult brain cortex of a wild-type fly (a), a blue cheese mutant fly (b), and an autophagy mutant fly (c). The tissues are stained for Ref(2)P (red) and DNA (blue). Ref(2)P accumulates ubiquitously into large sphere-shaped inclusion bodies/aggregates in blue cheese and autophagy mutants compared to wild type.

may reach 100%, whereas, in flies with restrictive alleles the infection rate drops to 0.01%, at least for some viral strains [23, 28]. It appears that the restrictive allele appeared several thousands of years ago and spread in the population as a result of natural selection since it confers a selective advantage [29]. The appearance of the sigma virus strain capable of infecting *Drosophila* flies carrying the restrictive ref(2)P alleles occurred much more recently (25 years ago) and rapidly spread in natural population across Europe [30]. Homozygous Ref(2)P null flies are fully viable but the males are sterile. The molecular mechanisms of male sterility are not clear [19, 20]. Electron microscopy studies revealed that in the testes of ref(2)P[od1] and ref(2)P[od3] loss-of-function mutants (where Ref(2)P protein lacks the UBA domain) and ref(2)P^{od2} loss-of-function mutant (where Ref(2)P protein lacks the PB1 domain), characteristics of degeneration were frequently observed, such as the appearance of large myelin figures around the spermatids [20]. Additionally, the most striking difference was observed in the mitochondria, which varied in size and appeared degenerated [20]. Mammalian p62 has been shown to contribute to autophagic degradation of ubiquitinated mitochondria and to their clustering [31]. Therefore, it would be interesting to test this scenario in Ref(2)P mutant testis.

One open question is how Ref(2)P controls sigma virus multiplication at the molecular and cellular level. Work from Contamine's group suggests a direct interaction between Ref(2)P and a sigma virus protein, since Ref(2)P has been shown to interact with the sigma virus capsid P protein and to share conformation-dependent epitopes with the capsid N protein [32]. Additionally, Ref(2)P has been shown to interact genetically with DaPKC and the *Drosophila* homologue of TRAF6, dTRAF2, to participate in the Toll-signaling pathway, and to regulate the NF-κB proteins Dorsal and DIF [33, 34]. Interestingly, mammalian p62 was shown to interact with sindbis virus capsid protein, and genetic knockdown of p62 blocked the targeting of viral capsid to autophagosomes [35]. Taken together, these results suggest that Ref(2)P may target sigma virus capsid for autophagosomal degradation and also may function as a scaffolding protein during assembly of viral protein complexes. This scenario has to be tested experimentally. Intriguingly, Ref(2)P was shown to accumulate in rod-shaped structures in *Drosophila* egg chambers, structures that may represent aggregates of viruses or bacteria (Figure 3).

Another aspect of Ref(2)P function was recently reported in *Drosophila* hemocytes. Interestingly, Ref(2)P was shown to have a role in hemocyte spreading and protrusion formation [36]. This suggests that selective autophagy of an ubiquitinated substrate may function in an autophagy-dependent mechanism for cortical remodeling of hemocytes. Taken together, all the above information demonstrates that Ref(2)P, like its mammalian homologue p62, has diverse cellular functions whose molecular mechanisms have to be examined in detail.

2.1.2. Blue Cheese, the Drosophila homologue of the Mammalian Selective Aggregate Clearance Mediator Alfy. The mammalian phosphatidylinositol-3-phosphate-(PI3P-) binding protein Alfy was shown to be required for selective degradation of protein aggregates [8, 9, 37]. Alfy is a huge protein containing 3527 amino acids residues. It harbors several functional domains in the C terminus: a BEACH domain followed by a series of WD40 repeats and a PI(3)P-binding FYVE domain [8]. Despite its FYVE-domain which would suggest a localization to PI(3)P-rich endosomes, Alfy is not found on endosomes but instead localizes mainly to the nuclear envelope. Under conditions of starvation or proteasomal inhibition, Alfy relocalizes to cytoplasmic structures located close to autophagic membranes and ubiquitin-containing protein aggregates. Electron microscopy studies revealed that similar structures can be found within autophagosomes [8]. Importantly, Alfy was shown to be required for selective degradation of aggregated proteins such as polyQ-cotaining mutant huntingtin [9]. This function was proposed to be mediated by Alfy's physical interaction with PI(3)P, Atg5, and p62 [9, 37]. Therefore, Alfy functions as a scaffold receptor for recruitment of misfolded, ubiquitinated proteins to the autophagosomal membrane that become degraded by autophagy.

Blue cheese is the *Drosophila* homologue of Alfy and is highly conserved with its human homologue ($\sim$50% identity between fly and human homologs) [8, 38], and it contains similar functional domains at its C-terminal. *blue cheese* mutant flies exhibit a reduced adult life span and age-related neurodegeneration associated with accumulation of ubiquitin-conjugated protein aggregates throughout the adult central neruous system, neural atrophy, and cell death [38]. Ref(2)P accumulates in ubiquitinated inclusions in the brain of *blue cheese* mutant flies, suggesting that blue cheese is required for autophagic degradation of p62-associated ubiquitinated proteins *in vivo* [38] (Figure 2).

Finley and colleagues performed a genetic modifier screen for blue cheese genetic interactions based on alteration of the blue cheese eye phenotype. They found that recessive mutations in lysosomal trafficking genes and members of the ubiquitin and SUMO signaling pathways as well as in cytoskeletal and motor proteins have potential genetic interactions with Blue cheese [39]. They also showed that mutations of several lysosomal transport genes also alter high-molecular-weight UB-protein profiles and reduce adult life span [39]. Importantly, it was recently shown by Simonsen and Finley groups that overexpression of the C-terminal region of Blue cheese ameliorates neurodegeneation related phenotypes *in vivo* [9]. The authors tested the enhanced expression of Blue cheese in *Drosophila* eye model of polyglutamine toxicity, where UAS-polyQ127 transgene was expressed in the fly eye. It is well established that poly Q expression in the eye results in ommatidial disorganization, pigmentation loss, reduced eye size, and the appearance of necrotic regions. Enhanced expression of full-length Blue cheese (UAS-FL-Bchs) or C-terminal Blue cheese (UAS-bchs-C1000) with UAS-polyQ127 in the eye resulted in reduced number of necrotic areas and an overall improvement in eye size, morphology, and pigmentation. Taken together, these results suggest that the Alfy/Bchs proteins have a role in macroautophagic clearance of aggregation-prone proteins.

2.2. Mitophagy, Xenophagy, and Nucleophagy in Drosophila. Selective autophagy was recently shown to play an important role in the quality control of organelles and intracellular pathogens [31, 40]. However, mitophagy (selective autophagy of mitochondria), xenophagy (selective autophagy of bacteria and viruses), and nucleophagy (selective autophagy of nuclear fragments) are largely unexplored in *Drosophila*. Moreover, pexophagy (selective autophagy of peroxisomes) is not described yet in *Drosophila*. In the following lines, I will summarize what is reported so far in the literature about the processes above in *Drosophila*.

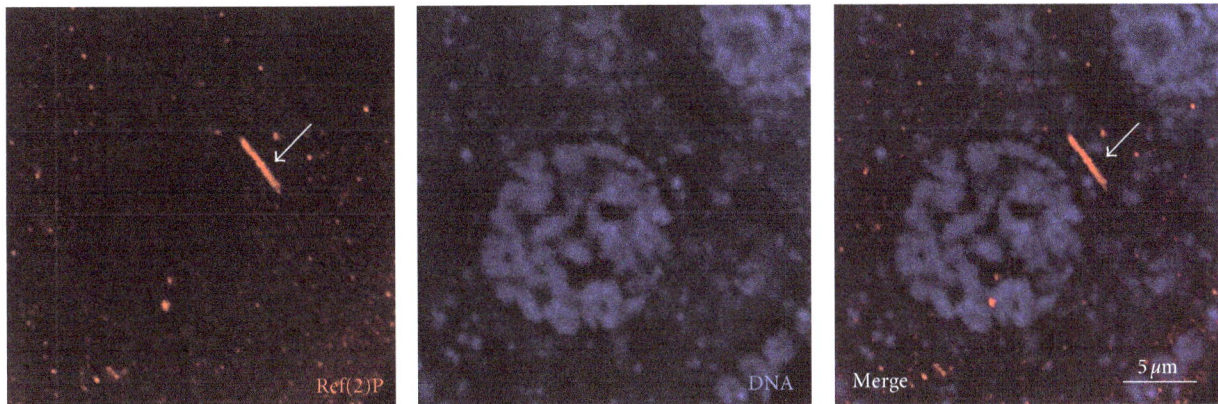

FIGURE 3: Ref(2)P localization in *Drosophila* egg chamber. Confocal micrograph of a middle section of a stage 8 egg chamber of wild-type fly, illustrating a portion of a nurse cell. The tissue is stained for Ref(2)P (red) and DNA (blue). Note the rod-like structure stained for Ref(2)P (arrow).

2.2.1. Mitophagy. Mitophagy has been recently described in yeast and mammals [31]. In yeast, the outer mitochondrial membrane protein Atg32 binds to the autophagosomal membrane protein Atg8 through its LIR motif [41]. In mammals, mitophagy was described during the physiological process of red blood cell differentiation and it requires the outer mitochondrial membrane protein NIP3-like protein NIX, which is also binds to LC3 through its LIR motif [42, 43]. Additionally, when mitochondria are damaged and depolarized, the kinase PTEN-induced putative kinase protein 1 (PINK1) accumulates to mitochondria and recruits the E3 ubiquitin ligase Parkin from the cytoplasm specifically to the damaged mitochondria. Subsequently, Parkin ubiquitylates mitochondrial proteins and promotes mitochondrial degradation by autophagy [31].

Genetic studies in *Drosophila* showed that the PINK1-Parkin pathway promotes mitochondrial fission or alternatively inhibit their fusion [44, 45]. It was recently shown in S2 cells that *Drosophila* PINK1 localizes to depolarized mitochondria and recruits Parkin and this promotes mitochondria degradation by autophagy [46]. Importantly, the profusion factor mitofusin (Mfn; also known as marf in *Drosophila*) was shown to be a novel substrate of Parkin [46]. Interestingly, it was also reported that activation of autophagy through Atg1 overexpression rescues PINK1 mutant phenotypes in *Drosophila* [47]. These studies suggest that, like in mammals, mitophagy also occurs in *Drosophila* and is dependent on PINK1 and Parkin, although the molecular details have to be further clarified.

Finally, it was recently reported that mitochondrial dynamics are abnormal in autophagy deficient egg chamber [48]. Dying atg1 germline mutant egg chambers exhibit abnormal mitochondrial remodeling that included the presence of mitochondrial islands suggesting that there is a crosstalk between autophagy, mitochondrial dynamics, and cell death during *Drosophila* oogenesis [48].

2.2.2. Xenophagy. Autophagy has been associated with the elimination of intracellular pathogens during mammalian innate immune responses, a process called xenophagy [40]. In *Drosophila*, xenophagy is largely unexplored. There are two reports that provide evidence for conserved mechanisms of xenophagy in *Drosophila*. In the first one, Kurata and colleagues reported that, in primary *Drosophila* hemocytes and S2 cells, autophagy prevented the intracellular growth of *Listeria monocytogenes* and promoted host survival after this infection [49]. Additionally, recognition of diaminopimelic acid-type peptidoglycan by the pattern-recognition receptor PGRP-LE was required for the induction of autophagy. Importantly, autophagy induction occurred independently of the Toll and IMD innate-signaling pathways [49].

In a second study, it was found that autophagy implements an antiviral role against the mammalian viral pathogen vesicular stomatitis virus (VSV) in *Drosophila* S2 cells as well as in adult flies [50]. The surface glycoprotein of VSV, VSVG, was shown to be the pathogen-associated molecular pattern that initiates the autophagic response. Autophagy was shown to restrain viral replication, and repression of autophagy resulted in increased viral replication and pathogenesis. Importantly, it was shown that this response was regulated by the phosphatidylinositol 3-kinase (PI3K)/Akt signaling pathway which controls autophagy in response to nutrient availability [50]. These data suggest that xenophagy occurs in *Drosophila*, and the molecular mechanisms are well conserved compared to mammals.

2.2.3. Nucleophagy. Nucleophagy is the process where parts of the nucleus can be specifically degraded by autophagy [51]. Nucleophagy is best characterized in yeast *Saccharomyces cerevisiae*, and is called piecemeal microautophagy [52]. During piecemeal microautophagy of the nucleus there is formation of nucleus-vacuole junctions where parts of the nucleus are sequestered into invaginations of the vacuolar membrane, followed by fission of nuclear fragments, and its release into the vacuolar lumen, where they are degraded. A direct interaction of the nuclear membrane protein Nvj1p with that vacuole protein Vac8p of the vacuole are required for this process [51, 52]. Recently, nucleophagy was also reported in mammals in nuclear envelopathies caused by mutations in the genes encoding A-type lamins (LMNA) and emerin (EMD) [53]. Nucleophagy was also observed rarely in wild-type cells [53].

In *Drosophila*, nuclear autophagy has been recently described during the cell death of nurse cells in late oogenesis [54]. Immunofluorescence analysis of mCherry-DrAtg8a autophagy marker in the nurse cells during the late stages of oogenesis revealed the presence of large autolysosomes adjacent to or attached to the condensed and fragmented nurse cell nucleus. Ultrastructural analysis revealed the presence of large autolysosomes which contained condensed material resembling the material of the fragmented nurse cell nucleus, suggesting that the nurse cell nuclear fragments are removed by autophagy [54].

2.3. Selective Degradation of Proteins in Drosophila. Autophagy has been shown to be responsible for the selective degradation of proteins in mammals and yeast like beta-synuclein [55], catalase [56], and acetaldehyde dehydrogenase [57]. In *Drosophila*, there is also a growing number of cases in which proteins can be preferentially degraded by autophagy.

2.3.1. Degradation of Survival Factors. Degradation of survival factors is a way of cell to die [58]. There are two recent reports that support this hypothesis in *Drosophila*. In the first study, we have demonstrated that the inhibitor of apoptosis protein dBruce was degraded by autophagy in the nurse cells during cell death in late oogenesis [54]. Genetic inhibition of autophagy in the female germline resulted in late stage egg chambers containing persistent nurse cell nuclei that did not contain fragmented DNA and in attenuation of caspase-3 activation. Importantly, we found that *Drosophila* inhibitor of apoptosis dBruce is degraded by autophagy, and this is responsible to control DNA fragmentation [54]. A second report showed that degradation of inhibitor of apoptosis protein DIAP1 during developmental dendrite pruning of *Drosophila* class IV dendritic arborization neurons is depended on Valosin-containing protein (VCP), a ubiquitin-selective AAA chaperone involved in endoplasmic reticulum-associated degradation and the maturation of autophagosomes [59, 60]. These results suggest that autophagic degradation of survival factors can cause cell death during development in *Drosophila*.

2.3.2. Degradation of Rhodopsin and Retinal Degeneration. Activated rhodopsin is degraded in endosomal pathways in normal photoreceptor cells in *Drosophila*, and accumulation of activated rhodopsin in some *Drosophila* mutants leads to retinal degeneration [61]. In a recent study, it was reported that activated rhodopsin is degraded by autophagy in order to prevent retinal degeneration [62]. Light-dependent retinal degeneration in the *Drosophila* eye is caused by silencing or mutation of autophagy genes, such as autophagy-related protein 7 and 8, or genes essential for PE (phosphatidylethanolamine) biogenesis and autophagosome formation, including phosphatidylserine decarboxylase (Psd)

and CDP-ethanolamine:diacylglycerol ethanolaminephosphotransferase (Ept). Silencing of atg-7/8 or Psd/Ept resulted in an increase in the amount of rhodopsin localized to Rab7-positive late endosomes [62]. These results suggest that autophagic and endosomal/lysosomal pathways suppress light-dependent retinal degeneration and that rhodopsin is a substrate for autophagic degradation in this context.

2.3.3. Degradation of Highwire. Beyond its role in cellular homeostasis, autophagy is implicated in the regulation of developmental growth and remodeling of various cells and tissues during development [63]. One such example in *Drosophila* is the synaptic development of the larval neuromuscular junction. Shen and Ganetzky showed that autophagy promotes the synaptic development of the *Drosophila* larval neuromuscular junction, by downregulating an E3 ubiquitin ligase, Highwire, which restrains neuromuscular junction growth via a MAPKKK pathway [64, 65]. Autophagy mutants exhibit neuromuscular junction undergrowth and Atg1 overexpression, resulting in neuromuscular junction overgrowth. Moreover, overgrowth associated with Atg1 overexpression is suppressed by mutations in *atg18*, demonstrating that this overgrowth is due to elevated levels of autophagy [64, 65]. In a recent paper, *Drosophila* Rae1 was identified as a component of the Highwire complex. Loss of *Rae1* function in neurons results in morphological defects at the neuromuscular junction that are similar to those seen in Highwire mutants [66]. The authors found that Rae1 physically and genetically interacts with Highwire and limits synaptic terminal growth by regulating the MAP kinase kinase kinase Wallenda. Moreover, they found that the Rae1 is sufficient to promote Highwire protein abundance by binding to Highwire and protecting it from autophagic degradation [66]. Together, these findings indicate that Rae1 prevents autophagy-mediated degradation of Highwire and that selectively controls Highwire protein abundance during synaptic development.

3. Concluding Remarks and Future Directions

From the literature analyzed above, it is obvious that the molecular mechanisms of selective autophagy in *Drosophila* remain largely unexplored. The precise mechanisms of selective autophagy of organelles and proteins has not been directly shown in *Drosophila*, and the molecular details of the interaction of selective autophagy receptors Ref(2)P and blue cheese with the autophagic machinery have to be shown experimentally. The presence of putative LIR motif in Ref(2)P offers a fertile ground for further functional analysis *in vivo*. p62 and Ref(2)P have been proposed to collect ubiquitinated proteins and target them for degradation [4]. It would therefore be interesting to test whether induced expression of Ref(2)P ameliorates phenotypes related to neurodegeneration *in vivo*. It will also be important to elucidate in details how small or large aggregates are removed per se. Elucidation of these processes may have applications in fighting aggregation-related diseases, such as neurodegenerative diseases as well as cancer. There is emerging evidence that mammalian p62 directly interacts with Keap1 and that p62 is a target gene for Nrf2 transcription factor implicated in oxidative stress signaling [4, 13]. It would be interesting to test the interaction of Ref(2)P with the *Drosophila* homologue of Keap1, dkeap1 [67]. It would also be interesting to test whether the *Drosophila* homologues of BNIP3-like proteins play a role in selective degradation of mitochondria.

In conclusion, *Drosophila* offers a fertile ground for studying the molecular mechanisms of selective autophagy. Future studies will hopefully uncover the molecular details of this process.

Acknowledgments

I. P. Nezis acknowledges support from the European Research Council. I. P. Nezis would like to thank the anonymous reviewers for their helpful comments on the improvement of the manuscript and Dr. Kim Finley for providing the blue cheese mutant flies.

References

[1] Z. Yang and D. J. Klionsky, "Eaten alive: a history of macroautophagy," *Nature Cell Biology*, vol. 12, no. 9, pp. 814–822, 2010.

[2] D. C. Rubinsztein, G. Mariño, and G. Kroemer, "Autophagy and aging," *Cell*, vol. 146, no. 5, pp. 682–695, 2011.

[3] F. Madeo, N. Tavernarakis, and G. Kroemer, "Can autophagy promote longevity?" *Nature Cell Biology*, vol. 12, no. 9, pp. 842–846, 2010.

[4] T. Johansen and T. Lamark, "Selective autophagy mediated by autophagic adapter proteins," *Autophagy*, vol. 7, no. 3, pp. 279–296, 2011.

[5] C. K. McPhee and E. Baehrecke, "Autophagy in *Drosophila melanogaster*," *Biochimica et Biophysica Acta*, vol. 1793, no. 9, pp. 1452–1460, 2009.

[6] P. Wild, H. Farhan, D. G. McEwan et al., "Phosphorylation of the autophagy receptor optineurin restricts Salmonella growth," *Science*, vol. 333, no. 6039, pp. 228–233, 2011.

[7] S. Jiang, C. D. Wells, and P. J. Roach, "Starch-binding domain-containing protein 1 (Stbd1) and glycogen metabolism: identification of the Atg8 family interacting motif (AIM) in Stbd1 required for interaction with GABARAPL1," *Biochemical and Biophysical Research Communications*, vol. 413, no. 3, pp. 420–425, 2011.

[8] A. Simonsen, H. C. Birkeland, D. J. Gillooly et al., "Alfy, a novel FYVE-domain-containing protein associated with protein granules and autophagic membranes," *Journal of Cell Science*, vol. 117, no. 18, pp. 4239–4251, 2004.

[9] M. Filimonenko, P. Isakson, K. D. Finley et al., "The selective macroautophagic degradation of aggregated proteins requires the PI3P-binding protein alfy," *Molecular Cell*, vol. 38, no. 2, pp. 265–279, 2010.

[10] G. Bjørkøy, T. Lamark, A. Brech et al., "p62/SQSTM1 forms protein aggregates degraded by autophagy and has a protective effect on huntingtin-induced cell death," *Journal of Cell Biology*, vol. 171, no. 4, pp. 603–614, 2005.

[11] S. Pankiv, T. H. Clausen, T. Lamark et al., "p62/SQSTM1 binds directly to Atg8/LC3 to facilitate degradation of ubiquitinated protein aggregates by autophagy," *Journal of Biological Chemistry*, vol. 282, no. 33, pp. 24131–24145, 2007.

[12] J. Moscat and M. T. Diaz-Meco, "p62 at the crossroads of autophagy, apoptosis, and cancer," *Cell*, vol. 137, no. 6, pp. 1001–1004, 2009.

[13] I. P. Nezis and H. Stenmark, "p62 at the interphaseof autophagy, oxidative stress signaling and cancer," *Antioxidant Redox & Signaling*. In press.

[14] J. Moscat, M. T. Diaz-Meco, and M. W. Wooten, "Of the atypical PKCs, Par-4 and p62: recent understandings of the biology and pathology of a PB1-dominated complex," *Cell Death and Differentiation*, vol. 16, no. 11, pp. 1426–1437, 2009.

[15] S. Pankiv, T. Lamark, J. A. Bruun, A. Øvervatn, G. Bjørkøy, and T. Johansen, "Nucleocytoplasmic shuttling of p62/SQSTM1 and its role in recruitment of nuclear polyubiquitinated proteins to promyelocytic leukemia bodies," *Journal of Biological Chemistry*, vol. 285, no. 8, pp. 5941–5953, 2010.

[16] M. Komatsu, H. Kurokawa, S. Waguri et al., "The selective autophagy substrate p62 activates the stress responsive transcription factor Nrf2 through inactivation of Keap1," *Nature Cell Biology*, vol. 12, no. 3, pp. 213–223, 2010.

[17] A. Jain, T. Lamark, E. Sjøttem et al., "p62/SQSTM1 is a target gene for transcription factor NRF2 and creates a positive feedback loop by inducing antioxidant response element-driven gene transcription," *Journal of Biological Chemistry*, vol. 285, no. 29, pp. 22576–22591, 2010.

[18] I. P. Nezis, A. Simonsen, A. P. Sagona et al., "Ref(2)P, the *Drosophila melanogaster* homologue of mammalian p62, is required for the formation of protein aggregates in adult brain," *Journal of Cell Biology*, vol. 180, no. 6, pp. 1065–1071, 2008.

[19] D. Contamine, A. M. Petitjean, and M. Ashburner, "Genetic resistance to viral infection: the molecular cloning of a *Drosophila* gene that restricts infection by the rhabdovirus sigma," *Genetics*, vol. 123, no. 3, pp. 525–533, 1989.

[20] S. Dezelee, F. Bras, D. Contamine, M. Lopez-Ferber, D. Segretain, and D. Teninges, "Molecular analysis of ref(2)P, a *Drosophila* gene implicated in sigma rhabdovirus multiplication and necessary for male fertility," *The EMBO Journal*, vol. 8, no. 11, pp. 3437–3446, 1989.

[21] A. Carré-Mlouka, S. Gaumer, P. Gay et al., "Control of sigma virus multiplication by the *ref(2)P* gene of *Drosophila melanogaster*: an in vivo study of the PB1 domain of Ref(2)P," *Genetics*, vol. 176, no. 1, pp. 409–419, 2007.

[22] S. A. Hogenhout, M. G. Redinbaugh, and D. Ammar, "Plant and animal rhabdovirus host range: a bug's view," *Trends in Microbiology*, vol. 11, no. 6, pp. 264–271, 2003.

[23] T. Huszar and J. L. Imler, "*Drosophila* viruses and the study of antiviral host-defense," *Advances in Virus Research*, vol. 72, pp. 227–265, 2008.

[24] A. Fleuriet, "Comparison of various physiological traits in flies (*Drosophila melanogaster*) of wild origin, infected or uninfected by the hereditary rhabdovirus sigma," *Archives of Virology*, vol. 69, no. 3-4, pp. 261–272, 1981.

[25] A. Fleuriet, "Effect of overwintering on the frequency of flies infected by the Rhabdovirus sigma in experimental populations of *Drosophila* melanogaster," *Archives of Virology*, vol. 69, no. 3-4, pp. 253–260, 1981.

[26] P. Dru, F. Bras, S. Dezelee et al., "Unusual variability of the *Drosophila melanogaster* ref(2)P protein which controls the multiplication of sigma rhabdovirus," *Genetics*, vol. 133, no. 4, pp. 943–954, 1993.

[27] M. L. Wayne, D. Contamine, and M. Kreitman, "Molecular population genetics of ref(2)P, a locus which confers viral

resistance in *Drosophila*," *Molecular Biology and Evolution*, vol. 13, no. 1, pp. 191–199, 1996.

[28] P. Gay, "*Drosophila* genes which intervene in multiplication of sigma virus," *Molecular and General Genetics*, vol. 159, no. 3, pp. 269–283, 1978.

[29] J. Bangham, D. J. Obbard, K. W. Kim, P. R. Haddrill, and F. M. Jiggins, "The age and evolution of an antiviral resistance mutation in *Drosophila* melanogaster," *Proceedings of the Royal Society B*, vol. 274, no. 1621, pp. 2027–2034, 2007.

[30] J. A. Carpenter, D. J. Obbard, X. Maside, and F. M. Jiggins, "The recent spread of a vertically transmitted virus through populations of *Drosophila* melanogaster," *Molecular Ecology*, vol. 16, no. 18, pp. 3947–3954, 2007.

[31] R. J. Youle and D. P. Narendra, "Mechanisms of mitophagy," *Nature Reviews Molecular Cell Biology*, vol. 12, no. 1, pp. 9–14, 2011.

[32] F. Wyers, P. Dru, B. Simonet, and D. Contamine, "Immunological cross-reactions and interactions between the *Drosophila* melanogaster ref(2)P protein and sigma rhabdovirus proteins," *Journal of Virology*, vol. 67, no. 6, pp. 3208–3216, 1993.

[33] A. Avila, N. Silverman, M. T. Diaz-Meco, and J. Moscat, "The *Drosophila* atypical protein kinase C-Ref(2)P complex constitutes a conserved module for signaling in the toll pathway," *Molecular and Cellular Biology*, vol. 22, no. 24, pp. 8787–8795, 2002.

[34] A. Goto, S. Blandin, J. Royet, J. M. Reichhart, and E. A. Levashina, "Silencing of Toll pathway components by direct injection of double-stranded RNA into *Drosophila* adult flies," *Nucleic Acids Research*, vol. 31, no. 22, pp. 6619–6623, 2003.

[35] A. Orvedahl, S. MacPherson, R. Sumpter, Z. Tallóczy, Z. Zou, and B. Levine, "Autophagy protects against sindbis virus infection of the central nervous system," *Cell Host and Microbe*, vol. 7, no. 2, pp. 115–127, 2010.

[36] P. Kadandale, J. D. Stender, C. K. Glass, and A. A. Kiger, "Conserved role for autophagy in Rho1-mediated cortical remodeling and blood cell recruitment," *Proceedings of the National Academy of Sciences of the United States of America*, vol. 107, no. 23, pp. 10502–10507, 2010.

[37] T. H. Clausen, T. Lamark, P. Isakson et al., "p62/SQSTM1 and ALFY interact to facilitate the formation of p62 bodies/ALIS and their degradation by autophagy," *Autophagy*, vol. 6, no. 3, pp. 330–344, 2010.

[38] K. D. Finley, P. T. Edeen, R. C. Cumming et al., "Blue cheese mutations define a novel, conserved gene involved in progressive neural degeneration," *Journal of Neuroscience*, vol. 23, no. 4, pp. 1254–1264, 2003.

[39] A. Simonsen, R. C. Cumming, K. Lindmo et al., "Genetic modifiers of the drosophila blue cheese gene link defects in lysosomal transport with decreased life span and altered ubiquitinated-protein profiles," *Genetics*, vol. 176, no. 2, pp. 1283–1297, 2007.

[40] V. Deretic, "Autophagy as an innate immunity paradigm: expanding the scope and repertoire of pattern recognition receptors," *Current Opinions in Immunology*, vol. 24, no. 1, pp. 21–31, 2012.

[41] T. Kanki, K. Wang, Y. Cao, M. Baba, and D. J. Klionsky, "Atg32 is a mitochondrial protein that confers selectivity during mitophagy," *Developmental Cell*, vol. 17, no. 1, pp. 98–109, 2009.

[42] I. Novak, V. Kirkin, D. G. E. McEwan et al., "Nix is a selective autophagy receptor for mitochondrial clearance," *EMBO Reports*, vol. 11, no. 1, pp. 45–51, 2010.

[43] R. L. Schweers, J. Zhang, M. S. Randall et al., "NIX is required for programmed mitochondrial clearance during reticulocyte maturation," *Proceedings of the National Academy of Sciences of the United States of America*, vol. 104, no. 49, pp. 19500–19505, 2007.

[44] I. E. Clark, M. W. Dodson, C. Jiang et al., "*Drosophila* pink1 is required for mitochondrial function and interacts genetically with parkin," *Nature*, vol. 441, no. 7097, pp. 1162–1166, 2006.

[45] H. Deng, M. W. Dodson, H. Huang, and M. Guo, "The Parkinson's disease genes pink1 and parkin promote mitochondrial fission and/or inhibit fusion in *Drosophila*," *Proceedings of the National Academy of Sciences of the United States of America*, vol. 105, no. 38, pp. 14503–14508, 2008.

[46] E. Ziviani, R. N. Tao, and A. J. Whitworth, "*Drosophila* parkin requires PINK1 for mitochondrial translocation and ubiquitinates Mitofusin," *Proceedings of the National Academy of Sciences of the United States of America*, vol. 107, no. 11, pp. 5018–5023, 2010.

[47] S. Liu and B. Lu, "Reduction of protein translation and activation of autophagy protect against PINK1 pathogenesis in *Drosophila* melanogaster," *Plos Genetics*, vol. 6, no. 12, Article ID e1001237, pp. 1–12, 2010.

[48] E. A. Tanner, T. A. Blute, C. B. Brachmann, and K. McCall, "Bcl-2 proteins and autophagy regulate mitochondrial dynamics during programmed cell death in the *Drosophila* ovary," *Development*, vol. 138, no. 2, pp. 327–338, 2011.

[49] T. Yano, S. Mita, H. Ohmori et al., "Autophagic control of listeria through intracellular innate immune recognition in *Drosophila*," *Nature Immunology*, vol. 9, no. 8, pp. 908–916, 2008.

[50] S. Shelly, N. Lukinova, S. Bambina, A. Berman, and S. Cherry, "Autophagy is an essential component of *Drosophila* immunity against vesicular stomatitis virus," *Immunity*, vol. 30, no. 4, pp. 588–598, 2009.

[51] D. Mijaljica, M. Prescott, and R. J. Devenish, "The intricacy of nuclear membrane dynamics during nucleophagy," *Nucleus*, vol. 1, no. 3, pp. 213–223, 2010.

[52] R. Krick, Y. Mühe, T. Prick et al., "Piecemeal microautophagy of the nucleus: genetic and morphological traits," *Autophagy*, vol. 5, no. 2, pp. 270–272, 2009.

[53] Y. E. Park, Y. K. Hayashi, G. Bonne et al., "Autophagic degradation of nuclear components in mammalian cells," *Autophagy*, vol. 5, no. 6, pp. 795–804, 2009.

[54] I. P. Nezis, B. V. Shravage, A. P. Sagona et al., "Autophagic degradation of dBruce controls DNA fragmentation in nurse cells during late *Drosophila* melanogaster oogenesis," *Journal of Cell Biology*, vol. 190, no. 4, pp. 523–531, 2010.

[55] J. L. Webb, B. Ravikumar, J. Atkins, J. N. Skepper, and D. C. Rubinsztein, "α-synuclein is degraded by both autophagy and the proteasome," *Journal of Biological Chemistry*, vol. 278, no. 27, pp. 25009–25013, 2003.

[56] L. Yu, F. Wan, S. Dutta et al., "Autophagic programmed cell death by selective catalase degradation," *Proceedings of the National Academy of Sciences of the United States of America*, vol. 103, no. 13, pp. 4952–4957, 2006.

[57] J. Onodera and Y. Ohsumi, "Ald6p is a preferred target for autophagy in yeast, saccharomyces cerevisiae," *Journal of Biological Chemistry*, vol. 279, no. 16, pp. 16071–16076, 2004.

[58] L. Yu, L. Strandberg, and M. J. Lenardo, "The selectivity of autophagy and its role in cell death and survival," *Autophagy*, vol. 4, no. 5, pp. 567–573, 2008.

[59] S. Rumpf, S. B. Lee, L. Y. Jan, and Y. N. Jan, "Neuronal remodeling and apoptosis require VCP-dependent degradation of the

apoptosis inhibitor DIAP1," *Development*, vol. 138, no. 6, pp. 1153–1160, 2011.

[60] J. S. Ju, R. A. Fuentealba, S. E. Miller et al., "Valosin-containing protein (VCP) is required for autophagy and is disrupted in VCP disease," *Journal of Cell Biology*, vol. 187, no. 6, pp. 875–888, 2009.

[61] Y. Chinchore, A. Mitra, and P. J. Dolph, "Accumulation of rhodopsin in late endosomes triggers photoreceptor cell degeneration," *Plos Genetics*, vol. 5, no. 2, Article ID e1000377, 2009.

[62] R. Midorikawa, M. Yamamoto-Hino, W. Awano et al., "Autophagy-dependent rhodopsin degradation prevents retinal degeneration in *Drosophila*," *Journal of Neuroscience*, vol. 30, no. 32, pp. 10703–10719, 2010.

[63] N. Mizushima and M. Komatsu, "Autophagy: renovation of cells and tissues," *Cell*, vol. 147, no. 4, pp. 728–741, 2011.

[64] W. Shen and B. Ganetzky, "Autophagy promotes synapse development in *Drosophila*," *Journal of Cell Biology*, vol. 187, no. 1, pp. 71–79, 2009.

[65] W. Shen and B. Ganetzky, "Nibbling away at synaptic development," *Autophagy*, vol. 6, no. 1, pp. 168–169, 2010.

[66] X. Tian, J. Li, V. Valakh, A. Diantonio, and C. Wu, "*Drosophila* Rae1 controls the abundance of the ubiquitin ligase Highwire in post-mitotic neurons," *Nature Neuroscience*, vol. 14, no. 10, pp. 1267–1275, 2011.

[67] G. P. Sykiotis and D. Bohmann, "Keap1/Nrf2 signaling regulates oxidative stress tolerance and lifespan in *Drosophila*," *Developmental Cell*, vol. 14, no. 1, pp. 76–85, 2008.

5

Antiproliferative Factor-Induced Changes in Phosphorylation and Palmitoylation of Cytoskeleton-Associated Protein-4 Regulate Its Nuclear Translocation and DNA Binding

David A. Zacharias,[1] Matthew Mullen,[2] and Sonia Lobo Planey[2]

[1] *Whitney Laboratory, Department of Neuroscience, University of Florida, St. Augustine, FL 32080, USA*
[2] *Department of Basic Sciences, The Commonwealth Medical College, Scranton, PA 18509, USA*

Correspondence should be addressed to David A. Zacharias, daz@whitney.ufl.edu and Sonia Lobo Planey, splaney@tcmedc.org

Academic Editor: Jerome Rattner

Cytoskeleton-associated protein 4 (CKAP) is a reversibly palmitoylated and phosphorylated transmembrane protein that functions as a high-affinity receptor for antiproliferative factor (APF)—a sialoglycopeptide secreted from bladder epithelial cells of patients with interstitial cystitis (IC). Palmitoylation of CKAP4 by the palmitoyl acyltransferase, DHHC2, is required for its cell surface localization and subsequent APF signal transduction; however, the mechanism for APF signal transduction by CKAP4 is unknown. In this paper, we demonstrate that APF treatment induces serine phosphorylation of residues S3, S17, and S19 of CKAP4 and nuclear translocation of CKAP4. Additionally, we demonstrate that CKAP4 binds gDNA in a phosphorylation-dependent manner in response to APF treatment, and that a phosphomimicking, constitutively nonpalmitoylated form of CKAP4 localizes to the nucleus, binds DNA, and mimics the inhibitory effects of APF on cellular proliferation. These results reveal a novel role for CKAP4 as a downstream effecter for APF signal transduction.

1. Introduction

Cytoskeleton-associated protein 4 (CKAP4; also known as CLIMP-63, ERGIC-63, and p63) is a 63 kDa, reversibly palmitoylated and phosphorylated, type II transmembrane (TM) protein, originally identified as a resident of the endoplasmic reticulum/Golgi intermediate complex (ERGIC) [1–5]. The first report describing CKAP4 [1] (referring to it as "p62") demonstrated that its palmitoylation peaked during mitosis and suggested that palmitoylation may be an important regulator of vesicular transport between various membranous compartments. Soon thereafter, Schweizer and colleagues cloned CKAP4 and identified membrane-proximal cysteine 100 (C100) as the site for palmitoylation [4]. More recently, DHHC2 was identified as the palmitoyl acyltransferase (PAT) that palmitoylates CKAP4 at C100 [6].

CKAP4 is localized prominently to the endoplasmic reticulum (ER). One major function of CKAP4 is to anchor rough ER to microtubules, organizing the overall structure of ER with respect to the microtubule network [3–5, 7, 8].

The binding of CKAP4 to microtubules is regulated by phosphorylation of three critical serine residues (S3, S17, and S19) located in its cytosolic, N-terminal domain (Figure 1) [2]. CKAP4 is unique among microtubule binding proteins in at least one respect: it is a TM protein. However, it is similar to many other microtubule-binding proteins in that phosphorylation blocks its ability to bind microtubules [9]. Overexpression of a mutant version of CKAP4 that mimics phosphorylation of three serine residues (S3E, S17E, and S19E) within the microtubule binding domain results in a restructuring or "collapse" of the ER around the nucleus without any observable effect on the microtubule network. Similar effects on the ER structure occurred when a deletion mutant of CKAP4 lacking the same serine residues was overexpressed in cells. Conversely, overexpression of a full-length, phosphorylation-incompetent (S3A, S17A, and S19A) mutant of CKAP4 colocalized with and was able to bundle microtubules similar to wild-type CKAP4 [7].

CKAP4 is also expressed on the plasma membrane (PM). Palmitoylation by DHHC2 is required for expression of

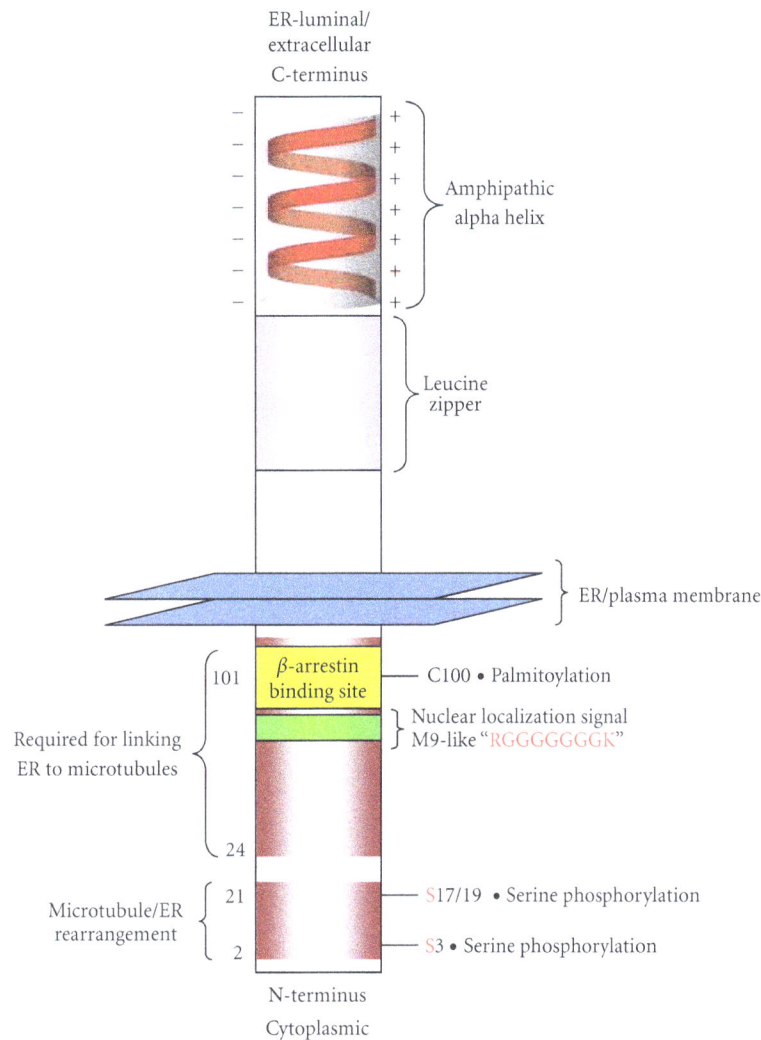

FIGURE 1: *CKAP4 domains*. CKAP4 is a 63 kDa, oligomeric, type II, single-pass TM domain protein that is palmitoylated and phosphorylated. The luminal/extracellular domain contains an amphipathic alpha helical region (506-KVQEQVHTLLSQDQAQAARLPPQD FLDRLSSLDNLKASVSQVEADLKMLRTAVDSLVAYSVKIETNENNLESAKGLLDDLRNDLDRLFVKVEKIHEKV-602) that is important for oligomerization and a leucine zipper (468-LASTVRSLGETQLVLYGDVEELKRSVGELPSTVESL-504). Together, these regions are homologous to the DNA-binding domain of bZIP transcription factors. The cytoplasmic, N-terminus contains a cysteine residue adjacent to the TM domain at position 100 that is palmitoylated (a modification that is important for trafficking from the ER to the PM), and three serine residues (S3, S17, and S19) that are required for phosphorylation-dependent binding of CKAP4 to the microtubule cytoskeleton. Two regions in the cytoplasmic N-terminus are required for binding to and bundling microtubules, thereby maintaining the connection between the ER and the cytoskeleton [3]. There is also a PQ protein-protein interaction domain (49-PHPQQHPQQHPQNQ-63) and a putative glycine-rich nuclear localization signal sequence (65-GKGGHRGGGGGGGGK-79).

CKAP4 on the cell surface [6], where it functions as a high-affinity receptor for antiproliferative factor (APF) [10]. APF is a lipophilic, nine-residue sialoglycopeptide with an amino acid sequence identical to the putative 6th TM domain of Frizzled 8 [11]. APF is present in the urine of patients with interstitial cystitis (IC), a chronic, painful bladder disease with a poorly understood etiology [12]. Exposing cells to APF *in vitro* dramatically alters gene expression and blocks proliferation of normal bladder epithelial cells and cancer cell lines including bladder (T24) and cervical (HeLa) adenocarcinoma, mimicking critical aspects of the pathology of the bladder epithelium in IC patients [6, 11, 13, 14]. The IC_{50} of synthetic and purified APF in proliferation assays is

~1 nM [14, 15], indicating that the affinity of APF for CKAP4 is high.

Previously, we observed an increase in the nuclear abundance of CKAP4 in HeLa cells following exposure to APF [6]. Importantly, this APF-induced change in CKAP4 localization was dependent on palmitoylation by DHHC2. Concurrent with the increased nuclear abundance of CKAP4, the expression level of several genes, (i.e., vimentin, zonula occludens-1, and E-cadherin) changed significantly in HeLa and normal bladder epithelial cells following APF exposure [6, 10, 13, 16]. These genes are among thirteen others shown to have significantly altered expression in bladder tissue from IC patients and known to be involved in the

TABLE 1: *Mutant constructs used to study CKAP4 phosphorylation and palmitoylation.* CKAP4 mutants that mimic constitutive depalmitoylation and various states of serine (S3, S17 and S19) phosphorylation were generated to determine their effect on the subcellular distribution of CKAP4 in response to APF. CKAP4SΔE translocated from the PM to the nucleus in response to APF; CKAP4SΔA localized to the PM but did not translocate to the nucleus in response to APF; none of the C100S mutants were expressed on the PM; CKAP4C100S/SΔE was expressed in the nucleus. The results indicate that CKAP4 must be palmitoylated for PM expression and depalmitoylated and phosphorylated for translocation to the nucleus. A summary of these results are provided in the table.

Name	Mutations	Comments posttranslational modifications	Resting distribution	Distribution after APF
wild-type (WT) CKAP4	none	expressed or endogenous	PM, ER	> nucleus
CKAP4 C100S	C100S	constitutively depalmitoylated	ER	= ER
CKAP4 SΔA	S3A, S17A, S19A	constitutively dephosphorylated	PM, ER	= PM, ER
CKAP4 C100S/SΔA	C100S, S3A, S17A, S19A	constitutively depalmitoylated, constitutively dephosphorylated	ER	ER
CKAP4 SΔE	S3E, S17E, S19E	mimics phosphorylation	PM, ER	> nucleus
CKAP4 C100S/SΔE	C100S, S3E, S17E, S19E	constitutively depalmitoylated, mimics phosphorylation	Nucleus	= nucleus

regulation of proliferation, cell adhesion, and tumorigenesis [17–19]. The redistribution of CKAP4 to the nucleus and the concurrent changes in gene expression suggest that APF induces specific changes in the palmitoylation and/or phosphorylation state of CKAP4, and that these changes affect its subcellular distribution and function within the cell. Consequently, we generated CKAP4 mutants that mimic constitutive depalmitoylation and various states of serine phosphorylation to determine their effect on the subcellular distribution of CKAP4 in response to APF.

Our results show that APF increases serine phosphorylation of CKAP4, and that phosphorylation of residues S3, S17, and S19 are required for its nuclear translocation; moreover, a phosphomimicking, constitutively nonpalmitoylated form of CKAP4 localizes to the nucleus, binds DNA, and mimics the inhibitory effects of APF on cellular proliferation. These results reveal a novel role for CKAP4 in APF-dependent signaling from the plasma membrane to the nucleus and suggest that CKAP4 may regulate transcription by binding directly to DNA.

2. Experimental Procedures

2.1. DNA Constructs. A vector construct containing wild-type CKAP4 (WT CKAP4) fused in-frame to the N-terminus of the V5 and 6xHis epitope tags in the mammalian expression vector pcDNA3.1V5His6-TOPO (Invitrogen, Carlsbad, CA) was generated by PCR using CKAP4-specific primers and cDNA from HeLa cells. A palmitoylation-incompetent form of CKAP4 (CKAP4C100S) was created by site-directed mutagenesis of the wild-type vector construct (Stratagene, La Jolla, CA) to change the cysteine at position 100 to serine. A series of CKAP4 mutants that mimic depalmitoylation and/or constitutive dephosphorylation and phosphorylation were constructed by site-directed mutagenesis of the WT CKAP4 or CKAP4C100S vector constructs as described in

Table 1. cDNA encoding the extracellular domain of CKAP4 (C-terminal residues 126–501) was generated by PCR and cloned in-frame with the His6 epitope tag of pRSET-B (Invitrogen) at the 5′ end. The accuracy of all constructs was verified by DNA sequencing.

2.2. Cell Culture and Transfections. HeLa (ATCC #CCL-2; American Type Culture Collection, Manassas, VA) cells were maintained in Dulbecco's modified Eagle's medium(DMEM) containing 10% fetal bovine serum (FBS), 100 U/mL penicillin, and 100 μg/mL streptomycin and 1 μg/mL fungizone (all from Invitrogen). Cells were trans-fected using FuGENE6 reagent (Roche, Basel, Switzerland) according to the manufacturer's instructions.

2.3. Immunocytochemistry. HeLa cells expressing WT CKAP4 or CKAP4 mutants that mimic depalmitoylation and/or constitutive dephosphorylation and phosphorylation (see Table 1) were seeded at a density of 2×10^4 cells/well in 8-well LabTek chamber slides (Nalge Nunc, Rochester, NY) and grown to semiconfluence in DMEM medium containing 10% FBS, 100 U/mL penicillin, 100 μg/mL streptomycin, 1 μg/mL fungizone, and 0.4 mg/mL G418 (all from Invitrogen). Cells were fixed for 20 minutes with 3% paraformaldehyde in phosphate-buffered saline (PBS), permeabilized with 0.1% Triton X-100 in PBS, and blocked in PBS/5% normal goat serum (NGS). The following primary antibodies were used: mouse mAb G1/296 against CKAP4 ("anti-CLIMP-63", diluted 1:100, Alexis Biochemicals, San Diego, CA), rabbit pAb against tubulin (diluted 1:100, Abcam, Cambridge, MA), rabbit pAb against fibrillarin (diluted 1:100, Abcam), and fluorescein isothiocyanate- (FITC-) conjugated mouse mAb against the V5 epitope (diluted 1:500, Invitrogen). Secondary antibodies were FITC-labeled goat anti-rabbit or goat anti-mouse (diluted 1:1000, Invitrogen) and tetramethyl rhodamine isothiocyanate- (TRITC-) labeled goat

Antiproliferative Factor-Induced Changes in Phosphorylation and Palmitoylation of Cytoskeleton-Associated Protein-4 Regulate Its Nuclear Translocation and DNA Binding

45

anti-mouse (diluted 1 : 1000, Jackson ImmunoResearch Laboratories, West Grove, PA). Slides were mounted in SlowFade Antifade reagent (Invitrogen) and imaged using a Nikon TE2000 epifluorescence microscope or Leica SP5II confocal.

2.4. Nuclear and Nucleolar Protein Isolation.

HeLa cells were serum-starved overnight and then treated with 20 nM synthetic APF (Peptides International) for 24 hours. Nuclear extracts were generated using the NE-PER kit (ThermoFisher Scientific, Waltham, MA) according to the manufacturer's instructions. Nucleoli were isolated using a variation on the method used by Busch and coworkers in 1963 [20] as described by the Lamond lab (http://www.lamondlab.com/f7protocols.htm). Briefly, APF-treated or untreated HeLa cells were washed with cold PBS (pH 7.4), trypsinized, and collected by centrifugation at 500 ×g for 5 min. The cell pellets were resuspended in 15 volumes of a hypotonic buffer (10 mM Tris-HCl, pH 7.4, 10 mM NaCl, and 1 mM $MgCl_2$), incubated on ice for 30 min, and then lysed by adding Nonidet P-40 to a final concentration of 0.3%. The samples were homogenized 10 times using a glass dounce homogenizer (0.4-mm clearance; Kimberly/Kontes, Owens, IL) while keeping the homogenizer on ice. Nuclei were collected by centrifugation at 218 ×g for 5 min and resuspended in 10 volumes of 0.25 M sucrose containing 10 mM $MgCl_2$. Nuclei were then purified by centrifugation at 1430 ×g for 5 min through a 0.88 M sucrose cushion containing 0.05 mM $MgCl_2$. Purified nuclei were resuspended in 10 volumes of 0.34 M sucrose containing 0.05 mM $MgCl_2$ and sonicated on ice for six, 10-second bursts with 10-second intervals between each burst. Nucleoli were then purified from the resulting homogenate by centrifugation at 3000 ×g for 10 min through a 0.88 M sucrose cushion containing 0.05 mM $MgCl_2$. Purified nucleoli were resuspended in 0.34 M sucrose containing 0.05 mM $MgCl_2$ for further analyses.

2.5. Immunoprecipitation.

Lysates (500 μg) from serum-starved HeLa cells treated with 20 nM APF or serum-starved controls were immunoprecipitated with an mAb antibody against CKAP4 (G1/296; 1 μg mAb/100 μg total protein lysate; Alexis Biochemicals) overnight at 4°C. Samples were then bound to Protein A, washed (4X with RIPA/Empigen buffer), eluted (4X LDS sample buffer; Invitrogen), boiled at 95°C for 5 min and resolved on a 4–12% Bis-Tris gel and transferred to a nitrocellulose membrane under reducing conditions. Western Blot analysis detected phosphoserine on CKAP4 using primary mouse mAb, α-phosphoserine antibody (Invitrogen; diluted 1 : 1000) and secondary, HRP-conjugated goat anti-mouse (diluted 1 : 20,000, Jackson ImmunoResearch Laboratories). The signal was detected on film by ECL (ThermoFisher Scientific). The membrane was stripped and reprobed with a mouse mAb G1/296 against CKAP4 (Alexis Biochemicals) to normalize the phosphoserine signal to the amount of immunoprecipitated CKAP4. Densitometric analysis of the immunoreactive bands was performed using ImageJ, and the ratios of phosphorylated to nonphosphorylated CKAP4 were determined.

2.6. Western Blot Analysis.

Cells were lysed in ice-cold RIPA buffer containing protease inhibitors (ThermoFisher Scientific), sonicated, and centrifuged for 15 minutes at 4°C. The supernatant protein concentration was measured using the Micro BCA protein assay kit (ThermoFisher Scientific). Proteins were separated by electrophoresis using 4–12% NuPAGE Novex Bis-Tris polyacrylamide gels in MOPS running buffer (Invitrogen) and then transferred to nitrocellulose. Membranes were blocked for 2 hours at room temperature in TBST buffer (Tris-buffered saline, pH 7.4, with 0.1% Tween 20) containing 5% nonfat milk and incubated with an mAb antibody against CKAP4 (G1/296; diluted 1 : 1000; Alexis Biochemicals) overnight at 4°C. The membrane was subsequently washed with TBST, incubated for 1 hour at room temperature in HRP-conjugated goat, anti-mouse (diluted 1 : 20000; ThermoFisher Scientific) secondary antibody and developed by enhanced chemiluminescence (ThermoFisher Scientific). To assess equal loading of protein, the membranes were stripped and reprobed for β-tubulin (diluted 1 : 1000; Abcam) and also for fibrillarin (diluted 1 : 1000; Abcam) to ensure the purity of the nuclear fraction. The membranes were exposed to film (BioMax AR, Kodak, Rochester, NY) and the resulting images scanned at 300 dpi. The protein bands of interest were quantified using ImageJ and the integrated signal densities normalized to β-tubulin (cytosol) or fibrillarin (nuclear) and subsequently expressed in terms of the fractional abundance relative to untreated control cells.

2.7. ^{32}P Metabolic Labeling.

HeLa cells grown to ~80% confluence in 10 cm dishes were serum starved for three hours and then metabolically labeled with 150 μCi $γ^{32}P$-ATP for one hour. The cells were then exposed to APF (20 nM; 24 hours) or left in serum-free medium (control) for 24 hours. The cells were then harvested, washed three times in ice-cold PBS, and the nuclear fraction was isolated using the NE-PER kit (ThermoFisher Scientific). Equal quantities of each fraction were separated by SDS-PAGE and transferred to nitrocellulose for Western Blot analysis using a CKAP4 mAb (G1/296; diluted 1 : 500; Alexis Biochemicals) as described above. Following the Western Blot, the membrane was rinsed in 1% H_2O_2 for one minute to eliminate the chemiluminescent signal and then wrapped in plastic wrap and exposed to film for 12 hours to detect the phosphorylated proteins.

2.8. Cell Surface Biotinylation.

Sulfo-NHS-biotin (ThermoFisher Scientific) was dissolved in serum-free DMEM to a final concentration of 0.5 mg/mL. HeLa cells were grown in serum-free medium for 36 hours (10 cm dishes at ~80% confluence), washed three times in PBS, and the medium replaced with serum-free DMEM/Biotin. The labeling reaction proceeded at 4°C for 60 minutes. Next, the cells were washed three times in PBS/100 mM gylcine to quench unreacted biotin. Serum-free DMEM containing APF (20 nM) or without APF (control) was added back to the cells, which were then incubated at 37°C for four hours. Cells were harvested by trypsinization, washed three times in PBS, and the nuclear protein fraction was isolated using the NE-PER kit (ThermoFisher Scientific) according to the manufacturer's protocol. Equal quantities of protein from each nuclear fraction were separated by SDS-PAGE and then blotted to nitrocellulose. The membrane was blocked overnight at 4°C in TBST buffer containing 10% milk and subsequently washed three times in TBS, once in TBST and once with PBS. The membrane was then incubated two hours at room temperature in PBS/1% BSA containing Streptavidin HRP (ThermoFisher Scientific) at a concentration of 1:500 dilution (0.02 μg/mL). The membrane was then washed three times (10 minutes each) at 4°C in TBS and once in TBST. Biotinylated proteins on the membrane were detected using ECL (ThermoFisher Scientific; Super Signal West Pico). The same membrane was subsequently Western Blotted simultaneously with antibodies to CKAP4 (G1/296; diluted 1:1000; Alexis Biochemicals) and fibrillarin (diluted 1:100, Abcam). The signal from the biotinylated CKAP4 in the nuclear fraction was normalized to the nuclear abundance of CKAP4 as determined by Western Blot and the total amount of nuclear protein loaded into each lane as was determined by the signal from fibrillarin.

2.9. Genomic DNA Affinity Chromatography (GDAC). HeLa cells treated with APF (or control) or expressing CKAP4C100S/SΔE (for 24–48 hours) were scraped into gDNA binding buffer (20 mM Tris [pH 7.5], 100 mM KCl, 10% Glycerol, 1 mM EDTA, 1 mM DTT, 1 mg/mL BSA, 0.1% SDS) with 1X phosphatase inhibitor and 1X phosphatase inhibitor cocktail 1 (both from ThermoFisher Scientific). Proteins were extracted for 2 hours on ice and then the insoluble fraction was pelleted by centrifugation at 15,000 ×g for 10 minutes at 4°C. Next, 15–100 mg of gDNA cellulose or cellulose alone (control; both from Sigma) was added to the cleared supernatant and mixed by constant inversion on a rotating wheel at 4°C for two hours to overnight. To capture the bacterially expressed C-terminus of CKAP4, 0.4–40 μg of purified protein (as determined by the Micro BCA assay, ThermoFisher Scientific) was added to 15–50 mg gDNA-cellulose in 500 μL of gDNA-binding buffer and mixed as above at 4°C. After proteins were allowed to bind, the gDNA-cellulose or cellulose alone was pelleted by centrifugation and washed with PBS three times. This wash protocol was sufficient to remove nonspecifically bound proteins and nonphosphorylated CKAP4 from the cellulose alone and from the gDNA-cellulose. After the final wash, SDS-PAGE sample buffer was added to 10 μL of the supernatant previously removed from the gDNA pellet, the cellulose, or gDNA-cellulose pellet. The samples were heated for 5 minutes at 95°C, separated by SDS-PAGE, transferred to nitrocellulose, and probed with α-V5 antibody (mAb; 1:5000; Invitrogen) followed by goat-α-mouse-HRP (1:10,000, ThermoFisher Scientific) to detect the transiently expressed CKAPC100S/SΔE or α-CKAP4 (mAb G1/296; 1:1000; Alexis Biochemicals) followed by goat-α-mouse-HRP (1:10,000, ThermoFisher Scientific) to detect endogenous and/or transiently expressed CKAP4. The signal was detected on film using chemiluminescence (SuperSignal West Pico, ThermoFisher Scientific).

3. Results

3.1. CKAP4 Translocates from the Plasma Membrane to the Nucleus in Response to APF. Previously, we observed by immunolocalization that CKAP4 is expressed on the surface of HeLa cells and became more abundant in the nucleus following APF exposure [6]. To confirm that CKAP4 translocates to the nucleus in response to APF, we used three different techniques—surface labeling, cellular fractionation, and immunocytochemistry. First, we surface-labeled HeLa cell proteins with sulfo-NHS biotin, a water soluble form of biotin that does not pass through the PM and binds only to primary amines of the extracellular portions of proteins. We then treated the cells with APF for three hours, isolated the nuclear protein fraction, and separated the proteins by SDS PAGE followed by transfer to nitrocellulose and detection with streptavidin-HRP. As demonstrated in Figure 2(a), biotinylated CKAP4 protein was detected in the nuclear fraction of APF-treated cells but not in the nuclear protein fraction isolated from untreated cells. This demonstrates that CKAP4 is present on the cell surface and that it is internalized and translocates to the nucleus in response to APF treatment.

To determine whether there is a fractional change in nuclear abundance of CKAP4 in response to APF stimulation, we performed Western Blot analysis on nuclear and cytosolic protein fractions extracted from APF-treated and untreated cells. We found that exposure of HeLa cells to APF resulted in a ~4-fold increase in the relative abundance of CKAP4 in the nucleus (Figures 2(b) and 2(c)) confirming our earlier observations [6]. Lastly, we show by immunocytochemistry that transiently expressed, wild-type (WT) CKAP4 localizes to the nucleus after APF treatment (Figure 3(a)). Collectively, these data demonstrate that CKAP4 translocates to the nucleus of HeLa cells after APF treatment.

3.2. APF-Induced Nuclear Translocation of CKAP4 Is Regulated by Serine Phosphorylation. Previously we showed that DHHC2-mediated palmitoylation of CKAP4 on cysteine-100 (C100) is required for its nuclear localization in response to APF treatment [6]. Work by others has shown that phosphorylation of serines 3, 17, and 19 (S3, S17, and S19) also affects CKAP4 subcellular distribution, promoting disengagement from microtubules [7]. These findings suggest that APF binding may alter the palmitoylation and/or phosphorylation state of CKAP4, affecting its subcellular distribution and function within the cell. Consequently, we generated CKAP4 mutants that mimic constitutive depalmitoylation and various states of serine phosphorylation to examine the effect of APF on their subcellular distribution in HeLa cells using immunocytochemistry (Table 1). The CKAP4 mutations included C100 to serine (C100S) to block palmitoylation; mutation of S3, S17, and S19 to alanine (SΔA) to block phosphorylation; mutation of the same three serines to glutamic acid (SΔE) to mimic phosphorylation [7]. Each mutant was fused to a V5 epitope tag to immunologically distinguish it from endogenous CKAP4. Also, since phosphorylation regulates the association of

Antiproliferative Factor-Induced Changes in Phosphorylation and Palmitoylation of Cytoskeleton-Associated Protein-4
Regulate Its Nuclear Translocation and DNA Binding

47

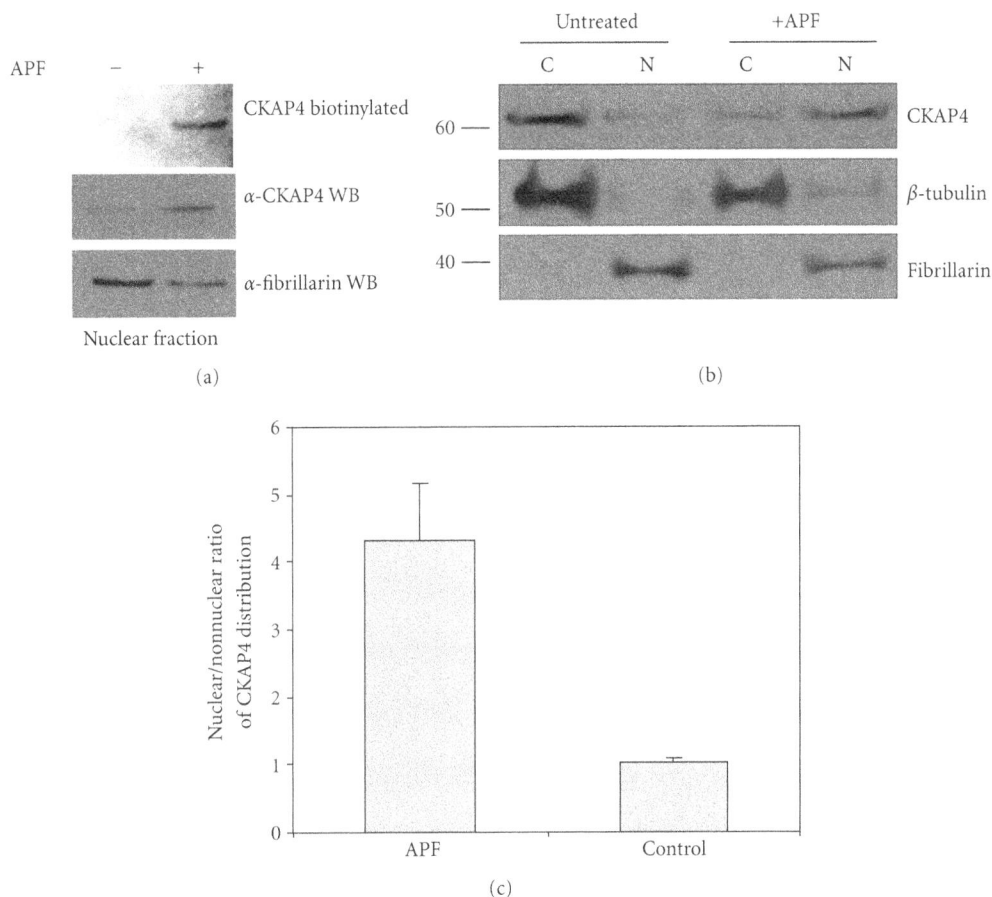

FIGURE 2: *Surface-labeled CKAP4 translocates from the plasma membrane into the nucleus following APF exposure.* (a) HeLa cell-surface proteins were labeled with Sulfo NHS-biotin as described in Section 2. Following exposure to 20 nM APF for 24 hours (or no treatment), the cells were harvested and the nuclear protein fraction was isolated (Pierce NE-PER), separated by SDS-PAGE, and transferred to nitrocellulose. The membrane was then probed with streptavidin-HRP (1 : 5000; Pierce) to bind biotinylated proteins, and the signal was detected by ECL (Pierce). Following detection of the biotinylated proteins from the nucleus, the (streptavidin) HRP on the membrane was inactivated by incubating the blot in PBS containing 3% H_2O_2 and 1% sodium azide. The same membrane was then reprobed with antibodies to CKAP4 ("anti-CLIMP-63", diluted 1 : 1000, Alexis Biochemicals) and fibrillarin (a nuclear marker and loading control; Abcam; diluted 1 : 1000). (b) HeLa cells were treated with APF (20 nM) for 24 hours, which resulted in a significant increase in the abundance of CKAP4 in the nucleus compared to control samples. Treated cells were harvested and the nuclear and cytosolic fractions were isolated and separated by SDS-PAGE as described in Section 2. Protein expression was analyzed by Western Blotting with antibodies for β-tubulin (diluted 1 : 1000, Abcam; loading control for the nonnuclear fraction), CKAP4 ("anti-CLIMP-63", diluted 1 : 1000, Alexis Biochemicals), and fibrillarin (diluted 1 : 1000, Abcam; loading control and specific marker for the nuclear fraction), and then with an HRP-conjugated anti-mouse secondary antibody (1 : 20000; ThermoFisher Scientific). The proteins were detected by ECL (Pierce) with multiple exposures to film. The integrated density of the bands on the film was measured using ImageJ. Exposure times were controlled to ensure that the signals on film were not saturated. (c) The nuclear/cytosolic ratio represents the relative distribution of CKAP4 in the nuclear versus cytosolic fractions extracted from cells treated with or without APF. CKAP4 abundance in the APF-treated and control samples were normalized for loading to β-tubulin for the nonnuclear fractions and to fibrillarin for the nuclear fractions. The nuclear/cytosolic ratio for CKAP4 in the APF and control samples was determined from these normalized values. The standard deviation describes the variability among the normalized, nuclear, and cytosolic ratios from three independent experiments. A two tailed, paired t-test of the two data arrays (plus APF and control) indicate that the difference between these ratios is significant ($P = 0.01$; $n = 3$). Cells treated with APF stop dividing, so the 10 cm dishes containing control and APF treated cells contained fewer cells (and protein) at the end of the experiment, normalizing the CKAP4 signals to loading controls corrected for this disparity. Fibrillarin is a well-characterized nuclear marker that is also known to localize to nucleoli. The data shown are representative of four independent experiments.

CKAP4 with microtubules, we coimmunolabeled tubulin and CKAP4 to visualize differences in the relative localization of CKAP4 within the microtubule network. Table 1 provides a summary of the data.

As shown in Figure 3, we confirmed that palmitoylation was required for CKAP4 expression on the PM regardless of the state of phosphorylation (see Figures 3(b), 3(d), 3(f)). Importantly, when CKAP4 is not expressed on the PM, it

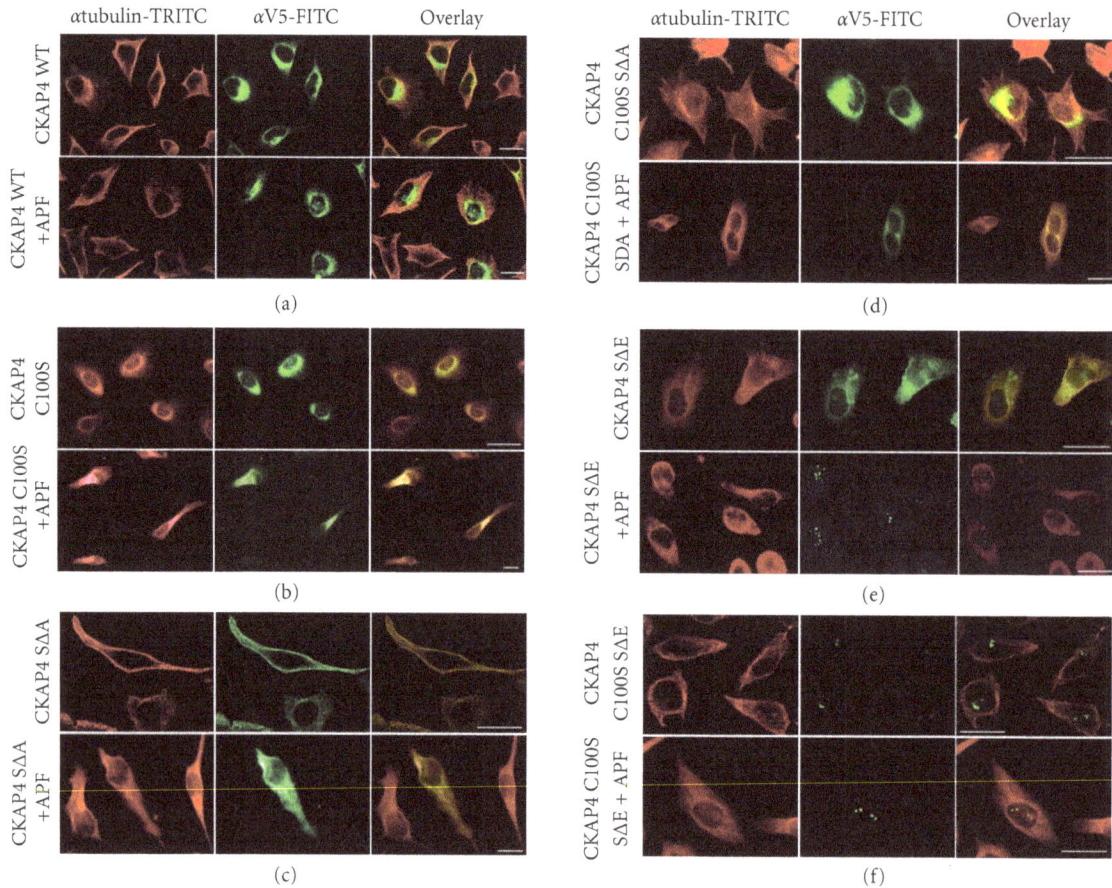

FIGURE 3: *CKAP4 phosphorylation and palmitoylation regulate CKAP4 trafficking.* CKAP4 mutants that mimic constitutive depalmitoylation and various states of serine [3, 17, 19] phosphorylation were generated to determine the effect of these two posttranslational modifications on the subcellular distribution of CKAP4 in response to APF (see Table 1). HeLa cells were transfected for 24–36 hours with the construct indicated to the left of each panel. Cells were serum starved for 6 hours and then treated with APF (20 nM) for 18–24 hours. Subsequently, the cells were fixed in 4% buffered paraformaldehyde, permeabilized with 0.1% Triton X-100, and immunostained for (1) β-tubulin (red; TRITC) and (2) V5 (green: FITC) (as described in Section 2) to distinguish the transfected, V5-epitope tagged WT (a) and mutant versions of CKAP4 from endogenous CKAP4 (b)–(d). Mutant versions of CKAP4 that cannot be palmitoylated (C100S) or phosphorylated (S∆A) do not translocate to the nucleus in response to APF. (e) Those that mimic phosphorylation (S∆E) translocate to the nucleus in response to APF. (f) CKAP4C100SS∆E, which is constitutively depalmitoylated and phosphomimicking, is expressed primarily in the nucleus. Images taken in each channel were superimposed to illustrate the distribution of mutant CKAP4 with respect to the cytoskeleton. The cells were imaged by epifluorescence or confocal microscopy at 60X and 63X, respectively, (Scale bars = 25 microns).

does not respond to APF by a change in subcellular distribution (Figures 3(b), 3(d), and 3(f)). CKAP4C100S/S∆A expression was restricted to internal membranes and did not change in response to APF (Figure 3(d)). CKAP4S∆A was expressed throughout the cell, including the PM, and resembled that of WT CKAP4 (Figures 3(a) and 3(c)). Treatment of cells expressing CKAP4S∆A with APF did not cause translocation of the mutant form of CKAP4 into the nucleus, suggesting that CKAP4 must be phosphorylated on serine residues 3, 17, and 19, thus disengaged from the microtubule network, to translocate into the nucleus. CKAP4S∆E was also distributed throughout the cells including the PM, but in contrast to CKAP4S∆A, APF treatment caused translocation of CKAP4S∆E into the nucleus (Figure 3(e)). Surprisingly, CKAP4C100S/S∆E was expressed primarily if not exclusively in the nucleus (even in the absence of APF), and its localization did not change in response to APF treatment

(Figure 3(f)). These data suggest that phosphorylation of CKAP4 on serine residues 3, 17, and 19 are required for its nuclear translocation in response to APF.

To test the idea that APF treatment promotes serine phosphorylation of CKAP4, we used an antiphosphoserine antibody to measure any change in phosphorylation of endogenous CKAP4 from whole cell lysates in response to APF. Serine phosphorylation of CKAP4 immunoprecipitated from APF-treated HeLa cells was increased ~5.0-fold compared with CKAP4 immunoprecipitated from whole cell lysates of untreated cells (Figure 4), demonstrating that APF increases CKAP4 phosphorylation on serine residues.

3.3. Nuclear CKAP4 Is Phosphorylated following APF-Induced Translocation. The observed increase in serine phosphorylation of endogenous CKAP4 and nuclear localization of the

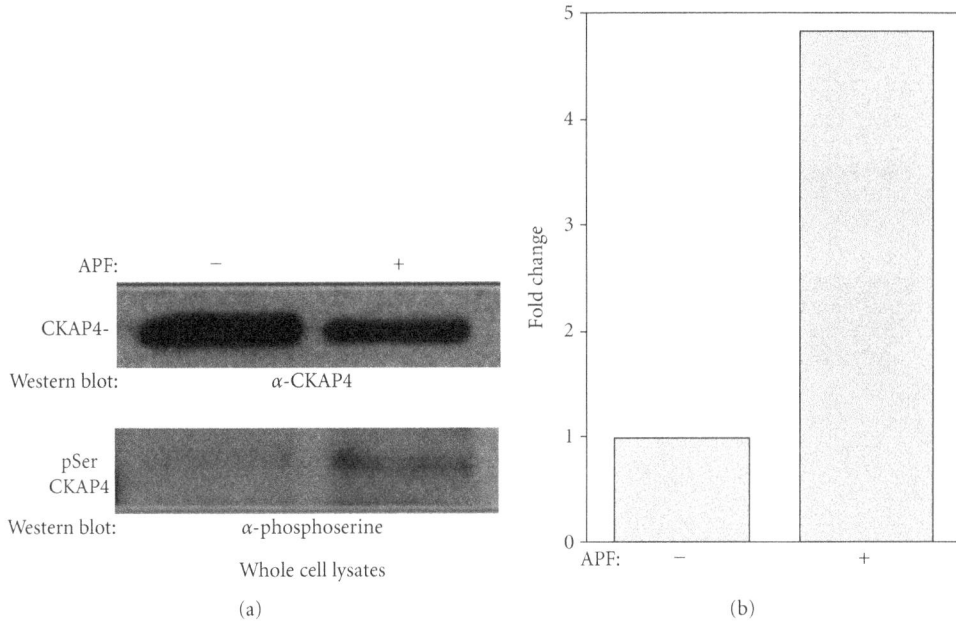

FIGURE 4: *APF induces serine phosphorylation of CKAP4.* (a) APF treatment induces a significant increase in serine phosphorylation of CKAP4 as demonstrated by immunoprecipitation of CKAP4 followed immunoblotting to detect phosphoserine. Whole cell lysates (500 μg) from HeLa cells treated with 20 nM APF or serum-starved controls were immunoprecipitated with CKAP4 antibody (Alexis) overnight at 4°C. Samples were then bound to Protein A, washed (4X with RIPA/Empigen buffer), eluted (4X LDS sample buffer; Invitrogen), boiled at 95°C for 5 min and resolved on a 4–12% Bis-Tris gel and transferred to a nitrocellulose membrane under reducing conditions. Western Blot analysis for pSer detected phospho-serine using primary (Invitrogen and secondary antibodies (goat anti-rabbit HRP-labeled antibody; Pierce)) developed with Enhanced Chemiluminescence reagent (Pierce) and exposed to film. The membrane was stripped with Restore Stripping Buffer (Pierce) and reprobed for CKAP4 (Alexis) to normalize the phosphoserine signal to the amount of immunoprecipitated CKAP4. (b) Densitometric analysis of the immunoreactive bands was done using ImageJ, and the ratios of phosphorylated to nonphosphorylated CKAP4 were determined.

phosphomimicking CKAP4 mutants in response to APF suggested that endogenous CKAP4 should be phosphorylated if it is in the nucleus. To determine if nuclear CKAP4 is indeed phosphorylated, we metabolically labeled cells with γ^{32}P-ATP, exposed them to APF, and isolated the nuclear fraction. The level of phosphorylation was assessed by Western Blot after separation of proteins by SDS-PAGE. In agreement with previous experiments, the abundance of CKAP4 in the nucleus increased following exposure to APF (Figure 5(a), top panel). Nuclear CKAP4 was phosphorylated to an equal degree in both APF-stimulated and control cells (Figure 5(a), bottom panel). The ratio of the CKAP4 phosphorylation signal over the corresponding CKAP4 Western Blot signal was the same for both (1.00 versus 0.975; APF treated versus control) indicating that nuclear CKAP4 is phosphorylated.

In HeLa cells exposed to APF or cells transfected with CKAP4C100S/SΔE, we observed by immunolabeling that CKAP4 was not evenly distributed within the nucleus, but rather appeared to be concentrated on nucleoli and other unidentified subnuclear structures (data not shown). To determine whether CKAP4 was associated with nucleoli, we treated HeLa cells with APF for 24 hours and measured the abundance of CKAP4 in the nucleolar and nuclear fractions (nonnucleolar fraction of the nucleus) by Western Blot (Figure 5(b)). Our results demonstrate that APF induces

an association of CKAP4 with nucleoli (Figures 5(b) and 5(c)). Immunolocalization of CKAP4C100S/SΔE suggested that it too associated with nucleoli (independent of APF treatment). To confirm this biochemically, we expressed CKAP4C100S/SΔE in HeLa cells, isolated the nucleolar fraction, and measured its association with nucleoli by Western Blot. Figure 5 shows that CKAP4C100S/SΔE associates with nucleoli to a greater degree than endogenous CKAP4 in APF-treated cells. These data support the idea that CKAP4 becomes phosphorylated after binding to APF, and that phosphorylation enhances its nuclear translocation.

3.4. CKAP4C100S/SΔE Behaves as an APF Mimetic. The similarity between the nuclear localization of endogenous CKAP4 following APF treatment and transiently expressed CKAP4 C100S/SΔE suggested that the mutant may mimic APF activity, such as inhibiting cellular proliferation. To determine if this hypothesis was correct, we transfected HeLa cells with varying quantities of CKAP4C100S/SΔE cDNA and measured their proliferation over the course of 48 hours. Figure 6 shows that that transient expression of CKAP4C100S/SΔE significantly inhibits HeLa cell proliferation in a dose-dependent manner (versus mock transfected cells), a result that is characteristic of cells exposed to APF [6, 11, 21]. These results indicate that CKAP4C100S/SΔE may

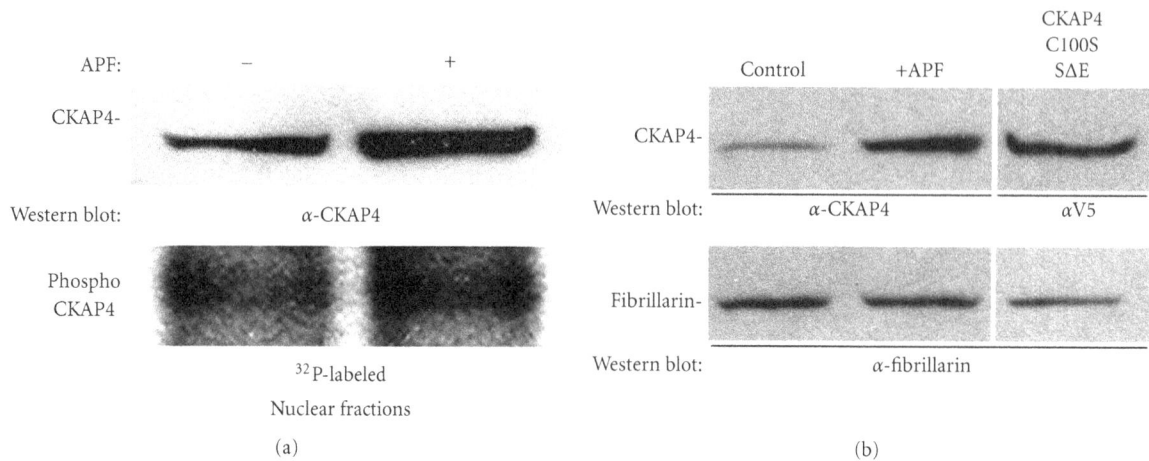

(a)

(b)

(c)

FIGURE 5: *Nuclear CKAP4 is phosphorylated following APF-induced translocation.* (a) CKAP4 in the nuclear fraction was phosphorylated to the same degree in APF-treated and control cells as demonstrated by metabolic labeling with γ^{32}P-ATP. The ratio of the CKAP4 phosphorylation signal over the corresponding CKAP4 Western Blot signal was approximately the same for both (1.00 versus 0.975; APF treated versus control). HeLa cells at ~80% confluence in 10 cm dishes were serum starved for 3 hours then 150 μCi γ^{32}P-ATP was added to each dish for 1 additional hour. Then the cells were either exposed to APF (20 nM; 24 hours) or left in serum-free medium (control) for 24 hours. At the end of 24 hours, the cells were harvested, washed three times in ice-cold PBS, and the nuclear and cytoplasmic fractions were isolated using the NE-PER (Pierce). Equal quantities of each fraction were separated by SDS-PAGE and transferred to nitrocellulose. The membrane was incubated overnight at 4°C in α-CKAP4 antibody (1 : 500) in TBST and 1% milk, washed and incubated for six hours at 4°C in a-mouse, HRP-conjugated secondary antibody (Pierce; 1 : 20,000) in TBST and 1% milk. CKAP4 bands were detected by enhanced chemiluminescence (ECL; Pierce; 20 second exposure; left panel). Following the Western Blot, the membrane was rinsed in 1% H_2O_2 for 1 minute to eliminate the chemiluminescent signal then wrapped in plastic wrap and exposed to film for 12 hours to detect the phosphorylated proteins. Densitometric analysis of the immunoreactive bands was done using ImageJ, and the ratios of phosphorylated to nonphosphorylated CKAP4 were determined. *APF increases the association of CKAP4 with nucleoli.* (b) and (c) Western Blot analysis shows that treatment of HeLa cells with APF (20 nM) increased the association of endogenous CKAP4 with the nucleolar fraction, and that the constitutively depalmitoylated and phosphomimicking CKAP4 mutant, CKAP4C100S/SΔE, associated with the nucleolar fraction to a greater extent than endogenous CKAP4 isolated from APF-treated cells. Nucleoli were isolated using a variation on the method published by Busch and coworkers [20] as described by the Angus Lamond lab (University of Dundee, UK). The nucleolar proteins were separated by SDS-PAGE and Western Blotted for CKAP4 (α-V5 in the case of CKAP4C100S/SΔE) and fibrillarin. The bands were detected on film by ECL and quantified using ImageJ. The CKAP4 signal in each lane of the Western Blot was normalized to the fibrillarin band in the same lane. The normalized value for CKAP4 from control cells was set to 1 and the other values were set relative to control. The values in the graph are means and SD. "*" indicates that the means of the values were significantly different than serum-starved control when evaluated using the students *t*-test (two-tailed. Serum starved versus APF treated $P = 0.018$; serum starved versus CKAP4C100S/SΔE $P = 0.002$; CKAP4C100S/SΔE versus APF treated, $P = 0.008$; $n = 3$ for each).

Antiproliferative Factor-Induced Changes in Phosphorylation and Palmitoylation of Cytoskeleton-Associated Protein-4
Regulate Its Nuclear Translocation and DNA Binding

51

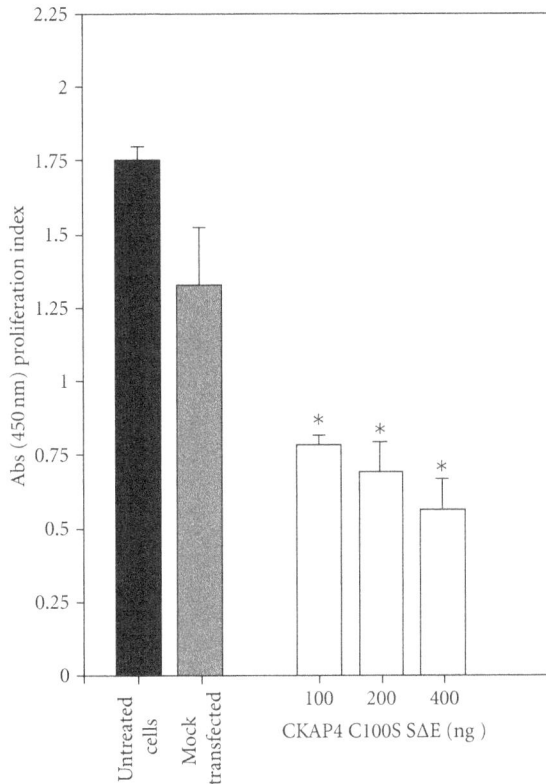

FIGURE 6: *CKAP4C100S/SΔE expression inhibits HeLa cell prolif-eration.* HeLa cells were mock transfected (Fugene; control) or transfected with 100, 200, or 400 ng of CKAP4C100S/SΔE DNA. After an additional 24 hours of growth, proliferation was measured using the WST-1 proliferation assay (Biovision) according to the manufacturer's protocol. Each data point is the mean and SD from 12 independent wells. Statistically significant differences in the rate of cellular proliferation were detected for HeLa cells transfected with 100, 200, and 400 ng of CKAP4C100S/SΔE DNA versus mock-transfected cells using the students t-test ($^*P < 0.001$).

function as an APF mimetic and support the argument that endogenous CKAP4 is depalmitoylated and phosphorylated following APF binding.

3.5. Phosphorylated CKAP4 Binds to gDNA.

Because CKAP4 localizes to the nucleus/nucleolus in response to APF treat-ment and because its predicted structure includes a region homologous to bZIP transcription factors (Figure 1), we assessed whether CKAP4 could bind to DNA using genomic DNA (gDNA) affinity chromatography (GDAC)[22]. For this technique, nuclear lysates from APF- or mock-treated HeLa cells were incubated with gDNA-bound cellulose beads in the presence of nonspecific competitor (dI-dC), and proteins that bound DNA were isolated and separated by SDS-PAGE followed by Western Blot with an anti-CKAP4 antibody. As shown in Figure 7(a), CKAP4 binds gDNA in an APF-dependent manner (Lane 3) when compared to mock-treated control (Lane 4). The CKAP4-gDNA binding is phosphorylation dependent as CKAP4 isolated from cells that were treated without phosphatase inhibitors prior to

APF treatment failed to bind gDNA (Lane 6). Furthermore, in nuclear lysates from cells transfected with the V5-tagged, phosphomimicking CKAP4 mutant (CKAP4 C100S/SΔE), we observed enhanced binding to gDNA (Figure 7(b)). In Figure 7(c), we show that purified rCKAP4 (C-terminal residues 126–501, which includes the bZIP-like DNA-binding domain) also binds gDNA. Collectively, these data suggest that APF induces phosphorylation of CKAP4 and that phosphorylation is required for CKAP4 to bind gDNA.

4. Discussion

CKAP4 has an established role in maintaining the struc-ture of the ER relative to the cytoskeleton by binding to microtubules—a function which is dependent on three N-terminal serine residues (S3, S17, and S19) that when phos-phorylated, cause CKAP4 to disengage from microtubules (see Figure 1) [3, 7]. Expression of mutant versions of CKAP4 that do not bind microtubules profoundly changes ER morphology [2]. More recently, our knowledge of CKAP4 function has expanded to include it being a cell surface receptor for APF [10], surfactant protein A (SPA) [23], and tissue plasminogen activator (tPA) [24]. In a previous study, we showed that palmitoylation of CKAP4 by DHHC2 is required for its localization on the PM and for mediating APF-induced signaling events, including inhibition of cellu-lar proliferation and changes in gene/protein expression [6]. However, prior to the data presented here, there was nothing known about the mechanism for APF signal transduction by CKAP4.

Previously, we observed by immunocytochemistry that exposing cells to APF resulted in an increased abundance of CKAP4 in the nucleus [6]. These data suggested that APF binding to CKAP4 on the cell surface induced its translocation into the nucleus. In this study, we used cell-surface labeling with Sulfo-NHS-biotin (which reacts only with extracellular primary amines but does not bind to proteins inside of the cell during the labeling process) to monitor CKAP4 localization in response to APF binding. Our results showed that biotinylated CKAP4 from the cell surface is abundant in the nuclear fraction of HeLa cells treated with APF, but not in untreated cells. Detection of a relatively small amount of nonbiotinylated CKAP4 in the nucleus of untreated cells in this experiment is consistent with data from other experiments, indicating that CKAP4 may have some inherent nuclear function not related to APF-induced signal transduction, moreover, it suggests that in order to block proliferation, CKAP4 in the nucleus must have bound to APF at some point.

Using a series of mutants that mimic combined states of palmitoylation and phosphorylation, we determined that APF promotes serine phosphorylation of CKAP4; more-over, that phosphorylation of serines 3, 17, and 19 is required for APF-induced nuclear translocation of CKAP4, as mutation of these residues to alanine (SΔA) inhibited this process. Additionally, mutation of these serine residues to glutamic acid to create a phosphomimicking form of CKAP4 (CKAP4SΔE) promoted the nuclear translocation of CKAP4

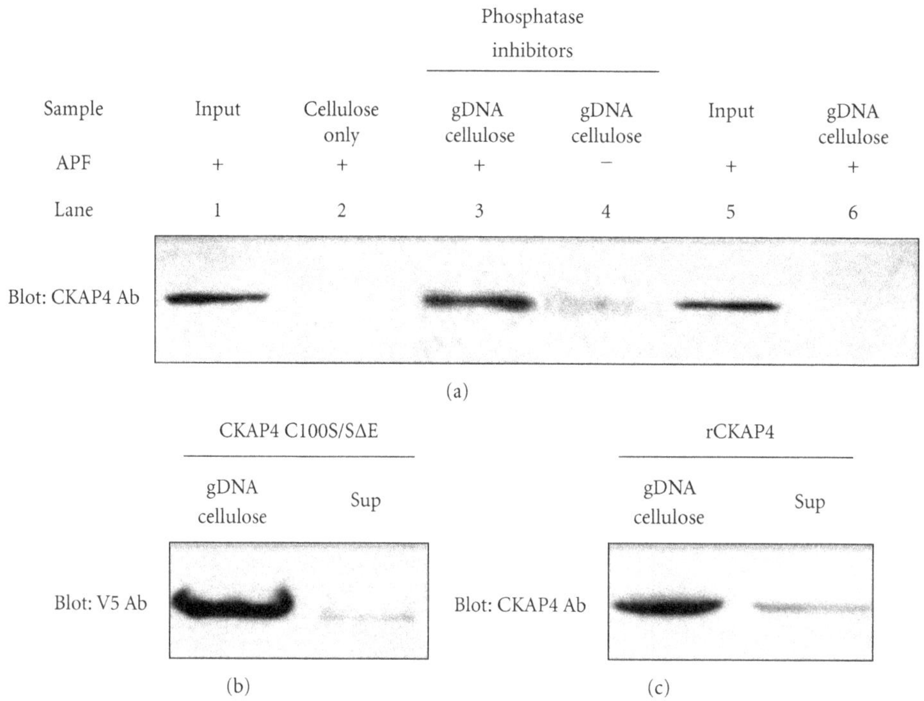

Sample	Input	Cellulose only	gDNA cellulose	gDNA cellulose	Input	gDNA cellulose
APF	+	+	+	−	+	+
Lane	1	2	3	4	5	6

Blot: CKAP4 Ab

(a)

CKAP4 C100S/SΔE

gDNA cellulose	Sup

Blot: V5 Ab

(b)

rCKAP4

gDNA cellulose	Sup

Blot: CKAP4 Ab

(c)

FIGURE 7: *CKAP4 binds directly to genomic DNA.* (a) CKAP4 binds gDNA in an APF-dependent manner (lane 3) when compared to mock-treated control (lane 4). The CKAP4-gDNA binding is phosphorylation dependent as CKAP4 isolated from cells treated without phosphatase inhibitors prior to APF treatment failed to bind gDNA (lane 6). As controls, we loaded APF treated nuclear lysates in lane 1 and phosphatase/APF treated lysates in lane 5. Lane 2 is a no gDNA cellulose control. (b) The ability of transiently transfected V5-tagged, CKAP4 C100S/SΔE to bind gDNA in the absence of APF treatment (c) and the ability of purified rCKAP4 (residues 126–501, which includes the bZIP-like DNA-binding domain) to bind gDNA were also assessed by comparing the amount of CKAP4 captured relative to what remained in the cell lysate after binding (Sup).

in response to APF. Interestingly, a phosphomimicking form of CKAP4 that was also constitutively depalmitoylated localized in the nucleus without APF stimulation. These data suggest that when CKAP4 is bound by APF on the PM, it becomes depalmitoylated and phosphorylated and translocates into the nucleus. In direct support of the idea that CKAP4 must be phosphorylated to enter the nucleus, we show by metabolic labeling with γ^{32}P-ATP that nuclear CKAP4 is phosphorylated. Despite using multiple methods, we did not observe palmitoylated CKAP4 in the nucleus. The mechanism by which CKAP4 or any other palmi-toyl-protein becomes depalmitoylated is a matter of great interest. However, there is very little known about how depalmitoylation is regulated. Serines 3 and 19 reside within a consensus site for PKC and serine 17 within a CKII site [7]; however, whether these kinases are responsible for phosphorylation of these serines remains to be determined.

CKAP4 has a ~106-amino acid cytosolic tail, a single TM domain, and a large, extracellular/ER-luminal domain consisting of 474 amino acids (see Figure 1). Much of the extracellular domain, including the portion required for oligomerization, is predicted to fold into a coiled-coil confor-mation, with an amphipathic alpha helix rich in basic residues comprising the C-terminal (~80) residues [3]. These structures are among the most common and well-characterized domains involved in protein-protein

interactions. Moreover, amino acid residues 468–602 are homologous to the DNA-binding domain of bZIP transcription factors. Many details about the predicted structure of CKAP4, its nonrandom distribution in the nucleus, and enhanced association with nucleoli (see Figure 5), suggest that it might bind to DNA. Genomic DNA affinity chromatography is a straight-forward method to determine if a protein binds to DNA without *a priori* knowledge of the exact binding sites [25]. Using GDAC, we demonstrated that CKAP4 isolated from APF-treated HeLa cells bound genomic DNA (gDNA). Furthermore, our results show that inclusion of phosphatase inhibitors was required for endogenous CKAP4 to bind to gDNA, but not for binding of the purified CKAP4 C-terminal fragment to gDNA. This indicates that the C-terminal fragment alone, separated from the phosphoserines near the N-terminus, is free to reside in a conformational state that favors gDNA binding—the same conformation that exists in the endogenous protein when phosphorylated. Future experiments will be carried out to identify the specific DNA sequences to which CKAP4 binds.

CKAP4 is 63 kDa, homo-oligomeric, and is embedded in membranes; these are not typical, physical properties of proteins that enter the nucleus by diffusion. Generally, proteins that diffuse into the nucleus are cytosolic and have a molecular weight less than 40 kDa. Proteins too large to diffuse through the nuclear pore translocate with the assistance of a shuttle protein (or karyopherin) that is physically bound to a nuclear localization signal (NLS) sequence of the imported protein. CKAP4 contains a glycine-rich domain that may act as an NLS. Similar domains found in several families of proteins are known to mediate nuclear import [22, 26]. Ongoing work indicates that this domain of CKAP4 is sufficient to drive nuclear localization of a nonrelated protein (data not shown).

Because of its size and TM domain, it would seem unlikely that CKAP4 would have a nuclear function. However, there is precedent for large, TM-domain proteins entering the nucleus. The EGF (epidermal growth factor) receptor family are glycosylated, single-pass TM domain proteins of $\sim$140 kDa that also translocate, unproteolyzed, from the plasma membrane into the nucleus after ligand binding [27]. Within the nucleus, they regulate transcription and participate, enzymatically, in signal transduction pathways. The idea that these proteins translocated to and had functional activity in the nucleus, beyond their tyrosine kinase activity at the plasma membrane, was controversial; however, more recently, the mechanistic details about how they make their way into the nucleus and how they function within the nucleus are becoming clearer (reviewed in [27]).

In summary, the data presented here provide new insight into the mechanism by which CKAP4 mediates APF signal transduction. At the same time, we have added to a small but growing body of literature demonstrating that a receptor in the plasma membrane can translocate into the nucleus after binding ligand. Finally, our data demonstrate that CKAP4C100S/S∆E has APF-like activity as it localizes to the nucleus and inhibits proliferation without binding to APF. This finding suggests that CKAP4C100S/S∆E could be used to treat hyperproliferative diseases such as cancer. APF itself has been proposed as an anticancer therapy [28].

Abbreviations

APF: Antiproliferative factor
CKAP4:   Cytoskeleton-associated protein 4/p63
DHHC: Asp-His-His-Cys
DMEM: Dulbecco's modified Eagle's medium
ER: Endoplasmic reticulum
FBS: Fetal bovine serum
FITC: Fluorescein isothiocyanate
GDAC: Genomic DNA affinity chromatography
gDNA: Genomic DNA
IC: Interstitial cystitis
LZ: Leucine zipper
mAb: Monoclonal antibody
MT: Microtubule
NGS: Normal goat serum
NLS: Nuclear localization signal
PAGE: Polyacrylamide gel electrophoresis
PAT: Palmitoyl acyltransferase
PBS: Phosphate-buffered saline
PM: Plasma membrane
SDS: Sodium dodecyl sulfate
SP-A: Surfactant protein-A
TM: Transmembrane
tPA: Tissue plasminogen activator
TRITC: Tetramethyl rhodamine isothiocyanate.

Acknowledgments

The authors would like to thank Karen Overstreet for technical assistance. This paper was supported by funding from the Chris DeMarco, American Cancer Society Institutional Research Grant (Agency no. IRG-01-188-04), The Climb for Cancer Foundation, The Whitney Laboratory, University of Florida, and The Commonwealth Medical College. The antibody to fibrillarin was a generous gift from Gerry Shaw (UF and EnCor Bioscience).

References

[1] D. I. Mundy and G. Warren, "Mitosis and inhibition of intracellular transport stimulate palmitoylation of a 62-kD protein," *Journal of Cell Biology*, vol. 116, no. 1, pp. 135–146, 1992.

[2] D. R. C. Klopfenstein, F. Kappeler, and H. P. Hauri, "A novel direct interaction of endoplasmic reticulum with microtubules," *EMBO Journal*, vol. 17, no. 21, pp. 6168–6177, 1998.

[3] A. Schweizer, J. Rohrer, H. P. Hauri, and S. Kornfeld, "Retention of p63 in an ER-Golgi intermediate compartment depends on the presence of all three of its domains and on its ability to form oligomers," *Journal of Cell Biology*, vol. 126, no. 1, pp. 25–39, 1994.

[4] A. Schweizer, J. Rohrer, P. Jeno, A. De Maio, T. G. Buchman, and H. P. Hauri, "A reversibly palmitoylated resident protein (p63) of an ER-Golgi intermediate compartment is related to a circulatory shock resuscitation protein," *Journal of Cell Science*, vol. 104, no. 3, pp. 685–694, 1993.

[5] A. Schweizer, M. Ericsson, T. Bachi, G. Griffiths, and H. P. Hauri, "Characterization of a novel 63 kDa membrane protein. Implications for the organization of the ER-to-Golgi pathway," *Journal of Cell Science*, vol. 104, no. 3, pp. 671–683, 1993.

[6] S. L. Planey, S. K. Keay, C. O. Zhang, and D. A. Zacharias, "Palmitoylation of cytoskeleton associated protein 4 by DHHC2 regulates antiproliferative factor-mediated signaling," *Molecular Biology of the Cell*, vol. 20, no. 5, pp. 1454–1463, 2009.

[7] C. Vedrenne, D. R. Klopfenstein, and H. P. Hauri, "Phosphorylation controls CLIMP-63-mediated anchoring of the endoplasmic reticulum to microtubules," *Molecular Biology of the Cell*, vol. 16, no. 4, pp. 1928–1937, 2005.

[8] D. R. Klopfenstein, J. Klumperman, A. Lustig, R. A. Kammerer, V. Oorschot, and H. P. Hauri, "Subdomain-specific localization of CLIMP-63 (p63) in the endoplasmic reticulum is mediated by its luminal α-helical segment," *Journal of Cell Biology*, vol. 153, no. 6, pp. 1287–1299, 2001.

[9] N. Sergeant, A. Bretteville, M. Hamdane et al., "Biochemistry of Tau in Alzheimer's disease and related neurological disorders," *Expert Review of Proteomics*, vol. 5, no. 2, pp. 207–224, 2008.

[10] T. P. Conrads, G. M. Tocci, B. L. Hood et al., "CKAP4/p63 is a receptor for the frizzled-8 protein-related antiproliferative factor from interstitial cystitis patients," *Journal of Biological Chemistry*, vol. 281, no. 49, pp. 37836–37843, 2006.

[11] S. K. Keay, Z. Szekely, T. P. Conrads et al., "An antiproliferative factor from interstitial cystitis patients is a frizzled 8 protein-related sialoglycopeptide," *Proceedings of the National Academy*

of Sciences of the United States of America, vol. 101, no. 32, pp. 11803–11808, 2004.

[12] S. Keay, "Cell signaling in interstitial cystitis/painful bladder syndrome," *Cellular Signalling*, vol. 20, no. 12, pp. 2174–2179, 2008.

[13] S. Keay, F. Seillier-Moiseiwitsch, C. O. Zhang, T. C. Chai, and J. Zhang, "Changes in human bladder epithelial cell gene expression associated with interstitial cystitis or antiproliferative factor treatment," *Physiological Genomics*, vol. 14, pp. 107–115, 2003.

[14] S. Keay, G. Tocci, K. Koch et al., "The frizzled 8-related antiproliferative factor from IC patients inhibits bladder and kidney carcinoma cell proliferation in vitro," *European Journal of Cancer Supplements*, vol. 4, no. 12, pp. 87–88, 2006.

[15] P. Kaczmarek, S. K. Keay, G. M. Tocci et al., "Structure-activity relationship studies for the peptide portion of the bladder epithelial cell antiproliferative factor from interstitial cystitis patients," *Journal of Medicinal Chemistry*, vol. 51, no. 19, pp. 5974–5983, 2008.

[16] J. Kim, S. K. Keay, J. D. Dimitrakov, and M. R. Freeman, "p53 mediates interstitial cystitis antiproliferative factor (APF)-induced growth inhibition of human urothelial cells," *FEBS Letters*, vol. 581, no. 20, pp. 3795–3799, 2007.

[17] M. Gamper, V. Viereck, V. Geissbühler et al., "Gene expression profile of bladder tissue of patients with ulcerative interstitial cystitis," *BMC Genomics*, vol. 10, article 199, 2009.

[18] L. H. Tseng, I. Chen, M. Y. Chen, C. L. Lee, Y. H. Lin, and L. K. Lloyd, "Isolation of nucleoli," *International Urogynecology Journal and Pelvic Floor Dysfunction*, 2009.

[19] L. H. Tseng, I. Chen, M. Y. Chen, C. L. Lee, Y. H. Lin, and L. K. Lloyd, "Retention of p63 in an ER-Golgi intermediate compartment depends on the presence of all three of its domains and on its ability to form oligomers," *International Urogynecology Journal and Pelvic Floor Dysfunction*, 2009.

[20] H. Busch, M. Muramatsu, H. Adams, W. J. Steele, M. C. Liau, and K. Smetana, "Isolation of nucleoli," *Experimental Cell Research*, vol. 24, pp. 150–163, 1963.

[21] H. H. Rashid, J. E. Reeder, M. J. O'Connell, C. O. Zhang, E. M. Messing, and S. K. Keay, "Interstitial cystitis antiproliferative factor (APF) as a cell-cycle modulator," *BMC Urology*, vol. 4, article 1, 2004.

[22] C. M. Van Dusen, L. Yee, L. M. McNally, and M. T. McNally, "A glycine-rich domain of hnRNP H/F promotes nucleocytoplasmic shuttling and nuclear import through an interaction with transportin 1," *Molecular and Cellular Biology*, vol. 30, no. 10, pp. 2552–2562, 2010.

[23] N. Gupta, Y. Manevich, A. S. Kazi, J. Q. Tao, A. B. Fisher, and S. R. Bates, "Identification and characterization of p63 (CKAP4/ERGIC-63/CLIMP-63), a surfactant protein A binding protein, on type II pneumocytes," *American Journal of Physiology*, vol. 291, no. 3, pp. L436–L446, 2006.

[24] T. M. Razzaq, R. Bass, D. J. Vines, F. Werner, S. A. Whawell, and V. Ellis, "Functional regulation of tissue plasminogen activator on the surface of vascular smooth muscle cells by the type-II transmembrane protein p63 (CKAP4)," *Journal of Biological Chemistry*, vol. 278, no. 43, pp. 42679–42685, 2003.

[25] J. Kumaran and E. N. Fish, "Genomic DNA affinity chromatography: a technique to isolate interferon-inducible DNA binding factors," *Methods in Molecular Medicine*, vol. 116, pp. 57–67, 2005.

[26] M. Cokol, R. Nair, and B. Rost, "Finding nuclear localization signals," *EMBO Reports*, vol. 1, no. 5, pp. 411–415, 2000.

[27] S. C. Wang and M. C. Hung, "Nuclear translocation of the epidermal growth factor receptor family membrane tyrosine kinase receptors," *Clinical Cancer Research*, vol. 15, no. 21, pp. 6484–6489, 2009.

[28] J. J. Barchi Jr. and P. Kaczmarek, "Short and sweet: evolution of a small glycopeptide from a bladder disorder to an anticancer lead," *Molecular Interventions*, vol. 9, no. 1, pp. 14–17, 2009.

6

TCA Cycle Defects and Cancer: When Metabolism Tunes Redox State

Simone Cardaci[1] and Maria Rosa Ciriolo[1, 2]

[1] *Department of Biology, University of Rome "Tor Vergata", Via della Ricerca Scientifica, 00133 Rome, Italy*
[2] *IRCCS San Raffaele Pisana, Via di Val Cannuta, 00166 Rome, Italy*

Correspondence should be addressed to Maria Rosa Ciriolo, ciriolo@bio.uniroma2.it

Academic Editor: Giuseppe Filomeni

Inborn defects of the tricarboxylic acid (TCA) cycle enzymes have been known for more than twenty years. Until recently, only recessive mutations were described which, although resulted in severe multisystem syndromes, did not predispose to cancer onset. In the last ten years, a causal role in carcinogenesis has been documented for inherited and acquired alterations in three TCA cycle enzymes, succinate dehydrogenase (SDH), fumarate hydratase (FH), and isocitrate dehydrogenase (IDH), pointing towards metabolic alterations as the underlying hallmark of cancer. This paper summarizes the neoplastic alterations of the TCA cycle enzymes focusing on the generation of pseudohypoxic phenotype and the alteration of epigenetic homeostasis as the main tumor-promoting effects of the TCA cycle affecting defects. Moreover, we debate on the ability of these mutations to affect cellular redox state and to promote carcinogenesis by impacting on redox biology.

1. Introduction

Cancer cells differ from normal ones due to a plethora of oncogenes-driven biochemical changes designed to sustain an high rate of growth and proliferation [1]. The first tumor-specific alteration in metabolism was reported at the beginning of the 20th century by Warburg [2]. His observations demonstrated that cancer cell metabolism relies on an increased glycolytic flux maintained even in the presence of oxygen ("aerobic glycolysis" or "Warburg effect"), without an associated increase in oxidative phosphorylation rate. The switch from respiration to glycolysis has usually been considered a consequence, rather than a cause, of cancer. However, in the last decade, the discovery that inherited and acquired alterations in some enzymes of tricarboxylic acid (TCA) cycle have a causal role in carcinogenesis has changed this viewpoint, pointing towards altered metabolism as the underlying hallmark of neoplastic transformation. These alterations consist of germline defects in genes encoding subunits of SDH and FH, as well as somatic mutations in coding sequence for IDH. Together with metabolomics studies documenting the alteration of HIF-dependent signaling pathway and epigenetic dynamics as main tumor-promoting effects of these mutations, a mounting body of evidence also supports how alterations in the TCA cycle enzymes may favor tumorigenesis by impacting on cellular redox state. Therefore, in this paper, we summarize the prooncogenic defects in the TCA cycle enzymes discussing their involvement in the tuning of redox environment and the engagement of redox-dependent tumorigenic signaling.

2. Fundamentals of the TCA Cycle

The TCA cycle is a core pathway for the metabolism of sugars, lipids, and amino acids [3]. It is usually presented in a naive perspective of a cyclic mitochondrial route constantly oxidizing the acetyl moiety of acetyl-coenzyme A to CO_2, generating NADH and $FADH_2$, whose electrons fuel the mitochondrial respiratory chain for ATP generation. The TCA cycle begins with the condensation of acetyl-CoA with oxaloacetate to form citrate, catalyzed by citrate synthase. Citrate can be exported to the cytoplasm, where it is used as precursor for lipid biosynthesis or remains in the mitochondria, where it is converted to isocitrate by aconitase.

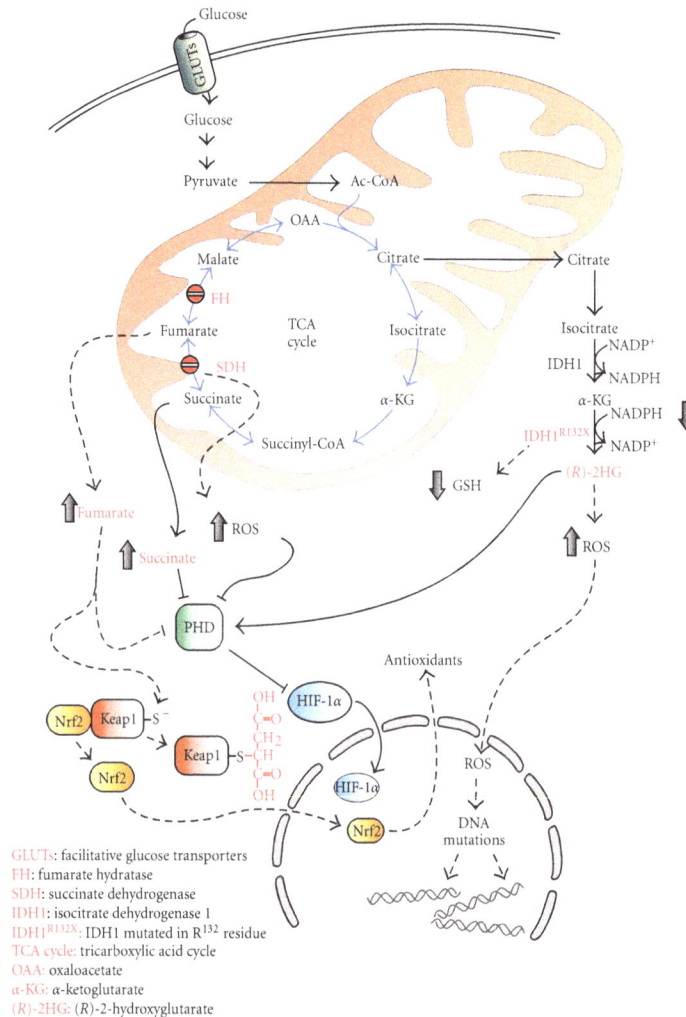

FIGURE 1: Redox alterations induced by TCA cycle defects. Redox alterations induced by mutations in SDH, FH, and IDH are shown. Loss of function of SDH increases ROS levels leading to DNA mutations and HIF-1α stabilization. IDH1 and IDH2 (not shown) mutations decrease GSH and NADPH levels. (R)-2-HG, produced by oncogenic mutations in IDH1 and IDH2, triggers ROS accumulation. Defects in FH stimulate nuclear translocation of Nrf2 and the transcription of antioxidant enzymes through the succination of Keap1. Enzymes and metabolites involved in tumor formation and redox alterations are in red. Blue arrows indicate TCA cycle reactions. Dotted arrows indicate pathways modulating cell redox state.

In the next step, α-ketoglutarate (α-KG), formed by the oxidative decarboxylation of isocitrate catalyzed by IDH, is converted to succinyl-CoA by a further decarboxylation by the α-KG dehydrogenase complex. Succinyl-CoA is then transformed to succinate by the succinyl-CoA synthetase. Fumarate, produced by succinate oxidation catalyzed by the SDH complex, is hydrated to malate by FH. Oxidation of malate, catalyzed by malate dehydrogenase, finally regenerates oxaloacetate, thus ensuring the completion of the cycle (Figure 1). On the mere biochemical viewpoint, the TCA cycle in nontumor cells has been divided into two stages: (i) decarboxylating, in which citrate is converted to succinyl-CoA releasing two CO_2 molecules; (ii) reductive, which comprises the successive oxidations of succinate to oxaloacetate. Interestingly, emerging findings from the last year support the hypothesis that, in several cell systems

such as (i) cancer cells containing mutations in complex I or complex III of the electron transport chain (ETC), (ii) patient-derived renal carcinoma cells with mutations in FH, (iii) cells with normal mitochondria subjected to acute pharmacological ETC, inhibition, as well as (iv) tumor cells exposed to hypoxia, the first stage of the cycle can proceed in the opposite direction through the reductive carboxylation of α-KG to form citrate. This allows cells to produce acetyl-coenzyme A to support *de novo* lipogenesis and their viability [4–6]. Although in physiological and resting conditions mitochondria are necessary and sufficient to perform the cycle, isoforms of some of its enzymes have been also found in the cytosol. This ensures a dual compartmentalization (cytosolic and mitochondrial) of reactions and metabolites which, being free to diffuse through the outer and the inner mitochondrial membranes by channels and active carriers, respectively, allows the cycle to respond to environmental and developmental signals, thus sustaining anabolic reactions as well as fueling the ATP-producing machinery. The TCA cycle is also a major pathway for interconversion of metabolites arising from transamination and deamination of amino acids and provides the substrates for amino acids synthesis by transamination, as well as for gluconeogenesis and fatty acid synthesis. Regulation of the TCA cycle depends primarily on a supply of oxidized cofactors: in tissues where its primary role is energy production, a respiratory control mediated by respiratory chain and oxidative phosphorylation is operative. This activity relies on availability of NAD^+ and ADP, which in turn depends on the rate of utilization of ATP in chemical and physical work.

3. Genetic Defects in the TCA Cycle

Genetic defects affecting the TCA cycle enzymes have been known for more than two decades. Until recently, only recessive mutations were documented whose clinical consequences were similar to alterations in the electron transport chain (ETC) and oxidative phosphorylation [7]. These defects were associated with multisystem disorders and severe neurological damage, but no cancer predisposition, as a result of very considerably impaired ATP formation in the central nervous system. In the last ten years, dominant defects associated with oncogenesis were described in cytoplasmic and mitochondrial isoforms of three nuclear-encoded enzymes, SDH, FH, and IDH, allowing to investigate the extrametabolic roles of the TCA cycle metabolites and their signaling to tumor formation.

3.1. Succinate Dehydrogenase. The SDH complex (also known as succinate:ubiquinone oxidoreductase or mitochondrial complex II) is a highly conserved heterotetrameric tumor suppressor, composed by two catalytic subunits (SDHA and SDHB), which protrude into the mitochondrial matrix, and two hydrophobic subunits (SDHC and SDHD), which anchor the catalytic components to the inner mitochondrial membrane and provide the binding site for the ubiquinone, as well [8]. All the subunits are encoded by nuclear genome and, unlike most of the TCA cycle enzymes, have no cytosolic counterparts. SDH catalyzes the oxidation of succinate to fumarate in the TCA cycle with the simultaneously reduction of ubiquinone to ubiquinol in the ETC. A decade ago, mutations in SDHB, SDHC, and SDHD subunits were identified in patients with hereditary paragangliomas (hPGLs) and pheochromocytomas (PCCs), a rare neuroendocrine neoplasm of the chromaffin tissue of the adrenal medulla or derived from the parasympathetic tissue of the head and neck paraganglioma, respectively [9–12]. More recently, mutations in SDHA and the SDH assembly factor 2 (SDHAF2), required for flavination of SDH [13, 14], have been associated with hPGL/PCC syndrome [15]. The genetic defects in the SDH genes predisposing to the hPGL as well as PCC are heterozygous germline mutations, inducing the inactivation of the protein and the neoplastic transformation develops as result of loss of heterozygosity, caused by the complete loss of the enzyme function by a second mutagenic hit (usually deletion) [16]. In addition to hPGL and PCC, a number of other neoplasms have been associated with mutations in SDH genes, including gastrointestinal stromal tumors, renal cell cancers, thyroid tumors, neuroblastomas, and testicular seminoma [8].

3.2. Fumarate Hydratase. FH is homotetrameric TCA cycle enzyme which catalyzes the stereospecific and reversible hydration of fumarate to L-malate. Homozygous FH deficiencies result in fumaric aciduria [17], characterized by early onset of severe encephalopathy and psychomotor retardation; on the contrary, heterozygous FH mutations predispose to multiple cutaneous and uterine leiomyomas (MCUL), as well as to hereditary leiomyomatosis and renal cell cancer (HLRCC) [18, 19]. In particular, the kidney tumors in HLRCC, whose morphological spectrum include papillary type II, tubulopapilar, tubular, collecting duct, and clear cell carcinoma, are particularly aggressive. Growing evidence suggests that *FH* mutations may also be involved in the pathogenesis of breast, bladder, as well as Leydig cell tumors [20, 21]. The most common types of tumor predisposing genetic defects are missense mutations (57%), followed by frameshift and nonsense mutations (27%), as well as large-scale deletions, insertions, and duplications [22]. Like SDH, enzymatic activity of FH is completely absent in HLRCC as result of the loss of the wild-type allele in the transformed cell.

3.3. Isocitrate Dehydrogenase. IDH is a member of the β-decarboxylating dehydrogenase family of enzymes and catalyzes the oxidative decarboxylation of isocitrate to produce 2-oxoglutarate (α-KG) and CO_2 in the TCA cycle. Nuclear genome encodes three isoforms of IDH: IDH1 and IDH2 are $NADP^+$-dependent homodimers, whereas IDH3 is a NAD^+-reliant heterotetrameric enzyme. Whereas IDH1 is found into cytoplasm and peroxisomes, IDH2 and IDH3 are exclusively localized into the mitochondrial matrix, and, although all three isoforms are able to decarboxylate isocitrate, IDH3 is the main form of IDH functioning in the TCA cycle under physiological conditions whereas IDH1 and IDH2 are mainly involved in the reductive glutamine metabolism, under hypoxia and ETC alterations [4, 5, 23]. Though it plays a central role in energy production, to date there have been no reports of cancer-associated mutations in any of the IDH3 subunits. Conversely, genomewide mutation analyses and high-throughput deep sequencing revealed the presence of mutations in either IDH1 or its mitochondrial counterpart IDH2 in 70% of grade II-III gliomas and secondary glioblastomas [24, 25]. Since these initial reports, mutations in IDH1 and IDH2 have been identified in 16-17% of patients with acute myeloid leukemia, in 20% of angioimmunoblastic T-cell lymphomas [26], and spotted in a variety of other malignancies at lower frequencies [27, 28] such as B-acute lymphoblastic leukemias, thyroid, colorectal, and prostate cancer [29, 30]. Unlike *SDH* and *FH* mutations in hPGL and HLRCC, respectively, IDH1 and IDH2 mutations are somatic and monoallelic. Moreover, whereas mutations in *SDH* and *FH* occur throughout the gene, the majority of IDH mutations identified in gliomas and AML are changes in the amino acid residues R132 in IDH1 and either R172 or R140 in IDH2 [31]. As result of these alterations, mutated IDHs are unable to efficiently catalyze the oxidative decarboxylation of isocitrate and acquire a neomorphic catalytic activity that allows a NADPH-dependent reduction of α-KG into the oncometabolite (R)-2-hydroxyglutaric acid $((R)$-2HG) [31, 32].

4. Mechanisms of Tumorigenesis Caused by the TCA Cycle Defects

The finding that many tumors arousing from mutations in both *SDH* and *FH* genes are characterized by hypoxic features has suggested that the activation of the hypoxia-inducible transcription factor-1α (HIF-1α) could play a supportive role in the tumorigenic processes induced by TCA cycle dysfunctions. Indeed, HIF-1α is known to coordinate the biochemical reprogramming of cancer cells aimed to sustain their growth and proliferation as well as tumor vascularization [33–35]. The causal link between TCA cycle dysfunction and HIF-1α activation was initially suggested by Selak and coworkers demonstrating that the accumulation of succinate in SDH-deficient cells causes the inhibition of prolyl 4-hydroxylases (PHDs), a negative regulators of the stability of the α subunit of HIF [36]. The PHDs are members of the superfamily of α-KG-dependent hydroxylases, which couple the hydroxylation of the substrates with the oxidation of α-KG to succinate in reactions that are dependent on O_2 and Fe^{2+} [37]. In normoxic conditions, PHDs hydroxylate two proline residues in the oxygen-dependent degradation domain of HIF-1α, allowing it to be polyubiquitinated and degraded *via* proteasome. The accumulated succinate in SDH-deficient or SDH-inactive cells impairs PHDs activity leading to HIF-1α stabilization under normoxic conditions (pseudohypoxia) [36]. Similarly to succinate, also fumarate, which accumulates in tumors harboring loss of FH function, has been demonstrated to be potent inhibitors of PHDs [38]. Interestingly, fumarate-mediated stabilization of HIF was observed to induce the upregulation of several HIF-target genes, including those that stimulate cell growth and angiogenesis, allowing to hypothesize pseudohypoxia response as a plausible mechanism for HLRCC onset [38]. Despite that this large body of evidence showed a direct link between HIF-1α expression and tumorigenesis, recent findings have raised some questions about the protumorigenic role of pseudohypoxic adaptation in all types of tumors arousing from TCA cycle defects. The first question was raised from the study of Adam and colleagues. They demonstrated that neither the presence of HIF nor the absence of PHDs is required for hyperplastic renal cysts formation (typical hallmark of HLRCC) in a kidney-specific *Fh1* (the ortholog of human FH) knockout mice that recapitulates many features of the human disease [39], suggesting that alternative oncogenic actions of fumarate could be responsible for HLRCC generation (see next paragraph). In addition to this study, depicting HIF as a sort of "bystander player" in the onset of tumors harboring FH mutations, another report indicated this transcription factor as a tumor-suppressor

protein in tumors carrying IDH1/2 mutations. Indeed, as demonstrated by Koivunen and colleagues, contrarily to succinate and fumarate, (R)-2HG stimulates PHDs activity, driving, in such a way, HIF-1α for proteasome-mediated degradation [40]. Moreover, they pointed out that HIF-1α downregulation enhances the proliferation of human astrocytes and promotes their transformation, providing a justification for exploring PHDs inhibition as a potential treatment strategy for tumors harboring IDH1/2 mutations [40].

As member of the α-KG-dependent hydroxylases, PHDs catalyze the hydroxylation of a wide range of substrates, besides HIF-1α [37]. Therefore, the reduced hydroxylation of PHD targets may contribute to tumorigenesis regardless of HIF-1α activity and the acquisition of a hypoxic signature. For instance, it has been proposed that SDH deficiency could impair PHD-dependent programmed cell death of neurons, therefore setting the stage for neoplastic transformation of neuronal cells. This hypothesis finds support in the recent studies demonstrating that the proapoptotic activity of the prolyl hydroxylase EglN3 requires a functional SDH, being feedback inhibited by succinate [41, 42]. Since EglN3 is required during development to allow the programmed cell death of some sympathetic neuronal precursor cells, its inhibition, elicited by the elevation of succinate levels, could play a role in the pathogenesis of tumors arousing from a defective developmental apoptosis, such as pheochromocytomas.

Highlighting the HIF-1α-independent tumorigenic mechanisms, growing body of evidence clearly places the alteration of TCA flux upstream of the epigenetic dynamics as well. Histone methylation is an important epigenetic modification which has been demonstrated to regulate gene expression by modifying chromatin structure and, thereby, fine-tuning the binding of transcription factors [43, 44]. One of the most studied enzymes regulating histone methylation signature are the Jumonji C-terminal domain (JmjC) family of histone demethylases [45]. As they remove the methyl groups on the arginine and lysine residues of histones after performing an α-KG- and oxygen-dependent hydroxylation, they have been included in the α-KG-dependent hydroxylases family. It was shown that succinate accumulation, in SDH-deficient cells, negatively affects the activity of many members of such class of histone demethylase. For instance, succinate-mediated JMJD3 inhibition leads to changes in the methylation mark of histone H3 on arginine [46]. Furthermore, in a yeast model of paraganglioma, the histone demethylase, Jhd1, was found to be inhibited by succinate accumulation in an α-KG-competitive manner [47]. Similarly, recent studies demonstrate that, besides SDH alterations, also IDH1/2 defects are associated with hypermethylated phenotype. Indeed, in cells harboring *IDH1/2* mutations, intracellular (R)-2-HG levels can reach the value of 10 mM. These concentrations promote the competitive inhibition of the α-KG-dependent histone N^ε-lysine demethylase JMJD2A, and the ten-eleven translocation (TET) family of 5-methylcytosine (5mC) hydroxylases, a class of protein mediating the α-KG-dependent removal of methyl mark from 5-methylcytosines, resulting in an enhanced histone and DNA methylation, respectively [48, 49]. Interestingly, although fumarate is able to inhibit HIF-regulating PHDs similarly to succinate, no evidence attesting its putative capability to mirror its cognate metabolite succinate in affecting histone methylation has been documented so far. In addition, as TET enzymes are members of the α-KG-dependent hydroxylases family, a putative ability of both succinate and fumarate in their inhibition can be reasonably argued. On the basis of the ability of the epigenetic alterations to affect lineage-specific differentiation and to result in the activation of oncogenes or silencing of tumor suppressors [44, 50, 51], the competitive inhibition of histone and DNA demethylases elicited by defects in fluxes of TCA cycle metabolites may drive tumorigenesis by promoting cell transformation and uncontrolled proliferation.

5. Redox-Dependent Tumorigenic Alterations Elicited by the TCA Cycle Defects

Apart from the mere metabolic viewpoint, compelling evidence suggests that the reactive oxygen species (ROS), produced by a deregulated mitochondrial functioning, might trigger the oncogenic signal or, at least, participate in the progression of tumors characterized by defects in the TCA cycle enzymes (Figure 1). This assumption finds support in the observation that, compared with their normal counterparts, many types of cancer cells have increased levels of ROS generated by a defective mitochondrial electron-transport chain [52–54]. By exploiting their chemical reactiveness with biomolecules, such as nucleic acids, ROS are known to induce several types of DNA damages, including depurination and depyrimidination, single- and double-stranded DNA breaks, base and sugar modifications, and DNA-protein crosslinks. In such a way, permanent modifications of DNA, resulting from sustained prooxidant conditions, drive the mutagenic events underlying carcinogenesis.

The observation that specific SDHC mutant (*mev-1*) of the *C. elegans* nematode was able to generate superoxide $O_2^{\bullet}$ [55, 56] suggested the possibility that ROS could have a causal role in the pathogenesis of tumors bearing defects in the TCA cycle. This hypothesis was further strengthened by the evidence that mouse fibroblasts transfected with a murine equivalent of the *mev-1* mutant were featured by a sustained ROS production and a significantly higher DNA mutation frequency than wild-type counterparts [57]. Although these lines of evidence supported the mutagenic role of ROS generated by defective SDH complex, no detectable DNA damages, despite an increased production of ROS and protein oxidation, was described in a *S. cerevisiae* strain lacking Sdh2 (the yeast ortholog of mammalian SDHB) [47]. To link the prooxidant state elicited by SDH dysfunctions to tumorigenesis, Guzy and colleagues proposed that the ROS could play a supportive role in the oncogenic process by contributing to the activation of HIF-1α [58]. Indeed, relying on a previously characterized role of respiratory chain-derived ROS as signals for HIF-1α stabilization under hypoxia [59, 60], it has been shown that cells expressing mutant SDHB, but not mutant SDHA, are

characterized by significant mitochondrial ROS production required, together with succinate, for a complete inactivation of PHDs and HIF-1α stabilization [58]. Therefore, these results reinforce the role of ROS as amplifier of the pseudo-hypoxic response, observed in all cells carrying SDH defects, providing a biochemical rationale for the severity of SDHB mutations which are usually associated with aggressive PCC.

The capabilities of the TCA cycle defects in the tuning of cellular redox state have been supported by the evidence that also oncogenic mutations in IDH1/2 genes are associated with the oxidation of intracellular milieu. Normally, in aerobic organisms, the control of cellular redox state is ensured by the balance between the prooxidant species, mainly produced by mitochondria, NADPH oxidases or as byproduct of the intermediate metabolism, and their clearance through the synergistic action of the antioxidant enzymes and the thiol-containing antioxidants. Among the latter, the tripeptide glutathione (GSH) plays a pivotal role in determining the steady-state value of the intracellular redox potential. Indeed, its intracellular abundance (1–10 mM) allows GSH to participate, as electron donor, in the enzymatic reduction of hydrogen peroxide and lipid peroxides and in the generation of reversible S-glutathionylated adducts with protein thiols, preventing them to undergo irreversible forms of oxidation [61]. The capability of IDH mutations to induce oxidative intracellular conditions is linked to a decrease in GSH levels, observed both in IDH1-R132H—and IDH2-R172K—expressing glioma cells with respect to their *wt* counterparts [62]. GSH is synthesized in two ATP-dependent steps: (i) synthesis of γ-glutamylcysteine, from L-glutamate and cysteine *via* the rate-limiting enzyme glutamate-cysteine ligase (GCL); (ii) addition of glycine to the C-terminal of γ-glutamylcysteine *via* the enzyme glutathione synthetase. Intracellular glutamate, required for the first reaction of GSH biosynthesis, is mainly produced by the oxidative deamination of glutamine catalyzed by the enzyme glutaminase [63]. As IDH1/2 mutant cells are characterized by lower levels of glutamate with respect to their *wt* matching parts [62], it is possible that oncogenic defects in IDH result in impaired GSH synthesis due to a lower glutamate availability, thus phenocopying the prooxidant conditions observed in glutaminase deficient cells [64]. The dampened glutamate levels could be the result of an enhanced α-KG demand of IDH1/2 mutant cells allowing the biosynthesis of the oncometabolite (R)-2-HG. This assumption is supported by the evidence that treatment of glioma cells with (R)-2-HG does not deplete neither glutamate nor glutathione levels [62], suggesting that many metabolic changes observed in IDH-mutated cells are not due to the direct action of (R)-2-HG but a consequence of its oncogenic production. The involvement of IDH1/2 mutations in the generation of prooxidant conditions is not only related to the alteration of intracellular GSH content. Indeed, the oxidative decarboxylation of isocitrate, which is impaired in all mutants of IDH1 and IDH2 proteins, is coupled to a reduced ability to generate NADPH. Moreover, the failure to sustain intracellular NADPH production is associated with an increased NADPH oxidation, necessary to allow the reductive biosynthesis of (R)-2-HG [31, 32, 65]. As GSH and the thiol-based antioxidant protein thioredoxin require NADPH as a source of reducing equivalents for their own regeneration [61], the altered equilibrium of NADP$^+$/NADPH elicited by IDH1/2 mutations could contribute to the shift of the intracellular redox state towards more oxidizing conditions. Although, these lines of evidence bring about the ability of mutant IDH1/2 to elicit prooxidant conditions independently on the direct action of (R)-2-HG on human redox metabolome, it has been proposed that this oncometabolite could contribute itself to oxidize intracellular environment. Indeed, some reports demonstrate its ability to induce oxidative damages in cerebral cortex of young rats [66] and to elicit ROS generation through the stimulation of NMDA receptor [67]. Although these findings support prooxidant capability of (R)-2-HG, to date no striking evidence has been provided attesting its mutagenic role.

Whereas accumulating pieces of evidence support the capability of oncogenic mutations in SDH as well as IDH genes to oxidize intracellular milieu, conflicting findings do not allow for defining a clear role of FH deficiency in cellular redox state modulation. The most convincing evidence showing the capability of FH-deficient cells to promote intracellular ROS accumulation comes from the work of Sudarshan and colleagues [68]. This study demonstrated that inactivating mutations of *FH* in an HLRCC-derived cell line result in glucose-induced NADPH oxidases-mediated generation of $O_2^{\bullet-}$ and ROS-dependent HIF-1α stabilization. On the contrary, O'Flaherty and colleagues provided clear evidence that accumulation of fumarate, due to the absence of a functional FH, is the sole mechanism responsible for the inhibition of HIF-1α prolyl hydroxylation, independently on defect in mitochondrial oxidative metabolism [69]. Indeed, the complete correction of HIF-1α pathway activation in Fh1$^{-/-}$ MEFs by extra-mitochondrial FH expression suggests that, at least in tumors harboring FH defects, neither impaired mitochondrial function nor the consequent dependence of energy metabolism on glycolysis contributes significantly to HIF-1α engagement. The most substantial pieces of evidence, depicting the elevation of fumarate levels as a condition linked to the reduction of intracellular redox state, came from two recent studies demonstrating that FH loss results in the activation of nuclear factor erythroid 2-related factor 2 (Nrf2) [39, 70], the pivotal transcription factor responsible for the induction of the antioxidant-responsive-element- (ARE-) driven genes, which codify for phase II detoxification enzymes and antioxidant proteins such as glutathione S-transferases and GCL [71]. Both studies demonstrated that reconstitution of FH-deficient cells with wild-type FH or an extra-mitochondrial FH decreased fumarate levels and restored Nrf2 regulation [39, 70]. In addition, elevation of intracellular fumarate content by a membrane-permeable fumarate ester was found sufficient to induce Nrf2 and its orchestrated antioxidant program [70]. According to the current view, in resting conditions, Nrf2 is retained in the cytoplasm through its interaction with Keap1 which prevents its nuclear translocation and rules its ubiquitin-proteasome-mediated turnover, as well. However, in the presence of electrophiles as well as during redox unbalance, Keap1 is modified at several reactive cysteine residues, resulting in Nrf2 stabilization and the activation of the protective gene expression program [71, 72]. In line with this accepted model, both groups revealed by mass spectroscopy analyses that fumarate was able to succinate several cysteine residues previously shown to be electrophile targets, including Cys^{151} and Cys^{288}, thereby providing a mechanistic explanation of the fumarate-induced Nrf2 activation [39, 70]. Although ROS can promote carcinogenesis by inducing oxidative damages to DNA, a recent outstanding study demonstrates that oncogene-induced Nrf2 activation promotes tumorigenesis by lowering ROS levels and conferring a more reduced intracellular environment [73]. Therefore, on the basis of these evidence, it is possible to hypothesize that the fumarate-mediated activation of the Nrf2-antioxidant pathway might drive the oncogenic signal for tumors characterized by defects in the FH enzyme. Although this assumption has not been demonstrated yet, the observation that heme oxygenase 1, one of the best defined target genes of Nrf2, is upregulated in FH-deficient cells allowing their survival [74] supports the putative causal role of Keap1 succination in the onset of tumors carrying FH defects. Furthermore, mounting bodies of evidence show that Nrf2 and its downstream genes are overexpressed in many cancer cell lines and human cancer tissues conferring them advantage for survival and growth as well as acquired chemoresistance [75, 76]. Therefore, it is possible to speculate that besides driving renal tumorigenesis, fumarate-induced succination of Nrf2 could contribute to the reduced sensitivity of particularly aggressive and recurrent forms of kidney cancer, such as HLRCC [77], to many chemotherapeutic approaches. The enhanced activation of Nrf2 observed both by Pollard and Furge groups contributes to explain the results obtained by Raimundo and coworkers in nontumor cells [78]. Indeed, they documented that FH-deficient diploid human fibroblasts are characterized by a highly reduced redox state with increased GSH levels, as result of increased expression of the GSH biosynthetic enzyme GCL. As highly reducing environment has been shown to stimulate cell proliferation [79], it is possible to hypothesize that the reduced redox state elicited by FH mutations could favor the doublings of stem-cell-like populations promoting thus the initial event of tumor formation. This assumption finds support in the observation that lower ROS levels have been found in many cancer stem cells with respect to the nontumorigenic counterparts, allowing them to maintain a high proliferative status and to prevent their differentiation [80].

6. Concluding Remarks

The direct involvement of TCA cycle enzymes in tumor formation has been arousing from a decade. In tumors associated with defects of SDH, FH, and IDH enzymes, the underlying mechanisms of tumorigenesis involve the accumulation of metabolites (succinate, fumarate, and (R)-2-HG) that convey oncogenic signals (oncometabolites). Large amount of evidence points towards the generation of pseudohypoxic phenotype and the alteration of epigenetic homeostasis as the main cancer-promoting effects of the TCA cycle affecting mutations. Besides inhibiting the α-KG-dependent hydroxylases, mounting body of evidence supports the ability of these oncometabolites to alter cellular redox state in precancerous as well as transformed cells. Therefore, alternatively or concomitantly to the generation of pseudohypoxic phenotype and the alteration of epigenetic dynamics, the oncometabolites-induced engagement of redox-dependent signaling pathways could contribute both to the neoplastic transformation of healthy cells as well as to the progression of malignancies characterized by germline mutations in SDH and FH and of somatic defects in IDH. These emerging findings reveal a dynamic interaction between the genetic profile, the metabolic status, and the redox tuning of the cell. Moreover, the different impact of oncogenic mutations of the TCA cycle on cellular redox state could contribute to explain the differences in the clinical phenotype and outcome of their associated tumors, opening new perspectives in the comprehension of the molecular mechanisms of oncogenesis and therapeutic targeting of these neoplastic alterations.

Acknowledgments

This work was partially supported by Grants from AIRC (no. IG 10636) and from Ministero dell'Università e della Ricerca (MIUR).

References

[1] G. Kroemer and J. Pouyssegur, "Tumor cell metabolism: Cancer's Achilles' heel," *Cancer Cell*, vol. 13, no. 6, pp. 472–482, 2008.

[2] O. Warburg, "On the origin of cancer cells," *Science*, vol. 123, no. 3191, pp. 309–314, 1956.

[3] I. E. Scheffler, *Mitochondria*, John Wiley & Sons, 2nd edition, 2008.

[4] D. R. Wise, P. S. Ward, J. E. Shay et al., "Hypoxia promotes isocytrate dehydrogenase-dependent carboxylation of α-ketoglutarate to citrate to support cell growth and viability," *Proceedings of the National Academy of Sciences USA*, vol. 108, no. 49, 1961119616 pages, 1961.

[5] C. M. Metallo, P. A. Gameiro, E. L. Bell et al., "Reductive glutamine metabolism by IDH1 mediates lipogenesis under hypoxia," *Nature*, vol. 481, no. 7381, pp. 380–384, 2011.

[6] A. R. Mullen, W. W. Wheaton, E. S. Jin et al., "Reductive carboxylation supports growth in tumour cells with defective mitochondria," *Nature*, vol. 481, no. 7381, pp. 385–388, 2011.

[7] D. C. Wallace, W. Fan, and V. Procaccio, "Mitochondrial energetics and therapeutics," *Annual Review of Pathology*, vol. 5, pp. 297–348, 2010.

[8] C. Bardella, P. J. Pollard, and I. Tomlinson, "SDH mutations in cancer," *Biochimica et Biophysica Acta*, vol. 1807, no. 11, pp. 1432–1443, 2011.

[9] B. E. Baysal, R. E. Ferrell, J. E. Willett-Brozick et al., "Mutations in SDHD, a mitochondrial complex II gene, in hereditary paraganglioma," *Science*, vol. 287, no. 5454, pp. 848–851, 2000.

[10] S. Niemann and U. Müller, "Mutations in SDHC cause autosomal dominant paraganglioma, type 3," *Nature Genetics*, vol. 26, no. 3, pp. 268–270, 2000.

[11] D. Astuti, F. Latif, A. Dallol et al., "Gene mutations in the succinate dehydrogenase subunit SDHB cause susceptibility to familial pheochromocytoma and to familial paraganglioma," *American Journal of Human Genetics*, vol. 69, no. 1, pp. 49–54, 2001.

[12] B. E. Baysal, J. E. Willett-Brozick, E. C. Lawrence et al., "Prevalence of SDHB, SDHC, and SDHD germline mutations in clinic patients with head and neck paragangliomas," *Journal of Medical Genetics*, vol. 39, no. 3, pp. 178–183, 2002.

[13] H. X. Hao, O. Khalimonchuk, M. Schraders et al., "SDH5, a gene required for flavination of succinate dehydrogenase, is mutated in paraganglioma," *Science*, vol. 325, no. 5944, pp. 1139–1142, 2009.

[14] J. P. Bayley, H. P. M. Kunst, A. Cascon et al., "SDHAF2 mutations in familial and sporadic paraganglioma and phaeochromocytoma," *The Lancet Oncology*, vol. 11, no. 4, pp. 366–372, 2010.

[15] N. Burnichon, J. J. Brière, R. Libé et al., "SDHA is a tumor suppressor gene causing paraganglioma," *Human Molecular Genetics*, vol. 19, no. 15, pp. 3011–3020, 2010.

[16] E. Gottlieb and I. P. M. Tomlinson, "Mitochondrial tumour suppressors: a genetic and biochemical update," *Nature Reviews Cancer*, vol. 5, no. 11, pp. 857–866, 2005.

[17] A. B. Zinn, D. S. Kerr, and C. L. Hoppel, "Fumarase deficiency: a new cause of mitochondrial encephalomyopathy," *The New England Journal of Medicine*, vol. 315, no. 8, pp. 469–475, 1986.

[18] V. Launonen, O. Vierimaa, M. Kiuru et al., "Inherited susceptibility to uterine leiomyomas and renal cell cancer," *Proceedings of the National Academy of Sciences of the United States of America*, vol. 98, no. 6, pp. 3387–3392, 2001.

[19] I. P. M. Tomlinson, N. A. Alam, A. J. Rowan et al., "Germline mutations in FH predispose to dominantly inherited uterine fibroids, skin leiomyomata and papillary renal cell cancer the multiple leiomyoma consortium," *Nature Genetics*, vol. 30, no. 4, pp. 406–410, 2002.

[20] L. G. Carvajal-Carmona, N. A. Alam, P. J. Pollard et al., "Adult leydig cell tumors of the testis caused by germline fumarate hydratase mutations," *The Journal of Clinical Endocrinology and Metabolism*, vol. 91, no. 8, pp. 3071–3075, 2006.

[21] H. J. Lehtonen, M. Kiuru, S. K. Ylisaukko-Oja et al., "Increased risk of cancer in patients with fumarate hydratase germline mutation," *Journal of Medical Genetics*, vol. 43, no. 6, pp. 523–526, 2006.

[22] A. King, M. A. Selak, and E. Gottlieb, "Succinate dehydrogenase and fumarate hydratase: linking mitochondrial dysfunction and cancer," *Oncogene*, vol. 25, no. 34, pp. 4675–4682, 2006.

[23] L. Dang, S. Jin, and S. M. Su, "IDH mutations in glioma and acute myeloid leukemia," *Trends in Molecular Medicine*, vol. 16, no. 9, pp. 392–397, 2010.

[24] D. W. Parsons, S. Jones, X. Zhang et al., "An integrated genomic analysis of human glioblastoma multiforme," *Science*, vol. 321, no. 5897, pp. 1807–1812, 2008.

[25] H. Yan, D. W. Parsons, G. Jin et al., "IDH1 and IDH2 mutations in gliomas," *The New England Journal of Medicine*, vol. 360, no. 8, pp. 765–773, 2009.

[26] R. A. Cairns, J. Iqbal, F. Lemonnier et al., "IDH2 mutations are frequent in angioimmunoblastic T-cell lymphoma," *Blood*, vol. 119, no. 8, pp. 1901–1903, 2012.

[27] S. Abbas, S. Lugthart, F. G. Kavelaars et al., "Acquired mutations in the genes encoding IDH1 and IDH2 both are recurrent aberrations in acute myeloid leukemia: prevalence and prognostic value," *Blood*, vol. 116, no. 12, pp. 2122–2126, 2010.

[28] P. Paschka, R. F. Schlenk, V. I. Gaidzik et al., "IDH1 and IDH2 mutations are frequent genetic alterations in acute myeloid leukemia and confer adverse prognosis in cytogenetically normal acute myeloid leukemia with NPM1 mutation without FLT3 internal tandem duplication," *Journal of Clinical Oncology*, vol. 28, no. 22, pp. 3636–3643, 2010.

[29] M. R. Kang, M. S. Kim, J. E. Oh et al., "Mutational analysis of IDH1 codon 132 in glioblastomas and other common cancers," *International Journal of Cancer*, vol. 125, no. 2, pp. 353–355, 2009.

[30] K. E. Yen, M. A. Bittinger, S. M. Su, and V. R. Fantin, "Cancer-associated IDH mutations: biomarker and therapeutic opportunities," *Oncogene*, vol. 29, no. 49, pp. 6409–6417, 2010.

[31] L. Dang, D. W. White, S. Gross et al., "Cancer-associated IDH1 mutations produce 2-hydroxyglutarate," *Nature*, vol. 462, no. 7274, pp. 739–744, 2009.

[32] P. S. Ward, J. Patel, D. R. Wise et al., "The common feature of leukemia-associated IDH1 and IDH2 mutations is a neomorphic enzyme activity converting α-ketoglutarate to 2-hydroxyglutarate," *Cancer Cell*, vol. 17, no. 3, pp. 225–234, 2010.

[33] A. P. Gimenez-Roqueplo, J. Favier, P. Rustin et al., "The R22X mutation of the SDHD gene in hereditary paraganglioma abolishes the enzymatic activity of complex II in the mitochondrial respiratory chain and activates the hypoxia pathway," *American Journal of Human Genetics*, vol. 69, no. 6, pp. 1186–1197, 2001.

[34] P. L. M. Dahia, K. N. Ross, M. E. Wright et al., "A HIf1α regulatory loop links hypoxia and mitochondrial signals in pheochromocytomas," *PLoS Genetics*, vol. 1, no. 1, pp. 72–80, 2005.

[35] S. Vanharanta, P. J. Pollard, H. J. Lehtonen et al., "Distinct expression profile in fumarate-hydratase-deficient uterine fibroids," *Human Molecular Genetics*, vol. 15, no. 1, pp. 97–103, 2006.

[36] M. A. Selak, S. M. Armour, E. D. MacKenzie et al., "Succinate links TCA cycle dysfunction to oncogenesis by inhibiting HIF-α prolyl hydroxylase," *Cancer Cell*, vol. 7, no. 1, pp. 77–85, 2005.

[37] A. Ozer and R. K. Bruick, "Non-heme dioxygenases: cellular sensors and regulators jelly rolled into one?" *Nature Chemical Biology*, vol. 3, no. 3, pp. 144–153, 2007.

[38] J. S. Isaacs, J. J. Yun, D. R. Mole et al., "HIF overexpression correlates with biallelic loss of fumarate hydratase in renal cancer: novel role of fumarate in regulation of HIF stability," *Cancer Cell*, vol. 8, no. 2, pp. 143–153, 2005.

[39] J. Adam, E. Hatipoglu, L. O'Flaherty et al., "Renal cyst formation in Fh1-deficient mice is independent of the Hif/Phd pathway: roles for fumarate in KEAP1 succination and Nrf2 signaling," *Cancer Cell*, vol. 20, no. 4, pp. 524–537, 2011.

[40] P. Koivunen, S. Lee, C. G. Duncan, C. G. Kaelin Jr. et al., "Transformation by the (R)-enantiomer of 2-hydroxyglutarate linked to EGLN activation," *Nature*, vol. 483, pp. 484–488, 2012.

[41] S. Lee, E. Nakamura, H. Yang et al., "Neuronal apoptosis linked to EglN3 prolyl hydroxylase and familial pheochromocytoma genes: developmental culling and cancer," *Cancer Cell*, vol. 8, no. 2, pp. 155–167, 2005.

[42] S. Schlisio, R. S. Kenchappa, L. C. W. Vredeveld et al., "The kinesin KIF1Bβ acts downstream from EglN3 to induce apoptosis and is a potential 1p36 tumor suppressor," *Genes and Development*, vol. 22, no. 7, pp. 884–893, 2008.

[43] P. Chi, C. D. Allis, and G. G. Wang, "Covalent histone modifications-miswritten, misinterpreted and mis-erased in human cancers," *Nature Reviews Cancer*, vol. 10, no. 7, pp. 457–469, 2010.

[44] E. Caffarelli and P. Filetici, "Epigenetic regulation in cancer development," *Frontiers in Bioscience*, vol. 1, no. 17, pp. 2682–2694, 2011.

[45] H. Hou and H. Yu, "Structural insights into histone lysine demethylation," *Current Opinion in Structural Biology*, vol. 20, no. 6, pp. 739–748, 2010.

[46] A. M. Cervera, J. P. Bayley, P. Devilee, and K. J. McCreath, "Inhibition of succinate dehydrogenase dysregulates histone modification in mammalian cells," *Molecular Cancer*, vol. 8, article 89, 2009.

[47] E. H. Smith, R. Janknecht, and J. L. Maher III, "Succinate inhibition of α-ketoglutarate-dependent enzymes in a yeast model of paraganglioma," *Human Molecular Genetics*, vol. 16, no. 24, pp. 3136–3148, 2007.

[48] R. Chowdhury, K. K. Yeoh, Y. M. Tian et al., "The oncometabolite 2-hydroxyglutarate inhibits histone lysine demethylases," *EMBO Reports*, vol. 12, no. 5, pp. 463–469, 2011.

[49] M. E. Figueroa, O. Abdel-Wahab, C. Lu et al., "Leukemic IDH1 and IDH2 mutations result in a hypermethylation phenotype, disrupt TET2 function, and impair hematopoietic differentiation," *Cancer Cell*, vol. 18, no. 6, pp. 553–567, 2010.

[50] J. Füllgrabe, E. Kavanagh, and B. Joseph, "Histone onco-modifications," *Oncogene*, vol. 30, no. 31, pp. 3391–3403, 2011.

[51] P. Sarkies and J. E. Sale, "Cellular epigenetic stability and cancer," *Trends in Genetics*, vol. 28, no. 3, pp. 118–127, 2012.

[52] T. P. Szatrowski and C. F. Nathan, "Production of large amounts of hydrogen peroxide by human tumor cells," *Cancer Research*, vol. 51, no. 3, pp. 794–798, 1991.

[53] S. Toyokuni, "Persistent oxidative stress in cancer," *FEBS Letters*, vol. 358, no. 1, pp. 1–3, 1995.

[54] S. Kawanishi, Y. Hiraku, S. Pinlaor, and N. Ma, "Oxidative and nitrative DNA damage in animals and patients with inflammatory diseases in relation to inflammation-related carcinogenesis," *Biological Chemistry*, vol. 387, no. 4, pp. 365–372, 2006.

[55] N. Ishii, M. Fujii, P. S. Hartman et al., "A mutation in succinate dehydrogenase cytochrome b causes oxidative stress and ageing in nematodes," *Nature*, vol. 394, no. 6694, pp. 694–697, 1998.

[56] N. Senoo-Matsuda, K. Yasuda, M. Tsuda et al., "A defect in the cytochrome b large subunit in complex II causes both superoxide anion overproduction and abnormal energy metabolism in caenorhabditis elegans," *The Journal of Biological Chemistry*, vol. 276, no. 45, pp. 41553–41558, 2001.

[57] T. Ishii, K. Yasuda, A. Akatsuka, O. Hino, P. S. Hartman, and N. Ishii, "A mutation in the SDHC gene of complex II increases oxidative stress, resulting in apoptosis and tumorigenesis," *Cancer Research*, vol. 65, no. 1, pp. 203–209, 2005.

[58] R. D. Guzy, B. Sharma, E. Bell, N. S. Chandel, and P. T. Schumacker, "Loss of the SdhB, but not the SdhA, subunit of complex II triggers reactive oxygen species-dependent hypoxia-inducible factor activation and tumorigenesis," *Molecular and Cellular Biology*, vol. 28, no. 2, pp. 718–731, 2008.

[59] N. S. Chandel, E. Maltepe, E. Goldwasser, C. E. Mathieu, M. C. Simon, and P. T. Schumacker, "Mitochondrial reactive oxygen species trigger hypoxia-induced transcription," *Proceedings of the National Academy of Sciences of the United States of America*, vol. 95, no. 20, pp. 11715–11720, 1998.

[60] N. S. Chandel, D. S. McClintock, C. E. Feliciano et al., "Reactive oxygen species generated at mitochondrial Complex III stabilize hypoxia-inducible factor-1α during hypoxia: a mechanism of O₂ sensing," *Journal of Biological Chemistry*, vol. 275, no. 33, pp. 25130–25138, 2000.

[61] G. Filomeni, G. Rotilio, and M. R. Ciriolo, "Disulfide relays and phosphorylative cascades: partners in redox-mediated signaling pathways," *Cell Death and Differentiation*, vol. 12, no. 12, pp. 1555–1563, 2005.

[62] Z. J. Reitman, G. Jin, E. D. Karoly et al., "Profiling the effects of isocitrate dehydrogenase 1 and 2 mutations on the cellular metabolome," *Proceedings of the National Academy of Sciences of the United States of America*, vol. 108, no. 8, pp. 3270–3275, 2011.

[63] J. M. Matés, J. A. Segura, J. A. Campos-Sandoval et al., "Glutamine homeostasis and mitochondrial dynamics," *International Journal of Biochemistry and Cell Biology*, vol. 41, no. 10, pp. 2051–2061, 2009.

[64] W. Hu, C. Zhang, R. Wu, Y. Sun, A. Levine, and Z. Feng, "Glutaminase 2, a novel p53 target gene regulating energy metabolism and antioxidant function," *Proceedings of the National Academy of Sciences of the United States of America*, vol. 107, no. 16, pp. 7455–7460, 2010.

[65] S. Zhao, Y. Lin, W. Xu et al., "Glioma-derived mutations in IDH1 dominantly inhibit IDH1 catalytic activity and induce HIF-1α," *Science*, vol. 324, no. 5924, pp. 261–265, 2009.

[66] A. Latini, K. Scussiato, R. B. Rosa et al., "D-2-hydroxyglutaric acid induces oxidative stress in cerebral cortex of young rats," *European Journal of Neuroscience*, vol. 17, no. 10, pp. 2017–2022, 2003.

[67] S. Kölker, V. Pawlak, B. Ahlemeyer et al., "NMDA receptor activation and respiratory chain complex V inhibition contribute to neurodegeneration in D-2-hydroxyglutaric aciduria," *European Journal of Neuroscience*, vol. 16, no. 1, pp. 21–28, 2002.

[68] S. Sudarshan, C. Sourbier, H. S. Kong et al., "Fumarate hydratase deficiency in renal cancer induces glycolytic addiction and hypoxia-inducible transcription factor 1α stabilization by glucose-dependent generation of reactive oxygen species," *Molecular and Cellular Biology*, vol. 29, no. 15, pp. 4080–4090, 2009.

[69] L. O'Flaherty, J. Adam, L. C. Heather et al., "Dysregulation of hypoxia pathways in fumarate hydratase-deficient cells is independent of defective mitochondrial metabolism," *Human Molecular Genetics*, vol. 19, no. 19, pp. 3844–3851, 2010.

[70] A. Ooi, J. C. Wong, D. Petillo et al., "An antioxidant response phenotype shared between hereditary and sporadic type 2 papillary renal cell carcinoma," *Cancer Cell*, vol. 20, no. 4, pp. 511–523, 2011.

[71] N. F. Villeneuve, A. Lau, and D. D. Zhang, "Regulation of the Nrf2-keap1 antioxidant response by the ubiquitin proteasome system: an insight into cullin-ring ubiquitin ligases," *Antioxidants and Redox Signaling*, vol. 13, no. 11, pp. 1699–1712, 2010.

[72] J. D. Hayes and M. McMahon, "NRF2 and KEAP1 mutations: permanent activation of an adaptive response in cancer," *Trends in Biochemical Sciences*, vol. 34, no. 4, pp. 176–188, 2009.

[73] G. M. Denicola, F. A. Karreth, T. J. Humpton et al., "Oncogene-induced Nrf2 transcription promotes ROS detoxification and tumorigenesis," *Nature*, vol. 475, no. 7354, pp. 106–110, 2011.

[74] C. Frezza, L. Zheng, O. Folger et al., "Haem oxygenase is synthetically lethal with the tumour suppressor fumarate hydratase," *Nature*, vol. 477, no. 7363, pp. 225–228, 2011.

[75] A. Lau, N. F. Villeneuve, Z. Sun, P. K. Wong, and D. D. Zhang, "Dual roles of Nrf2 in cancer," *Pharmacological Research*, vol. 58, no. 5-6, pp. 262–270, 2008.

[76] Y. Inami, S. Waguri, A. Sakamoto et al., "Persistent activation of Nrf2 through p62 in hepatocellular carcinoma cells," *The Journal of Cell Biology*, vol. 193, no. 2, pp. 275–284, 2011.

[77] E. C. Pfaffenroth and W. M. Linehan, "Genetic basis for kidney cancer: opportunity for disease-specific approaches to therapy," *Expert Opinion on Biological Therapy*, vol. 8, no. 6, pp. 779–790, 2008.

[78] N. Raimundo, J. Ahtinen, K. Fumić et al., "Differential metabolic consequences of fumarate hydratase and respiratory chain defects," *Biochimica et Biophysica Acta*, vol. 1782, no. 5, pp. 287–294, 2008.

[79] F. Q. Schafer and G. R. Buettner, "Redox environment of the cell as viewed through the redox state of the glutathione disulfide/glutathione couple," *Free Radical Biology and Medicine*, vol. 30, no. 11, pp. 1191–1212, 2001.

[80] M. Diehn, R. W. Cho, N. A. Lobo et al., "Association of reactive oxygen species levels and radioresistance in cancer stem cells," *Nature*, vol. 458, no. 7239, pp. 780–783, 2009.

7

Zinc Protoporphyrin Upregulates Heme Oxygenase-1 in PC-3 Cells via the Stress Response Pathway

Simon C. M. Kwok

ORTD, Albert Einstein Medical Center, 5501 Old York Road, Korman 214, Philadelphia, PA 19141-3098, USA

Correspondence should be addressed to Simon C. M. Kwok; kwoks@einstein.edu

Academic Editor: Afshin Samali

Zinc protoporphyrin IX (ZnPP), a naturally occurring molecule formed in iron deficiency or lead poisoning, is a potent competitive inhibitor of heme oxygenase-1 (HO-1). It also regulates expression of HO-1 at the transcriptional level. However, the effect of ZnPP on HO-1 expression is controversial. It was shown to induce HO-1 expression in some cells, but suppress it in others. The objective of this study is to investigate the effect of ZnPP on HO-1 expression in prostate cancer PC-3 cells. Incubation of PC-3 cells with $10 \mu M$ ZnPP for 4 h showed only a slight induction of HO-1 mRNA and protein, but the induction was high after 16 h and was maintained through 48 h of incubation. Of all the known responsive elements in the HO-1 promoter, ZnPP activated mainly the stress response elements. Of the various protein kinase inhibitors and antioxidant tested, only Ro 31-8220 abrogated ZnPP-induced HO-1 expression, suggesting that activation of HO-1 gene by ZnPP may involve protein kinase C (PKC). The involvement of PKC α, β, δ, η, θ, and ζ isoforms was ruled out by the use of specific inhibitors. The isoform of PKC involved and participation of other transcription factors remain to be studied.

1. Introduction

Heme oxygenase-1 (HO-1), also known as heat shock protein 32 (Hsp32), is an inducible enzyme that catalyzes the breakdown of heme, producing carbon monoxide, iron, and biliverdin. It is known to be a cytoprotective enzyme against oxidative stress [1]. It is often upregulated in tumor tissues, and its inhibition is considered as a means of sensitizing the tumors to anticancer drugs [2]. Although HO-1 expression is increased in malignant prostate tissues [3], its expression in prostate cancer cell line, PC-3, is low [4]. Induction of HO-1 expression by hemin in PC-3 cells resulted in decreased cell proliferation and migration [4]. Overexpression of HO-1 also led to nuclear location [5] and was associated with downregulation of matrix metalloprotease 9 (MMP9), which plays an important role in tumor cell invasion and angiogenesis [4]. The real function of HO-1 in tumor cells remains to be studied.

HO-1 expression can be induced by many inducers, and many regulatory pathways have been proposed [6, 7]. A number of antioxidant response element (ARE)-like motifs are present in the promoter of HO-1 gene. Six of these sites were found as clusters at E1 (-3928 bp) and E2 (-9069 bp) regions of the human HO-1 promoter; they are termed StRE1 through StRE6 [6]. Besides these StRE sites, other response elements, such as HSE [8], SREBP binding site [9], and an intronic SP1 enhancer [10] have also been reported to be present in HO-1 promoter. In addition, an Egr-1 binding site in mouse HO-1 promoter that is inducible by zinc protoporphyrin IX (ZnPP) has also been reported [11].

ZnPP, a naturally occurring molecule formed in iron deficiency or lead poisoning, is a potent competitive inhibitor of HO-1. Inhibition of HO-1 by ZnPP led to suppression of tumor cell growth [12] and ZnPP has been suggested to be a useful agent for antitumor therapy [13]. However, ZnPP has also been shown to regulate expression of HO-1 at the transcriptional level, and the effect of ZnPP on HO-1 expression is controversial. For example, it is shown to induce HO-1 expression in hamster fibroblast (HA-1) cells [11] but not in Neuro-2A mouse neuroblastoma cells and primary cultures of rat cortical neurons [14]. In fact, it even suppressed the induction of HO-1 by statins or lipopolysaccharide [14]. In our earlier study, ZnPP was found to induce HO-1 expression in human prostate adenocarcinoma PC-3 and breast adenocarcinoma MCF-7 cells [15]. It is a much stronger inducer of HO-1 than atorvastatin, one of the statins. In this study, we used prostate cancer PC-3 cells to investigate the mechanism of action of ZnPP. We herein report that ZnPP upregulates HO-1 in PC-3 cells via the antioxidant response pathway.

2. Materials and Methods

2.1. Reagents. N-acetyl cysteine (NAC) was product of Sigma-Aldrich (St. Louis, MO, USA). ZnPP and protein kinase inhibitors were purchased through EMD Chemicals Inc. (Gibbstown, NJ, USA). Antibodies against human β-actin and HO-1 were purchased from Cell Signaling Technology (Danvers, MA, USA) and Enzo Life Sciences (Plymouth Meeting, PA, USA), respectively. Antibodies against Keap1 and phospho-Nrf2 (pS40) were products of Proteintech Group, Inc. (Chicago, IL, USA) and Epitomics, Inc. (Burlingame, CA, USA), respectively. Antibodies against Bach1 were purchased from both Poteintech Group, Inc. and Epitomics, Inc.

2.2. Cell Line and Cell Culture. Human prostate adenocarcinoma PC-3 cell line was purchased from the American Type Culture Collection (Manassas, VA, USA). These cells were maintained as monolayer cultures in DMEM/F12 medium (Invitrogen, Carlsbad, CA, USA) supplemented with 10% fetal bovine serum, 100 units/mL of penicillin, 100 μg/mL of streptomycin, and 0.25 μg/mL of amphotericin B (complete medium) and were kept at 37°C in a humidified atmosphere containing 5% CO_2.

2.3. Construction of Reporter Plasmids. The enhancer-luciferase reporter plasmids were constructed by inserting sequences of various synthetic response elements into the filled-in *Nhe*I/*Bgl*II sites of pGL3-promoter vector (Promega, Madison, WI, USA) or *Eco*RV site of pGL4-promoter vector via blunt-end ligation as described in our earlier study [15]. Internal control plasmid, pGL4.74[hRluc/TK], was purchased from Promega (Madison, WI, USA).

2.4. RT-PCR and qPCR. PC-3 cells were grown to 80% confluence in T25 flasks and treated with 10 μM ZnPP or equal amount of DMSO (vehicle) for various time intervals up to 48 h. Total RNA was extracted using NucleoSpin Nucleic Acid Purification Kits (Clontech, Palo Alto, CA, USA). First-strand cDNA was synthesized from 5 μg of total RNA using Thermo-Script (Invitrogen, Carlsbad, CA, USA) in a volume of 20 μL. PCR was done for 30 cycles (denaturation at 94°C for 30 sec, annealing at 59°C for 30 sec, and extension at 72°C for 60 sec) using 1 μL of the first-strand cDNA, 10 pmol of gene specific primers and 2.5 units of JumpStart Taq DNA polymerase (Sigma-Aldrich, St. Louis, MO) in a volume of 50 μL. Primers for β-actin: 5'-CCTCGCCTTTGCCGATCC-3' and 5'-GGATCTTCATGAGGTAGTCAGTC-3'. Primers for HO-1: 5'-AGAAGAGCTGCACCGCAAGG-3' and 5'-CCTCTGAAGTTTAGGCCATTGC-3'.

Real-time PCR (RT-qPCR) was performed with StepOne real-time PCR system (Applied Biosystems, Foster City, CA, USA) using TaqMan Gene Expression Master Mix and ready-made human HMOX1 (Hs 01110250_m1), ACTB (Hs 99999903_m1), NQO1 (Hs00168547_m1), GSTP1 (Hs00943351_g1), BACH1 (Hs00230917_m1), NFE2L2 (Hs00975960_m1), and KEAP1 (Hs00202227_m1) Gene Expression Assays (Applied Biosystems, Foster City, CA, USA). Reactions were done in triplicates, and average C_T values were used to calculate "fold-induction" over vehicle-treated control using Comparative C_T method. ACTB was used as the internal control gene.

2.5. Luciferase Reporter Assay. Luciferase reporter assays were carried out as described in our previous study [16]. Briefly, cells grown to 90% confluence in 24-well plates were cotransfected in triplicates with 250 ng of enhancer-luciferase reporter plasmid and 25 ng of pGL4.74[hRluc/TK] internal control plasmid, using Lipofectamine 2000 (Invitrogen, Carlsbad, CA, USA). Six hours after transfection, the medium was replaced with fresh one containing 10 μM ZnPP or same amount of DMSO (vehicle). At 30–48 h after-transfection, the growth medium was removed, and the cells were rinsed twice with ice-cold phosphate buffered saline and were lysed by shaking for 15 min at 25°C with 100 μL of Passive Lysis Reagent (Promega, Madison, WI, USA). Aliquots of 20 μL of the cell lysates were assayed for firefly and renilla luciferase activities using a 20/20 Luminometer (Turner Biosystems, Sunnyvale, CA, USA) and Dual-Luciferase Reporter Assay System (Promega, Madison, WI, USA). The results were expressed as Relative Luciferase Activity (a ratio of the activities of firefly luciferase/renilla luciferase).

2.6. Measurement of Cell Survival. Cells were seeded in triplicates at 0.5–1 × 10^4 cells/well in 48-well plate in complete medium. At about 25% confluence, cells were treated with various concentrations of ZnPP or vehicle (DMSO) for 48 h. Cell survival was determined using CellTiter 96 Nonradioactive Cell Proliferation Assay (Promega, Madison, WI, USA) according to the protocol provided by the manufacturer. The color developed was measured at 490 nm.

2.7. Western Blot Analysis. PC-3 cells grown to 80% confluence in T25 flasks were treated with 10 μM ZnPP or vehicle for various time intervals up to 48 h. Cells were lysed with 0.5 mL of 1X Laemmli sample buffer containing 1% Halt protease inhibitor and phosphatase inhibitor cocktails (Thermo Scientific, Rockford, IL, USA), sonicated for 2 × 15 sec, and centrifuged at 10,000 rpm for 15 min at 4°C. Aliquots of 50 μg protein extract were analyzed on 10% SDS-polyacrylamide gel and transferred to PVDF membranes. The blots were analyzed by western blot according to the procedure provided by WesternDot 625 kit (Invitrogen, Carlsbad, CA, USA). Briefly, the blots were incubated in 8 mL of Blocking Buffer in a small plastic dish for 1 h at room temperature with gentle agitation. Then they were incubated with the diluted primary antibody (1 : 1000 dilution) at 4°C overnight. After washing 3 times with 50 mL of 1X Wash Buffer, 5 min each, blots were incubated with 8 mL of Biotin-XX-Goat anti-rabbit anti-body (1 : 2000 dilution) in Blocking Buffer for 1 h. They were washed 3 times with 50 mL of 1X Wash Buffer for 5 min each, and then incubated with 8 mL of Qdot 625 streptavidin conjugate (1 : 2000 dilution) in Blocking Buffer for 60 min at room temperature. Finally, the blots were washed 3 times with 50 mL of 1X Wash Buffer for 5 min each, and once with 20 mL of ultrapure water. The wet blots were placed on a UV trans-illuminator and pictures were taken with a Polaroid camera and orange filter.

2.8. Data Analysis. Data points shown represent mean ± standard error. Statistically significant differences between data points of two groups were determined by Student's *t*-test. By convention, a *P* value of <0.05 was considered statistically significant.

3. Results

ZnPP is relatively nontoxic to PC-3 cells. In fact, it induced significant cell proliferation at a concentration of 0.6–10 μM, and only suppressed cell growth above 10 μM (Figure 1(a)). Therefore, 10 μM ZnPP was used for all subsequent experiments. Basal expression level of HO-1 protein in PC-3 was undetectable. ZnPP induced HO-1 protein expression in a dose-dependent manner, with the highest induction level at 10 μM (Figure 1(b)). For a time course study, incubation of the cells with 10 μM ZnPP for 4 h showed only a slight induction of HO-1 protein, but the induction was high after 16 h and was maintained through 48 h of incubation (Figure 2(a)). The HO-1 mRNA level as determined by RT-PCR also showed similar profile (Figure 2(b)).

Of all the known responsive elements in the HO-1 pro-moter, ZnPP activated mainly the ARE-like elements (StREs). As shown in Figure 3, StRE3 showed the highest (6.6-fold) induction level by ZnPP, although these elements had differ-ent basal expression levels of relative luciferase activities due to different copy number of the response elements present in the luciferase-reporter constructs. ZnPP did not activate the HSE, SREBP, and SP1 elements (Figure 3).

A number of protein kinases are known to be involved with the activation of antioxidant response element. To inves-tigate the effect of various protein kinase inhibitors and anti-oxidant on the activation of StRE by ZnPP, cells trans-fected with StRE3-pGL3 were pretreated for 2 h with SB203580 (p38-MAPK inhibitor), LY294002 (phosphatidyli-nositol 3-kinase inhibitor), U0126 (MEK inhibitor), SP600125 (JNK inhibitor), IPA-3 (p21-Activated Kinase Inhibitor III), NAC (antioxidant), rottlerin (PKC-δ inhibitor), Ro 31-8220 (pan PKC inhibitor), or Ro 32-0432 (PKC-α inhibitor) prior to treatment with ZnPP for 24 h. As shown in Figure 4, SP600125 and Ro 32-0432 had little effect on the activation of StRE3 by ZnPP. SB203580, NAC, and IPA-3 reduced the activation of StRE3 by ZnPP to 72.9%, 62.4%, and 83.6%, res-pectively, of the level by ZnPP alone. However, LY294002, U0126, and Ro 31-8220 attenuated the activation to 39.4%, 40.2%, and 41.5%, respectively, of the ZnPP-alone control. On the other hand, rottlerin activated StRE3 element by itself and

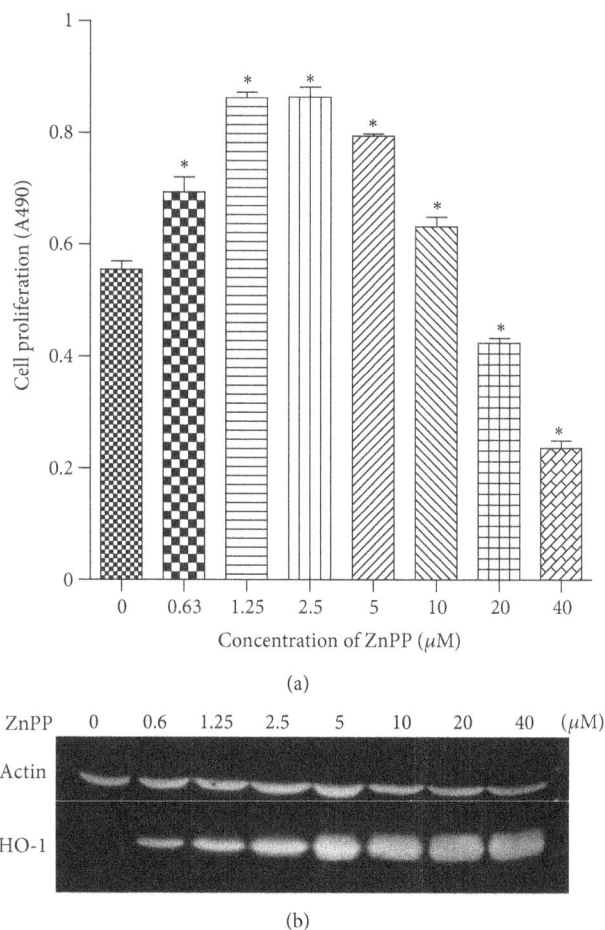

(a)

(b)

Figure 1: Effect of ZnPP on cell proliferation and HO-1 expression of PC-3 cells. (a) PC-3 cells were treated with various concentrations of ZnPP for 48 h, and number of live cells was estimated by MTS cell proliferation assay as described in Section 2. Results were expressed as Absorbance at 490 nm (mean ± S.E.). $N = 3$; $^*P < 0.05$ compared with untreated control. (b) PC-3 cells were treated with various concentrations of ZnPP for 24 h, and HO-1 expression was determined by western blot analysis as described in Section 2 using specific antibodies against β-actin and HO-1. Immunoreactive protein bands detected by WesternDot 625 appeared as fluorescent bands.

had a synergistic effect with ZnPP (2-fold over the level by ZnPP alone).

To confirm the effect of LY294002, U0126, Ro 31-8220, and rottlerin on ZnPP-activation of StRE3 element, HO-1 mRNA levels were determined by real-time PCR and protein levels by western blot analyses. For real-time PCR, the relative levels of HO-1 mRNA were determined in cells pretreated with LY294002, U0126, Ro 31-8220, or rottlerin for 1 h prior to ZnPP treatment for 3 h. The results showed that LY294002 and U0126 did not attenuate ZnPP-induction of HO-1 mRNA, but Ro 31-8220 completely suppressed the effect of ZnPP (Figure 5). On the other hand, real-time PCR also confirmed the synergistic effect of rottlerin with ZnPP; ZnPP alone upregulated HO-1 by 8.2-fold over vehicle-treated con-trol and rottlerin plus ZnPP upregulated HO-1 by 36.0-fold

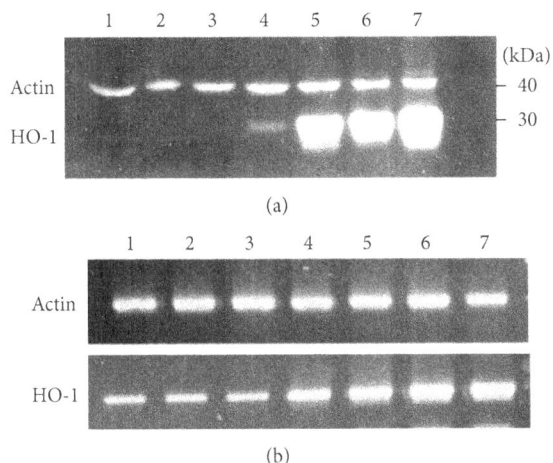

(a)

(b)

Figure 2: Induction of HO-1 by ZnPP in PC-3 for various lengths of time. PC-3 cells were treated with 10 μM ZnPP for 0.5, 1, 4, 16, 24, and 48 h (lanes 2, 3, 4, 5, 6, and 7, resp.). Control cells were treated with equal amount of DMSO for 48 h (lane 1). (a) Western blot analysis was done as described in Section 2, using specific antibodies against β-actin and HO-1. Immunoreactive protein bands detected by WesternDot 625 appeared as fluorescent bands. (b) RT-PCR analysis using gene-specific primers. PCR fragments were visualized by ethidium bromide staining.

(Figure 5). Western blot analyses basically confirmed the results of real-time PCR. Ro 31-8220 significantly suppressed the upregulation of HO-1 protein by ZnPP, while LY294002, U0126, and Ro 32-0432 had little effect on ZnPP-induction of HO-1 (Figure 6(a)). However, the synergistic effect of rottlerin with ZnPP was not evident due to the high level of upregulation of HO-1 by ZnPP alone. Western blot analysis also showed that SB203580 and IPA-3 had no effect on ZnPP-induction of HO-1 expression (data not shown).

Since Ro 31-8220 is a pan PKC inhibitor, this suggests that PKC may be involved in ZnPP-upregulation of HO-1. There are many PKC isoforms, but the involvement of PKC-α can be excluded by the lack of suppression of Ro 32-0432 (PKC-α inhibitor) on ZnPP-upregulation of HO-1 protein (Figure 6(a), lane 7). To determine if other PKC isoforms may be involved in ZnPP-activation of HO-1, PC-3 cells were treated with myristoylated pseudosubstrates of PKC-θ, PKC-ζ and PKC-η, or PKC-β inhibitor prior to ZnPP treatment. Western blot analysis showed no effect of these inhibitors on HO-1 protein level (Figure 6(b)). Hence, involvement of PKC-θ, PKC-ζ and PKC-η, and PKC-β can also be excluded.

Since Bach1, Nrf2 and Keap1 proteins have been shown to interact with antioxidant response element, the effect of ZnPP on the mRNA and protein levels of these proteins were investigated by real-time PCR and Western blot analyses. Real-time PCR analyses showed that there were no drastic changes in the expression of BACH1, NFE2L2 (NRF2), and KEAP1 genes when PC-3 cells were treated with ZnPP. The expression levels of BACH1, NFE2L2, and KEAP1 in cells treated with 10 μM ZnPP for 3 h were 73.5, 81.8, and 98.8%, respectively, of those in control cells. Western blot analysis showed no change in phospho-Nrf2(pS40) protein but a gradual decrease in Keap1 protein in cells treated with ZnPP

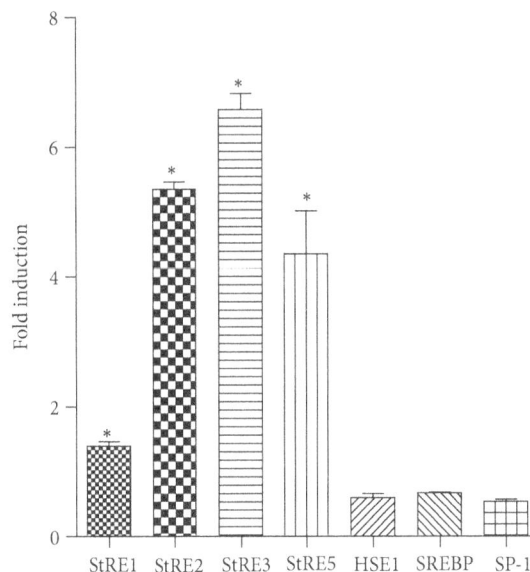

Figure 3: Effect of ZnPP on StRE1, StRE2, StRE3, StRE5, HSE, SREBP, and SP1 elements of human HO-1 promoter in PC-3 cells. PC-3 cells were transfected with enhancer-luciferase reporter plasmid harboring one of these elements, treated with 10 μM ZnPP for 48 h, and luciferase activities were determined as described in Section 2. Results were expressed as "Fold Induction" over vehicle (DMSO)-treated control of the corresponding responsive element (mean $\pm$ S.E.). N = 3; $^*P$ < 0.05 compared with vehicle-treated control of the corresponding responsive element.

for 2–24 h (Figure 7). Only a faint band of Bach1 protein was detected in PC-3 control cells, but none at all in cells treated with ZnPP, despite antibodies from two companies were used. To investigate the effect of ZnPP on the expression of other antioxidant responsive genes, the expression levels of NQO1 and GSTP1 in cells treated with 10 μM ZnPP for 3 h were determined by real-time PCR. The results showed that the expression levels of NQO1 and GSTP1 were 89.8 and 77.7%, respectively, of those in control cells.

To determine if rottlerin and ZnPP can act as prooxidant, cells were pretreated with 10 mM NAC prior to incubation with 10 μM ZnPP or 2 μM rottlerin, and relative levels of HO-1 mRNA were determined by real-time PCR. The results showed that NAC, even at 10 mM, reduced ZnPP-induction of HO-1 expression by only 43% (Figure 8). Rottlerin by itself did not induce HO-1 mRNA expression after 5 h incubation (Figure 8), and only a faint band of HO-1 protein was detected on western blot after 24 h incubation with rottlerin (data not shown). Therefore, the effect of NAC on rottlerin was not determined.

4. Discussion

In this study, we demonstrated that ZnPP upregulated HO-1 expression in PC-3 cells in a dose-dependent manner, and that upregulation was done mainly through activation of the ARE-like response elements (StREs) of HO-1 promoter. Although similar response elements are present in the pro-

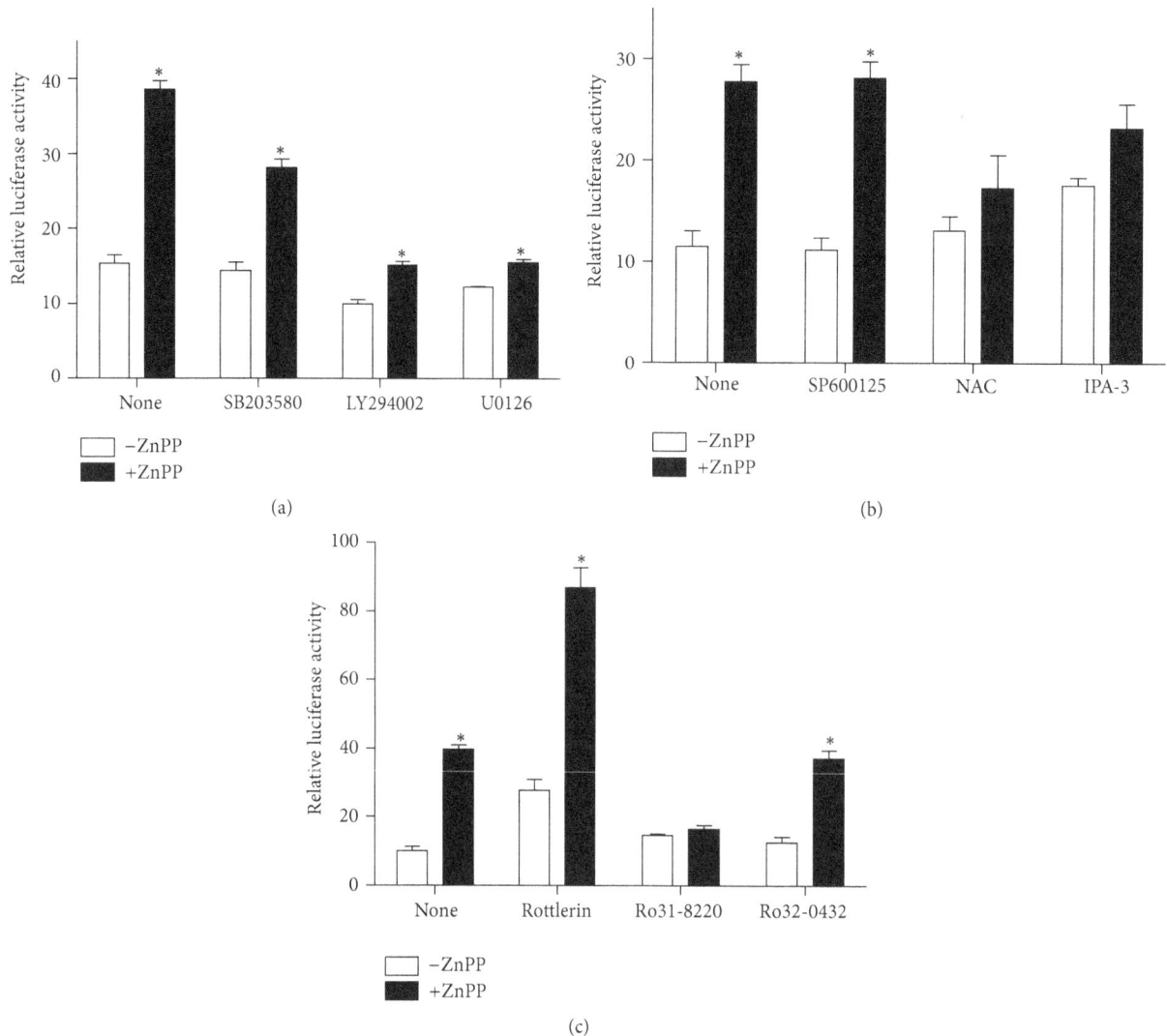

FIGURE 4: Effect of protein kinase inhibitors and antioxidant on ZnPP-activation of StRE3 element of human HO-1 promoter in PC-3 cells. PC-3 cells were transfected with StRE3-pGL3 luciferase reporter plasmid, pretreated with $3\,\mu M$ SB203580, $5\,\mu M$ LY294002, $10\,\mu M$ U0126, $3\,\mu M$ SP600125, $500\,\mu M$ NAC, $10\,\mu M$ IPA-3, $2\,\mu M$ rottlerin, $5\,\mu M$ Ro 31-8220, or $5\,\mu M$ Ro 32-0432 for 2 h, and then treated with $10\,\mu M$ ZnPP for 24 h, and luciferase activities were determined as described in Section 2. Results were expressed as "relative luciferase activity (ratio of activities of firefly luciferase/renilla luciferase)" (mean $\pm$ S.E.). $N = 3$; $^{*}P < 0.05$ compared with the untreated control.

moter of other antioxidant responsive genes, such as NQO1 and GSTP1 [17, 18], real-time PCR analysis showed no upregulation of these genes by ZnPP. It should be noted that the induction of HO-1 protein is much stronger than that of HO-1 mRNA, suggesting ZnPP may also affect other pathways that stabilize the HO-1 protein level. However, no known protease inhibitor activity of ZnPP has been reported. We also demonstrated that preincubation of the cells with Ro 31-8220, a pan inhibitor of PKC, abrogated ZnPP-induction of HO-1 expression, suggesting that activation of HO-1 gene by ZnPP may involve PKC.

It is intriguing to note that induction of HO-1 expression in PC-3 cells by hemin inhibited cell proliferation [4], whereas our results showed that although ZnPP also induced HO-1 expression in PC-3 cells, proliferation was not inhibited until ZnPP concentration exceeded $10\,\mu M$ (Figure 1(a)). This

may be due to the fact that ZnPP is also a well-known HO-1 inhibitor. HO-1 enzymatic activity may be required for the inhibition of cell proliferation. When ZnPP level was increased to above $10\,\mu M$, HO-1 level finally overwhelmed the inhibitory effect of ZnPP.

The ability of ZnPP to activate the ARE-like response elements (StREs) of HO-1 promoter suggests that ZnPP upregulates HO-1expression via the Nrf2-ARE signaling pathway. The level of Nrf2 is regulated by Keap1. The binding of Nrf2 by Keap1 results in the ubiquitination and degradation of Nrf2. Phosphorylation of Nrf2 and/or modification of cysteine residues of Keap1 results in decreased Nrf2 ubiquitination and degradation, and hence activation of the ARE signaling pathway [19]. Phosphorylation of Nrf2 at Ser^{40} residue by PKC results in dissociation of Nrf2 from the Nrf2-Keap1 complex, translocation of Nrf2 into nucleus, and activation of ARE-like

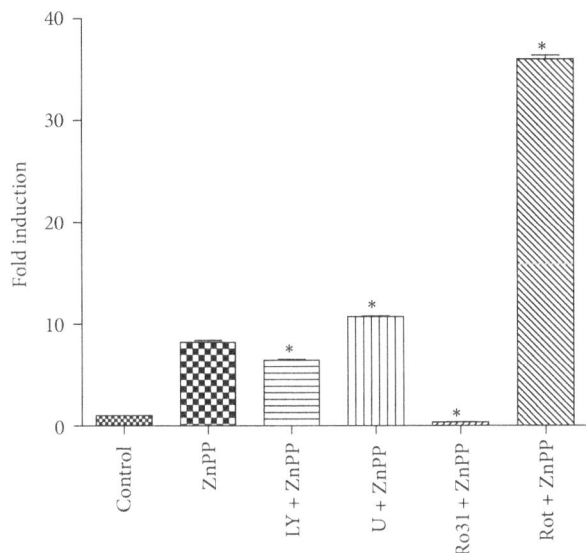

FIGURE 5: Real-time PCR analyses of HO-1 mRNA levels in PC-3 cells treated with ZnPP in the presence of various kinase inhibitors. PC-3 cells were pretreated with $5\,\mu M$ LY294002, $10\,\mu M$ U0126, $5\,\mu M$ Ro 31-8220, or $2\,\mu M$ rottlerin for 1 h, and then treated with $10\,\mu M$ ZnPP for 3 h, and relative HO-1 mRNA levels were determined by real-time PCR as described in Section 2. Results were expressed as "Fold Induction" over vehicle (DMSO-) treated control. $N = 3$; $^{*}P <$ 0.05 compared with ZnPP alone.

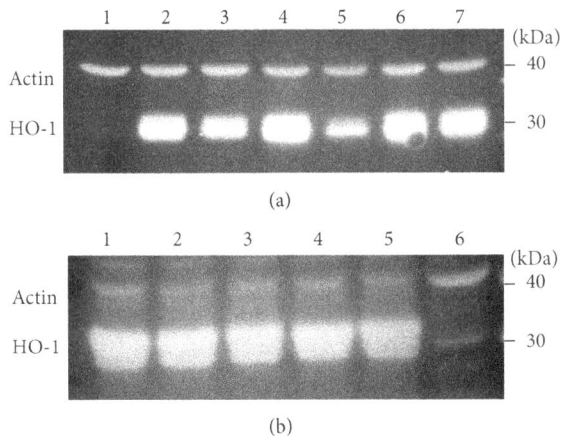

(a)

(b)

FIGURE 6: Western blot analysis of HO-1 protein induced by ZnPP in PC-3 in the presence of various kinase inhibitors or antioxidant. (a) PC-3 cells were pretreated with nothing (lane 2), $5\,\mu M$ LY294002 (lane 3), $10\,\mu M$ U0126 (lane 4), $5\,\mu M$ Ro 31-8220 (lane 5), $2\,\mu M$ rottlerin (lane 6), or $5\,\mu M$ Ro 32-0432 (lane 7) for 2 h, and then treated with $10\,\mu M$ ZnPP for 24 h. Control cells were treated with equal amount of DMSO for 24 h (lane 1). Total proteins ($50\,\mu g$) were analyzed by western blot as described in Section 2, using specific antibodies against β-actin and HO-1. Immunoreactive protein bands detected by WesternDot 625 appeared as fluorescent bands. (b) PC-3 cells were pretreated with $5\,\mu M$ myristoylated PKC-θ pseudosubstrate (lane 1), $5\,\mu M$ myristoylated PKC-ζ pseudosubstrate (lane 2), $5\,\mu M$ myristoylated PKC-η pseudosubstrate (lane 3), $5\,\mu M$ PKC-β inhibitor (lane 4), or nothing (lane 5) for 2 h, and then treated with $10\,\mu M$ ZnPP for 24 h. Control cells were treated with equal amount of DMSO for 24 h (lane 6). Western blot analysis was done as described above.

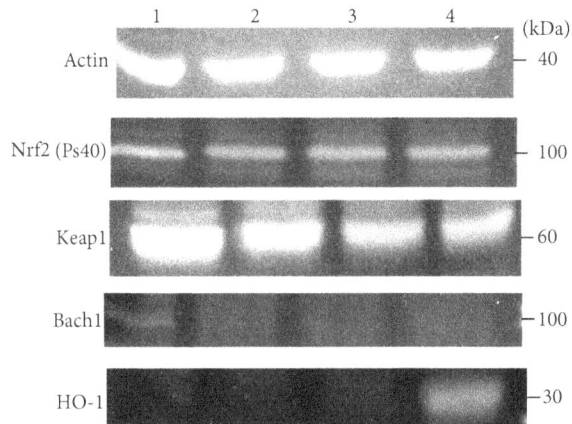

FIGURE 7: Effect of ZnPP on the level of phospho-Nrf2(pS40), Keap1, Bach1, and HO-1 in PC-3. PC-3 cells were treated with $10\,\mu M$ ZnPP for 2 h (lane 2), 4 h (lane 3), and 24 h (lane 4). Vehicle-treated cells were served as control (lane 1). Total proteins ($50\,\mu g$) were analyzed by western blot as described in Section 2, using specific antibodies against β-actin, phospho-Nrf2(pS40), Keap1, Bach1, and HO-1. Immunoreactive protein bands detected by WesternDot 625 appeared as fluorescent bands.

element of genes responsive to oxidative stress [20, 21]. Our results showed no change in phospho-Nrf2 (pS40) level in ZnPP-treated cells as compared to control (Figure 7). However, our results did show a progressive decrease in Keap1 level in ZnPP-treated cells (Figure 7). The significance of the decrease in Keap1 level remains to be studied. On the other hand, ARE is known to be bound and repressed by a transcription factor called Bach1. Inactivation of the repressor Bach1 will also lead to the activation of ARE [22]. However, Bach1 is unlikely involved in the upregulation of HO-1 by ZnPP for two reasons. First, basal expression level of Bach1 in PC-3 cells is very low; it was barely detectable in control cells by western blot (Figure 7), and no downregulation of Bach1 mRNA induced by ZnPP was detected by real-time PCR. Second, preincubation of the cells with 10 mM NAC prior to ZnPP treatment did not completely abrogate the ZnPP-induction of HO-1 (Figure 8). This suggests that ZnPP does not act as a prooxidant that would inactivate Bach1.

Since upregulation of HO-1 by ZnPP can be suppressed by Ro 31-8220, ZnPP-induction may be mediated by PKC. The involvement of PKC-δ in the upregulation of HO-1 by many phyto-chemicals has been demonstrated, as rottlerin and PKC-δ small interfering RNA were able to attenuate HO-1 induction by these compounds [23–27]. It should be noted that although rottlerin is not an efficient PKC-δ inhibitor [28], the involvement of PKC-δ was confirmed by the use of PKC-δ small interfering RNA in these studies. On the other hand, involvement of PKC-ζ [29] and yet unidentified atypical PKC [30] has also been demonstrated. Our results showed that preincubation of PC-3 cells with Ro 32-0432, PKC-β inhibitor, rottlerin, and pseudosubstrates of PKC-ζ, PKC-θ, and PKC-η did not suppress ZnPP-induction of HO-1 expression, and hence the involvement of α, β, δ, η, θ, and ζ isoforms of PKC can be ruled out. PC-3 cells are known to express PKC α, δ, ε, η, and μ isoforms as determined by nuclease protection assay [31] and α, ε, ζ, and ι as determined by western blot [32].

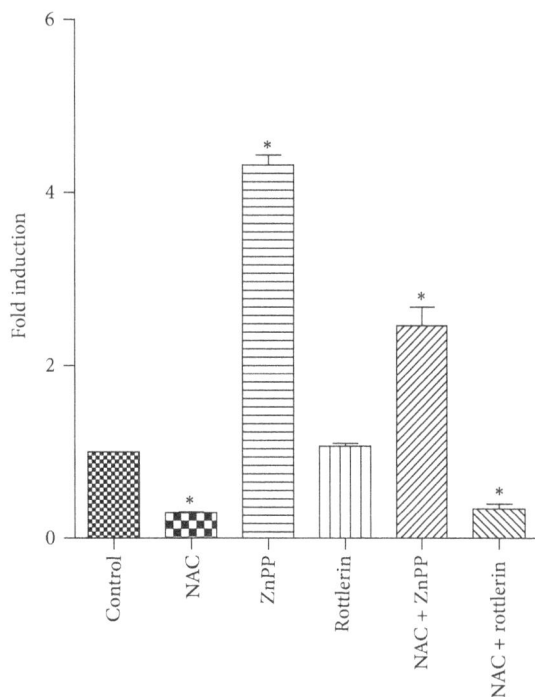

FIGURE 8: Effect of NAC on ZnPP- or rottlerin-induction of HO-1. PC-3 cells were treated with nothing (control); 10 mM NAC (5 h); 10 μM ZnPP (5 h); 2 μM rottlerin (5 h); 10 mM NAC (1 h pretreatment) plus 10 μM ZnPP (5 h); or 10 mM NAC (1 h pretreatment) plus 2 μM rottlerin (5 h). Relative HO-1mRNA levels were determined by real-time PCR as described in Section 2. Results were expressed as "Fold Induction" over control. $N = 3$; $^{*}P < 0.05$ compared with control.

Involvement of PKC ε, ι, and μ isoforms has not been tested in this study, because specific inhibitors for these isoforms are not commercially available. The PKC isoform involved in ZnPP-induction of HO-1 remains to be determined.

Our results showed that rottlerin did not attenuate ZnPP-induction of HO-1. Instead, rottlerin had a synergistic effect with ZnPP, and this was confirmed at the mRNA level (Figure 5). There is at least one report showing that rottlerin was able to upregulate HO-1 through reactive oxygen species (ROS) dependent and PKC-δ-independent pathway in human colon cancer HT29 cells, as its induction was abrogated by antioxidant NAC but not by suppression of PKC-δ expression by small interfering RNA technology [33]. In PC-3 cells, rottlerin did not induce HO-1 to any significant level by itself, but had a synergistic effect with ZnPP.

Ro 31-8220, a pan inhibitor of PKC, has been reported to induce apoptosis independent of PKC activity [34]. However, our results showed that Ro 31-8220 was able to suppress ZnPP-induction of HO-1 mRNA after only 3 h of incubation (Figure 5). Furthermore, PC-3 cells treated with 5 μM Ro 31-8220 for 24 h retained 93% viability (data not shown). These suggest that the suppression of ZnPP-induction of HO-1 by Ro 31-8220 was not due to induction of apoptosis. On the other hand, Ro 31-8220 has been shown to inhibit other kinases, such as MAPKAP kinase-1β (also known as Rsk-2) and p70 S6 kinase [35]. Furthermore, Ro 31-8220 has been

shown to activate JNK1 [36]. Involvement of these kinases in ZnPP-activation has not been ruled out.

In conclusion, ZnPP upregulates HO-1 in PC-3 cells via the activation of StRE of HO-1 promoter. The pathway through which the StRE is activated remains to be determined.

Abbreviations

ARE:	Antioxidant response element
Bach1:	Broad-complex, tramtrack and bric à brac and cap'n'collar homology 1
DMSO:	Dimethyl sulfoxide
Egr-1:	Early growth response 1
GSTP1:	Glutathione S-transferase P1
HO-1:	Heme oxygenase-1
HSE:	Heat-shock response element
JNK:	C-Jun N-terminal kinase
Keap1:	Kelch-like ECH-associated protein 1
MAPK:	Mitogen-activated protein kinase
MAPKAPK:	MAPK-activated protein kinase
MEK:	Mitogen-activated protein/extracellular signal-regulated kinase kinase
NAC:	N-acetyl cysteine
NQO1:	NAD(P)H quinone oxidoreductase 1
Nrf2:	Nuclear factor erythroid-derived 2 related factor 2
PKC:	Protein kinase C
ROS:	Reactive oxygen species
SREBP:	Sterol regulatory element binding transcription factor 1
SP1:	SP1 transcription factor
StRE:	Stress response element
ZnPP:	Zinc protoporphyrin IX.

Acknowledgment

This project was funded by a Grant from Albert Einstein Society, AEHN, Philadelphia, PA, USA.

References

[1] M. Bauer, K. Huse, U. Settmacher, and R. A. Claus, "The heme oxygenase-carbon monoxide system: regulation and role in stress response and organ failure," *Intensive Care Medicine*, vol. 34, no. 4, pp. 640–648, 2008.

[2] A. Jozkowicz, H. Was, and J. Dulak, "Heme oxygenase-1 in tumors: is it a false friend?" *Antioxidants and Redox Signaling*, vol. 9, no. 12, pp. 2099–2117, 2007.

[3] M. D. Maines and P. A. Abrahamsson, "Expression of heme oxygenase-1 (HSP32) in human prostate: normal, hyperplastic, and tumor tissue distribution," *Urology*, vol. 47, no. 5, pp. 727–733, 1996.

[4] G. Gueron, A. De Siervi, M. Ferrando et al., "Critical role of endogenous heme oxygenase 1 as a tuner of the invasive potential of prostate cancer cells," *Molecular Cancer Research*, vol. 7, no. 11, pp. 1745–1755, 2009.

[5] P. Sacca, R. Meiss, G. Casas et al., "Nuclear translocation of haeme oxygenase-1 is associated to prostate cancer," *British Journal of Cancer*, vol. 97, no. 12, pp. 1683–1689, 2007.

[6] J. Alam and J. L. Cook, "Transcriptional regulation of the heme oxygenase-1 gene via the stress response element pathway," *Current Pharmaceutical Design*, vol. 9, no. 30, pp. 2499–2511, 2003.

[7] J. Alam and J. L. Cook, "How many transcription factors does it take to turn on the heme oxygenase-1 gene?" *American Journal of Respiratory Cell and Molecular Biology*, vol. 36, no. 2, pp. 166–174, 2007.

[8] S. Okinaga, K. Takahashi, K. Takeda et al., "Regulation of human heme oxygenase-1 gene expression under thermal stress," *Blood*, vol. 87, no. 12, pp. 5074–5084, 1996.

[9] A. Kallin, L. E. Johannessen, P. D. Cani et al., "SREBP-1 regulates the expression of heme oxygenase 1 and the phosphatidylinositol-3 kinase regulatory subunit p55γ," *Journal of Lipid Research*, vol. 48, no. 7, pp. 1628–1636, 2007.

[10] J. Deshane, J. Kim, S. Bolisetty, T. D. Hock, N. Hill-Kapturczak, and A. Agarwal, "Sp1 regulates chromatin looping between an intronic enhancer and distal promoter of the human heme oxygenase-1 gene in renal cells," *Journal of Biological Chemistry*, vol. 285, no. 22, pp. 16476–16486, 2010.

[11] G. Yang, X. Nguyen, J. Ou, P. Rekulapelli, D. K. Stevenson, and P. A. Dennery, "Unique effects of zinc protoporphyrin on HO-1 induction and apoptosis," *Blood*, vol. 97, no. 5, pp. 1306–1313, 2001.

[12] K. Hirai, T. Sasahira, H. Ohmori, K. Fujii, and H. Kuniyasu, "Inhibition of heme oxygenase-1 by zinc protoporphyrin IX reduces tumor growth of LL/2 lung cancer in C57BL mice," *International Journal of Cancer*, vol. 120, no. 3, pp. 500–505, 2007.

[13] J. Fang, K. Greish, H. Qin et al., "HSP32 (HO-1) inhibitor, co-poly(styrene-maleic acid)-zinc protoporphyrin IX, a water-soluble micelle as anticancer agent: *in vitro* and *in vivo* anticancer effect," *European Journal of Pharmaceutics and Biopharmaceutics*, vol. 81, no. 3, pp. 540–547, 2012.

[14] C. H. Hsieh, J. C. Y. Jeng, M. W. Hsieh et al., "Involvement of the p38 pathway in the differential induction of heme oxygenase-1 by statins in Neuro-2A cells exposed to lipopolysaccharide," *Drug and Chemical Toxicology*, vol. 34, no. 1, pp. 8–19, 2011.

[15] S. C. M. Kwok, S. P. Samuel, and J. Handal, "Atorvastatin activates heme oxygenase-1 at the stress response elements," *Journal of Cellular and Molecular Medicine*, vol. 16, no. 2, pp. 394–400, 2012.

[16] S. C. M. Kwok and I. Daskal, "Brefeldin A activates CHOP promoter at the AARE, ERSE and AP-1 elements," *Molecular and Cellular Biochemistry*, vol. 319, no. 1-2, pp. 203–208, 2008.

[17] A. K. Jaiswal, "Regulation of genes encoding NAD(P)H:quinone oxidoreductases," *Free Radical Biology and Medicine*, vol. 29, no. 3-4, pp. 254–262, 2000.

[18] T. Nishinaka, Y. Ichijo, M. Ito et al., "Curcumin activates human glutathione S-transferase P1 expression through antioxidant response element," *Toxicology Letters*, vol. 170, no. 3, pp. 238–247, 2007.

[19] A. Giudice, C. Arra, and M. C. Turco, "Review of molecular mechanisms involved in the activation of the Nrf2-ARE signaling pathway by chemopreventive agents," *Methods in Molecular Biology*, vol. 647, pp. 37–74, 2010.

[20] H. C. Huang, T. Nguyen, and C. B. Pickett, "Phosphorylation of Nrf2 at Ser-40 by protein kinase C regulates antioxidant response element-mediated transcription," *Journal of Biological Chemistry*, vol. 277, no. 45, pp. 42769–42774, 2002.

[21] D. A. Bloom and A. K. Jaiswal, "Phosphorylation of Nrf2 at Ser40 by Protein Kinase C in Response to Antioxidants Leads to the Release of Nrf2 from INrf2, but Is Not Required for Nrf2 Stabilization/Accumulation in the Nucleus and Transcriptional Activation of Antioxidant Response Element-mediated NAD(P)H:Quinone Oxidoreductase-1 Gene Expression," *Journal of Biological Chemistry*, vol. 278, no. 45, pp. 44675–44682, 2003.

[22] J. F. Reichard, G. T. Motz, and A. Puga, "Heme oxygenase-1 induction by NRF2 requires inactivation of the transcriptional repressor BACH1," *Nucleic Acids Research*, vol. 35, no. 21, pp. 7074–7086, 2007.

[23] S. A. Rushworth, R. M. Ogborne, C. A. Charalambos, and M. A. O'Connell, "Role of protein kinase C δ in curcumin-induced antioxidant response element-mediated gene expression in human monocytes," *Biochemical and Biophysical Research Communications*, vol. 341, no. 4, pp. 1007–1016, 2006.

[24] B. C. Kim, W. K. Jeon, H. Y. Hong et al., "The anti-inflammatory activity of Phellinus linteus (Berk. & M.A. Curt.) is mediated through the PKCδ/Nrf2/ARE signaling to up-regulation of heme oxygenase-1," *Journal of Ethnopharmacology*, vol. 113, no. 2, pp. 240–247, 2007.

[25] R. M. Ogborne, S. A. Rushworth, and M. A. O'Connell, "Epigallocatechin activates haem oxygenase-1 expression via protein kinase Cδ and Nrf2," *Biochemical and Biophysical Research Communications*, vol. 373, no. 4, pp. 584–588, 2008.

[26] H. Zhang and H. J. Forman, "Acrolein induces heme oxygenase-1 through PKC-δ and PI3K in human bronchial epithelial cells," *American Journal of Respiratory Cell and Molecular Biology*, vol. 38, no. 4, pp. 483–490, 2008.

[27] S. E. Lee, S. I. Jeong, H. Yang et al., "Fisetin induces Nrf2-mediated HO-1 expression through PKC-δ and p38 in human umbilical vein endothelial cells," *Journal of Cellular Biochemistry*, vol. 112, no. 9, pp. 2352–2360, 2011.

[28] S. P. Soltoff, "Rottlerin: an inappropriate and ineffective inhibitor of PKCδ," *Trends in Pharmacological Sciences*, vol. 28, no. 9, pp. 453–458, 2007.

[29] A. I. Rojo, M. Salina, M. Salazar et al., "Regulation of heme oxygenase-1 gene expression through the phosphatidylinositol 3-kinase/PKC-ζ pathway and Sp1," *Free Radical Biology and Medicine*, vol. 41, no. 2, pp. 247–261, 2006.

[30] S. Numazawa, M. Ishikawa, A. Yoshida, S. Tanaka, and T. Yoshida, "Atypical protein kinase C mediates activation of NF-E2-related factor 2 in response to oxidative stress," *American Journal of Physiology-Cell Physiology*, vol. 285, no. 2, pp. C334–C342, 2003.

[31] C. T. Powell, N. J. Brittis, D. Stec, H. Hug, W. D. W. Heston, and W. R. Fair, "Persistent membrane translocation of protein kinase C α during 12-O- tetradecanoylphorbol-13-acetate-induced apoptosis of LNCaP human prostate cancer cells," *Cell Growth and Differentiation*, vol. 7, no. 4, pp. 419–428, 1996.

[32] J. R. Stewart and C. A. O'Brian, "Resveratrol antagonizes EGFR-dependent Erk1/2 activation in human androgen-independent prostate cancer cells with associated isozyme-selective PKCα inhibition," *Investigational New Drugs*, vol. 22, no. 2, pp. 107–117, 2004.

[33] E. J. Park, J. H. Lim, S. I. Nam, J. W. Park, and T. K. Kwon, "Rottlerin induces heme oxygenase-1 (HO-1) up-regulation through reactive oxygen species (ROS) dependent and PKC δ-independent pathway in human colon cancer HT29 cells," *Biochimie*, vol. 92, no. 1, pp. 110–115, 2010.

[34] Z. Han, P. Pantazis, T. S. Lange, J. H. Wyche, and E. A. Hendrickson, "The staurosporine analog, Ro-31-8220, induces apop-

tosis independently of its ability to inhibit protein kinase C," *Cell Death and Differentiation*, vol. 7, no. 6, pp. 521–530, 2000.

[35] D. R. Alessi, "The protein kinase c inhibitors Ro 318220 and GF 109203X are equally potent inhibitors of MAPKAP kinase-1β (Rsk-2) and p70 S6 kinase," *FEBS Letters*, vol. 402, no. 2-3, pp. 121–123, 1997.

[36] J. Beltman, F. McCormick, and S. J. Cook, "The selective protein kinase C inhibitor, Ro-31-8220, inhibits mitogen- activated protein kinase phosphatase-1 (MKP-1) expression, induces c-Jun expression, and activates Jun N-terminal kinase," *Journal of Biological Chemistry*, vol. 271, no. 43, pp. 27018–27024, 1996.

Identification and Characterization of Novel Perivascular Adventitial Cells in the Whole Mount Mesenteric Branch Artery Using Immunofluorescent Staining and Scanning Confocal Microscopy Imaging

Chandra Somasundaram,[1,2,3] **Rahul K. Nath,**[2,3] **Richard D. Bukoski,**[1] **and Debra I. Diz**[4]

[1] Cardiovascular Disease Research Program, JLC-Biomedical/Biotechnology Research Institute, North Carolina Central University, Durham, NC 27707, USA
[2] Research Division, Texas Nerve and Paralysis Institute, Houston, TX 77030, USA
[3] Intron Pharmaceuticals, Houston, TX 77005, USA
[4] Hypertension & Vascular Research Center, Wake Forest University School of Medicine, Winston-Salem, NC 27157, USA

Correspondence should be addressed to Chandra Somasundaram, chandra@drnathmedical.com

Academic Editor: G. S. Stein

A novel perivascular adventitial cell termed, adventitial neuronal somata (ANNIES) expressing the neural cell adhesion molecule (NCAM) and the vasodilator neuropeptide, calcitonin gene-related peptide (CGRP), exists in the adult rat mesenteric branch artery (MBA) in situ. In addition, we have previously shown that ANNIES coexpress CGRP and NCAM. We now show that ANNIES express the neurite growth marker, growth associated protein-43(Gap-43), palladin, and the calcium sensing receptor (CaSR), that senses changes in extracellular Ca(2+) and participates in vasodilator mechanisms. Thus, a previously characterized vasodilator, calcium sensing autocrine/paracrine system, exists in the perivascular adventitia associated with neural-vascular interface. Images of the whole mount MBA segments were analyzed under scanning confocal microscopy. Confocal analysis showed that the Gap-43, CaSR, and palladin were present in ANNIES about $37 \pm 4\%$, $94 \pm 6\%$, and $80 \pm 10\%$ respectively, comparable to CGRP (100%). Immunoblots from MBA confirmed the presence of Gap-43 (48 kD), NCAM (120 and 140 kD), and palladin (90–92 and 140 kD). In summary, CGRP, and NCAM-containing neural cells in the perivascular adventitia also express palladin and CaSR, and coexpress Gap-43 which may participate in response to stress/injury and vasodilator mechanisms as part of a perivascular sensory neural network.

1. Introduction

Vascular growth and remodeling occur in association with certain physiological and pathological conditions. In addition, vascular regeneration and repair are essential for the survival of blood vessels. These processes involve numerous cell types. There are still uncharacterized and less characterized cell types in vascular adventitia, which include vascular stem/progenitor cells [1–10] and adventitial neuronal somata (ANNIES) [11]. The vascular adventitia is a complicated tissue [12], which is found to be the most active layer in terms of cell turnover [13]. In addition, within the vascular adventitia

resides amyelinated nerves known as "nerva vasorum" [14]. Using fluorescence confocal microscopy (FCM) to visualize vascular wall 3D organization of different cellular and extracellular elements of the intact artery with minimal 3D distortion [11, 13], we recently demonstrated ANNIES coexpressing neural cell adhesion molecule (NCAM) and calcitonin gene-related peptide (CGRP) in the adult rat mesenteric branch artery (MBA). These cells can be enzymatically dispersed and maintained in culture [11].

The present study was designed to further characterize ANNIES in adult rat MBA as cells with axonal and neurite growth markers, such as growth-associated protein-43

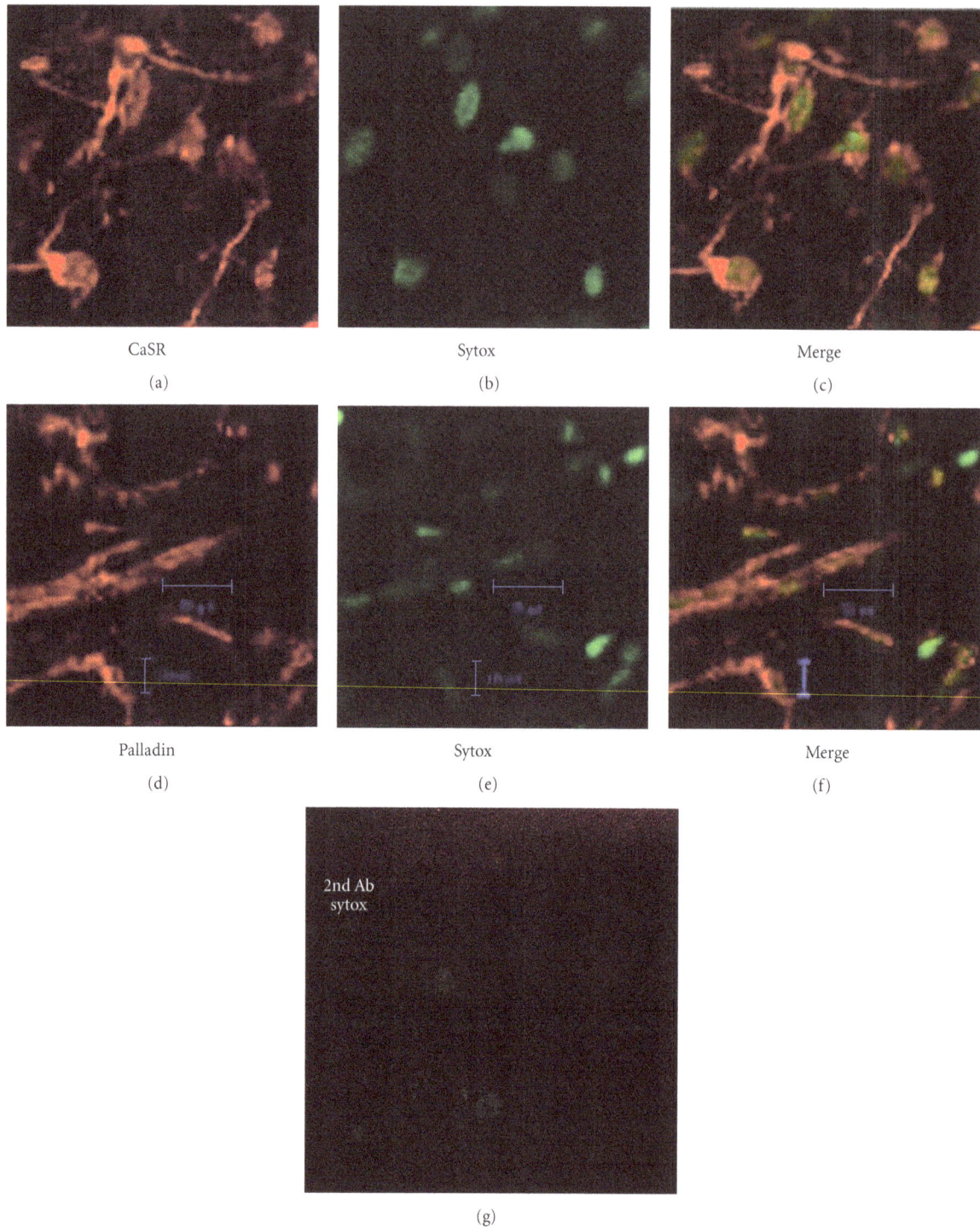

FIGURE 1: Immunofluorescent confocal analysis of the whole mount rat MBA showing the CaSR and palladin expression. ANNIES and the nerve fibers strongly express (94 ± 6%) the CaSR protein (top panel—in red, alexa fluor—647). Fibroblast like cells with no processes, and cells with nerve fibers (ANNIES) showing palladin expression (80 ± 10%) in red (lower panel). Nuclei stained in green (sytox)—(100x), (g). Rat MBA stained for secondary antibody alone, tagged with alexa fluor-647, and nuclear stain sytox—(100x). (Representative of 3 experiments.)

(Gap-43) and palladin, and to determine whether they express the calcium sensing receptor (CaSR). Recently, CaSR mRNA and protein expression has been demonstrated in rat whole MBA and other vascular tissue extracts, also shown in vascular smooth muscle cells in culutre [15, 16]. None of these studies show the specific cell(s) in the artery in situ expressing the CaSR. We demonstrate here that nucleated ANNIES cells together with nerve processes clearly express

Identification and Characterization of Novel Perivascular Adventitial Cells in the Whole Mount Mesenteric Branch Artery
Using Immunofluorescent Staining and Scanning Confocal Microscopy Imaging

75

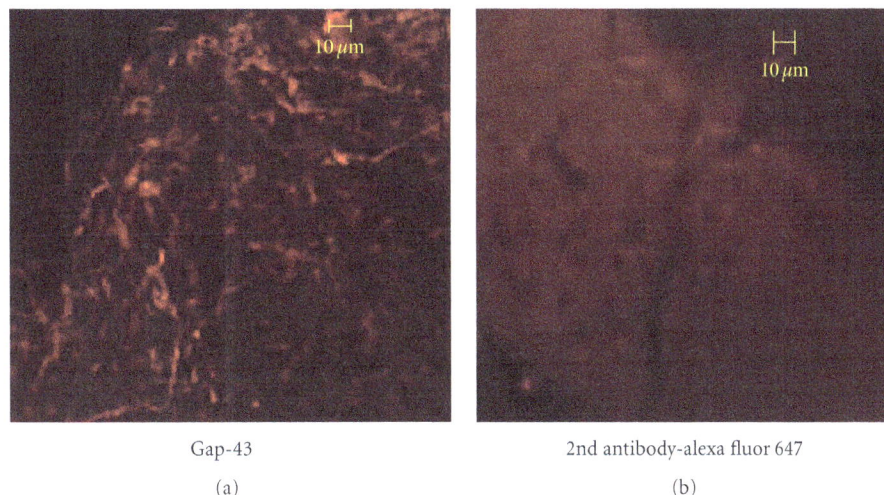

Gap-43	2nd antibody-alexa fluor 647
(a)	(b)

FIGURE 2: Immunofluorescent confocal analysis of the whole mount rat MBA expressing Gap-43. The whole mount of adult rat mesenteric arterial adventitia expresses $37 \pm 4\%$ Gap-43 (in red, alexa fluor-647) under scanning fluorescent confocal microscopy—20x. (Representative of 3 experiments.)

palladin, CaSR, and coexpress Gap-43 protein in the whole mount MBA by immunofluorescent staining followed by laser confocal analysis. In addition, we confirm the expression of these markers using Western blot analysis.

2. Materials and Methods

2.1. Animals. All procedures using laboratory animals were reviewed and approved by the Institutional Animal Care and Use Committee of North Carolina Central University. Male Wistar rats, 10–12 weeks of age, were obtained from Harlan Sprague Dawley (Indianapolis, USA). All animals were continually monitored, and upon arrival, they were maintained in colony rooms with fixed light: dark cycles and constant temperature and humidity and provided with Purina rodent chow (Harlan Teklad, Madison, Wis, USA) and water ad libitum.

2.2. Preparation and Isolation of Vessels. Mesenteric arteries were dissected from rats ($n = 10$) as previously described [11]. In brief, rats were deeply anesthetized with 2% isoflurane and then sacrificed by open chest cardiac puncture. The small intestine and all vessels feeding it were removed in block and placed in cold physiological salt solution (PSS). Branch I and II arteries were carefully dissected from the surrounding fat and mesenterium, taking care to leave a portion of the omental membrane attached to the vessel. A $12 \mu m$-diameter stainless steel wire was inserted into the lumen to remove blood and to serve as a handle for moving the vessel segment between solutions.

2.3. Immunostaining and Confocal Analysis. Vessels of MBA were fixed in buffered formalin for 20 min and washed three times in TBS (Tris-buffered saline). The vessels were kept in blocking solution containing 8% BSA in TBS and incubated with primary antibody such as polyclonal anti-Gap-43, anti-CaSR (Molecular probes), anti-palladin (gift from Dr. Otey,

UNC-Chapel Hill) overnight at 4°C. For dual immunofluorescence staining anti-CGRP (Phoenix Pharmaceuticals, Calif), and anti-Gap-43 (Molecular probes) were incubated for 2 h at room temperature (RT). After incubation, the vessels were washed thrice in TBS and incubated with appropriate secondary antibody tagged with alexa fluor 647 alone, and with 488 in case of co-staining for 1 h at RT. After washing thrice in TBS, vessels were incubated with the nuclear stain Sytox (Petticoat Junction, OR) to identify the nuclei of ANNIES. Vessels were mounted on glass slides in a glycerol-based Antifade medium (Molecular probes) after washing. Segments were viewed with a Zeiss LSM 510 confocal microscope (Zeiss Instruments) with 100x, 40x, and 20x oil immersion objectives.

We considered the number of CGRP staining cells as 100% of ANNIES, and calculated the percentage of ANNIES expressing the other markers (Gap-43, CaSR and palladin) in comparison to the CGRP expressing cells.

2.4. Western Blotting. Protein was extracted from minced MBA ($n = 5$) by homogenizing in a ground-glass homogenizer (MBA) in buffer containing 10 mM Tris, pH 7.5, 0.25 M sucrose, 3 mM $MgCl_2$ containing 1% (v/v) Triton X-100, dithiothreitol (1 mM), Pe-fabloc (1 mM), leupeptin (10 μM), bestatin (130 μM), pepstatin (1 μM), and calpain inhibitor II (10 $\mu g/mL$). The homogenate was then centrifuged at 16,000 g for 10 min and the pellet was used for NCAM, Gap-43 and palladin. The pellet was dissolved in buffer containing 10 mM Tris, pH 7.5, and 1% Triton X-100, size separated using 8% SDS-PAGE, and electroblotted onto nitrocellulose membrane (Bio-Rad Laboratories, CA) as described [17]. The membrane was then probed with antibodies separately (Chemicon International Inc., Calif) and visualized using an HRP-conjugated secondary antibody, and the chemiluminescence method (Amersham Pharmacia Biotech, NJ).

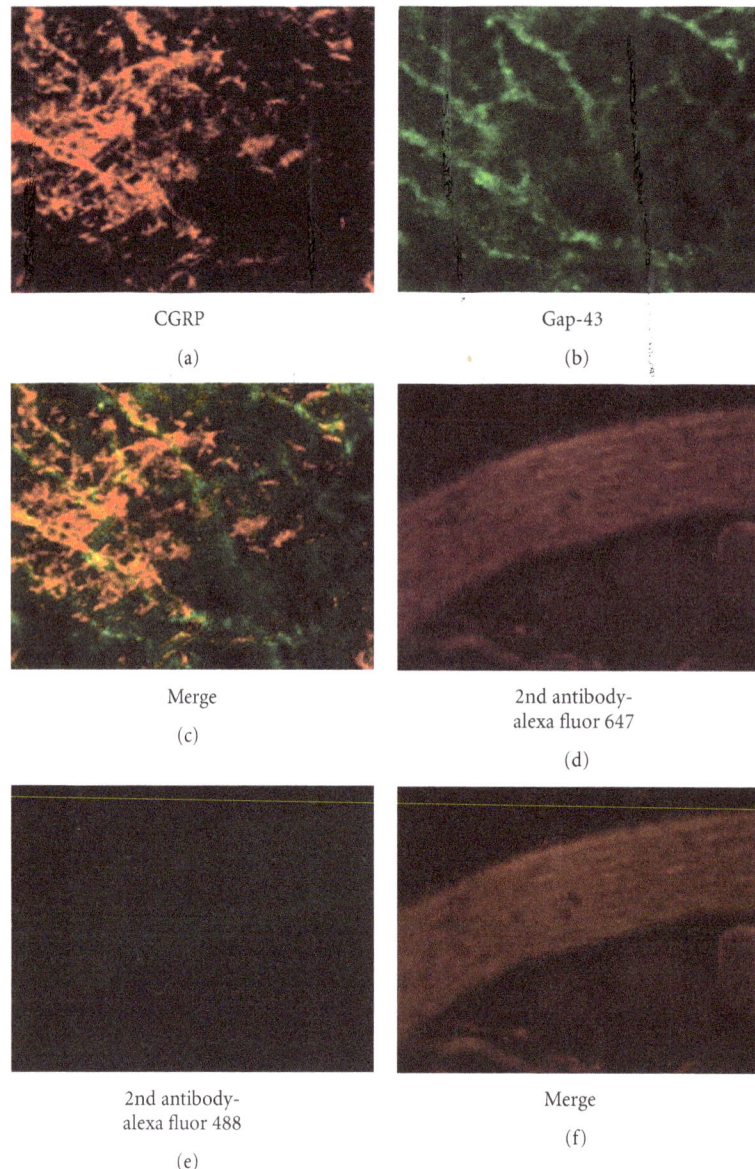

FIGURE 3: Immunofluorescent confocal analysis of the whole mount rat MBA expressing CGRP (in red, alexa fluor-647), Gap-43 (in green, alexa fluor-488), and the coexpression of these two markers (yellow). ANNIES and the nerve fibers strongly express CGRP (100%), and Gap-43. (Representative of 3 experiments.)

3. Results and Discussion

3.1. Immunofluorescent Analysis of the Whole Mount MBA by Confocal Microscopy. We used the whole mounts of the adult rat MBA and immunofluorescence confocal microscopy imaging to further characterize ANNIES in arterial adventitia. The use of whole-mount MBA from rats enabled us to have an accurate reconstruction of the cellular and nerve interrelationships and innervation patterns of perivascular adventia and vasa nervorum within the MBA. This also facilitates the ability to study the expression pattern of the neurochemical markers in situ, without the need to cut them as with conventional histology. Furthermore, this approach allows us to localize and visualize the morphology/phenotype and quantify these cells (ANNIES) and nerve

structures *in situ* coexpressing vasodilator trophic factor CGRP and NCAM [11], and axonal growth markers Gap-43 (Figure 3) as well as CaSR (Figure 1). CaSR is shown to play significant role in blood pressure regulation by releasing an unknown vasodilator [18]. More recently, Weston et al. [19] reported that this vasodilator is still unidentified.

3.2. ANNIES and Perivascular Nerve Fibers in Rat MBA Express CaSR. We hypothesize that the vasodilator released by CaSR, which is still unidentified [19], might be the potent vasodilator neuropeptide, CGRP. Our present finding demonstrates that $94 \pm 6\%$ ($n = 3$) ANNIES cell bodies express the CaSR (Figure 1), compared with the ANNIES identified by the expression of CGRP as previously demonstrated [11]. The CaSR is shown to regulate the production

Identification and Characterization of Novel Perivascular Adventitial Cells in the Whole Mount Mesenteric Branch Artery
Using Immunofluorescent Staining and Scanning Confocal Microscopy Imaging

77

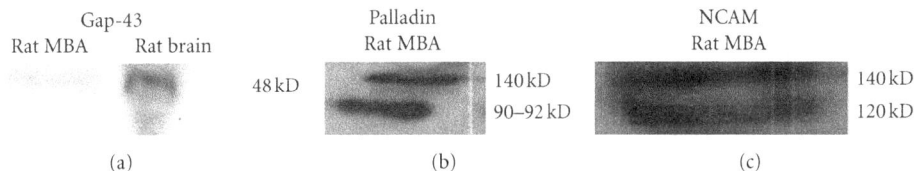

FIGURE 4: Western blot analysis of Gap-43, palladin, NCAM in adult rat MBA. Immunoblots from the rat MBA protein lysate also confirmed the presence of Gap-43 palladin and NCAM. Palladin expresses as two isoforms of 90–92 and 140 kD proteins. Gap-43 antibody detected a ~48 kDa protein, whereas anti-NCAM detected 120 and 140 kD proteins. (Representative of 3 experiments.)

of CGRP [20] that is involved in the regulation of the release of the neurotransmitter in the synaptic space [20]. Our laboratory has demonstrated that the CaSR mediates Ca2+ induced relaxation of isolated mesenteric arteries [18] via an unknown vasodilator substance, which is independent of endothelium and NO [18]. These previous findings support our hypothesis that this putative vasodilator released by the CaSR might be the CGRP expressed in ANNIES and perivascular nerve fibers [11]. We demonstrate here that nucleated ANNIES cells and the nerve processes strongly express the CaSR.

3.3. Rat MBA and ANNIES Express Gap-43, Palladin and Coexpress CGRP.

Gap-43 is a marker of neural outgrowth and regeneration [21]. It is shown to locate in the growth cones of growing neurites, where it interacts with F-actin associated adhesion molecule and/or extracellular matrix complexes to promote neurite extension [22]. In addition, it has been shown that the phosphorylated Gap-43 stabilizes long actin filaments and has the ability to directly influence the structure of the actin cytoskeleton response to extracellular signals [23]. Similarly, palladin has been shown to express in the growth cone and colocalize with focal adhesion [24] and respond to vascular [25] and dermal injury [26].

Confocal analysis of the whole mount MBA in our experiment showed that Gap-43 is strongly distributed in 37 ± 4% of the perivascular nerve fibers (Figure 2, $n = 3$), and coexpressing CGRP (Figure 3), also identified by the expression of CaSR (Figure 1). Palladin is present in both SMC [25] and ANNIES in the rat MBA (Figure 1). Palladin expression is apparent in a mixed population of fibroblasts and in cells with nerve processes, ANNIES (Figure 1) that were shown to coexpress CGRP and NCAM [11]. The CaSR and palladin were present in ANNIES about 94 ± 6% and 80 ± 10% ($n = 3$), respectively. Strong immunofluorescence for NCAM [11], and axonal growth markers such as Gap-43 as reported in this paper, revealed the presence of many neuronal cell bodies within the vasa nervorum of rat MBA. In addition, the presence of Gap-43 in ANNIES indicates that ANNIES shows features of sensory neurons of the DRG, undergoing responses to stress/injury, given that the expression of Gap-43 in DRG cells is increased in response to injury [27, 28]. In addition, CGRP induces schwann cell proliferation, and thus thought to be involved in peripheral nerve injury and repair [29].

3.4. Western Blot Analysis of MBA Expressing NCAM, Gap-43, and Palladin.

Immunoblots from the rat MBA also confirmed the presence of Gap-43 (Figure 4) and palladin as two isoforms of 90–92 and 140 kD ($n = 3$). This protein pattern is different from other adult tissues such as brain that expresses three isoforms (90–92, 140, and 200 kD) or SMC where only one isoform (90–92 kDa) is expressed [30]. CaSR protein expression by western blot in the protein extract of the whole mesenteric artery was reported [31]. We previously demonstrated beta-CGRP expression in rat MBA by RT-PCR, cloning, and sequence analysis [11].

Cells that are not with nerve fibers, probably fibroblasts, were also stained for palladin as it has been reported by other investigators [30]. There is no nerve like structures around some of the fibroblast like cells that express palladin. In addition, NCAM, CGRP [11], and Gap-43 are expressed in ANNIES, but these markers are not shown to express in the other adventitial fibroblasts, smooth muscle, and endothelial cells.

4. Conclusion

This is the first report that cells in the peripheral vasculature with a neuronal phenotype express markers of active neurite growth. The presence of CGRP-containing neural cells in the vascular adventitia may participate in response to injury and vasodilator mechanisms as part of a perivascular sensory neural network. ANNIES cell morphology with nerve fibers and expression of neural and neurite growth markers reveal that these cells are distinct from the other known cells in blood vessel. The additional finding that the CaSR is associated with ANNIES suggests that these cells may participate in the regulation of myogenic tone.

Acknowledgments

The authors thank Dr. Carol A. Otey (University of North Carolina, Chapel Hill) for providing them with monoclonal and polyclonal antibodies for palladin. This work was partly supported by Texas Nerve and Paralysis Institute, Houston, Texas. (Support: HL-64761 R. D.) Bukoski passed away.

References

[1] A. Pacilli and G. Pasquinelli, "Vascular wall resident progenitor cells. A review," *Experimental Cell Research*, vol. 315, no. 6, pp. 901–914, 2009.

[2] E. Zengin, F. Chalajour, U. M. Gehling et al., "Vascular wall resident progenitor cells: a source for postnatal vasculogenesis," *Development*, vol. 133, no. 8, pp. 1543–1551, 2006.

[3] C. W. Chen, E. Montelatici, M. Crisan et al., "Perivascular multi-lineage progenitor cells in human organs: regenerative units, cytokine sources or both?" *Cytokine and Growth Factor Reviews*, vol. 20, no. 5-6, pp. 429–434, 2009.

[4] M. Corselli, C. W. Chen, M. Crisan, L. Lazzari, and B. Péault, "Perivascular ancestors of adult multipotent stem cells," *Arteriosclerosis, Thrombosis, and Vascular Biology*, vol. 30, no. 6, pp. 1104–1109, 2010.

[5] M. Crisan, C. W. Chen, M. Corselli, G. Andriolo, L. Lazzari, and B. Péault, "Perivascular multipotent progenitor cells in human organs," *Annals of the New York Academy of Sciences*, vol. 1176, pp. 118–123, 2009.

[6] M. Crisan, B. Deasy, M. Gavina et al., "Purification and long-term culture of multipotent progenitor cells affiliated with the walls of human blood vessels: myoendothelial cells and pericytes," *Methods in Cell Biology*, vol. 86, pp. 295–309, 2008.

[7] M. Crisan, J. Huard, B. Zheng et al., "Purification and culture of human blood vessel-associated progenitor cells," *Current Protocols in Stem Cell Biology*, chapter 2, pp. 2B.2.1–2B.2.13, 2008.

[8] M. Crisan, S. Yap, L. Casteilla et al., "A perivascular origin for mesenchymal stem cells in multiple human organs," *Cell Stem Cell*, vol. 3, no. 3, pp. 301–313, 2008.

[9] D. Klein, H. P. Hohn, V. Kleff, D. Tilki, and S. Ergün, "Vascular wall-resident stem cells," *Histology and Histopathology*, vol. 25, no. 5, pp. 681–689, 2010.

[10] M. Tavian, B. Zheng, E. Oberlin et al., "The vascular wall as a source of stem cells," *Annals of the New York Academy of Sciences*, vol. 1044, pp. 41–50, 2005.

[11] C. Somasundaram, D. I. Diz, T. Coleman, and R. D. Bukoski, "Adventitial neuronal somata," *Journal of Vascular Research*, vol. 43, no. 3, pp. 278–288, 2006.

[12] B. van der Loo and J. F. Martin, "The adventitia, endothelium and atherosclerosis," *International Journal of Microcirculation-Clinical and Experimental*, vol. 17, no. 5, pp. 280–288, 1997.

[13] S. M. Arribas, C. J. Daly, M. C. González, and J. C. Mcgrath, "Imaging the vascular wall using confocal microscopy," *Journal of Physiology*, vol. 584, no. 1, pp. 5–9, 2007.

[14] W. Warwick, *DB: Gray's Anatomy*, Churchill Livingstone, Philadelphia, Pa, USA, 37th edition, 1989.

[15] E. Harno, G. Edwards, A. R. Geraghty et al., "Evidence for the presence of GPRC6A receptors in rat mesenteric arteries," *Cell Calcium*, vol. 44, no. 2, pp. 210–219, 2008.

[16] S. Smajilovic, J. L. Hansen, T. E. H. Christoffersen et al., "Extracellular calcium sensing in rat aortic vascular smooth muscle cells," *Biochemical and Biophysical Research Communications*, vol. 348, no. 4, pp. 1215–1223, 2006.

[17] Y. Wang, E. K. Awumey, P. K. Chatterjee et al., "Molecular cloning and characterization of a rat sensory nerve Ca2+-sensing receptor," *American Journal of Physiology—Cell Physiology*, vol. 285, no. 1, pp. C64–C75, 2003.

[18] R. D. Bukoski, K. Bian, Y. Wang, and M. Mupanomunda, "Perivascular sensory nerve Ca2+ receptor and Ca2+-induced relaxation of isolated arteries," *Hypertension*, vol. 30, no. 6, pp. 1431–1439, 1997.

[19] A. H. Weston, A. Geraghty, I. Egner, and G. Edwards, "The vascular extracellular calcium-sensing receptor: an update," *Acta Physiologica*, vol. 203, no. 1, pp. 127–137, 2011.

[20] M. Ruat, M. E. Molliver, A. M. Snowman, and S. H. Snyder, "Calcium sensing receptor: molecular cloning in rat and localization to nerve terminals," *Proceedings of the National Academy of Sciences of the United States of America*, vol. 92, no. 8, pp. 3161–3165, 1995.

[21] S. Mechsner, J. Schwarz, J. Thode, C. Loddenkemper, D. S. Salomon, and A. D. Ebert, "Growth-associated protein 43-positive sensory nerve fibers accompanied by immature vessels are located in or near peritoneal endometriotic lesions," *Fertility and Sterility*, vol. 88, no. 3, pp. 581–587, 2007.

[22] Y. Shen, S. Mani, S. L. Donovan, J. E. Schwob, and K. F. Meiri, "Growth-associated protein-43 is required for commissural axon guidance in the developing vertebrate nervous system," *Journal of Neuroscience*, vol. 22, no. 1, pp. 239–247, 2002.

[23] Q. He, E. W. Dent, and K. F. Meiri, "Modulation of actin filament behavior by GAP-43 (neuromodulin) is dependent on the phosphorylation status of serine 41, the protein kinase C site," *Journal of Neuroscience*, vol. 17, no. 10, pp. 3515–3524, 1997.

[24] S. M. Goicoechea, D. Arneman, and C. A. Otey, "The role of palladin in actin organization and cell motility," *European Journal of Cell Biology*, vol. 87, no. 8-9, pp. 517–525, 2008.

[25] L. Jin, Q. Gan, B. J. Zieba et al., "The actin associated protein palladin is important for the early smooth muscle cell differentiation," *PLoS ONE*, vol. 5, no. 9, Article ID e12823, pp. 1–13, 2010.

[26] M. J. Rönty, S. K. Leivonen, B. Hinz et al., "Isoform-specific regulation of the actin-organizing protein palladin during TGF-β1-induced myofibroblast differentiation," *Journal of Investigative Dermatology*, vol. 126, no. 11, pp. 2387–2396, 2006.

[27] P. A. Vo and D. R. Tomlinson, "Effects of nerve growth factor on expression of GAP-43 in right atria after sympathectomy in diabetic rats," *Diabetes, Obesity and Metabolism*, vol. 3, no. 5, pp. 350–359, 2001.

[28] S. Y. Tsai, L. Y. Yang, C. H. Wu et al., "Injury-induced Janus kinase/protein kinase C-dependent phosphorylation of growth-associated protein 43 and signal transducer and activator of transcription 3 for neurite growth in dorsal root ganglion," *Journal of Neuroscience Research*, vol. 85, no. 2, pp. 321–331, 2007.

[29] C. C. Toth, D. Willis, J. L. Twiss et al., "Locally synthesized calcitonin gene-related peptide has a critical role in peripheral nerve regeneration," *Journal of Neuropathology and Experimental Neurology*, vol. 68, no. 3, pp. 326–337, 2009.

[30] M. M. Parast and C. A. Otey, "Characterization of palladin, a novel protein localized to stress fibers and cell adhesions," *Journal of Cell Biology*, vol. 150, no. 3, pp. 643–655, 2000.

[31] A. H. Weston, M. Absi, D. T. Ward et al., "Evidence in favor of a calcium-sensing receptor in arterial endothelial cells: studies with calindol and Calhex 231," *Circulation Research*, vol. 97, no. 4, pp. 391–398, 2005.

9

Endothelial Cells and Astrocytes:
A *Concerto en Duo* in Ischemic Pathophysiology

Vincent Berezowski,[1,2,3] **Andrew M. Fukuda,**[4] **Roméo Cecchelli,**[1,2,3] **and Jérôme Badaut**[4,5]

[1] *Université Lille Nord de France, 59000 Lille, France*
[2] *UArtois, LBHE, EA 2465, 62300 Lens, France*
[3] *IMPRT-IFR114, 59000 Lille, France*
[4] *Departments of Physiology, Loma Linda University School of Medicine, Loma Linda, CA 92354, USA*
[5] *Departments of Pediatrics, Loma Linda University School of Medicine, Loma Linda, CA 92354, USA*

Correspondence should be addressed to Jérôme Badaut, jbadaut@llu.edu

Academic Editor: Carola Förster

The neurovascular/gliovascular unit has recently gained increased attention in cerebral ischemic research, especially regarding the cellular and molecular changes that occur in astrocytes and endothelial cells. In this paper we summarize the recent knowledge of these changes in association with edema formation, interactions with the basal lamina, and blood-brain barrier dysfunctions. We also review the involvement of astrocytes and endothelial cells with recombinant tissue plasminogen activator, which is the only FDA-approved thrombolytic drug after stroke. However, it has a narrow therapeutic time window and serious clinical side effects. Lastly, we provide alternative therapeutic targets for future ischemia drug developments such as peroxisome proliferator-activated receptors and inhibitors of the c-Jun N-terminal kinase pathway. Targeting the neurovascular unit to protect the blood-brain barrier instead of a classical neuron-centric approach in the development of neuroprotective drugs may result in improved clinical outcomes after stroke.

1. Introduction: Current Clinical Overview of Stroke

In the United States, stroke is the number one cause of chronic disability and the fourth leading cause of death, with approximately 7 million adults affected [1]. Annually there are approximately 800,000 strokes in the US, of which 87% are ischemic strokes, 10% are primary hemorrhages, and 3% are subarachnoid hemorrhages [1]. Together they cause the country a financial burden of approximately 62.7 billion dollars [2]. Cerebral ischemic stroke is caused by an occlusion of a cerebral blood vessel, typically by a thrombus, which causes a decrease in cerebral blood flow and thus limits the supply of oxygen and nutrients globally (in global ischemia) or to certain regions of the brain (in focal brain ischemia). This absence of blood flow in a brain region causes neuronal death in addition to damaging the vascular tree; the vascular tree is usually made more fragile during the ischemic period and damaged during reperfusion. Time is

an important parameter in the evolution of brain injury. In 2006, Saver et al. have estimated the impact of stroke on the brain tissue [3] to be immense; the brain may lose up to 120 million neurons, 830 billion synapses and 714 km of myelinated fibers for each hour after stroke onset [3]. Ischemic stroke seems to accelerate aging of the brain at a rate of 3.6 years each time when the symptoms are not treated [3]. Therefore, the clinical goal of acute stroke treatment is to reduce brain damage by limiting the time of ischemia through thrombectomy (mechanical endovascular approach) or thrombolytic therapy, which consists of in lysing the blood clot in order to restore cerebral blood flow.

Recombinant tissue plasminogen activator (rtPA) is currently the only thrombolytic molecule administered during acute cerebral infarction that provides a clinical benefit in terms of survival and neurological outcome [4]. The rtPA administration must be within the first 4 hours 30 minutes after stroke onset to maintain the beneficial effects without substantially raising the side effects/risk [5, 6], which limits

its use. Based on the organization of emergency care, only 5% of stroke patients are eligible for this therapy in this narrow time window, which leaves the remaining 95% of patients without any beneficial treatment available. The major risk of rtPA is the extension of the damage due to potential bleeding [7]. The need for drug development to prevent the neuronal loss has driven research on neuroprotective agents that aim to save viable neurons located in the ischemic penumbra area. However, all of the proposed neuroprotective treatments specifically targeting neurons that showed promise on the bench have failed in clinical trials [8].

In 2000, the neurovascular unit (NVU) was proposed as a physiological unit composed by neurons, astrocytes, and endothelial cells [9]; there is a growing interest in studying the changes of the NVU after stroke. In addition to cell death, ischemic stroke is characterized by changes in the properties of the blood-brain barrier (BBB) with physical disruption of the tight junctions contributing to aggravation of cerebral edema and consequently neuronal death. The new strategy for drug development is to have molecules with a broader spectrum targeting not just the neurons but the NVU as a whole entity. In the present paper, we will focus on some molecular and cellular mechanisms of astrocytes and endothelial cells. We will look specifically at: (1) the ways astrocytes and endothelial cells work in concert in stroke pathophysiology such as BBB disruption and edema formation, (2) how they could be affected after rtPA treatment, and (3) new drug developments in the future.

2. Definition of the Neurovascular/ Gliovascular Unit

Several groups have proposed the NVU as a physiological unit composed of not only endothelial cells, astrocytes, and neurons but also pericytes, smooth muscle cells, and the interacting circulating peripheral immune cells [10–12]. The term "gliovascular" emphasizes the importance of the interactions between astrocytes and cerebral blood vessels within the NVU [13], which are critical in cerebral blood flow regulation [14], brain energy metabolism [15], and also the maintenance of the BBB properties [13].

The BBB is located in the endothelial cells of brain vessels, with the presence of tight junctions and adherens junctions between the cells (Figure 1) that prevent paracellular diffusion and act as a unit to regulate ions and other molecules between peripheral blood flow and brain parenchyma. Tight junctions are composed of several protein families: transmembrane proteins (claudins and occludins), cytoplasmic proteins, and zona occludens proteins. They bind the afore mentioned proteins with structural cytoskeletal proteins such as actin. Adherens junctions are formed by proteins such as platelet-endothelial cell adhesion molecule (PECAM) and vascular endothelial-cadherin, which contribute to the close physical contact between endothelial cells and facilitate the formation of tight junctions.

The brain endothelial cells of the BBB also present specific transport proteins located on the luminal and abluminal membranes for nutrients, ions, and toxins to cross the endothelial layer between the blood stream and brain [13, 16]. For example, energy molecules are transported by specific solute carriers such as glucose transporter 1 (GLUT 1) and monocarboxylate transporters 1 and 2 (MCT1, MCT2). Large molecular weight solutes (e.g., large proteins and peptides) are able to cross the BBB and enter the intact CNS via endocytotic mechanisms called receptor-mediated transcytosis, such as with insulin, or adsorptive-mediated transcytosis, exemplified by albumin. On the other hand, transport can also be achieved by the ATP-binding protein (ABC) family, which consumes ATP to effectively transport a wide range of lipid-soluble compounds from the brain endothelium. In the BBB examples of ABC transporters for efflux transport are P-glycoprotein (P-gp), multidrug resistance-associated protein (MRP), and breast cancer resistance protein (BCRP) [16]. These efflux transporters are understood as gatekeepers of the brain because they keep tight control over which substances are allowed to enter the CNS through the endothelial cell barrier (Figure 1). Endothelial cells also present a metabolic barrier of the BBB, which functions to inactivate molecules capable of penetrating cerebral endothelial cells.

Quite recently it has been proposed that the primary barrier of the BBB may extend to the basal lamina, thus preventing the entry of immune cells into the parenchyma under normal brain conditions [12]. Historically the brain was thought to be an immune cell deficient organ, and the BBB was thought to prevent passage of any immune cells into the brain. However, peripheral immune cells from the blood have been observed to enter and be present in the brain at multiple time points during embryonic development [17] and in normal physiological conditions in adults [12]. Therefore, the theory of the CNS as an immune-independent organ has recently started to be reexamined and revised. Engelhardt and collaborators elegantly compare the perivascular space as a castle moat with perivascular antigen presenting cells floating as guards, confined by the inner and outer wall, which is the basement membrane of the astrocytic endfeet and the endothelial cell, respectively [12]. Endothelial cells and other cells, such as the astrocytes, may also contribute to the tight regulation of the movement of immune cells between the peripheral blood stream and the brain. However, the exact mechanisms by which peripheral cells enter the brain are still a matter of discussion. Moreover, rather than the BBB being a rigid wall, it provides a dynamic interface between the brain and the rest of the body.

As mentioned previously, the presence and the maintenance of these barrier properties are important for brain homeostasis and for neuronal functioning [13]. In fact, disruption of tight junctions leads to BBB disruption and extravasation of blood components and water, which contribute to vasogenic edema formation. We will cover these in more detail in the following section.

3. Edema Process after Stroke: Endothelium and Astrocyte, *Concerto en Duo*

3.1. BBB Disruption and Edema Formation. Cerebral edema has been traditionally divided into 2 major classes: cytotoxic and vasogenic [18] for cerebrovascular diseases and other

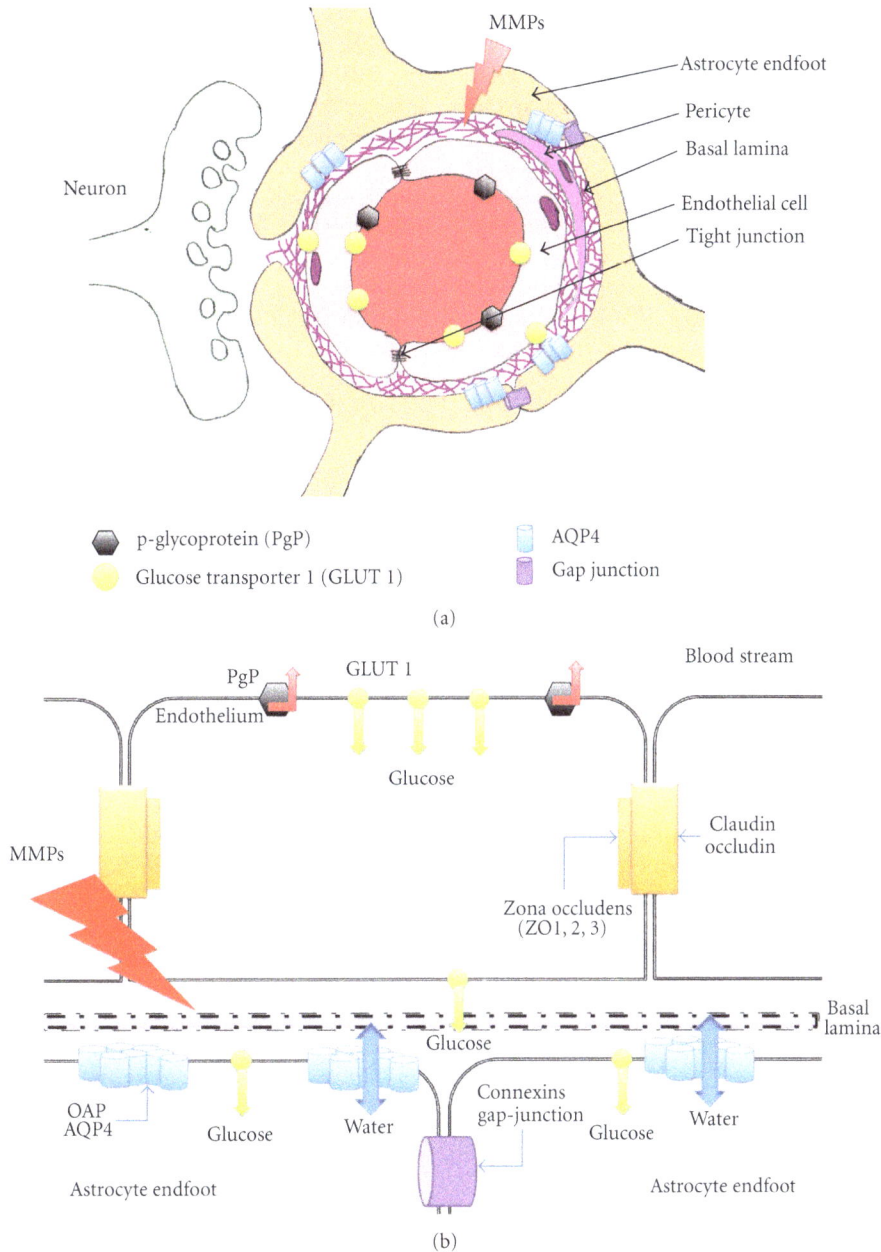

FIGURE 1: (a) Schematic drawing of the neurovascular unit (NVU) in the capillary bed composed by the neuron, astrocyte endfoot, basal lamina, pericyte, and endothelial cell. The endothelial cell is the first barrier between the blood stream and the nervous tissue. The presence of the tight junction composes the physical barrier and the movement of substrates is controlled by several transporters. The astrocyte endfeet are linked with the gap junction, allowing movement of several solutes in the astrocyte network. The basal lamina is composed of several proteins such as agrin, dystroglycan, and perlecan. (b) A close-up schematic drawing of the endothelial cells and astrocyte endfeet with some of the proteins involved in edema formation and resolution.

brain pathologies. Cytotoxic edema is defined by intracellular accumulation of water coming from the extracellular space without BBB disruption. Vasogenic edema appears after BBB disruption, leading to a diffusion of proteins from the blood to the tissue followed by water accumulation in the extracellular space [18]. However, this division alone does not explain fully the diversity and the complexity of the edema process in brain ischemia as well as in the other brain injuries and disorders. Based on several recent advances in

the understanding of the molecular mechanisms of edema formation and BBB properties, a third subtype of edematous processes was named *ionic edema* and described as a continuum between the cytotoxic to vasogenic edema in the cerebrovascular diseases [19, 20]. In fact, cytotoxic, or anoxic, edema occurs within the first few minutes after cerebral blood flow stoppage and is characterized as swelling of the astrocytes and neuronal dendrites [20, 21]. The cellular swelling within the first 10 minutes is a result of oxygen and glucose deprivation followed by a slow rise in extracellular [K$^+$] [22]. The absence of oxygen and energy nutrients induces a disruption of the cellular ionic gradients and leads to entry of ions into cells. Water follows this ionic gradient into the cells and induces cellular swelling. Cytotoxic/anoxic edema may evolve quickly to become ionic edema because the absence of oxygen and nutrients further alters the energy balance in endothelial cells and the ionic gradients, including transcapillary flux of Na$^+$ in these cells [19, 23]. The endothelial cells also require a large amount of ATP production, characterized by the high density of mitochondria, which are important for the regular homeostatic BBB functions such as maintenance of ionic gradients and membrane transporters [24, 25]. The absence of energy supplies for these cells would severely impair these functions. Reperfusion induces overpressure accompanied by shear stress on the nonperfused vascular tree that results in early transient leakage of the BBB [26, 27]. This leakage results in further entry of water through the endothelial cells resulting in brain swelling within 30 minutes after reperfusion [26, 27] and additional BBB permeability [27, 28]. This early opening of the BBB has also been described clinically in humans and is frequently associated with hemorrhagic transformation [29]. Early reperfusion probably mitigates the BBB alterations, but if it is delayed, reperfusion will exacerbate the amount of endothelial injury [30–32]. The final step is the development of vasogenic edema, in which there is disruption of cerebrovascular endothelial tight junctions leading to increased permeability to albumin and other plasma proteins [18]. Another contributing factor of brain edema formation in addition to tight junction disruption is brain endothelial transcytosis [33]. BBB disruption is usually coupled with the inflammatory response and activation of matrix metalloproteinases (MMP) [34, 35]. In fact, vasogenic edema development is aggravated by MMP-9, which degrades basal lamina, the connection between astrocytic endfeet and endothelial cells [36].

In the clinic, diffusion-weighted imaging (DWI) and T2-weighted imaging (T2WI) magnetic resonance imaging (MRI) modalities are used extensively to assess postischemic edema [20, 37, 38]. T2 values represent water content and apparent diffusion coefficient (ADC) values derived from DWI images represent water mobility in the tissue [20, 37]. ADC values decrease rapidly after stroke onset, indicating restricting water movement, and are interpreted as evidence of ionic edema with the characteristic swelling of the brain cells causing a decrease in extracellular space as proposed in our classification mentioned before. T2 values increase at later time points, which are associated with vasogenic edema [20, 39].

The molecular mechanisms and temporal development of edema after stroke have been well studied. However, the cellular and molecular mechanisms involved in edema resolution are not well understood in stroke and other brain diseases. The healing of the endothelial cells with stabilization of the tight junctions may be a critical step to limit the entry of blood components into the brain. Thus, stabilizing the NVU may be an essential component of controlling edema formation and BBB breakdown after stroke.

Postischemic BBB disruption has been commonly believed to be biphasic [40], but recent work suggests that the BBB disruption may be continuous for up to 5 weeks after ischemia in rats [28]. BBB leakage was demonstrated using gadolinium and magnetic resonance imaging (MRI) at 25 min; 2, 4, 6, 12, 18, 24, 36, 48, and 72 hours; and 1, 2, 3, 4, and 5 weeks after ischemia [28]. Similarly, albumin leakage through the BBB, especially in the hippocampus, has also been observed in spontaneously hypertensive stroke prone rats long term [41]. Although these data do not completely rule out the possibility of a biphasic pattern in the opening of the BBB, the long-term leakage of the BBB is important to note from the standpoint of postischemic edema because this disruption could account for a prolonged vasogenic edema.

3.2. *"Concerto en Duo": Astrocyte Network in Edema Formation and Resolution.* As part of the NVU the astrocyte endfeet in contact to the blood vessels are well known for to swell after stroke [42–44]. The recent knowledge on the transporters and channels in this astrocyte subdomain gives new perspectives on the understanding of astrocyte swelling. In fact, aquaporin 4 (AQP4), a member of the family of 13 water channel proteins, is proposed to have an important role in edema formation [20, 45]. AQP4 is the most abundant water channel in the brain, in part due to its high concentration on astrocytic endfeet which are in contact with all the cerebral blood vessels [46, 47]. More recently, AQP1 has also been described in a subpopulation of astrocytes within the nonhuman primate but not in rodents, suggesting interspecies differences and a possible role in brain water homeostasis [48]. AQP1 has also been reported to be present in peripheral endothelia and primary rat brain endothelial cell cultures [49]. Interestingly, Dolman and collaborators observed that mRNA AQP1 levels were lower in cultured brain endothelial cells when cocultured with astrocytes [49], suggesting an inhibition effect of the astrocytes on the AQP expression in endothelia. In fact, there are publications reporting a low level of AQP in endothelial cells *in vivo* [50], although AQP is more abundant in astrocytes [49, 51–54].

Currently, AQP4 is considered as a key player in the edema process by its location on the astrocyte endfeet [20, 55]. AQP4 is assembled in homotetramers where each individual aquaporin represents a water channel [56]. Interestingly, AQP4 is also organized in the astrocyte endfeet membrane in a larger geometric structure known as an orthogonal array of particles (OAPs), which has been described with freeze-fracture techniques and electron microscopy studies (Figure 2) [57]. OAPs are present in all astrocyte endfeet in contact with the blood vessels as well as the glia limitans. OAPs are formed with two isoforms of AQP4: long (AQP4-M1) and short splice variants (AQP4-M23). The ratio of AQP4-M1 to AQP4-M23 determines the size of these OAPs [57] in contact with the basal lamina of brain vessels (Figure 1). Experiments in oocytes showed that the AQP4-M23 isoform stabilizes the OAP structure [57, 58]. However, the exact functional roles of the OAPs remain unknown in normal and pathological conditions. Recently, AQP4-m1 mRNA and protein were found to increase quickly after stroke onset, while AQP4-m23 remained the same.

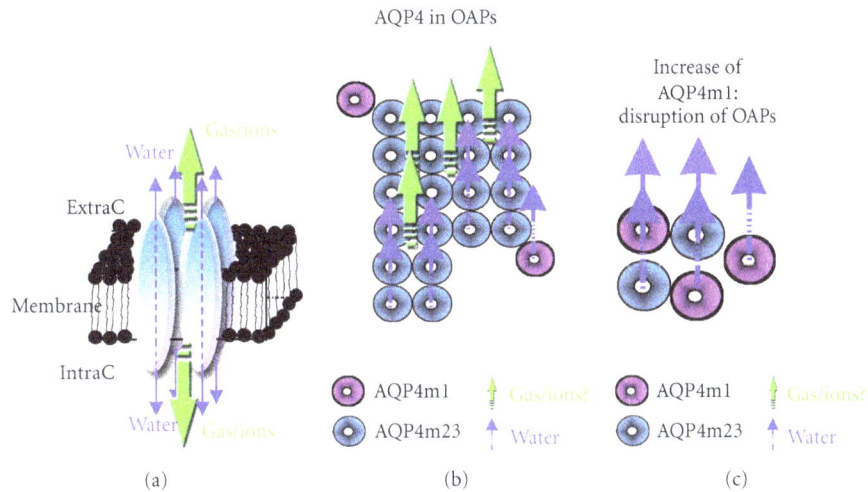

FIGURE 2: (a) Schematic drawing of the aquaporin homotetramer assembly within the lipid membrane: the central pore is proposed to be permeable to cations and gases (green arrows). Each individual aquaporin facilitates bidirectional water movement depending on the osmotic gradient (blue arrows). (b) AQP4 homotetramer is assembled in a higher structure named orthogonal array of particles (OAPs). Two isoforms of AQP4, AQP4-M1 (purple circles) and AQP4-M23 (blue circles) isoforms, contribute to the formation of OAPs. *In vitro* experiment showed that higher expression of AQP4-M23 contributes to the formation of larger OAPs. (c) Increase of AQP4-M1 induced disruption of OAPs. Recent knowledge on AQP leads us to hypothesize that the large OAPs contribute to gas and cation diffusion in the astrocyte membranes through central pores (green arrows).

The increase of AQP4-m1 early after ischemia could favor a shift toward M1 in the M1/M23 balance, which is known to favor small size OAPs [27]. In accordance with this work, previous studies have shown that early disorganization of OAPs on the astrocyte endfeet after global cerebral ischemia preceded astrocyte swelling [59]. Although a direct effect of the modification in the ratio of AQP4-M1 to AQP4-M23 on water permeability has not yet been directly investigated *in vivo*, in a preconditioning model, a strong increase in AQP4 expression and increase of AQP4-M1 were correlated with reduced edema and less water in the tissue, suggesting increased water diffusibility which resulted in the removal of excess liquid from the brain tissue [27]. Interestingly, it was recently proposed that the assemblage of 4 aquaporin molecules forms a central pore, through which water, ions, and gases may flow depending on the AQP subtype. For example, the central pore is permeable for O_2, CO_2, and possibly nitric oxide for AQP1, 4, and 5 [56, 60]. Thus, the disruption of the OAPs may also affect the diffusion of ions and gas through the central pore.

Due to its location in the astrocyte endfeet in contact with the blood vessels, AQP4 has been proposed to be linked with BBB integrity [52, 61, 62] and cell adhesion [63]. In the epithelial cells of the eye lens AQP0 is present in the OAPs and participates in epithelial cells linkage; however it does not facilitate water flux [64]. In this case, the presence of AQP4 in the astrocyte endfeet membrane was dependent on the presence of proteins in the basal lamina such as agrin, α-dystroglycan, and laminin [65, 66] in addition to syntrophin and dystrophin protein complexes [67, 68]. The connection of AQP4 to proteins in the basal lamina may explain the ability of astrocytes to maintain the integrity of the blood-brain barrier, suggesting a possible role for AQP4

as a structural molecule within the perivascular space [61]. However, reports using AQP4 knock out (AQP4-KO) mice show contradicting results regarding the modifications in the BBB structure suggesting that AQP4 may not be integral to the BBB structure [61, 69]. Similarly in our siRNA silencing studies, BBB permeability was not significantly changed at distance from the site of injection after injection of siRNA against AQP4, even though AQP4 expression was decreased [55]. We also showed that the upregulation of AQP4 in a preconditioning model did not prevent the early opening of the BBB after stroke [27].

Heparan sulfate proteoglycan is a large family of proteins with agrin and perlecan, involved in the basal lamina composition located between the astrocyte endfeet and endothelial cells [54, 70]. Agrin and dystroglycan seem to play an integral role in the maintenance of astrocyte polarity by the interaction with AQP4 in the astrocyte endfeet [54]. Specifically, agrin KO mice showed a significantly decreased density of OAP in the astrocyte endfeet when compared to wildype but overall immunoreactivity of AQP4 did not differ significantly [71]. Dysfunctions in the basal lamina are related to increase of the BBB disruption, promoting edema formation. In fact, a family of endopeptidases, matrix metalloproteinases (MMPs), has been shown to degrade the proteins of the basal lamina and contribute to vasogenic cerebral edema [36]. In the human brain, MMPs are usually very low in concentration under nonpathological conditions [72]. However, after injuries such as ischemic stroke, certain MMPs such as MMP-2, -3, and -7 and especially MMP-9 have been shown to be upregulated in the brain (reviewed in [72]). This layer between astrocytes and endothelial cells is a potential future target for the NVU protection. Recently, Dr. Bix and collaborators have shown that administration of perlecan domain V, which is the c-terminal fragment, administered 24 hours after ischemic stroke has beneficial effects by interacting with integrins [73]. Perlecan domain V increased expression of vascular endothelial growth factor (VEGF), thus promoting angiogenesis, and interestingly did not lead to increased BBB permeability [73] even though VEGF is known to increase BBB permeability after ischemia [74]. Perlecan has also been shown to modulate postischemic astrogliosis through interaction with dystroglycans and integrins in the astrocytes [75].

Astrocytic AQP4 is not only linked with the matrix proteins but also with several other channels present in higher concentration in the astrocyte endfeet such as potassium inner rectifying channel 4.1 (KIR4.1), connexins (Cx), and also chloride channel 2 (CIC-2) [76, 77]. Colocalization of AQP4 and KIR4.1 suggests that AQP4 may have a role in potassium homeostasis by facilitating water diffusion along the potassium gradient and AQP4-KO mice display a delay in potassium reuptake during electrical activity [76]. The decrease of AQP4 expression using siRNA showed an associative decrease of connexin 43 (Cx43), a protein involved in gap junction formation, and a decrease of CIC-2, involved in the regulatory volume decrease function of the astrocytes. Interestingly, gap junctions and AQP4 are morphologically closely associated [78] with the astrocyte endfeet. The gap junctions in the astrocyte contribute to the formation of a complex network named the astroglial network [79]. Intercellular and intracellular communication that facilitate the movement of second messengers, amino acids, nucleotides, energy metabolites, and small peptides [79–82] in astrocyte processes occur through gap junctions, which are made up of a family of channel proteins called connexins [83, 84]. In astrocytes, Cx30 and Cx43 are predominant [83–85]. However, it is also important to note that Cx43, along with Cx37, Cx40 [86, 87], and Cx45 [87], is also expressed in brain endothelial cells. The protein level of Cx40 and Cx45 was shown to increase in cerebral arteries, but no change in protein or mRNA was observed for brain endothelial Cx43 and Cx37 after a model of brain injury causing cerebral vascular dysfunction [87]. The effect of astrocytic Cx43 upregulation or downregulation after ischemia still remains controversial and there is no consensus as to what provides beneficial effects [88]. However, in humans, there are reports that show that Cx43 protein levels were increased in the penumbra [89]. And because Cx43 and Cx30 knockouts have been observed to be more edema prone [90], it is possible that the increase in Cx43 after ischemia may be a physiological response to decrease edema. The induction of Cx43 may be facilitating water flow throughout the astrocyte network to diversify and dissipate the accumulation of fluid from just one region. From these data we hypothesize that gap junction proteins, specifically Cx43 on astrocytes, are working with AQP4. Evidence for this also comes from a significant decrease of Cx43 observed in mouse astrocyte cell cultures after administration of small interference RNA against AQP4 [91]. Although direct functional data are still lacking, one possibility is that AQP4 and Cx43 is working together to direct water flow between astrocytes and could be controlling astrocytic swelling.

The role of AQP4 in cerebral edema formation and resolution has been studied in several models. However the precise role of AQP4 remains unclear and depends on the pathological model used [92, 93]. Indeed, the absence of AQP4 was shown to prevent the formation of edema in a permanent ischemia model in AQP4-KO mice [94]. Similarly, edema formation is prevented in α-syntrophin knockout mice at 24 h after stroke [68]. This decrease of brain swelling was correlated with the loss of the perivascular AQP4 domain in α-syntrophin-KO mice [68]. These results suggest that perivascular AQP4 has an important role in edema formation. However, the absence of AQP4 in AQP4-KO mice also prevents water clearance in an experiment of intrastriatal infusion of a saline solution, showing that AQP4 is critical for water removal from tissue [95]. Conversely, in a preconditioning stroke model, a higher induction of AQP4 was correlated with edema reduction [27]. However, this reduction of edema may be referring to vasogenic edema, in which case, AQP4 is said to aid in edema resolution by actively pumping out water from the cerebral tissue to peripheral blood [95]. The redistribution of the water in the astrocyte compartment through the astrocyte network would also be possible for the CSF compartments. This hypothesis is supported by a publication showing an increase of AQP4 in ependymal cells in the border of the ventricles in a traumatic brain injury model [96].

To summarize, the exact mechanism causing decreased edema formation is not yet fully understood, but AQP4 and the astrocyte network with the gap-junction proteins may certainly be contributing. Osmotic gradients can also play an important role, and recently, high AQP4 expression was observed in hypersaline treatment after stroke correlating with decreased edema formation at 48 hours [97].

4. rtPA: A Unique Drug for Stroke Treatment with Aversive Effects on the NVU

4.1. Clinical Evidence (from Bed to the Bench, Neurotoxicity of rtPA). As discussed in the introduction, recombinant tissue plasminogen activator (rtPA) is currently the only thrombolytic molecule FDA approved for treatment of acute ischemic stroke [4]. The intact BBB is usually an obstacle for most neuropharmacological agents in healthy patients. The dysfunction of the BBB after ischemia could cause problems for the therapeutic function of rtPA. This protease targets fibrin-bound plasminogens and converts them into plasmins, which then cut the fibrin clot and lyse it. Intravenously infused at a dose of 0.9 mg/kg over one hour, rtPA provides increased survival and better neurological outcomes [4]. To be beneficial for the patient, rtPA must be administered within the first 4 h 30 min after stroke onset [5, 6]. Despite the organization of emergency care, only 5% of stroke patients are eligible for this therapy. In fact, late administration of rtPA translated to a higher risk of bleeding and extension of the lesion [7]. Higher doses of rtPA do not bind only the fibrin clot but also activate the circulating plasminogen activator (tPA). This activation contributes to a generalized fibrinolysis and fibrinogenolysis, which is suspected to be a cause of bleeding. But the mechanisms of the hemorrhagic transformation after rtPA treatment seem to be more complex than can be accounted for by the affinity of rtPA for fibrin alone. In fact, the enhanced fibrin specificity of tenecteplase and reteplase, two rtPA derivatives, resulted in no significant difference in terms of cerebral hemorrhage [98, 99].

Interestingly, the comparison with myocardial infarction shows a low incidence of cerebral hemorrhage after rtPA administration [100] suggesting a direct link between bleeding and the ischemic pathophysiology. Clinical studies showed that 80% of bleeding after cerebral thrombolysis occur preferentially in the ischemic territory [7].

4.2. Aversive Effects of rtPA Treatment on the NVU after Stroke. To have a better understanding of the aversive effects of rtPA its neurotoxic effects were examined. It is well known that endogenous tPA is present in the blood stream, endothelial cells, neurons, and microglial cells [101]. In the brain parenchyma, tPA activity was found to be pleiotropic and associated with synaptic plasticity and cell death [102–104]. In fact, tPA interacts with several neuronal proteins such as N-methyl-D-aspartate (NMDA) receptors, one subtype of glutamatergic receptors, low-density lipoprotein-receptor-related protein (LRP), and Annexin-II [101, 105, 106]. tPA is synthesized in neurons, stored in presynaptic vesicles, and released following depolarization in synergy with the neurotransmitters. In the synaptic cleft, tPA binds and cleaves the NR1 subunit of NMDA receptors that causes an amplification of calcium influx in postsynaptic neurons and an increase of the glutamatergic response in physiological conditions. However, this physiological response becomes excitotoxic after ischemia and is magnified after rtPA injection [101, 107, 108]. The injection of antibodies against the NR1-subunit prevented these proexcitotoxic effects of endogenous tPA and reduced brain infarction and BBB leakage after stroke [109]. These data suggest that the NMDA receptor may be a protective drug target for the NVU after stroke and may provide a potential extension of the rtPA therapeutic window [109].

The presence of rtPA in the brain parenchyma has been explained by its passage through the BBB in several *in vitro* models with different proposed mechanisms.

(i) rtPA diffuses into the brain parenchyma through an already opened BBB as a consequence of the ischemic process. As we discussed previously, the kinetics of the BBB opening is complex in the early stages after stroke and it is difficult to observe this with clinical imaging [29]. Interestingly, *in vitro* endothelial monolayer cultured with astrocytes enables us to observe the ability of rtPA to cross the intact BBB [110], which is increased under oxygen-glucose deprivation (OGD) [111]. Therefore, as rtPA potentially diffuses through an open or closed BBB in early time points after stroke onset, it may aggravate neuronal cell death as described previously.

(ii) rtPA could cross the BBB by degrading the endothelium via its own proteolytic activity, but it is not a requirement in the intact BBB [110]. The ability

of rtPA to cross the intact BBB at a thrombolytic dose suggests that this protease may interact first with the endothelial cells before the BBB breakdown. In fact, rtPA promotes breakdown of the BBB [112] by stimulating the synthesis activity of MMP-9 [113–116] and other MMP isoforms [117] exacerbating the degradation of the basal lamina and subsequent vasogenic edema formation and hemorrhage. The thrombolytic products could exacerbate the proposed mechanism [118].

(iii) Finally, LRP potentially contributes in trans-endothelial transport of the exogenous rtPA [106, 119, 120] and then activates the astrocytic MMP-9 and nuclear factor NF-κB, which promotes the expression of inducible nitric oxide synthase (iNOS). This increase of NO results in increased BBB permeability [121].

With all these data together, Yepes and collaborators have proposed the following potential cellular and molecular events to explain the toxicity of the rtPA and tPA on the NVU [104].

(1) Circulating endogenous tPA and rtPA cross the BBB (intact or damaged endothelial layer) and increase MMP-9 activity in the basal lamina soon after stroke onset which compromises the NVU integrity and makes it fragile.

(2) Then tPA and rtPA bind to the astrocytic LRP, inducing the loss of the extracellular domain of LRP [122, 123] in the basal lamina, and release the intracellular domain of LRP in the astrocytic cytoplasm to activate NF-κB. This NF-κB activation increases iNOS and MMP9 expression and overall function in the whole NVU, causing separation of astrocytic endfeet from the basal lamina. This is usually observed at the later stages of BBB breakdown. However, it is tempting to speculate that this cascade, which involves the perivascular cells of the NVU, would be an accelerated pathological process resulting from the use of rtPA. It is possible that rtPA and tPA may also affect the phenotype of the astrocyte endfeet by changes in the level of expression of key proteins such as AQP4 and also Cx43.

4.3. New Therapeutic Strategies for rtPA Treatment after Stroke. The BBB is definitely not a barrier to rtPA in stroke but the BBB does become a serious barrier to the effective usage of this drug in clinic due to the neurotoxic effects and the risk of hemorrhagic transformation. Interestingly, tPA may be endogenously synthesized by the central nervous system in neurons and endothelial cells [124]. However, tPA and rtPA have effects on the endothelial cells, astrocytes, and neurons and possibly other glial cell types such as oligodendrocytes and microglia. In order to prevent the aversive effects of rtPA while maintaining the benefits of early reperfusion, several new therapeutic strategies have been examined to prevent the interaction of rtPA with the NMDA receptor within the NVU [104]. In fact, NMDA receptors are expressed not only in neurons but also in oligodendrocytes and endothelial cells [125, 126]. One of these strategies uses an LRP antagonist (RAP) to minimize the binding of rtPA with LRP in the endothelial cells. A second strategy uses the ATD-NR1 antibody to block rtPA binding of the NR1 subunit on neuronal NMDA receptors. The last one uses a mutation of the rtPA to decrease its adverse effects on the nervous tissue [104]. An example of a natural drug, desmoteplase, the vampire bat *Desmodus Rotundus* Salivary Plasminogen Activator (DSPA), is a thrombolytic agent under development. It shows little neurotoxicity and has the ability to interact with the BBB endothelium through the same receptor (LRP) as that of tPA [127, 128]. Unfortunately, the clinical trial of DIAS-2 (Desmoteplase In Acute ischemic Stroke) showed no benefit of the desmoteplase versus placebo [129]. Although the outcome of this clinical trial was disappointing, promising alternatives pathways are being investigated. In fact, Gleevec, a FDA approved drug for treatment of chronic myelogenous leukemia, was recently proposed to prevent the complications associated with rtPA treatment [130]. Gleevec inhibits the activation of platelet-derived growth factor alpha receptor (PDGFR). It was shown that tPA increases BBB permeability through the indirect activation of perivascular astrocytic PDGFR [130].

MMP inhibition is a good strategy based on reports of easy monitoring of MMP blood levels, defining them as potential biomarkers of brain damage [131, 132]. But because endogenous MMPs are also key mediators in stroke recovery by contributing to inflammatory and remodeling responses, pharmacological targeting must be accurately applied for acute stroke phases so; their beneficial effects are not compromised [133, 134]. Despite efforts to understand the complex link between BBB integrity and the hemorrhage risk [112], a better definition and understanding of NVU kinetics and the mechanisms underlying their dysfunction is still needed to better define eligibility criteria for rtPA treatment. Thus, alternative approaches other than MMP inhibition as mentioned before in some recent developments will offer interesting treatment strategies after stroke.

5. NVU Protection May Be the Future instead of Neuroprotection in Stroke Treatment

5.1. Preconditioning for Future Development of New Drugs. Given the small number of patients eligible for thrombolysis, many pharmaceutical compounds have been developed to limit the progression of brain injury by targeting different mechanisms leading to neuronal death [135]. Despite promising protective effects observed in preclinical studies, no compound to date has demonstrated benefit against stroke-induced neuronal death after facing the rigorous wall of clinical trials [136].

As mentioned in Section 1, research on brain diseases has focused on neuronal damage, as it was thought to be the major cause of cognitive deficits. However, ischemic stroke is a complex brain disease characterized by sudden onset of disabilities related to brain damage with a vascular origin. Because the development of many neuroprotective molecules for treatment over the last twenty years has been unsuccessful, researchers have switched gears towards investigating the natural endogenous neuroprotection of ischemic tolerance [137]. The purpose of the ischemic tolerance preconditioning is to induce endogenous defense mechanisms prior to the ischemic event that will attenuate the eventual consequences of ischemia. This resistance to ischemic damage can be achieved experimentally by several stimuli including ischemic preconditioning [138]. The concept and protocols were adapted from previous studies done in myocardial infarction. In fact, a short duration of coronary occlusion is unable to cause myocyte necrosis. However, when carried out before a prolonged occlusion, a short occlusion significantly reduced the final infarct volume of the myocardium [139]. This initial nonharmful ischemic insult triggered endogenous mechanisms that made the organ more resistant to the next attack for up to two periods of ischemic tolerance [139]. The first period of ischemic tolerance resulted from posttranscriptional responses and began minutes after preconditioning. The second, longer period, began 24 hours after preconditioning and lasted up to 7 days with maximal protection found at 3 days.

As with the cardiac preconditioning, ischemic tolerance in the brain also has delayed mechanisms leading to neuroprotection [140]. However, the mechanisms are complex and not well understood. The induction of ischemic tolerance likely depends on the coordinated responses at the genomic, molecular, cellular, and tissue levels [141–143], which suggests the importance of the interactions between the astrocyte and endothelial cells in the NVU. Regarding neurovascular events in stroke pathophysiology, there has been a growing interest in vascular approaches to the preconditioning mechanisms. Protective effects of preconditioning were observed *in vivo*, demonstrating that endothelium function is preserved by improving cerebral blood flow during reperfusion in areas surrounding the lesion [144], and that BBB integrity is maintained with a reduction in edema formation [145]. The induced protection was again correlated not only with a decreased expression of MMP-9 [146] but also with a reduced neutrophil adhesion to endothelial cells through a decreased expression of ICAM-1 [147, 148]. These results were confirmed by *in vitro* studies that report a protective effect via preservation of BBB integrity, by both a decreased expression of the inflammatory molecules ICAM-1 and VCAM-1 [149, 150] and maintenance of tight junction structure [149]. Moreover, preconditioning also facilitates the increase of AQP4 expression at early time-points after stroke onset, which is associated with a decrease of the edema formation [27]. A recent study also reported the protective role of glial tissue preconditioning in severe stroke [151]. These recent observations suggest that future drug development must focus on drugs affecting the entire NVU instead of one cell type as was proposed in the 1990s with the development of calcium channel and NMDA inhibitors. Recently, some compounds like edaravone, an antioxidant, showed benefits in preclinical and clinical studies by protection of the NVU [152, 153]. But further trials are needed to confirm these promising preliminary results [154].

5.2. Protection of the NVU: Focus on PPARs. Preventive neuroprotection also involves management of risk factors, which is supported by studies showing that physical exercise [155] or lipid-lowering treatment reduces the occurrence and severity of stroke [156–158]. In this context, the involvement of pharmacological agents that are activators of nuclear receptors like peroxisome proliferator-activated receptors (PPARs) could be a promising study. Present in three isoforms, α, β/δ, and γ, these receptors exhibit pleiotropic activity in the sense that they can activate or repress the transcription of many genes involved in lipid and carbohydrate metabolism in addition to inflammation [159, 160]. PPARs are expressed in neurons, endothelial cells, and glial cells [161]. Activation of the PPARs has long-term effects lasting from hours to days, which correspond to an activation of gene transcription (named transactivation) as has been seen in lipid and carbohydrate metabolism. However, activation of the PPARs induce a cellular response within minutes to hours and this corresponds to an inhibition of gene transcription named transrepression [162]. The latter mechanism does not require binding to DNA, but rather protein-protein interaction involving other transcription factors like NF-κB of STAT-3 and AP-1, to inhibit their activity as reported for inflammatory genes [163].

Independent of its lipid-lowering activity, PPAR-α activation was found to be neuroprotective in several *in vivo* studies carried out in mice subjected to transient ischemia with preventive or curative treatments by agonists such as fenofibrate, WY-14643, and resveratrol (a polyphenol present in grapes) [164–166]. The observed protection is the result of an anti-inflammatory mechanism, which decreases the expression of adhesion molecules, ICAM-1 and VCAM-1, in brain endothelial cells. Effects of antioxidants were also observed. However, a study using a BBB *in vitro* model combining endothelial cells with glial cells from wild-type or PPAR-α knockout mice has demonstrated not only that the observed protection against OGD-induced hyperpermeability was dependent on this nuclear receptor activation but also that the ligand targeted specifically the endothelial cells without modulation of the classical PPAR-α target genes associated with inflammation or metabolism [167]. Moreover, protective effects of PPAR-γ were not only reported through similar mechanisms [168] but also via an inhibition of NFκB and TNF-α pathways [169, 170] and macrophages/microglial cells activation, thus preventing cytokine production [171]. One study also suggests that PPAR-γ agonists could inhibit excitotoxicity-induced neuronal death [172].

Statins are HMG-CoA reductase inhibitors. This enzyme catalyzes the conversion of HMG-CoA (3-hydroxy-3-methylglutaryl coenzyme A) to mevalonate, a precursor of cholesterol. As lipid lowering agents statins also exert pleiotropic effects at the vascular level [173]. In addition to protection against excitotoxicity in cultured neurons [174], statins have demonstrated preservation of BBB endothelial cells' integrity against glutamate excitotoxic challenge *in vitro* [175]. These compounds also enabled the reduction of MMP-9 synthesis in rtPA-activated astrocytes [176]. The effects of statins may involve nuclear receptors, through an increase in both expression and activity of PPAR-α [177–179]. More recently, brain endothelial PPAR-δ activation has proven to be protective against ischemia-induced cell death through inhibition of the miR-15a microRNA, thus strengthening the therapeutic concept based on activation of PPARs for the treatment of stroke-related microvascular dysfunction [180].

5.3. Inhibition of JNK Activation and NVU Protection. The c-Jun N-terminal kinases (JNKs) belong to the mitogen activated protein kinase (MAPK) family; the two other members being p38 and ERK [181, 182]. The isoforms JNK1 and JNK2 are ubiquitously distributed, while JNK3 is primarily expressed in the heart, brain, endocrine pancreas and testis [183]. JNKs are activated by phosphorylation, which is catalyzed by upstream kinases—MKK 4 and 7 [182–184]. JNK activation is essential for normal brain development and organogenesis during embryonic development [185]. However, the activation of JNKs plays several roles ranging from regulation of cell survival and apoptosis to cell proliferation [183, 185–187]. They are activated under pathological conditions both in the brain [188, 189] and in the periphery [190, 191]. In fact, JNK phosphorylation initially decreases after stroke and then starts to increase at 1.5 hours with a maximum at 9 hours after onset [192]. Phosphorylation of c-Jun, a JNK substrate, follows the same temporal pattern, peaking at 8 hours post-stroke [192, 193].

The development of the peptide named DJNKi, a competitive inhibitor of the JNK signaling pathway, has been shown to reduce lesion volume of mice with transient MCAO by 90% even when induced 6 hours after injury. This lesion volume decrease was accompanied by behavior improvements as well [193], suggesting an increase of the therapeutic time window almost 2 times longer than tPA. This positive outcome was also observed in a more severe model with a permanent occlusion model [194]. Moreover, DJNKi has been shown to be compatible for treatment of ischemic stroke even in the presence of rtPA and was shown to decrease lesion volume [195]. DJNKi also improved neurobehavior scores and decreased hemispheric swelling after a model of intra-cerebral hemorrhage [196]. Thus, DJNKi could possibly attenuate the highly probable side effect of hemorrhagic transformation caused by rtPA. Interestingly, in this model of intracerebral hemorrhage, DJNKi administration significantly increased AQP4 expression 48 hours after injury. This increase in AQP4 expression negatively correlated with decreased hemispheric swelling, thus pointing towards a possible role of DJNKi controlling edema as well. In fact, activation of the JNK pathway is present not only in the neurons but also in glial cells [197] and brain endothelial cells [198]. Such activation in nonneuronal cells may negatively impact neuronal cell death and function [197]. In the context of broad effects of this drug, Benakis et al. [199] showed that DJNKI-1, injected peripherally, is able to modulate some nonneuronal inflammatory processes. As discussed previously, the development of a drug targeting several cells such as in the NVU may help to move towards success in the clinic.

6. Summary and Perspectives in Stroke Research

In summary, the data found in the literature suggest that the failure of agents in protecting the brain against stroke may come from the fact that each developed compound targeted only one mechanism and one cell type of stroke pathophysiology. Ischemic preconditioning appears to be an attractive experimental strategy that would identify endogenous mechanisms of protection and regeneration. Recent evidence of such protective mechanisms supports a complex action on cells of the NVU, underlining the importance of the interactions between endothelial cells and astrocytes in the pathophysiology after stroke. As our knowledge of the NVU increases, molecules with pleiotropic activity will become increasing useful in the development of post-ischemic treatments in the clinics.

Conflict of Interest

The authors declare that they have no conflict of interests.

Acknowledgments

The authors thank Jacqueline Coats (Loma Linda University) for text revision. They also thank Françoise Dieterlen and the French Société de Biologie" for permitting the translation of some topics previously published in the French journal "Biologie Aujourd'hui. This paper was supported in part by the NIH R01HD061946 (to J. Badaut), the Swiss Science Foundation (FN 31003A-122166 to J. Badaut), and the European Union's Seventh Framework Programme (FP7/2007-2013) under Grant agreements nos. 201024 and 202213 (European Stroke Network).

References

[1] V. L. Roger, A. S. Go, D. M. Lloyd-Jones et al., "Heart disease and stroke statistics-2011 update: a report from the American Heart Association," *Circulation*, vol. 123, no. 4, pp. e18–e19, 2011.

[2] W. Rosamond, K. Flegal, G. Friday et al., "Heart disease and stroke statistics—2007 Update: a report from the American Heart Association Statistics Committee and Stroke Statistics Subcommittee," *Circulation*, vol. 115, no. 5, pp. e69–e171, 2007.

[3] J. L. Saver, "Time is brain—quantified," *Stroke*, vol. 37, no. 1, pp. 263–266, 2006.

[4] J. R. Marler, "Tissue plasminogen activator for acute ischemic stroke. The National Institute of Neurological Disorders and Stroke rt-PA Stroke study group," *New England Journal of Medicine*, vol. 333, no. 24, pp. 1581–1587, 1995.

[5] W. Hacke, M. Kaste, E. Bluhmki et al., "Thrombolysis with alteplase 3 to 4.5 hours after acute ischemic stroke," *New England Journal of Medicine*, vol. 359, no. 13, pp. 1317–1329, 2008.

[6] J. L. Saver, J. Gornbein, J. Grotta et al., "Number needed to treat to benefit and to harm for intravenous tissue plasminogen activator therapy in the 3- to 4.5-hour window Joint outcome table analysis of the ECASS 3 trial," *Stroke*, vol. 40, no. 7, pp. 2433–2437, 2009.

[7] J. P. Broderick, "Intracerebral hemorrhage after intravenous t-PA therapy for ischemic stroke," *Stroke*, vol. 28, no. 11, pp. 2109–2118, 1997.

[8] G. J. Del Zoppo, "The neurovascular unit in the setting of stroke," *Journal of Internal Medicine*, vol. 267, no. 2, pp. 156–171, 2010.

[9] C. Iadecola, "Neurovascular regulation in the normal brain and in Alzheimer's disease," *Nature Reviews Neuroscience*, vol. 5, no. 5, pp. 347–360, 2004.

[10] C. Iadecola and M. Nedergaard, "Glial regulation of the cerebral microvasculature," *Nature Neuroscience*, vol. 10, no. 11, pp. 1369–1376, 2007.

[11] E. A. Neuwelt, B. Bauer, C. Fahlke et al., "Engaging neuroscience to advance translational research in brain barrier biology," *Nature Reviews Neuroscience*, vol. 12, no. 3, pp. 169–182, 2011.

[12] B. Engelhardt and C. Coisne, "Fluids and barriers of the CNS establish immune privilege by confining immune surveillance to a two-walled castle moat surrounding the CNS castle," *Fluids and Barriers of the CNS*, vol. 8, no. 1, Article ID 4, 2011.

[13] N. J. Abbott, L. Rönnbäck, and E. Hansson, "Astrocyte-endothelial interactions at the blood-brain barrier," *Nature Reviews Neuroscience*, vol. 7, no. 1, pp. 41–53, 2006.

[14] E. Hamel, "Perivascular nerves and the regulation of cerebrovascular tone," *Journal of Applied Physiology*, vol. 100, no. 3, pp. 1059–1064, 2006.

[15] L. Pellerin, "Food for thought: the importance of glucose and other energy substrates for sustaining brain function under varying levels of activity," *Diabetes and Metabolism*, vol. 36, no. 3, pp. S59–S63, 2010.

[16] N. J. Abbott, A. A. K. Patabendige, D. E. M. Dolman, S. R. Yusof, and D. J. Begley, "Structure and function of the blood-brain barrier," *Neurobiology of Disease*, vol. 37, no. 1, pp. 13–25, 2010.

[17] T. Owens, I. Bechmann, and B. Engelhardt, "Perivascular spaces and the two steps to neuroinflammation," *Journal of Neuropathology and Experimental Neurology*, vol. 67, no. 12, pp. 1113–1121, 2008.

[18] A. W. Unterberg, J. Stover, B. Kress, and K. L. Kiening, "Edema and brain trauma," *Neuroscience*, vol. 129, no. 4, pp. 1021–1029, 2004.

[19] J. M. Simard, T. A. Kent, M. Chen, K. V. Tarasov, and V. Gerzanich, "Brain oedema in focal ischaemia: molecular pathophysiology and theoretical implications," *The Lancet Neurology*, vol. 6, no. 3, pp. 258–268, 2007.

[20] J. Badaut, S. Ashwal, and A. Obenaus, "Aquaporins in cerebrovascular disease: a target for treatment of brain edema?" *Cerebrovascular Diseases*, vol. 31, no. 6, pp. 521–531, 2011.

[21] Z. Zador, G. T. Manley, S. Stiver, and V. Wang, "Role of aquaporin-4 in cerebral edema and stroke," *Handbook of Experimental Pharmacology*, vol. 190, pp. 159–170, 2009.

[22] W. C. Risher, R. D. Andrew, and S. A. Kirov, "Real-time passive volume responses of astrocytes to acute osmotic and ischemic stress in cortical slices and in vivo revealed by two-photon microscopy," *GLIA*, vol. 57, no. 2, pp. 207–221, 2009.

[23] M. E. O'Donnell, T. I. Lam, L. Tran, and S. E. Anderson, "The role of the blood-brain barrier Na-K-2Cl cotransporter in stroke," *Advances in Experimental Medicine and Biology*, vol. 559, pp. 67–75, 2005.

[24] S. P. Duckles and D. N. Krause, "Mechanisms of cerebrovascular protection: oestrogen, inflammation and mitochondria," *Acta Physiologica*, vol. 203, no. 1, pp. 149–154, 2011.

[25] S. P. Duckles, D. N. Krause, C. Stirone, and V. Procaccio, "Estrogen and mitochondria: a new paradigm for vascular protection?" *Molecular Interventions*, vol. 6, no. 1, pp. 26–35, 2006.

[26] M. De Castro Ribeiro, L. Hirt, J. Bogousslavsky, L. Regli, and J. Badaut, "Time course of aquaporin expression after transient focal cerebral ischemia in mice," *Journal of Neuroscience Research*, vol. 83, no. 7, pp. 1231–1240, 2006.

[27] L. Hirt, B. Ternon, M. Price, N. Mastour, J. F. Brunet, and J. Badaut, "Protective role of early Aquaporin 4 induction against postischemic edema formation," *Journal of Cerebral Blood Flow and Metabolism*, vol. 29, no. 2, pp. 423–433, 2009.

[28] D. Strbian, A. Durukan, M. Pitkonen et al., "The blood-brain barrier is continuously open for several weeks following transient focal cerebral ischemia," *Neuroscience*, vol. 153, no. 1, pp. 175–181, 2008.

[29] E. C. Henning, L. L. Latour, and S. Warach, "Verification of enhancement of the CSF space, not parenchyma, in acute stroke patients with early blood-brain barrier disruption," *Journal of Cerebral Blood Flow and Metabolism*, vol. 28, no. 5, pp. 882–886, 2008.

[30] A. M. Romanic, R. F. White, A. J. Arleth, E. H. Ohlstein, and F. C. Barone, "Matrix metalloproteinase expression increases after cerebral focal ischemia in rats: inhibition of matrix metalloproteinase-9 reduces infarct size," *Stroke*, vol. 29, no. 5, pp. 1020–1030, 1998.

[31] A. Kastrup, T. Engelhorn, C. Beaulieu, A. De Crespigny, and M. E. Moseley, "Dynamics of cerebral injury, perfusion, and blood-brain barrier changes after temporary and permanent middle cerebral artery occlusion in the rat," *Journal of the Neurological Sciences*, vol. 166, no. 2, pp. 91–99, 1999.

[32] S. Nagahiro, S. Goto, K. Kogo, M. Sumi, M. Takahashi, and Y. Ushio, "Sequential changes in ischemic edema following transient focal cerebral ischemia in rats: magnetic resonance imaging study," *Neurologia Medico-Chirurgica*, vol. 34, no. 7, pp. 412–417, 1994.

[33] M. Plateel, E. Teissier, and R. Cecchelli, "Hypoxia dramatically increases the nonspecific transport of blood-borne proteins to the brain," *Journal of Neurochemistry*, vol. 68, no. 2, pp. 874–877, 1997.

[34] A. Rosell, A. Ortega-Aznar, J. Alvarez-Sabín et al., "Increased brain expression of matrix metalloproteinase-9 after ischemic and hemorrhagic human stroke," *Stroke*, vol. 37, no. 6, pp. 1399–1406, 2006.

[35] G. A. Rosenberg, E. Y. Estrada, and J. E. Dencoff, "Matrix metalloproteinases and TIMPs are associated with blood-brain barrier opening after reperfusion in rat brain," *Stroke*, vol. 29, no. 10, pp. 2189–2195, 1998.

[36] E. Candelario-Jalil, Y. Yang, and G. A. Rosenberg, "Diverse roles of matrix metalloproteinases and tissue inhibitors of metalloproteinases in neuroinflammation and cerebral ischemia," *Neuroscience*, vol. 158, no. 3, pp. 983–994, 2009.

[37] A. Obenaus and S. Ashwal, "Magnetic resonance imaging in cerebral ischemia: focus on neonates," *Neuropharmacology*, vol. 55, no. 3, pp. 271–280, 2008.

[38] C. A. Chastain, U. E. Oyoyo, M. Zipperman et al., "Predicting outcomes of traumatic brain injury by imaging modality and injury distribution," *Journal of Neurotrauma*, vol. 26, no. 8, pp. 1183–1196, 2009.

[39] J. Badaut, S. Ashwal, B. Tone, L. Regli, H. R. Tian, and A. Obenaus, "Temporal and regional evolution of aquaporin-4 expression and magnetic resonance imaging in a rat pup model of neonatal stroke," *Pediatric Research*, vol. 62, no. 3, pp. 248–254, 2007.

[40] D. R. Pillai, M. S. Dittmar, D. Baldaranov et al., "Cerebral ischemia-reperfusion injury in rats—A 3 T MRI study on biphasic blood-brain barrier opening and the dynamics of edema formation," *Journal of Cerebral Blood Flow and Metabolism*, vol. 29, no. 11, pp. 1846–1855, 2009.

[41] C. S. Ábrahám, N. Harada, M. A. Deli, and M. Niwa, "Transient forebrain ischemia increases the blood-brain barrier permeability for albumin in stroke-prone spontaneously hypertensive rats," *Cellular and Molecular Neurobiology*, vol. 22, no. 4, pp. 455–462, 2002.

[42] R. S. Bourke, H. K. Kimelberg, L. R. Nelson et al., "Biology of glial swelling in experimental brain edema," *Advances in Neurology*, vol. 28, pp. 99–109, 1980.

[43] H. K. Kimelberg, "Astrocytic swelling in cerebral ischemia as a possible cause of injury and target for therapy," *GLIA*, vol. 50, no. 4, pp. 389–397, 2005.

[44] J. M. Rutkowsky, B. K. Wallace, P. M. Wise, and M. E. O'Donnell, "Effects of estradiol on ischemic factor-induced astrocyte swelling and AQP4 protein abundance," *American Journal of Physiology*, vol. 301, no. 1, pp. C204–C212, 2011.

[45] M. J. Tait, S. Saadoun, B. A. Bell, and M. C. Papadopoulos, "Water movements in the brain: role of aquaporins," *Trends in Neurosciences*, vol. 31, no. 1, pp. 37–43, 2008.

[46] J. Badaut, A. Nehlig, J. M. Verbavatz, M. E. Stoeckel, M. J. Freund-Mercier, and F. Lasbennes, "Hypervascularization in the magnocellular nuclei of the rat hypothalamus: relationship with the distribution of aquaporin-4 and markers of energy metabolism," *Journal of Neuroendocrinology*, vol. 12, no. 10, pp. 960–969, 2000.

[47] J. Badaut, J. M. Verbavatz, M. J. Freund-Mercier, and F. Lasbennes, "Presence of aquaporin-4 and muscarinic receptors in astrocytes and ependymal cells in rat brain: a clue to a common function?" *Neuroscience Letters*, vol. 292, no. 2, pp. 75–78, 2000.

[48] I. I. Arciénega, J. F. Brunet, J. Bloch, and J. Badaut, "Cell locations for AQP1, AQP4 and 9 in the non-human primate brain," *Neuroscience*, vol. 167, no. 4, pp. 1103–1114, 2010.

[49] D. Dolman, S. Drndarski, N. J. Abbott, and M. Rattray, "Induction of aquaporin 1 but not aquaporin 4 messenger RNA in rat primary brain microvessel endothelial cells in culture," *Journal of Neurochemistry*, vol. 93, no. 4, pp. 825–833, 2005.

[50] M. Amiry-Moghaddam, R. Xue, F. M. Haug et al., "Alpha-syntrophin deletion removes the perivascular but not endothelial pool of aquaporin-4 at the blood-brain barrier and delays the development of brain edema in an experimental model of acute hyponatremia," *The FASEB Journal*, vol. 18, no. 3, pp. 542–544, 2004.

[51] S. Nielsen, E. A. Nagelhus, M. Amiry-Moghaddam, C. Bourque, P. Agre, and O. R. Ottersen, "Specialized membrane domains for water transport in glial cells: high- resolution immunogold cytochemistry of aquaporin-4 in rat brain," *Journal of Neuroscience*, vol. 17, no. 1, pp. 171–180, 1997.

[52] G. P. Nicchia, B. Nico, L. M. A. Camassa et al., "The role of aquaporin-4 in the blood-brain barrier development and integrity: studies in animal and cell culture models," *Neuroscience*, vol. 129, no. 4, pp. 935–945, 2004.

[53] H. Wolburg, S. Noell, A. Mack, K. Wolburg-Buchholz, and P. Fallier-Becker, "Brain endothelial cells and the glio-vascular complex," *Cell and Tissue Research*, vol. 335, no. 1, pp. 75–96, 2009.

[54] H. Wolburg, S. Noell, K. Wolburg-Buchholz, A. MacK, and P. Fallier-Becker, "Agrin, aquaporin-4, and astrocyte polarity

as an important feature of the blood-brain barrier," *Neuroscientist*, vol. 15, no. 2, pp. 180–193, 2009.

[55] J. Badaut, S. Ashwal, A. Adami et al., "Brain water mobility decreases after astrocytic aquaporin-4 inhibition using RNA interference," *Journal of Cerebral Blood Flow and Metabolism*, vol. 31, no. 3, pp. 819–831, 2011.

[56] J. Yu, A. J. Yool, K. Schulten, and E. Tajkhorshid, "Mechanism of gating and ion conductivity of a possible tetrameric pore in aquaporin-1," *Structure*, vol. 14, no. 9, pp. 1411–1423, 2006.

[57] J. E. Rash, K. G. V. Davidson, T. Yasumura, and C. S. Furman, "Freeze-fracture and immunogold analysis of aquaporin-4 (AQP4) square arrays, with models of AQP4 lattice assembly," *Neuroscience*, vol. 129, no. 4, pp. 915–934, 2004.

[58] C. S. Furman, D. A. Gorelick-Feldman, K. G. V. Davidson et al., "Aquaporin-4 square array assembly: opposing actions of M1 and M23 isoforms," *Proceedings of the National Academy of Sciences of the United States of America*, vol. 100, no. 23, pp. 13609–13614, 2003.

[59] M. Suzuki, Y. Iwasaki, and T. Yamamoto, "Disintegration of orthogonal arrays in perivascular astrocytic processes as an early event in acute global ischemia," *Brain Research*, vol. 300, no. 1, pp. 141–145, 1984.

[60] R. Musa-Aziz, L. M. Chen, M. F. Pelletier, and W. F. Boron, "Relative CO2/NH3 selectivities of AQP1, AQP4, AQP5, AmtB, and RhAG," *Proceedings of the National Academy of Sciences of the United States of America*, vol. 106, no. 13, pp. 5406–5411, 2009.

[61] J. Zhou, H. Kong, X. Hua, M. Xiao, J. Ding, and G. Hu, "Altered blood-brain barrier integrity in adult aquaporin-4 knockout mice," *NeuroReport*, vol. 19, no. 1, pp. 1–5, 2008.

[62] B. Nico, A. Frigeri, G. P. Nicchia et al., "Role of aquaporin-4 water channel in the development and integrity of the blood-brain barrier," *Journal of Cell Science*, vol. 114, no. 7, pp. 1297–1307, 2001.

[63] Y. Hiroaki, K. Tani, A. Kamegawa et al., "Implications of the aquaporin-4 structure on array formation and cell adhesion," *Journal of Molecular Biology*, vol. 355, no. 4, pp. 628–639, 2006.

[64] T. Gonen, P. Silz, J. Kistler, Y. Cheng, and T. Walz, "Aquaporin-0 membrane junctions reveal the structure of a closed water pore," *Nature*, vol. 429, no. 6988, pp. 193–197, 2004.

[65] A. Warth, S. Kröger, and H. Wolburg, "Redistribution of aquaporin-4 in human glioblastoma correlates with loss of agrin immunoreactivity from brain capillary basal laminae," *Acta Neuropathologica*, vol. 107, no. 4, pp. 311–318, 2004.

[66] E. Guadagno and H. Moukhles, "Laminin-induced aggregation of the inwardly rectifying potassium channel, Kir4.1, and the water-permeable channel, AQP4, via a dystroglycan-containing complex in astrocytes," *GLIA*, vol. 47, no. 2, pp. 138–149, 2004.

[67] J. D. Neely, M. Amiry-Moghaddam, O. P. Ottersen, S. C. Froehner, P. Agre, and M. E. Adams, "Syntrophin-dependent expression and localization of Aquaporin-4 water channel protein," *Proceedings of the National Academy of Sciences of the United States of America*, vol. 98, no. 24, pp. 14108–14113, 2001.

[68] M. Amiry-Moghaddam, T. Otsuka, P. D. Hurn et al., "An α-syntrophin-dependent pool of AQP4 in astroglial end-feet confers bidirectional water flow between blood and brain," *Proceedings of the National Academy of Sciences of the United States of America*, vol. 100, no. 4, pp. 2106–2111, 2003.

[69] S. Saadoun, M. J. Tait, A. Reza et al., "AQP4 gene deletion in mice does not alter blood-brain barrier integrity or brain morphology," *Neuroscience*, vol. 161, no. 3, pp. 764–772, 2009.

[70] G. Bix and R. V. Iozzo, "Novel interactions of perlecan: unraveling Perlecan's role in angiogenesis," *Microscopy Research and Technique*, vol. 71, no. 5, pp. 339–348, 2008.

[71] S. Noell, P. Fallier-Becker, U. Deutsch, A. F. MacK, and H. Wolburg, "Agrin defines polarized distribution of orthogonal arrays of particles in astrocytes," *Cell and Tissue Research*, vol. 337, no. 2, pp. 185–195, 2009.

[72] M. Ramos-Fernandez, M. F. Bellolio, and L. G. Stead, "Matrix metalloproteinase-9 as a marker for acute ischemic stroke: a systematic review," *Journal of Stroke and Cerebrovascular Diseases*, vol. 20, no. 1, pp. 47–54, 2011.

[73] B. Lee, D. Clarke, A. Al Ahmad et al., "Perlecan domain V is neuroprotective and proangiogenic following ischemic stroke in rodents," *Journal of Clinical Investigation*, vol. 121, no. 8, pp. 3005–3023, 2011.

[74] Z. G. Zhang, L. Zhang, Q. Jiang et al., "VEGF enhances angiogenesis and promotes blood-brain barrier leakage in the ischemic brain," *Journal of Clinical Investigation*, vol. 106, no. 7, pp. 829–838, 2000.

[75] A. J. Al-Ahmad, B. Lee, M. Saini, and G. J. Bix, "Perlecan domain V modulates astrogliosis In vitro and after focal cerebral ischemia through multiple receptors and increased nerve growth factor release," *GLIA*, vol. 59, no. 12, pp. 1822–1840, 2011.

[76] D. K. Binder, X. Yao, Z. Zador, T. J. Sick, A. S. Verkman, and G. T. Manley, "Increased seizure duration and slowed potassium kinetics in mice lacking aquaporin-4 water channels," *GLIA*, vol. 53, no. 6, pp. 631–636, 2006.

[77] V. Benfenati, G. P. Nicchia, M. Svelto, C. Rapisarda, A. Frigeri, and S. Ferroni, "Functional down-regulation of volume-regulated anion channels in AQP4 knockdown cultured rat cortical astrocytes," *Journal of Neurochemistry*, vol. 100, no. 1, pp. 87–104, 2007.

[78] J. E. Rash, "Molecular disruptions of the panglial syncytium block potassium siphoning and axonal saltatory conduction: pertinence to neuromyelitis optica and other demyelinating diseases of the central nervous system," *Neuroscience*, vol. 168, no. 4, pp. 982–1008, 2010.

[79] C. Giaume, A. Koulakoff, L. Roux, D. Holcman, and N. Rouach, "Astroglial networks: a step further in neuroglial and gliovascular interactions," *Nature Reviews Neuroscience*, vol. 11, no. 2, pp. 87–99, 2010.

[80] A. Tabernero, J. M. Medina, and C. Giaume, "Glucose metabolism and proliferation in glia: role of astrocytic gap junctions," *Journal of Neurochemistry*, vol. 99, no. 4, pp. 1049–1061, 2006.

[81] A. L. Harris, "Connexin channel permeability to cytoplasmic molecules," *Progress in Biophysics and Molecular Biology*, vol. 94, no. 1-2, pp. 120–143, 2007.

[82] G. Zoidl and R. Dermietzel, "On the search for the electrical synapse: a glimpse at the future," *Cell and Tissue Research*, vol. 310, no. 2, pp. 137–142, 2002.

[83] J. E. Rash, T. Yasumura, K. G. V. Davidson, C. S. Furman, F. E. Dudek, and J. I. Nagy, "Identification of cells expressing Cx43, Cx30, Cx26, Cx32 and Cx36 in gap junctions of rat brain and spinal cord," *Cell Communication and Adhesion*, vol. 8, no. 4-6, pp. 315–320, 2001.

[84] J. I. Nagy, F. E. Dudek, and J. E. Rash, "Update on connexins and gap junctions in neurons and glia in the mammalian nervous system," *Brain Research Reviews*, vol. 47, no. 1-3, pp. 191–215, 2004.

[85] J. E. Rash, T. Yasumura, F. E. Dudek, and J. I. Nagy, "Cell-specific expression of connexins and evidence of restricted gap junctional coupling between glial cells and between neurons," *Journal of Neuroscience*, vol. 21, no. 6, pp. 1983–2000, 2001.

[86] K. Nagasawa, H. Chiba, H. Fujita et al., "Possible involvement of gap junctions in the barrier function of tight junctions of brain and lung endothelial cells," *Journal of Cellular Physiology*, vol. 208, no. 1, pp. 123–132, 2006.

[87] M. A. Avila, S. L. Sell, B. E. Hawkins et al., "Cerebrovascular connexin expression: effects of traumatic brain injury," *Journal of Neurotrauma*, vol. 28, no. 9, pp. 1803–1811, 2011.

[88] R. Farahani, M. H. Pina-Benabou, A. Kyrozis et al., "Alterations in metabolism and gap junction expression may determine the role of astrocytes as "good Samaritans" or executioners," *GLIA*, vol. 50, no. 4, pp. 351–361, 2005.

[89] T. Nakase, Y. Yoshida, and K. Nagata, "Enhanced connexin 43 immunoreactivity in penumbral areas in the human brain following ischemia," *GLIA*, vol. 54, no. 5, pp. 369–375, 2006.

[90] W. M. Armstead, "Superoxide generation links protein kinase C activation to impaired ATP- sensitive K^+ channel function after brain injury," *Stroke*, vol. 30, no. 1, pp. 153–159, 1999.

[91] W. M. Armstead, "Age-dependent impairment of K(ATP) channel function following brain injury," *Journal of Neurotrauma*, vol. 16, no. 5, pp. 391–402, 1999.

[92] J. Badaut, J. F. Brunet, and L. Regli, "Aquaporins in the brain: from aqueduct to "multi-duct"," *Metabolic Brain Disease*, vol. 22, no. 3-4, pp. 251–263, 2007.

[93] A. S. Verkman, D. K. Binder, O. Bloch, K. Auguste, and M. C. Papadopoulos, "Three distinct roles of aquaporin-4 in brain function revealed by knockout mice," *Biochimica et Biophysica Acta*, vol. 1758, no. 8, pp. 1085–1093, 2006.

[94] G. T. Manley, M. Fujimura, T. Ma et al., "Aquaporin-4 deletion in mice reduces brain edema after acute water intoxication and ischemic stroke," *Nature Medicine*, vol. 6, no. 2, pp. 159–163, 2000.

[95] M. C. Papadopoulos, G. T. Manley, S. Krishna, and A. S. Verkman, "Aquaporin-4 facilitates reabsorption of excess fluid in vasogenic brain edema," *FASEB Journal*, vol. 18, no. 11, pp. 1291–1293, 2004.

[96] Q. Guo, I. Sayeed, L. M. Baronne, S. W. Hoffman, R. Guennoun, and D. G. Stein, "Progesterone administration modulates AQP4 expression and edema after traumatic brain injury in male rats," *Experimental Neurology*, vol. 198, no. 2, pp. 469–478, 2006.

[97] C. H. Chen, R. Xue, J. Zhang, X. Li, S. Mori, and A. Bhardwaj, "Effect of osmotherapy with hypertonic saline on regional cerebral edema following experimental stroke: a study utilizing magnetic resonance imaging," *Neurocritical Care*, vol. 7, no. 1, pp. 92–100, 2007.

[98] D. F. Chapman, P. Lyden, P. A. Lapchak, H. Nunez, H. Thibodeaux, and J. Zivin, "Comparison of TNK with wild-type tissue plasminogen activator in a rabbit embolic stroke model," *Stroke*, vol. 32, no. 3, pp. 748–752, 2001.

[99] P. A. Lapchak, D. M. Araujo, and J. A. Zivin, "Comparison of Tenecteplase with Alteplase on clinical rating scores following small clot embolic strokes in rabbits," *Experimental Neurology*, vol. 185, no. 1, pp. 154–159, 2004.

[100] J. H. Gurwitz, J. M. Gore, R. J. Goldberg et al., "Risk for intracranial hemorrhage after tissue plasminogen activator treatment for acute myocardial infarction," *Annals of Internal Medicine*, vol. 129, no. 8, pp. 597–604, 1998.

[101] O. Nicole, F. Docagne, C. Ali et al., "The proteolytic activity of tissue-plasminogen activator enhances NMDA receptor-mediated signaling," *Nature Medicine*, vol. 7, no. 1, pp. 59–64, 2001.

[102] K. Benchenane, J. P. López-Atalaya, M. Fernández-Monreal, O. Touzani, and D. Vivien, "Equivocal roles of tissue-type plasminogen activator in stroke-induced injury," *Trends in Neurosciences*, vol. 27, no. 3, pp. 155–160, 2004.

[103] A. L. Samson and R. L. Medcalf, "Tissue-type plasminogen activator: a multifaceted modulator of neurotransmission and synaptic plasticity," *Neuron*, vol. 50, no. 5, pp. 673–678, 2006.

[104] M. Yepes, B. D. Roussel, C. Ali, and D. Vivien, "Tissue-type plasminogen activator in the ischemic brain: more than a thrombolytic," *Trends in Neurosciences*, vol. 32, no. 1, pp. 48–55, 2009.

[105] C. J. Siao, S. R. Fernandez, and S. E. Tsirka, "Cell type-specific roles for tissue plasminogen activator released by neurons or microglia after excitotoxic injury," *Journal of Neuroscience*, vol. 23, no. 8, pp. 3234–3242, 2003.

[106] M. Yepes, M. Sandkvist, E. G. Moore, T. H. Bugge, D. K. Strickland, and D. A. Lawrence, "Tissue-type plasminogen activator induces opening of the blood-brain barrier via the LDL receptor-related protein," *Journal of Clinical Investigation*, vol. 112, no. 10, pp. 1533–1540, 2003.

[107] M. Fernández-Monreal, J. P. López-Atalaya, K. Benchenane et al., "Arginine 260 of the amino-terminal domain of NR1 subunit is critical for tissue-type plasminogen activator-mediated enhancement of N-methyl-D-aspartate receptor signaling," *Journal of Biological Chemistry*, vol. 279, no. 49, pp. 50850–50856, 2004.

[108] M. Fernández-Monreal, J. P. López-Atalaya, K. Benchenane et al., "Is tissue-type plasminogen activator a neuromodulator?" *Molecular and Cellular Neuroscience*, vol. 25, no. 4, pp. 594–601, 2004.

[109] R. MacRez, P. Obiang, M. Gauberti et al., "Antibodies preventing the interaction of tissue-type plasminogen activator with N-methyl-D-aspartate receptors reduce stroke damages and extend the therapeutic window of thrombolysis," *Stroke*, vol. 42, no. 8, pp. 2315–2322, 2011.

[110] K. Benchenane, V. Berezowski, C. Ali et al., "Tissue-type plasminogen activator crosses the intact blood-brain barrier by low-density lipoprotein receptor-related protein-mediated transcytosis," *Circulation*, vol. 111, no. 17, pp. 2241–2249, 2005.

[111] K. Benchenane, V. Berezowski, M. Fernández-Monreal et al., "Oxygen glucose deprivation switches the transport of tPA across the blood-brain barrier from an LRP-dependent to an increased LRP-independent process," *Stroke*, vol. 36, no. 5, pp. 1059–1064, 2005.

[112] C. S. Kidwell, L. Latour, J. L. Saver et al., "Thrombolytic toxicity: blood brain barrier disruption in human ischemic stroke," *Cerebrovascular Diseases*, vol. 25, no. 4, pp. 338–343, 2008.

[113] T. Aoki, T. Sumii, T. Mori, X. Wang, and E. H. Lo, "Blood-brain barrier disruption and matrix metalloproteinase-9 expression during reperfusion injury mechanical versus embolic focal ischemia in spontaneously hypertensive rats," *Stroke*, vol. 33, no. 11, pp. 2711–2717, 2002.

[114] M. A. Kelly, A. Shuaib, and K. G. Todd, "Matrix metalloproteinase activation and blood-brain barrier breakdown following thrombolysis," *Experimental Neurology*, vol. 200, no. 1, pp. 38–49, 2006.

[115] S. R. Lee, S. Z. Guo, R. H. Scannevin et al., "Induction of matrix metalloproteinase, cytokines and chemokines in rat cortical astrocytes exposed to plasminogen activators," *Neuroscience Letters*, vol. 417, no. 1, pp. 1–5, 2007.

[116] X. Wang, K. Tsuji, S. R. Lee et al., "Mechanisms of hemorrhagic transformation after tissue plasminogen activator reperfusion therapy for ischemic stroke," *Stroke*, vol. 35, no. 11, pp. 2726–2730, 2004.

[117] J.-C. Copin, D. J. Bengualid, R. F. Da Silva, O. Kargiotis, K. Schaller, and Y. Gasche, "Recombinant tissue plasminogen activator induces blood-brain barrier breakdown by a matrix metalloproteinase-9-independent pathway after transient focal cerebral ischemia in mouse," *European Journal of Neuroscience*, vol. 34, no. 7, pp. 1085–1092, 2011.

[118] S. Gautier, O. Petrault, P. Gele et al., "Involvement of thrombolysis in recombinant tissue plasminogen activator-induced cerebral hemorrhages and effect on infarct volume and postischemic endothelial function," *Stroke*, vol. 34, no. 12, pp. 2975–2979, 2003.

[119] P. A. Lapchak, D. F. Chapman, and J. A. Zivin, "Metalloproteinase inhibition reduces thrombolytic (tissue plasminogen activator)-induced hemorrhage after thromboembolic stroke," *Stroke*, vol. 31, no. 12, pp. 3034–3040, 2000.

[120] X. Wang, S. R. Lee, K. Arai et al., "Lipoprotein receptor-mediated induction of matrix metalloproteinase by tissue plasminogen activator," *Nature Medicine*, vol. 9, no. 10, pp. 1313–1317, 2003.

[121] X. Zhang, R. Polavarapu, H. She, Z. Mao, and M. Yepes, "Tissue-type plasminogen activator and the low-density lipoprotein receptor-related protein mediate cerebral ischemia-induced nuclear factor-κB pathway activation," *American Journal of Pathology*, vol. 171, no. 4, pp. 1281–1290, 2007.

[122] R. Polavarapu, J. An, C. Zhang, and M. Yepes, "Regulated intramembrane proteolysis of the low-density lipoprotein receptor-related protein mediates ischemic cell death," *American Journal of Pathology*, vol. 172, no. 5, pp. 1355–1362, 2008.

[123] R. Polavarapu, M. C. Gongora, H. Yi et al., "Tissue-type plasminogen activator-mediated shedding of astrocytic low-density lipoprotein receptor-related protein increases the permeability of the neurovascular unit," *Blood*, vol. 109, no. 8, pp. 3270–3278, 2007.

[124] J. O'Rourke, X. Jiang, Z. Hao, R. E. Cone, and A. R. Hand, "Distribution of sympathetic tissue plasminogen activator (tPA) to a distant microvasculature," *Journal of Neuroscience Research*, vol. 79, no. 6, pp. 727–733, 2005.

[125] R. Wong, "NMDA receptors expressed in oligodendrocytes," *BioEssays*, vol. 28, no. 5, pp. 460–464, 2006.

[126] A. Reijerkerk, G. Kooij, S. M. A. Van Der Pol et al., "The NR1 subunit of NMDA receptor regulates monocyte transmigration through the brain endothelial cell barrier," *Journal of Neurochemistry*, vol. 113, no. 2, pp. 447–453, 2010.

[127] G. T. Liberatore, A. Samson, C. Bladin, W. D. Schleuning, and R. L. Medcalf, "Vampire bat salivary plasminogen activator (desmoteplase): a unique fibrinolytic enzyme that does not promote neurodegeneration," *Stroke*, vol. 34, no. 2, pp. 537–543, 2003.

[128] J. P. López-Atalaya, B. D. Roussel, C. Ali et al., "Recombinant Desmodus rotundus salivary plasminogen activator crosses the blood-brain barrier through a low-density lipoprotein receptor-related protein-dependent mechanism without exerting neurotoxic effects," *Stroke*, vol. 38, no. 3, pp. 1036–1043, 2007.

[129] W. Hacke, A. J. Furlan, Y. Al-Rawi et al., "Intravenous desmoteplase in patients with acute ischaemic stroke selected by MRI perfusion-diffusion weighted imaging or perfusion CT (DIAS-2): a prospective, randomised, double-blind, placebo-controlled study," *The Lancet Neurology*, vol. 8, no. 2, pp. 141–150, 2009.

[130] E. J. Su, L. Fredriksson, M. Geyer et al., "Activation of PDGF-CC by tissue plasminogen activator impairs blood-brain barrier integrity during ischemic stroke," *Nature Medicine*, vol. 14, no. 7, pp. 731–737, 2008.

[131] J. Alvarez-Sabín, P. Delgado, S. Abilleira et al., "Temporal profile of matrix metalloproteinases and their inhibitors after spontaneous intracerebral hemorrhage: relationship to clinical and radiological outcome," *Stroke*, vol. 35, no. 6, pp. 1316–1322, 2004.

[132] S. Horstmann, P. Kalb, J. Koziol, H. Gardner, and S. Wagner, "Profiles of matrix metalloproteinases, their inhibitors, and laminin in stroke patients: influence of different therapies," *Stroke*, vol. 34, no. 9, pp. 2165–2170, 2003.

[133] J. Montaner, "Stroke biomarkers: can they help us to guide stroke thrombolysis?" *Drug News and Perspectives*, vol. 19, no. 9, pp. 523–532, 2006.

[134] A. Rosell and E. H. Lo, "Multiphasic roles for matrix metalloproteinases after stroke," *Current Opinion in Pharmacology*, vol. 8, no. 1, pp. 82–89, 2008.

[135] A. R. Green, "Pharmacological approaches to acute ischaemic stroke: reperfusion certainly, neuroprotection possibly," *British Journal of Pharmacology*, vol. 153, no. 1, pp. S325–S338, 2008.

[136] A. F. Ducruet, B. T. Grobelny, B. E. Zacharia, Z. L. Hickman, M. L. Yeh, and E. S. Connolly Jr., "Pharmacotherapy of cerebral ischemia," *Expert Opinion on Pharmacotherapy*, vol. 10, no. 12, pp. 1895–1906, 2009.

[137] U. Dirnagl, K. Becker, and A. Meisel, "Preconditioning and tolerance against cerebral ischaemia: from experimental strategies to clinical use," *The Lancet Neurology*, vol. 8, no. 4, pp. 398–412, 2009.

[138] J. Chen and R. Simon, "Ischemic tolerance in the brain," *Neurology*, vol. 48, no. 2, pp. 306–311, 1997.

[139] C. E. Murry, R. B. Jennings, and K. A. Reimer, "Preconditioning with ischemia: a delay of lethal cell injury in ischemic myocardium," *Circulation*, vol. 74, no. 5, pp. 1124–1136, 1986.

[140] T. Kirino, "Ischemic tolerance," *Journal of Cerebral Blood Flow and Metabolism*, vol. 22, no. 11, pp. 1283–1296, 2002.

[141] M. P. Stenzel-Poore, S. L. Stevens, and R. P. Simon, "Genomics of preconditioning," *Stroke*, vol. 35, no. 11, supplement 1, pp. 2683–2686, 2004.

[142] M. P. Stenzel-Poore, S. L. Stevens, Z. Xiong et al., "Effect of ischaemic preconditioning on genomic response to cerebral ischaemia: similarity to neuroprotective strategies in hibernation and hypoxia-tolerant states," *The Lancet*, vol. 362, no. 9389, pp. 1028–1037, 2003.

[143] J. L. Cadet and I. N. Krasnova, "Cellular and molecular neurobiology of brain preconditioning," *Molecular Neurobiology*, vol. 39, no. 1, pp. 50–61, 2009.

[144] A. Kunz, L. Park, T. Abe et al., "Neurovascular protection by ischemic tolerance: role of nitric oxide and reactive oxygen species," *Journal of Neuroscience*, vol. 27, no. 27, pp. 7083–7093, 2007.

[145] T. Masada, Y. Hua, G. Xi, S. R. Ennis, and R. F. Keep, "Attenuation of ischemic brain edema and cerebrovascular injury after ischemic preconditioning in the rat," *Journal of Cerebral Blood Flow and Metabolism*, vol. 21, no. 1, pp. 22–33, 2001.

[146] F. Y. Zhang, X. C. Chen, H. M. Ren, and W. M. Bao, "Effects of ischemic preconditioning on blood-brain barrier permeability and MMP-9 expression of ischemic brain," *Neurological Research*, vol. 28, no. 1, pp. 21–24, 2006.

[147] S. Zahler, C. Kupatt, and B. F. Becker, "Endothelial preconditioning by transient oxidative stress reduces inflammatory responses of cultured endothelial cells to TNF-α," *FASEB Journal*, vol. 14, no. 3, pp. 555–564, 2000.

[148] S. G. Zhou, X. Y. Lei, and D. F. Liao, "Effects of hypoxic preconditioning on the adhesion of neutrophils to vascular endothelial cells induced by hypoxia/reoxygenation," *Chinese Critical Care Medicine*, vol. 15, no. 3, pp. 159–162, 2003.

[149] P. An and Y. X. Xue, "Effects of preconditioning on tight junction and cell adhesion of cerebral endothelial cells," *Brain Research*, vol. 1272, no. C, pp. 81–88, 2009.

[150] A. V. Andjelkovic, S. M. Stamatovic, and R. F. Keep, "The protective effects of preconditioning on cerebral endothelial cells in vitro," *Journal of Cerebral Blood Flow and Metabolism*, vol. 23, no. 11, pp. 1348–1355, 2003.

[151] R. Gesuete, F. Orsini, E. R. Zanier et al., "Glial cells drive preconditioning-induced blood-brain barrier protection," *Stroke*, vol. 42, no. 5, pp. 1445–1453, 2011.

[152] T. Yamashita, K. Deguchi, S. Nagotani, and K. Abe, "Vascular protection and restorative therapy in ischemic stroke," *Cell Transplantation*, vol. 20, no. 1, pp. 95–97, 2011.

[153] P. A. Lapchak, "A critical assessment of edaravone acute ischemic stroke efficacy trials: is edaravone an effective neuroprotective therapy?" *Expert Opinion on Pharmacotherapy*, vol. 11, no. 10, pp. 1753–1763, 2010.

[154] S. Feng, Q. Yang, M. Liu et al., "Edaravone for acute ischaemic stroke," *Cochrane Database of Systematic Reviews*, vol. 12, p. CD007230, 2011.

[155] D. Deplanque and R. Bordet, "Physical activity: one of the easiest ways to protect the brain?" *Journal of Neurology, Neurosurgery and Psychiatry*, vol. 80, no. 9, p. 942, 2009.

[156] P. Amarenco, "Hypercholesterolemia, lipid-lowering agents, and the risk for brain infarction," *Neurology*, vol. 57, no. 5, pp. S35–S44, 2001.

[157] H. B. Rubins, J. Davenport, V. Babikian et al., "Reduction in stroke with gemfibrozil in men with coronary heart disease and low HDL cholesterol the veterans affairs HDL intervention trial (VA-HIT)," *Circulation*, vol. 103, no. 23, pp. 2828–2833, 2001.

[158] D. Deplanque, I. Masse, C. Lefebvre, C. Libersa, D. Leys, and R. Bordet, "Prior TIA, lipid-lowering drug use, and physical activity decrease ischemic stroke severity," *Neurology*, vol. 67, no. 8, pp. 1403–1410, 2006.

[159] B. Desvergne and W. Wahli, "Peroxisome proliferator-activated receptors: nuclear control of metabolism," *Endocrine Reviews*, vol. 20, no. 5, pp. 649–688, 1999.

[160] P. Lefebvre, G. Chinetti, J. C. Fruchart, and B. Staels, "Sorting out the roles of PPARα in energy metabolism and vascular homeostasis," *Journal of Clinical Investigation*, vol. 116, no. 3, pp. 571–580, 2006.

[161] S. Moreno, S. Farioli-vecchioli, and M. P. Cerù, "Immunolocalization of peroxisome proliferator-activated receptors and retinoid X receptors in the adult rat CNS," *Neuroscience*, vol. 123, no. 1, pp. 131–145, 2004.

[162] M. Ricote and C. K. Glass, "PPARs and molecular mechanisms of transrepression," *Biochimica et Biophysica Acta*, vol. 1771, no. 8, pp. 926–935, 2007.

[163] P. Delerive, K. De Bosscher, S. Besnard et al., "Peroxisome proliferator-activated receptor α negatively regulates the vascular inflammatory gene response by negative cross-talk with transcription factors NF-κB and AP-1," *Journal of Biological Chemistry*, vol. 274, no. 45, pp. 32048–32054, 1999.

[164] M. Collino, M. Aragno, R. Mastrocola et al., "Oxidative stress and inflammatory response evoked by transient cerebral ischemia/reperfusion: effects of the PPAR-α agonist WY14643," *Free Radical Biology and Medicine*, vol. 41, no. 4, pp. 579–589, 2006.

[165] D. Deplanque, P. Gelé, O. Pétrault et al., "Peroxisome proliferator-activated receptor-α activation as a mechanism of preventive neuroprotection induced by chronic fenofibrate treatment," *Journal of Neuroscience*, vol. 23, no. 15, pp. 6264–6271, 2003.

[166] H. Inoue, X. F. Jiang, T. Katayama, S. Osada, K. Umesono, and S. Namura, "Brain protection by resveratrol and fenofibrate against stroke requires peroxisome proliferator-activated receptor α in mice," *Neuroscience Letters*, vol. 352, no. 3, pp. 203–206, 2003.

[167] C. Mysiorek, M. Culot, L. Dehouck et al., "Peroxisome proliferator-activated receptor-α activation protects brain capillary endothelial cells from oxygen-glucose deprivation-induced hyperpermeability in the blood-brain barrier," *Current Neurovascular Research*, vol. 6, no. 3, pp. 181–193, 2009.

[168] Y. Luo, W. Yin, A. P. Signore et al., "Neuroprotection against focal ischemic brain injury by the peroxisome proliferator-activated receptor-γ agonist rosiglitazone," *Journal of Neurochemistry*, vol. 97, no. 2, pp. 435–448, 2006.

[169] T. Mabuchi, K. Kitagawa, T. Ohtsuki et al., "Contribution of microglia/macrophages to expansion of infarction and response of oligodendrocytes after focal cerebral ischemia in rats," *Stroke*, vol. 31, no. 7, pp. 1735–1743, 2000.

[170] L. Pantoni, C. Sarti, and D. Inzitari, "Cytokines and cell adhesion molecules in cerebral ischemia: experimental bases and therapeutic perspectives," *Arteriosclerosis, Thrombosis, and Vascular Biology*, vol. 18, no. 4, pp. 503–513, 1998.

[171] T. Kielian and P. D. Drew, "Effects of peroxisome proliferator-activated receptor-γ agonists on central nervous system inflammation," *Journal of Neuroscience Research*, vol. 71, no. 3, pp. 315–325, 2003.

[172] X. Zhao, Z. Ou, J. C. Grotta, N. Waxham, and J. Aronowski, "Peroxisome-proliferator-activated receptor-gamma (PPARγ) activation protects neurons from NMDA excitotoxicity," *Brain Research*, vol. 1073-1074, no. 1, pp. 460–469, 2006.

[173] M. Endres, "Statins and stroke," *Journal of Cerebral Blood Flow and Metabolism*, vol. 25, no. 9, pp. 1093–1110, 2005.

[174] A. Zacco, J. Togo, K. Spence et al., "3-hydroxy-3-methylglutaryl coenzyme a reductase inhibitors protect cortical neurons from excitotoxicity," *Journal of Neuroscience*, vol. 23, no. 35, pp. 11104–11111, 2003.

[175] C. R. W. Kuhlmann, M. Gerigk, B. Bender, D. Closhen, V. Lessmann, and H. J. Luhmann, "Fluvastatin prevents glutamate-induced blood-brain-barrier disruption in vitro," *Life Sciences*, vol. 82, no. 25-26, pp. 1281–1287, 2008.

[176] S. Wang, S. R. Lee, S. Z. Guo et al., "Reduction of tissue plasminogen activator-induced matrix metalloproteinase-9 by simvastatin in astrocytes," *Stroke*, vol. 37, no. 7, pp. 1910–1912, 2006.

[177] M. Jasińska, J. Owczarek, and D. Orszulak-Michalak, "Statins: a new insight into their mechanisms of action and consequent pleiotropic effects," *Pharmacological Reports*, vol. 59, no. 5, pp. 483–499, 2007.

[178] G. Martin, H. Duez, C. Blanquart et al., "Statin-induced inhibition of the rho-signaling pathway activates PPARα and induces HDL apoA-I," *Journal of Clinical Investigation*, vol. 107, no. 11, pp. 1423–1432, 2001.

[179] R. Paumelle and B. Staels, "Cross-talk between statins and PPARα in cardiovascular diseases: clinical evidence and basic mechanisms," *Trends in Cardiovascular Medicine*, vol. 18, no. 3, pp. 73–78, 2008.

[180] K. J. Yin, Z. Deng, M. Hamblin, J. Zhang, and Y. E. Chen, "Vascular PPARδ protects against stroke-induced brain injury," *Arteriosclerosis, Thrombosis, and Vascular Biology*, vol. 31, no. 3, pp. 574–581, 2011.

[181] A. Plotnikov, E. Zehorai, S. Procaccia, and R. Seger, "The MAPK cascades: signaling components, nuclear roles and mechanisms of nuclear translocation," *Biochimica et Biophysica Acta*, vol. 1813, no. 9, pp. 1619–1633, 2011.

[182] C. Widmann, S. Gibson, M. B. Jarpe, and G. L. Johnson, "Mitogen-activated protein kinase: conservation of a three-kinase module from yeast to human," *Physiological Reviews*, vol. 79, no. 1, pp. 143–180, 1999.

[183] V. Waetzig, Y. Zhao, and T. Herdegen, "The bright side of JNKs-Multitalented mediators in neuronal sprouting, brain development and nerve fiber regeneration," *Progress in Neurobiology*, vol. 80, no. 2, pp. 84–97, 2006.

[184] J. M. Kyriakis and J. Avruch, "Mammalian mitogen-activated protein kinase signal transduction pathways activated by stress and inflammation," *Physiological Reviews*, vol. 81, no. 2, pp. 807–869, 2001.

[185] W. Haeusgen, R. Boehm, Y. Zhao, T. Herdegen, and V. Waetzig, "Specific activities of individual c-Jun N-terminal kinases in the brain," *Neuroscience*, vol. 161, no. 4, pp. 951–959, 2009.

[186] H. Nishina, T. Wada, and T. Katada, "Physiological roles of SAPK/JNK signaling pathway," *Journal of Biochemistry*, vol. 136, no. 2, pp. 123–126, 2004.

[187] V. Waetzig and T. Herdegen, "MEKK1 controls neurite regrowth after experimental injury by balancing ERK1/2 and JNK2 signaling," *Molecular and Cellular Neuroscience*, vol. 30, no. 1, pp. 67–78, 2005.

[188] M. A. Bogoyevitch, "The isoform-specific functions of the c-Jun N-terminal kinases (JNKs): differences revealed by gene targeting," *BioEssays*, vol. 28, no. 9, pp. 923–934, 2006.

[189] S. Brecht, R. Kirchhof, A. Chromik et al., "Specific pathophysiological functions of JNK isoforms in the brain," *European Journal of Neuroscience*, vol. 21, no. 2, pp. 363–377, 2005.

[190] H. Chaudhury, M. Zakkar, J. Boyle et al., "C-Jun N-terminal kinase primes endothelial cells at atheroprone sites for apoptosis," *Arteriosclerosis, Thrombosis, and Vascular Biology*, vol. 30, no. 3, pp. 546–553, 2010.

[191] J. Cui, M. Zhang, Y. Q. Zhang, and Z. H. Xu, "JNK pathway: diseases and therapeutic potential," *Acta Pharmacologica Sinica*, vol. 28, no. 5, pp. 601–608, 2007.

[192] Y. Gao, A. P. Signore, W. Yin et al., "Neuroprotection against focal ischemic brain injury by inhibition of c-Jun N-terminal kinase and attenuation of the mitochondrial apoptosis-signaling pathway," *Journal of Cerebral Blood Flow and Metabolism*, vol. 25, no. 6, pp. 694–712, 2005.

[193] T. Borsellol, P. G. H. Clarkel, L. Hirt et al., "A peptide inhibitor of c-Jun N-terminal kinase protects against excitotoxicity and cerebral ischemia," *Nature Medicine*, vol. 9, no. 9, pp. 1180–1186, 2003.

[194] L. Hirt, J. Badaut, J. Thevenet et al., "D-JNKI1, a cell-penetrating c-Jun-N-terminal kinase inhibitor, protects against cell death in severe cerebral ischemia," *Stroke*, vol. 35, no. 7, pp. 1738–1743, 2004.

[195] K. Wiegler, C. Bonny, D. Coquoz, and L. Hirt, "The JNK inhibitor XG-102 protects from ischemic damage with delayed intravenous administration also in the presence of recombinant tissue plasminogen activator," *Cerebrovascular Diseases*, vol. 26, no. 4, pp. 360–366, 2008.

[196] D. Michel-Monigadon, C. Bonny, and L. Hirt, "C-Jun N-terminal kinase pathway inhibition in intracerebral hemorrhage," *Cerebrovascular Diseases*, vol. 29, no. 6, pp. 564–570, 2010.

[197] Z. Xie, C. J. Smith, and L. J. Van Eldik, "Activated glia induce neuron death Via MAP kinase signaling pathways involving JNK and p38," *GLIA*, vol. 45, no. 2, pp. 170–179, 2004.

[198] R. Kacimi, R. G. Giffard, and M. A. Yenari, "Endotoxin-activated microglia injure brain derived endothelial cells via NF-κB, JAK-STAT and JNK stress kinase pathways," *Journal of Inflammation*, vol. 8, article 7, 2011.

[199] C. Benakis, C. Bonny, and L. Hirt, "JNK inhibition and inflammation after cerebral ischemia," *Brain, Behavior, and Immunity*, vol. 24, no. 5, pp. 800–811, 2010.

10

WIPI-1 Positive Autophagosome-Like Vesicles Entrap Pathogenic *Staphylococcus aureus* for Lysosomal Degradation

Mario Mauthe,[1] Wenqi Yu,[2] Oleg Krut,[3] Martin Krönke,[3] Friedrich Götz,[2] Horst Robenek,[4] and Tassula Proikas-Cezanne[1]

[1] *Autophagy Laboratory, Interfaculty Institute for Cell Biology, Eberhard Karls University Tübingen, Auf der Morgenstelle 15, 72076 Tübingen, Germany*
[2] *Microbial Genetics, Interfaculty Institute for Microbiology and Infectious Medicine, University of Tübingen, 72076 Tübingen, Germany*
[3] *Institute for Medical Microbiology, Immunology and Hygiene, University of Cologne, 50935 Cologne, Germany*
[4] *Leibniz Institute for Arteriosclerosis Research, University of Münster, 48149 Münster, Germany*

Correspondence should be addressed to Tassula Proikas-Cezanne, tassula.proikas-cezanne@uni-tuebingen.de

Academic Editor: Anne Simonsen

Invading pathogens provoke the autophagic machinery and, in a process termed xenophagy, the host cell survives because autophagy is employed as a safeguard for pathogens that escaped phagosomes. However, some pathogens can manipulate the autophagic pathway and replicate within the niche of generated autophagosome-like vesicles. By automated fluorescence-based high content analyses, we demonstrate that *Staphylococcus aureus* strains (USA300, HG001, SA113) stimulate autophagy and become entrapped in intracellular PtdIns(3)P-enriched vesicles that are decorated with human WIPI-1, an essential PtdIns(3)P effector of canonical autophagy and membrane protein of both phagophores and autophagosomes. Further, *agr*-positive *S. aureus* (USA300, HG001) strains were more efficiently entrapped in WIPI-1 positive autophagosome-like vesicles when compared to *agr*-negative cells (SA113). By confocal and electron microscopy we provide evidence that single- and multiple-Staphylococci entrapped undergo cell division. Moreover, the number of WIPI-1 positive autophagosome-like vesicles entrapping Staphylococci significantly increased upon (i) lysosomal inhibition by bafilomycin A_1 and (ii) blocking PIKfyve-mediated PtdIns(3,5)P_2 generation by YM201636. In summary, our results provide evidence that the PtdIns(3)P effector function of WIPI-1 is utilized during xenophagy of *Staphylococcus aureus*. We suggest that invading *S. aureus* cells become entrapped in autophagosome-like WIPI-1 positive vesicles targeted for lysosomal degradation in nonprofessional host cells.

1. Introduction

Macroautophagy (hereafter autophagy) is a cytoprotective cellular degradation mechanism for long-lived proteins and organelles [1]. Autophagy is specific to eukaryotic cells and important for cellular survival by enabling a constitutive clearance and recycling of cytoplasmic material (basal autophagy). Crucial to the process of autophagy is the fact, that cytoplasmic material is stochastically degraded. Portions of the cytoplasm become randomly sequestered in unique, double-membrane vesicles, autophagosomes. Autophagosomes are generated by elongation and closure of a membrane precursor, the phagophore. Subsequently, autophagosomes fuse with lysosomes to acquire acidic hydrolases for cargo degradation [2]. This stochastic constitutive form of autophagy provides constant clearance of the cytoplasm. Upon stress, such as starvation, the autophagic activity is induced above basal level to compensate nutrient shortage by providing monomeric constituents, such as amino acids, and energy. Conversely, under nutrient-rich conditions autophagy is suppressed by the mTORC1 signaling circuit [3]. Importantly, autophagy is also activated in a specific manner and targets damaged organelles, protein aggregates, or pathogens for degradation [4]. Both, stochastic and specific autophagy are crucial to secure cellular homeostasis [5].

Prerequisite for the formation of autophagosomes is the generation of an essential phospholipid, phosphatidylinositol 3-phosphate (PtdIns(3)P), a result of the activity of the phosphatidylinositol 3-kinase class III (PtdIns3KC3) in complex with Beclin 1, p150, and Atg14L [6, 7]. The PtdIns(3)P signal is decoded through PtdIns(3)P-binding effectors specific to autophagy, such as the human WIPI proteins [8]. WIPI-1 (Atg18 in yeast) specifically binds PtdIns(3)P at the phagophore and fosters the recruitment of two ubiquitin-like conjugation systems, Atg12 and LC3, involved in phagophore elongation and closure [9]. Subsequently, WIPI-1 becomes a membrane protein of autophagosomes where it localizes at both the inner and outer membrane [10, 11]. Hence the specific localization of WIPI-1 at the phagophore and at autophagosomes upon the initiation of autophagy can monitor the process of canonical autophagy, as it is dependent on the PtdIns(3)P signal [11].

The process of autophagy is closely connected with a variety of diseases such as tumor development, neurodegeneration, and with cellular responses to pathogens, including viral infection and bacterial cell invasion [5, 12]. *Staphylococcus aureus*, a major pathogen for nosocomial infectious diseases was initially characterized as an extracellular pathogen, but was later found to also target nonprofessional host cells like keratinocytes, fibroblasts, endothelial cells, and epithelial cells where invading *S. aureus* liberates from the endosomal compartment [13]. In HeLa cells, *S. aureus* was found to become sequestered and to replicate in autophagosome-like vesicles as a result of autophagosome/lysosome fusion block, which ultimately leads to cell death [14].

Here, we visualized the invasion of mCherry-expressing *S. aureus* strains USA300, HG001, SA113 in human U2OS tumor cells that stably express GFP-WIPI-1 for automated fluorescence-based high content analyses, a procedure that monitors the autophagic process and that we have established earlier [15]. We provide evidence that *S. aureus* stimulates canonical autophagy in nonprofessional host cells and becomes entrapped in noncanonical WIPI-1 positive autophagosome-like vesicles. Time course experiments showed that the number of tumor cells that contain such WIPI-1 positive autophagosome-like vesicles with entrapped *S. aureus* cells increased over time (30 min–2 h). After an infection period of 2 h, 40–50% of the cells harbored WIPI-1 positive autophagosome-like vesicles sequestering *agr*-positive *S. aureus* (USA300, HG001), and 20% of the tumor cells contained entrapped *agr*-negative *S. aureus* (SA113). Importantly, we demonstrate that the number of WIPI-1 positive autophagosome-like vesicles harboring *S. aureus* significantly increased upon lysosomal inhibition, strongly arguing for the degradation of *S. aureus* through xenophagy. In addition, by employing GFP-FYVE and a selective PIKfyve inhibitor (YM201636) we further demonstrate the requirement of PtdIns(3)P-enriched membranes during the process of entrapping invading *S. aureus*.

2. Material and Methods

2.1. Eukaryotic Cell Culture. The human osteosarcoma cell line U2OS (ATCC) was cultured in DMEM (Invitrogen) supplemented with 10% FCS (PAA), 100 U/mL penicillin/100 μg/mL streptomycin (Invitrogen), 5 μg/mL plasmocin (Invivogen) at 37°C, 5% CO_2. Monoclonal human

U2OS cell clones stably expressing either GFP-WIPI-1 [15, 20] or GFP-2xFYVE [9] were cultured in DMEM (Invitrogen) supplemented with 10% FCS (PAA), 100 U/mL penicillin/100 μg/mL streptomycin (Invitrogen), 5 μg/mL plasmocin (Invivogen), 0.6 mg/mL G418 (Invitrogen) at 37°C, 5% CO_2. The following media were used for treatments: DMEM/FCS (DMEM supplemented with 10% FCS), DMEM (DMEM without FCS), and EBSS (Sigma-Aldrich).

2.2. Bacterial Strains. S. aureus USA300, HG001, SA113, or *S. carnosus* TM300 [21] (see Table 1) were electroporated with the pCtuf-*ppmch* plasmid. The pCtuf-*ppmch* plasmid encoded mCherry fused with the propeptide of lipase for fluorescence enhancement, and *ppmch* expression was controlled by the native constitutive EF-Tu promotor. Electroporated bacterial strains were grown in basic medium (1% peptone, 0.5% yeast extract, 0.5% NaCl, 0.1% glucose, 0.1% K_2HPO_4) at 37°C to an OD_{600} of 0.8 and harvested by centrifugation.

2.3. Bacterial Infection of Eukaryotic Host Cells. GFP-WIPI-1 expressing U2OS cells were seeded in 96-well plates (Brand) in DMEM/10% FCS 20 hours before bacterial infection. S. aureus (USA300, HG001, SA113) or *S. carnosus* carrying the pCtuf-*ppmch* plasmid, were diluted in DMEM, DMEM/10% FCS or EBSS (Sigma-Aldrich) to an m.o.i of 100, added to the GFP-WIPI-1 U2OS cells, and incubated for 0.5, 1, or 2 hours at 37°C, 5% CO_2. Alternatively, S. aureus USA300 cells were diluted (m.o.i of 100) in DMEM/FCS supplemented with either bafilomycin A_1 (200 nM, Sigma-Aldrich) or YM201636 (800 nM, Cayman Chemicals) or with both and used to infect GFP-WIPI-1 expressing U2OS cells for 2 hours at 37°C, 5% CO_2. Alternatively, GFP-2xFYVE expressing U2OS cells [9] were infected with S. aureus USA300 (in DMEM/FCS) for 2 hours at 37°C, 5% CO_2.

2.4. Autophagy Assay. GFP-WIPI-1 expressing U2OS cells, seeded in 96-well plates, were treated with nutrient-rich culture medium (DMEM/10% FCS), culture medium lacking serum (DMEM), or medium lacking serum and amino acids (EBSS) for 0.5, 1, or 2 hours. After fixation with 3.7% paraformaldehyde for 30 minutes, autophagy was accessed by WIPI-1 puncta formation analysis [11, 22] (see below).

2.5. Confocal Laser Scanning Microscopy. Confocal microscopy was conducted as previously described [8]. Images were acquired using an LSM510 microscope (Zeiss) and a 63 × 1.4 DIC Plan-Apochromat oil-immersion objective. For each image, 8–10 optical sections (0.5 μm) were acquired. Both, single optical sections as well as projections from 8–10 optical sections are presented.

2.6. Automated Fluorescence Image Acquisition and Analysis. Stable GFP-WIPI-1 U2OS cells were automatically imaged and analysed using the *In Cell Analyzer 1000* (GE Healthcare) as described earlier [9, 15]. Cells exposed to bacteria (see above) were stained with DAPI (5 μg/mL; Applichem). Fluorescence images were automatically acquired with a Nikon

TABLE 1: Bacterial strains used in this study.

Bacterial strain	Relevant properties	Relevant genotype	Reference
S. aureus USA300	Pathogenic, community-associated methicillin-resistant S. aureus (CA-MRSA)	agr$^+$	[16]
S. aureus HG001	Pathogenic, methicillin-sensitive S. aureus (MSSA)	agr$^+$	[17]
S. aureus SA113	Pathogenic, methicillin-sensitive S. aureus (MSSA)	agr$^-$	[18]
S. carnosus TM300	Apathogenic, food grade staphylococcal species		[19]

40x Plan Fluor objective and the excitation/emission filter D360_40X/HQ460_40M (DAPI), HQ535_50X/HQ620_60 M (mCherry), and S475_20X/HQ535_50M (GFP). GFP-WIPI-1 puncta were automatically analysed as previously described [15] and the number of GFP-WIPI-1 puncta-positive cells as well as the number of GFP-WIPI-1 puncta per cell was determined. Red fluorescent bacteria were automatically analysed by using the *dual area object analysis*. The algorithms *inclusion* and *multiscale top hat* were applied and the total area of bacterial fluorescence within the cell was determined. To determine the number of cells containing GFP-WIPI-1 positive autophagosome-like vesicles sequestering bacteria, automatically acquired fused images (DAPI, GFP, mCherry) of 100 individual cells for each treatment were analyzed.

2.7. Electron Microscopy. Stable GFP-WIPI-1 U2OS cells were infected with *S. aureus* USA300 (m.o.i of 100) in DMEM/FCS and fixed in 2% glutaraldehyde and 0.5% osmium tetroxide in 0.1 M PBS, dehydrated with ethanol, and embedded in Epon using standard procedures as previously described [23]. Thin sections were cut using an ultramicrotome and contrasted with uranyl acetate and lead citrate. Thin sections were examined in an EM410 electron microscope (Philips) and documented digitally (DITABIS).

2.8. Statistical Analysis. Statistical significance was evaluated using two-tailed heteroscedastic *t*-testing and *P* values were calculated.

3. Results

3.1. Visualizing Basal and Induced Autophagy by Automated GFP-WIPI-1 Image Acquisition and Analysis. The WIPI-1 puncta-formation assay allows the assessment of the evolutionarily conserved, PtdIns(3)P-dependent initiation of autophagy on the basis of fluorescence microscopy, previously employed by using confocal microscopy or automated image acquisition and analysis [11, 15]. Thereby, endogenous WIPI-1 can be visualized by indirect immunofluorescence or alternatively by introducing GFP-WIPI-1 as conducted in the present study. Fluorescent WIPI-1 puncta reflect the accumulation of WIPI-1 at membranes via its specific binding to PtdIns(3)P was found to represent phagophores and autophagosomes [10, 11]. In addition, WIPI-1 binds to PtdIns(3)P at the endoplasmic reticulum and at the plasma membrane upon the induction of autophagy, indicative for membrane origins where phagophore/autophagosome formation is initiated by unknown mechanisms [10]. Here, we employed automated GFP-WIPI-1 image acquisition

and analysis as follows. Human U2OS cells that stably express GFP-WIPI-1 were seeded in 96-well plates and basal autophagy, and starvation-induced autophagy was monitored in up to 3000 individual cells per treatment over time (Figure 1). After an incubation period of 0.5, 1, or 2 h with nutrient-rich culture medium (DMEM/FCS), basal autophagic activity was found in approximately 10% of the cells (Figures 1(a) and 1(d)). Serum starvation (DMEM) elevated the number of GFP-WIPI-1 puncta-positive cells to approximately 50% (Figures 1(b) and 1(d)), and both serum and amino acid starvation (EBSS) further elevated this number to approximately 85% (Figures 1(c) and 1(d)). In addition, we demonstrate that with regard to nutrient-rich medium (DMEM/FCS), the number of GFP-WIPI-1 puncta per cell also increased upon serum (DMEM) or upon both serum and amino acid starvation (EBSS) (Figure 1(e)). These culture media (DMEM/FCS, DMEM, EBSS) were used in the following experiments to infect GFP-WIPI-1 expressing U2OS cells with mCherry-expressing Staphylococci.

3.2. Formation of GFP-WIPI-1 Positive Autophagosome-Like Vesicles upon Staphylococcus aureus Infection. Upon infection of GFP-WIPI-1 U2OS cells with pathogenic Staphylococci, here *S. aureus* HG001, in nutrient-rich medium (DMEM/FCS), we identified canonical, autophagosomal GFP-WIPI-1 membranes (Figures 2(a) and 2(b)), and new GFP-WIPI-1 autophagosome-like vesicles that were larger in diameter with decreased fluorescence intensity (Figure 2(c)) when compared to the canonical GFP-WIPI-1 puncta. GFP-WIPI-1 autophagosome-like vesicles (Figure 2(c)) were rarely observed when starvation media (DMEM, EBSS) were used during the infection with *S. aureus* HG001 (Supplementary Figure 1 available online at doi:10.1155/2012/179207).

To monitor and quantify this particular GFP-WIPI-1 response upon mCherry-expressing Staphylococci infection in an automated fashion (Figure 3), cells were stained with DAPI and by using three different excitation/emission filters, DAPI, GFP, and mCherry fluorescence images were acquired (Figure 3). Up to 2723 individual cells per treatment were automatically recognized by both DAPI and the overall cellular GFP fluorescence. GFP images were used to automatically detect and determine the number of cells harboring GFP-WIPI-1 puncta by applying a decision tree as previously described [15]. Additionally, mCherry fluorescence was used to automatically determine the fluorescence area, reflecting the load of intracellular Staphylococci. For the quantification of cells harboring GFP-WIPI-1 positive autophagosome-like vesicles entrapping Staphylococci, fused images (DAPI, GFP, mCherry) of 100 individual cells were used (Figure 3).

GFP-WIPI-1

(a)

(b)

(c)

Nutrient-rich medium
(DMEM/FCS)

Serum starvation
(DMEM)

Amino acid and serum
starvation
(EBSS)

(d)

(e)

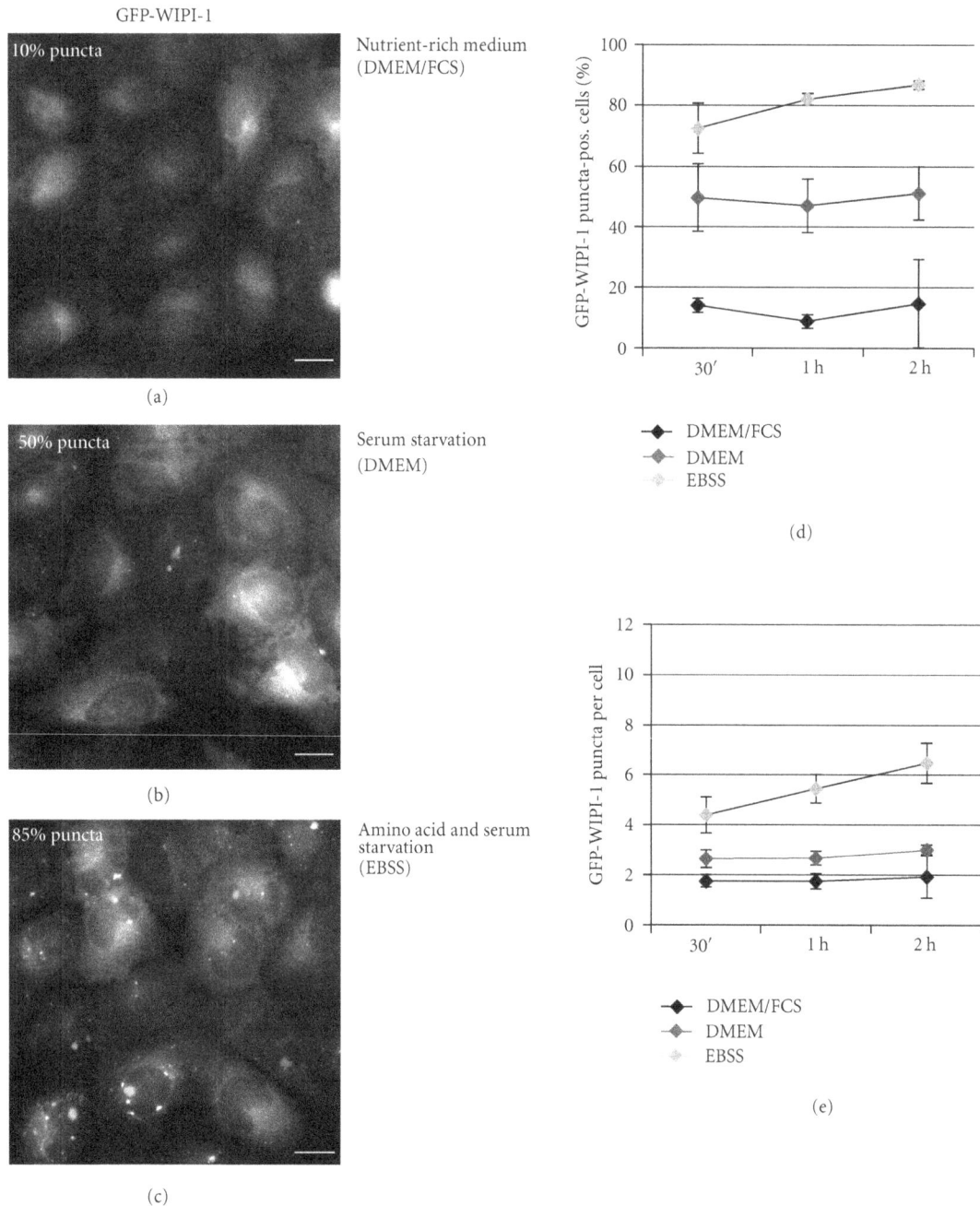

FIGURE 1: GFP-WIPI-1 puncta formation upon serum and amino acid starvation. GFP-WIPI-1 U2OS cells were treated with nutrient-rich culture medium (DMEM/FCS), serum-free culture medium (DMEM), or with medium lacking both serum and amino acids (EBSS) for 0.5, 1, and 2 h. Fluorescence images were automatically acquired and 2 h treatment images are shown ((a)–(c)). The number of GFP-WIPI-1 puncta-positive cells (d), and of GFP-WIPI-1 puncta per cell (e) was automatically determined. Each measure point represents mean value from up to 3000 individually analyzed cells per treatment condition $\pm$ SD ($n = 2$, each in triplicates). Scale bars: 20 μm.

3.3. Pathogenic Staphylococcus aureus USA300, HG001, and SA113 Stimulated Canonical Autophagosome Formation and Became Entrapped in GFP-WIPI-1 Positive Autophagosome-Like Vesicles. In the following experiment, GFP-WIPI-1 expressing U2OS cells were infected for 0.5, 1 and 2 h with mCherry-expressing *S. aureus* USA300 (Figure 4, Supplementary Figure 2), HG001 (Figure 5, Supplementary Figure 3), or SA113 (Figure 6, Supplementary Figure 4)

either in nutrient-rich medium (DMEM/FCS), serum-free medium (DMEM), or serum and amino acid-free medium (EBSS). Subsequently, fluorescence images (approximately 2000 individual cells per treatment) were automatically acquired and analyzed as described (Figure 3). Please note that the control experiments in Figure 1 were conducted in parallel to the experiments presented in Figures 4–7 hence provide the comparison for conditions without

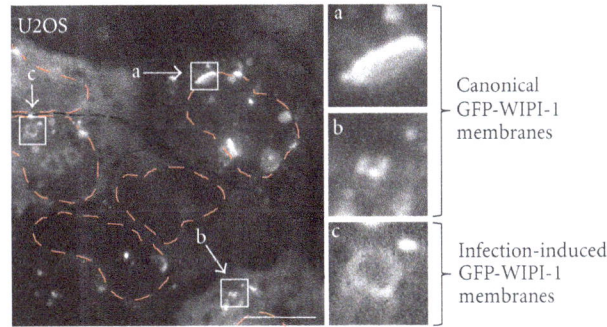

FIGURE 2: GFP-WIPI-1 images upon infection of U2OS cells with *S. aureus* HG001. GFP-WIPI-1 U2OS cells were infected with *S. aureus* HG001 in DMEM/FCS for 2 h and images were automatically acquired. GFP-WIPI-1 fluorescence of the cells (indicated with the black-dashed line) is shown, and cell nuclei are indicated (red-dashed line) according to DAPI staining (not shown). Highlighted are the different GFP-WIPI-1 structures observed: large perinuclear GFP-WIPI-1 positive membranes (a) and cytoplasmic GFP-WIPI-1 puncta (b), reflecting canonical autophagosomal membranes. In addition, GFP-WIPI-1 positive autophagosomal-like vesicles appeared specifically upon infection (c). Scale bars: 20 μm. Supplementary information is provided (Supplementary Figure 1).

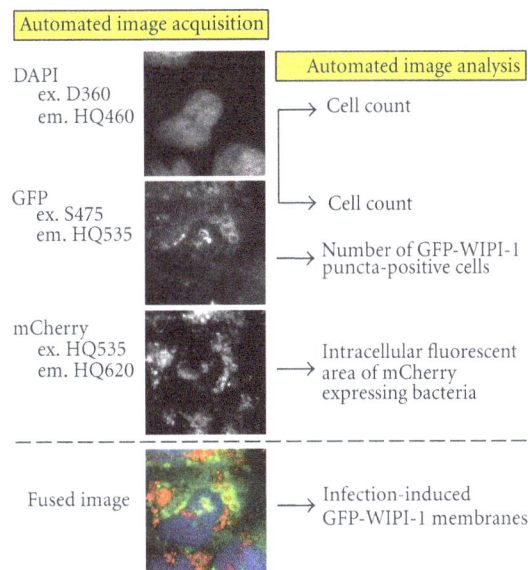

FIGURE 3: Automated image acquisition and analysis of stably expressing GFP-WIPI-1 U2OS cells with mCherry-expressing Staphylococci. Fluorescence images (middle panel) were automatically acquired using different emission/excitation filters for DAPI, GFP, and mCherry (left panel). DAPI and GFP images were used to automatically detect individual cells, and GFP images were used for detecting and analyzing GFP-WIPI-1 puncta formation (indicated in the right panel). Additionally, for each individual cell the bacterial area was determined (indicated in the right panel) and a fused image was further used to determine the number of cells harboring WIPI-1 positive autophagosome-like vesicles entrapping Staphylococci.

(Figure 1) and with (Figures 4–7, Supplementary Figures 2–5) Staphylococci.

As shown in Figure 1, under nutrient rich conditions (DMEM/FCS) the number of GFP-WIPI-1 puncta-positive cells is low (approximately 10%), reflecting cells that undergo basal autophagy. Interestingly, upon infection of GFP-WIPI-1 expressing U2OS cells with *S. aureus* USA300 in DMEM/FCS, a prominent increase of GFP-WIPI-1 puncta-positive cells (up to approximately 70% within 2 h of infection) was observed (Figure 4(a), in green). In addition, the number of GFP-WIPI-1 puncta per individual cell also increased upon *S. aureus* USA300 infection in DMEM/FCS (Supplementary Figure 6(B)). The elevated number of GFP-WIPI-1 puncta-positive cells and GFP-WIPI-1 puncta per cell correlated with an increase of intracellular *S. aureus* USA300 (Figure 4(a), in red). Using serum-free conditions either in the presence (DMEM, Figure 4(b), in red) or absence of amino acids (EBSS, Figure 4(c), in red), no increase of intracellular *S. aureus* USA300 was observed. However, infection of *S. aureus* USA300 in DMEM also resulted in an increase (up to approximately 70%) of GFP-WIPI-1 puncta-positive cells (Figure 4(b), in green), whereas *S. aureus* USA300 in EBSS (Figure 4(c)) did not trigger a further increase of the number of GFP-WIPI-1 puncta-positive cells when compared to EBSS treatment alone (Figure 1).

Next, we determined the number of cells displaying entrapped *S. aureus* USA300 within GFP-WIPI-1 positive

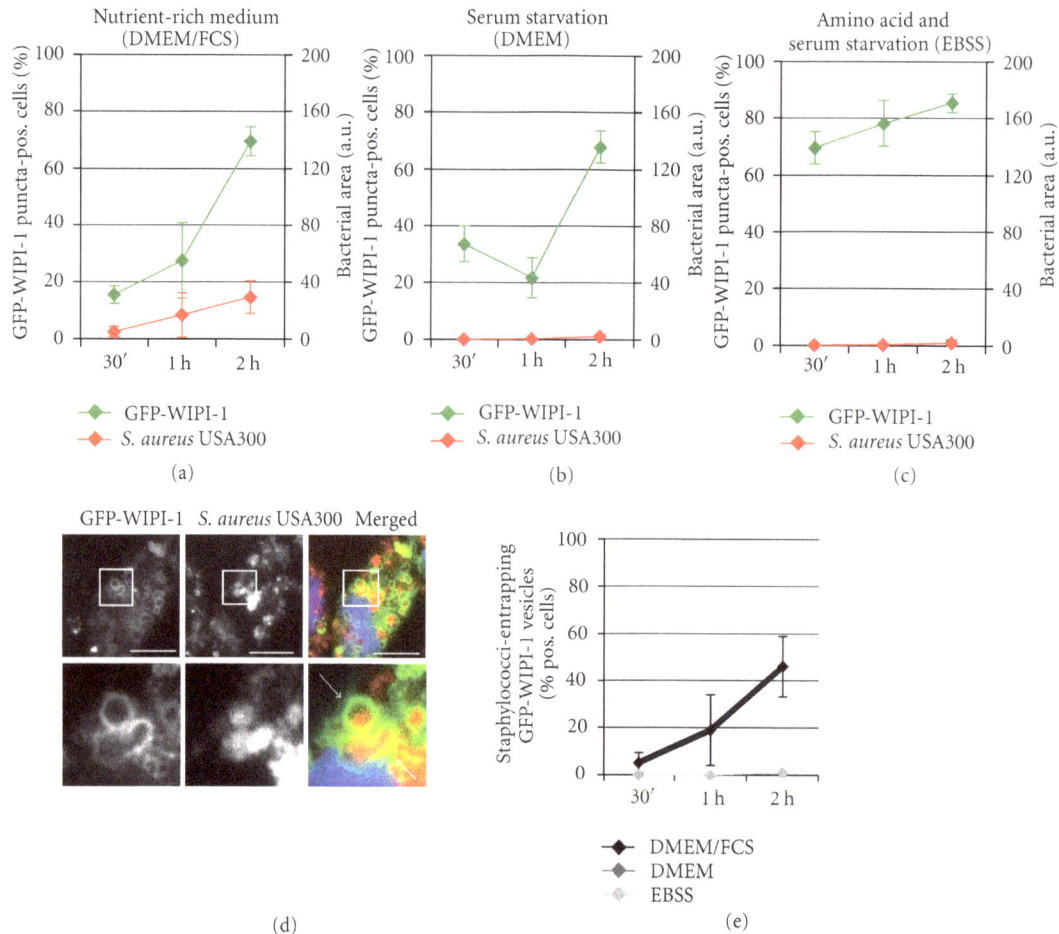

Figure 4: Pathogenic *S. aureus* USA300 induces GFP-WIPI-1 puncta formation and becomes entrapped in GFP-WIPI-1 positive autophagosome-like vesicles. GFP-WIPI-1 U2OS cells were infected with mCherry-expressing *S. aureus* USA300 for 0.5, 1, and 2 h in DMEM/FCS, DMEM, or EBSS. Automated image acquisition and analysis were conducted as described in Figure 3. The quantification of up to 2000 individual cells is presented for GFP-WIPI-1 (in green) and *S. aureus* USA300 (in red) using either DMEM/FCS (a), DMEM (b), or EBSS (c) for infection ± SD ($n = 2$, each in duplicates). Representative images (2 h infection in DMEM/FCS) are shown (d). Scale bars: 20 μm. From 100 infected cells for each of the treatment condition, the number of cells displaying GFP-WIPI-1 positive autophagosomal-like vesicles entrapping *S. aureus* USA300 was determined (e) ± SD ($n = 2$, each in duplicates).

autophagosome-like vesicles (Figures 4(d) and 4(e)). In line with the increased number of cells carrying intracellular *S. aureus* USA300 when nutrient-rich medium (DMEM/FCS) was used (Figure 4(a)), the number of cells with GFP-WIPI-1 positive autophagosome-like vesicles that entrap *S. aureus* USA300 (approximately 40%) also increased (Figure 4(e)). This was not observed by using DMEM or EBSS (Figure 4(e)). We also provide the control images corresponding to *S. aureus* USA300 infections using either DMEM or EBSS (Supplementary Figure 2).

The infection of stably expressing GFP-WIPI-1 U2OS cells with *S. aureus* HG001 in DMEM/FCS also triggered an elevation of GFP-WIPI-1 puncta-positive cells (up to 76%) (Figure 5(a), in green) and of GFP-WIPI-1 puncta per cell (Supplementary Figure 6(C)). Again, the increased number of GFP-WIPI-1 puncta-positive cells correlated with an increased bacterial load (Figure 5(a), in red) and the increase in the number of cells displaying GFP-WIPI-1

positive autophagosome-like vesicles entrapping *S. aureus* HG001 (approximately 40%) (Figures 5(d) and 5(e)). Also in this case, this feature was not observed by using DMEM or EBSS (Figure 5(e)), but DMEM conditions still triggered an increase of GFP-WIPI-1 puncta formation (Figure 5(b), Supplementary Figure 6(C)) when compared with control setting (Figure 1, Supplementary Figure 6(A)). Control images corresponding to *S. aureus* HG001 infections using either DMEM or EBSS are also provided (Supplementary Figure 3).

Next, we employed the *agr*-deficient *S. aureus* strain SA113 and infected stably expressing GFP-WIPI-1 U2OS cells. Clearly, upon infection in DMEM/FCS the number of GFP-WIPI-1 puncta-positive cells increased over time to up to 60% (Figure 6(a), in green), which correlated with an increasing bacterial load (Figure 6(a), in red). See also the increased number of GFP-WIPI-1 puncta per cell upon *S. aureus* SA113 infection in DMEM/FCS (Supplementary

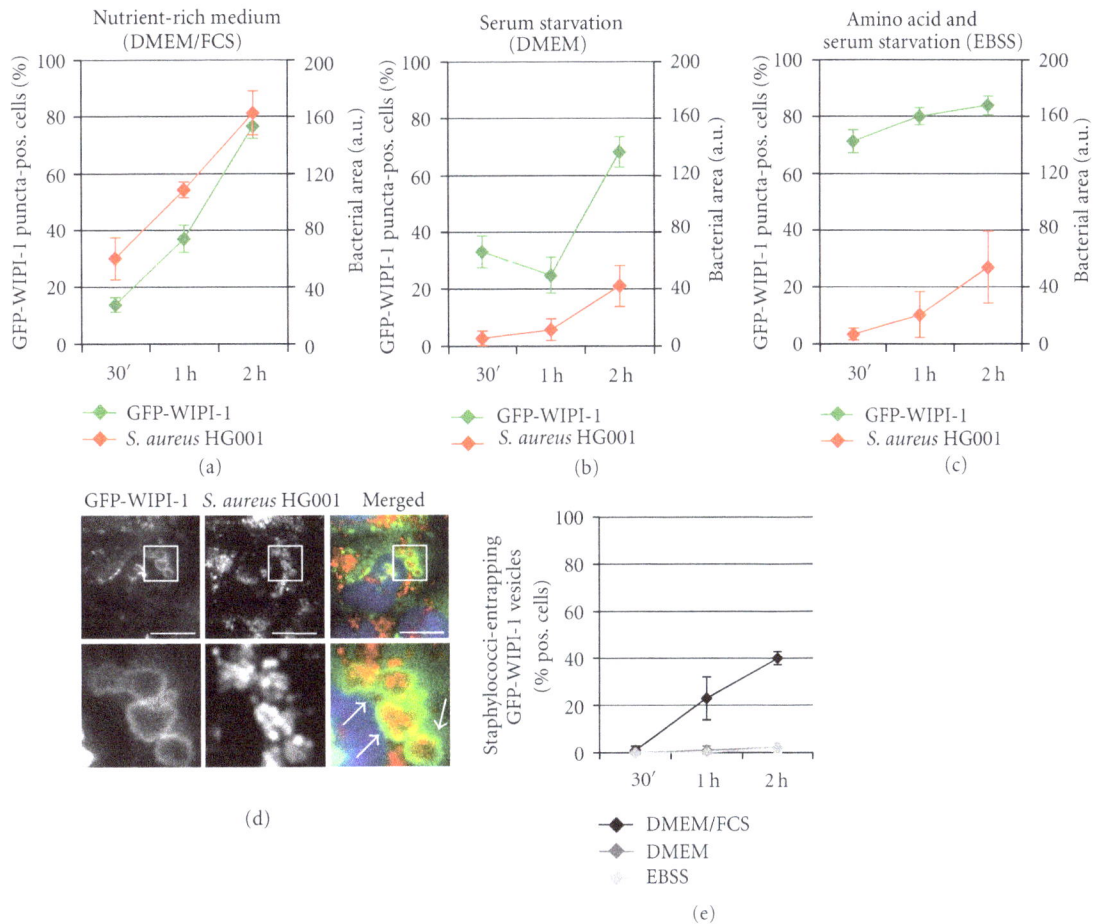

FIGURE 5: Pathogenic *S. aureus* HG001 induces GFP-WIPI-1 puncta formation and becomes entrapped in GFP-WIPI-1 positive autophagosome-like vesicles. According to Figure 4, GFP-WIPI-1 U2OS cells were infected with mCherry-expressing *S. aureus* HG001 in DMEM/FCS (a), DMEM (b), and EBSS (c), and up to 2000 individual cells were analyzed. Images (2 h, DMEM/FCS) are shown (d). Scale bars: 20 μm. The number of cells displaying GFP-WIPI-1 positive autophagosomal-like vesicles entrapping *S. aureus* HG001 was determined (e) $\pm$ SD ($n = 2$, each in duplicates).

Figure 6(D)). In contrast to the effect of the employed *agr*-positive *S. aureus* strains USA300 (Figure 4) and HG001 (Figure 5), the number of cells displaying *S. aureus* SA113 entrapped in GFP-WIPI-1 positive autophagosome-like vesicles was prominently lower (approximately 18%) (Figures 6(d) and 6(e)). However, the presence of *S. aureus* SA113 in DMEM also triggered an increase of GFP-WIPI-1 puncta-positive cells (Figure 6(b)) when compared to control settings (Figure 1), whereas in EBSS no further elevation was achieved (Figure 6(c)), and in both cases, cells did not display entrapped *S. aureus* SA113 (Figure 6(e)). Control images of *S. aureus* SA113 infections with either DMEM or EBSS are also provided (Supplementary Figure 4).

3.4. Apathogenic Staphylococcus carnosus TM300 Cells Were Not Entrapped in Intracellular GFP-WIPI-1 Positive Autophagosome-Like Vesicles.
In contrast to the pathogenic *S. aureus* strains (see above), infection of stably expressing GFP-WIPI-1 U2OS cells with the apathogenic *S. carnosus* TM300 did not result in an invasion of host cells in either

of the used media (Figures 7(a)–7(c)). In line, GFP-WIPI-1 positive autophagosome-like vesicles were not induced (Figures 7(d) and 7(e)). Control images for *S. carnosus* TM300 in DMEM or EBSS are provided (Supplementary Figure 5). Interestingly, within 2 h of incubation with *S. carnosus* TM300 in DMEM/FCS, the number of GFP-WIPI-1 puncta-positive cells increased (approximately 45%) (Figure 7(a)) when compared to the control settings (Figure 1), which was not observed by using DMEM (Figure 7(b)) or EBSS (Figure 7(c)). However, the number of GFP-WIPI-1 puncta per individual cell did not increase upon infection of *S. carnosus* TM300 in DMEM/FCS (Supplementary Figure 6(E)) when compared to uninfected conditions (Supplementary Figure 6(A)).

3.5. Inhibition of PtdIns(3,5)P$_2$ Production and Lysosomal Inhibition Increased the Number of WIPI-1 Positive Autophagosome-Like Vesicles Entrapping Staphylococcus aureus.
Next, we questioned whether pathogenic *S. aureus* cells entrapped in GFP-WIPI-1 positive autophagosomal-like vesicles are degraded in the lysosome. We employed

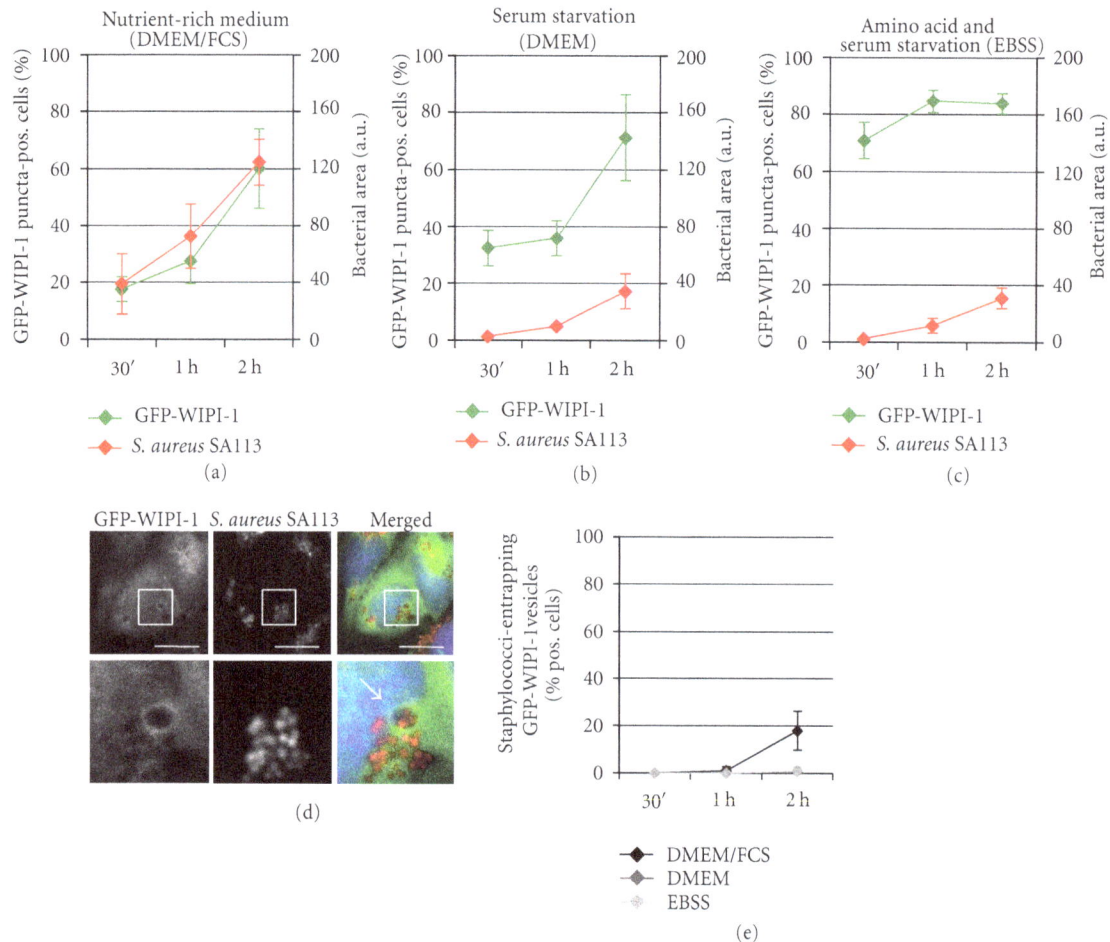

FIGURE 6: Pathogenic *S. aureus* SA113 induces GFP-WIPI-1 puncta formation and becomes entrapped in GFP-WIPI-1 positive autophagosome-like vesicles. According to Figures 4 and 5, GFP-WIPI-1 U2OS cells were infected with mCherry-expressing *S. aureus* SA113 in DMEM/FCS (a), DMEM (b), and EBSS (c) and analyzed (up to 2000 individual cells), representative images (2 h, DMEM/FCS) are shown ((d), scale bars: 20 μm), and the quantification of cells displaying GFP-WIPI-1 positive autophagosomal-like vesicles entrapping *S. aureus* SA113 is presented (e) $\pm$ SD ($n = 2$, each in duplicates).

the lysosomal inhibitor bafilomycin A$_1$ (Baf A$_1$) to block autophagosome/lysosome fusion events upon infection of GFP-WIPI-1 expressing U2OS cells with *S. aureus* USA300 in DMEM/FCS. Upon Baf A$_1$ addition the number of cells harboring GFP-WIPI-1 positive autophagosomal-like vesicles entrapping *S. aureus* USA300 (Figure 8(a), left panel) significantly increased. And, the number of GFP-WIPI-1 positive autophagosomal-like vesicles per individual cell also significantly increased (Figure 8(b), left panel). In this situation (Figure 8(a), left panel; Figure 8(b), left panel) we found that the bacterial load did not significantly change (Supplementary Figure 7).

Further, during infection of GFP-WIPI-1 expressing U2OS cells with *S. aureus* USA300 in DMEM/FCS we employed YM201636 (YM), a specific PIKfyve inhibitor that blocks PtdIns(3,5)P$_2$ production from PtdIns(3)P [24]. Upon YM treatment the number of cells harboring GFP-WIPI-1 positive autophagosomal-like vesicles (Figure 8(a), left panel) and the number of the vesicles per cell (Figure 8(b), left panel) significantly increased. Again, the

intracellular bacterial load within the cells did not change (Supplementary Figure 7). Baf A$_1$/YM cotreatment had an additive effect (Figures 8(a) and 8(b) left panels). The corresponding automated GFP-WIPI-1 puncta formation analysis is also provided (Figures 8(a) and 8(b) right panels).

3.6. Confocal and Electron Microscopy of Intracellular Staphylococcus aureus USA300.
To achieve more image resolution, we infected GFP-WIPI-1 expressing U2OS cells with *S. aureus* USA300 in DMEM/FCS followed by confocal laser scanning microscopy (Figure 9(a)). Clearly, GFP-WIPI-1 positive autophagosome-like vesicles harbored multiple *S. aureus* USA300 cells and the analysis of individual confocal sections confirmed that these vesicles are found in the cytoplasm (Figure 9(a), 1–4).

It has been shown that *S. aureus* invading HeLa cells become sequestered in Rab7-positive endosomes [14]. As Rab7 marks late endosomes, we here used GFP-2xFYVE to visualize early endosomes. We used GFP-2xFYVE expressing U2OS cells for infection with *S. aureus* USA300 in

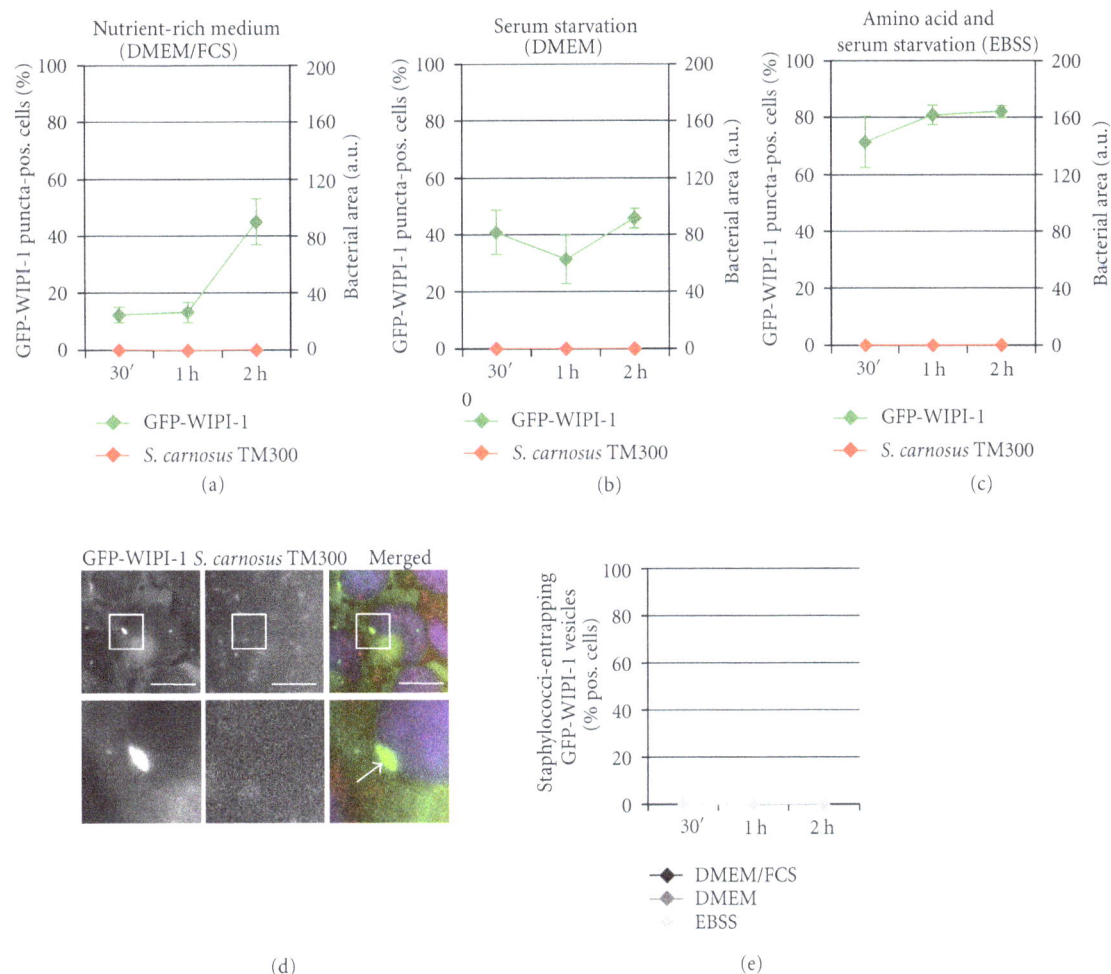

FIGURE 7: Apathogenic *S. carnosus* TM300 cells are not entrapped in GFP-WIPI-1 positive autophagosome-like vesicles. According to Figures 4–6, GFP-WIPI-1 U2OS cells were infected with mCherry-expressing *S. carnosus* TM300 in DMEM/FCS (a), DMEM (b), and EBSS (c) and analyzed (up to 2000 individual cells). Representative images (2 h, DMEM/FCS) are presented ((d), scale bars: 20 μm). The number of cells with GFP-WIPI-1 positive autophagosomal-like vesicles entrapping *S. carnosus* TM300 is presented (e) ± SD ($n = 2$, each in duplicates).

DMEM/FCS. Indeed, we also found that *S. aureus* USA300 cells were entrapped in GFP-2xFYVE positive endosomes (Figure 9(b), 1–4).

Further, by electron microscopy we found that intracellular *S. aureus* USA300 cells are entrapped in vesicles with a single *S. aureus* USA300 cell (Figure 3.6), or in vesicles harboring multiple *S. aureus* USA300 cells (Figure 3.6). In both cases, intracellular *S. aureus* USA300 cells showed clear signs of ongoing cell division (red arrows).

4. Discussion

Autophagy is considered an ancient eukaryotic pathway for cellular self-digestion that evolved with the endomembrane system [25]. As the endomembrane system provided an opportunity for invading pathogens to manipulate the host cell, it is further considered that the autophagic response to pathogen invasion may have also evolved as an early host defense program of eukaryotic cells [25, 26]. Interestingly enough, this hypothesis explains that (i) autophagy is in

part a stochastic degradation pathway to clear the cytoplasm, thereby securing the functionality of both proteins and the endomembrane system, but is also (ii) a specific response triggered by certain stress exposures, such as pathogen invasion. In fact, the autophagic response to pathogen invasion has been identified because autophagy-related proteins (ATG) essential to the stochastic process of autophagy, such as Atg5 and LC3, have also been found to decorate membranes harboring intracellular pathogens and to be functionally involved in the cellular response to pathogens [4, 27]. Still, molecular mechanisms of autophagic responses to pathogen exposure are insufficiently understood.

Bacterial pathogens employ a variety of mechanisms to manipulate host cell membranes [28, 29]. Commonly, many bacteria interfere with the phosphoinositide metabolism that is often targeted by bacterial virulence factors [30]. Among the phosphoinositides, PtdIns(3)P is the essential variant for the forming autophagosomal membrane, hence it can be anticipated that PtdIns(3)P might commonly interconnect bacterial infection with the autophagic pathway. In fact, it

(a)

(b)

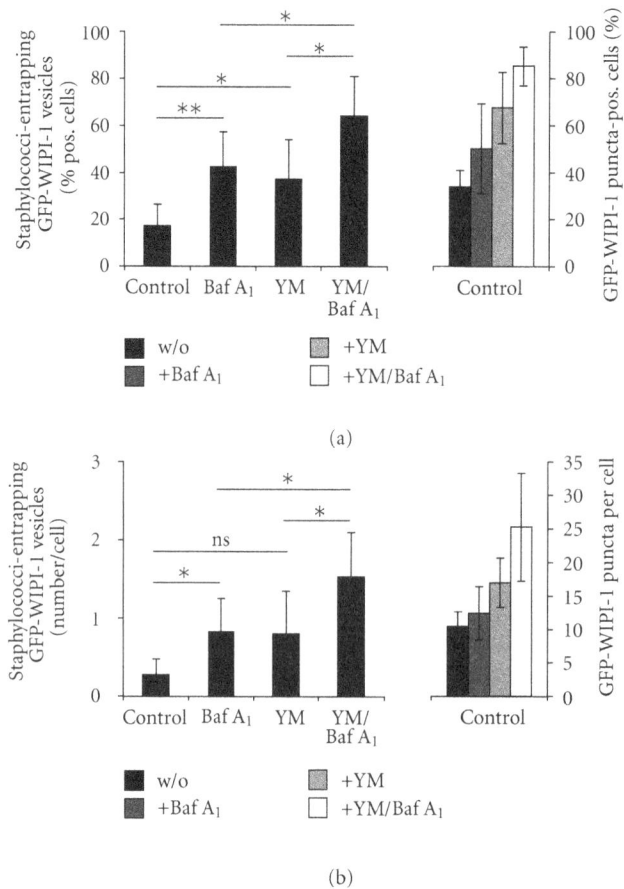

FIGURE 8: Bafilomycin A_1 and YM201636 treatments increased the number of GFP-WIPI-1 positive autophagosome-like vesicles entrapping Staphylococci. GFP-WIPI-1 U2OS cells were infected with *S. aureus* USA300 in DMEM/FCS in the absence (control) or presence of 200 nM bafilomycin A_1 (Baf A_1), 800 nM YM201636 (YM), or with both (Baf A_1/YM) for 2 h. Images were automatically acquired (not shown). The number of GFP-WIPI-1 puncta-positive cells ((a), right panel) and the number of GFP-WIPI-1 puncta per cell ((b), right panel) was determined. From 100 infected cells for each of the treatment condition, the number of cells displaying GFP-WIPI-1 positive autophagosomal-like vesicles entrapping *S. aureus* USA300 ((a), left panel) and the number of GFP-WIPI-1 autophagosomal-like vesicles entrapping *S. aureus* USA300 per cell ((b), left panel) was determined ($n = 3$). $*P < 0.05$, $**P < 0.01$, ns: not significant.

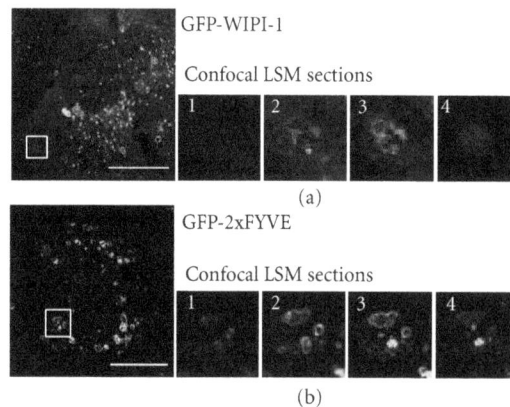

FIGURE 9: Confocal laser scanning microscopy of *S. aureus* USA300 infected GFP-WIPI-1 or GFP-2xFYVE expressing U2OS cells. GFP-WIPI-1 (a) or GFP-2xFYVE (b) expressing U2OS cells were infected with *S. aureus* USA300 for 2 h in DMEM/FCS. Representative images ($n = 3$) are shown. Magnifications display individual LSM sections (1–4). Scale bars: 20 μm.

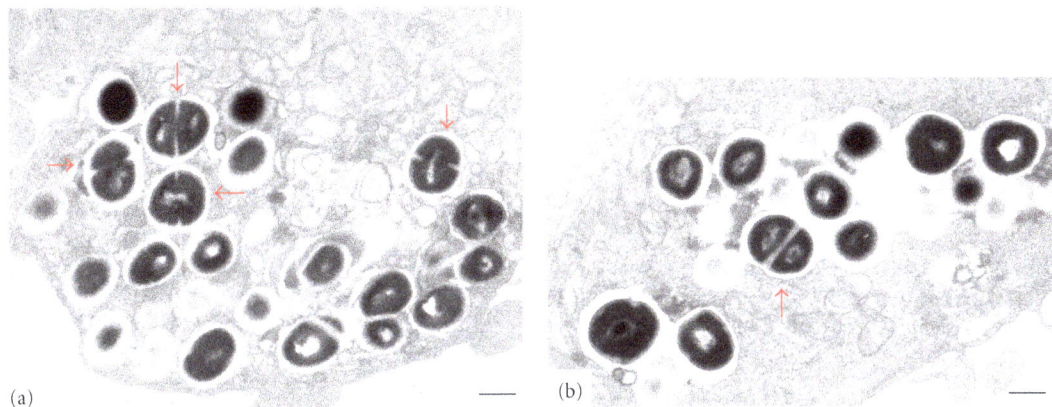

FIGURE 10: Electron microscopy of *S. aureus* USA300 infected GFP-WIPI-1 expressing U2OS cells. GFP-WIPI-1 U2OS cells were infected with *S. aureus* USA300 in DMEM/FCS followed by conventional electron microscopy. Either single *S. aureus* USA300 cells were found to reside within a vesicle (a), or multiple cells were found in enlarged vesicles (b). Red arrows indicate dividing Staphylococci. Scale bars: 500 nm.

has been shown that PtdIns(3)P is involved in the formation of Salmonella-containing vacuoles serving as a niche in host cells, and that PtdIns(3)P is targeted by *M. tuberculosis* to inhibit phagosome maturation [31]. Here, we addressed this question by investigating the process of *S. aureus* invasion of tumor cells.

A study by Schnaith and coworkers suggested a model that connected the autophagic response with *S. aureus* infection via the bacterial *agr*-virulence factor [14]. In this model, late phagosomes with (i) *agr*-positive *S. aureus* become entrapped in autophagosome-like vesicles, where *S. aureus* replicate and subsequently escape into the cytoplasm to promote host cell death, but (ii) *agr*-deficient *S. aureus* are subjected to lysosomal degradation [14].

We here provide evidence, that exposure of nonprofessional host cells (tumor cells) to Staphylococci stimulates the canonical WIPI-1 response at the onset of autophagy, which is to bind to PtdIns(3)P at the phagophore to foster the recruitment of downstream ATGs, such as Atg5 and LC3 [9, 32]. Interestingly, this response is attributable to the interaction of Staphylococci with the host cell membrane, as we found WIPI-1 to become stimulated upon both noninvasive and invasive Staphylococci. In line, WIPI-1 was also stimulated upon peptidoglycan treatment (data not shown). By further analyzing invasive *S. aureus* strains in this study, we identified new WIPI-1 positive autophagosome-like vesicles that entrapped multiple *S. aureus* particles. And, moreover, *agr*-positive *S. aureus* strains were more efficiently entrapped when compared to *agr*-deficient *S. aureus* cells. Our results demonstrate that WIPI-1, a principal PtdIns(3)P effector at the onset of stochastic, canonical autophagy, is also involved in selective engagement of the autophagic pathway, moreover underscored by the notion that Staphylococci prominently stimulated WIPI-1 in nutrient-rich conditions. And, our results demonstrate that *S. aureus* (i) stimulates autophagy and (ii) in addition, becomes entrapped in WIPI-1 positive autophagosome-like vesicles.

The most compelling explanation would be that WIPI-1 becomes stimulated upon *S. aureus* interaction with the plasma membrane, subsequently WIPI-1 positive phagophore membranes, for example, originated from the endoplasmic reticulum, are utilized to sequester *S. aureus* where bacterial replication occurs. In addition, we also found *S. aureus* particles sequestered in phagosomes, marked by the FYVE domain [33], which are intended for phagocytosis. Hence our results can be viewed as host cell response to *S. aureus*, critically involving PtdIns(3)P membranes that either serve as phagosome membranes, or that are utilized to further sequester *S. aureus*, thereby generating a replication niche. Evidence that bacterial replication occurs is given by our electron microscopy analysis showing dividing *S. aureus* cells within the sequestering vesicle. The importance of PtdIns(3)P-enriched membranes during sequestration of invading *S. aureus* is further emphasized by our finding that more WIPI-1 positive autophagosome-like vesicles entrap *S. aureus* cells when phosphorylation of PtdIns(3)P to PtdIns(3,5)P_2 by PIKfyve was specifically blocked.

PtdIns(3)P-enriched membranes promote vesicle fusion with lysosomes. In line, FYVE domain marked phagosomes that carry *S. aureus* would be subjected to phagocytosis as suggested [14]. If WIPI-1 positive autophagosome-like vesicles entrapping *S. aureus* identified in this study would reflect cytoplasmic sequestration of invaded *S. aureus* with PtdIns(3)P-enriched WIPI-1 positive phagophores, the resulting autophagosome-like vesicles should become subjected to fusion with the lysosomal compartment, because they are enriched in PtdIns(3)P. But it was shown that lysosomal fusion is blocked upon *S. aureus* invasion [14]. To address this question we employed bafilomycin A$_1$ to inhibit the functionality of the lysosomal compartment. Clearly, lysosomal inhibition significantly increased the number of WIPI-1 positive autophagosome-like vesicles harboring *agr*-positive Staphylococci. This demonstrates that nonprofessional host cells employ autophagy as a defense response with regards to *S. aureus* infection, in line with previous suggestions [34]. However, under some circumstances [14] bacterial replication and vesicle escape might override this cellular defense program.

Acknowledgments

The authors kindly thank Marianne Opalka for technical assistance. This work was supported by grants from the Federal Ministry for Education and Science (BMBF, BioProfile) and the German Research Society (DFG, SFB 773) to TP-C.

References

[1] Z. Yang and D. J. Klionsky, "Eaten alive: a history of macroautophagy," *Nature Cell Biology*, vol. 12, no. 9, pp. 814–822, 2010.

[2] Z. Yang and D. J. Klionsky, "Mammalian autophagy: core molecular machinery and signaling regulation," *Current Opinion in Cell Biology*, vol. 22, no. 2, pp. 124–131, 2010.

[3] J. Kim, M. Kundu, B. Viollet, and K. L. Guan, "AMPK and mTOR regulate autophagy through direct phosphorylation of Ulk1," *Nature Cell Biology*, vol. 13, no. 2, pp. 132–141, 2011.

[4] A. Van Der Vaart, M. Mari, and F. Reggiori, "A picky eater: exploring the mechanisms of selective autophagy in human pathologies," *Traffic*, vol. 9, no. 3, pp. 281–289, 2008.

[5] N. Mizushima, B. Levine, A. M. Cuervo, and D. J. Klionsky, "Autophagy fights disease through cellular self-digestion," *Nature*, vol. 451, no. 7182, pp. 1069–1075, 2008.

[6] T. Noda, K. Matsunaga, and T. Yoshimori, "Atg14L recruits PtdIns 3-kinase to the ER for autophagosome formation," *Autophagy*, vol. 7, no. 4, pp. 438–439, 2011.

[7] T. Proikas-Cezanne and P. Codogno, "Beclin 1 or not Beclin 1," *Autophagy*, vol. 7, no. 7, pp. 671–672, 2011.

[8] T. Proikas-Cezanne, S. Waddell, A. Gaugel, T. Frickey, A. Lupas, and A. Nordheim, "WIPI-1α (WIPI49), a member of the novel 7-bladed WIPI protein family, is aberrantly expressed in human cancer and is linked to starvation-induced autophagy," *Oncogene*, vol. 23, no. 58, pp. 9314–9325, 2004.

[9] M. Mauthe, A. Jacob, S. Freiberger et al., "Resveratrol-mediated autophagy requires WIPI-1 regulated LC3 lipidation in the absence of induced phagophore formation," *Autophagy*, vol. 7, no. 12, pp. 1448–1461, 2011.

[10] T. Proikas-Cezanne and H. Robenek, "Freeze-fracture replica immunolabelling reveals human WIPI-1 and WIPI-2 as membrane proteins of autophagosomes," *Journal of Cellular and Molecular Medicine*, vol. 15, no. 9, pp. 2007–2010, 2011.

[11] T. Proikas-Cezanne, S. Ruckerbauer, Y. D. Stierhof, C. Berg, and A. Nordheim, "Human WIPI-1 puncta-formation: a novel assay to assess mammalian autophagy," *FEBS Letters*, vol. 581, no. 18, pp. 3396–3404, 2007.

[12] B. Levine, N. Mizushima, and H. W. Virgin, "Autophagy in immunity and inflammation," *Nature*, vol. 469, no. 7330, pp. 323–335, 2011.

[13] I. Fedtke, F. Götz, and A. Peschel, "Bacterial evasion of innate host defenses—the Staphylococcus aureus lesson," *International Journal of Medical Microbiology*, vol. 294, no. 2-3, pp. 189–194, 2004.

[14] A. Schnaith, H. Kashkar, S. A. Leggio, K. Addicks, M. Krönke, and O. Krut, "Staphylococcus aureus subvert autophagy for induction of caspase-independent host cell death," *Journal of Biological Chemistry*, vol. 282, no. 4, pp. 2695–2706, 2007.

[15] S. G. Pfisterer, M. Mauthe, P. Codogno et al., "Ca^{2+}/calmodulin-dependent kinase signaling via CaMKI and AMPK contributes to the regulation of WIPI-1 at the onset of autophagy," *Molecular Pharmacology*, vol. 80, no. 6, pp. 1066–1075, 2011.

[16] L. K. McDougal, C. D. Steward, G. E. Killgore, J. M. Chaitram, S. K. McAllister, and F. C. Tenover, "Pulsed-field gel electrophoresis typing of oxacillin-resistant Staphylococcus aureus isolates from the United States: establishing a national database," *Journal of Clinical Microbiology*, vol. 41, no. 11, pp. 5113–5120, 2003.

[17] S. Herbert, A. K. Ziebandt, K. Ohlsen et al., "Repair of global regulators in *Staphylococcus aureus* 8325 and comparative analysis with other clinical isolates," *Infection and Immunity*, vol. 78, no. 6, pp. 2877–2889, 2010.

[18] S. Iordanescu and M. Surdeanu, "Two restrictions and modification systems in *Staphylococcus aureus* NCTC8325," *Journal of General Microbiology*, vol. 96, no. 2, pp. 277–281, 1976.

[19] R. Rosenstein, C. Nerz, L. Biswas et al., "Genome analysis of the meat starter culture bacterium *Staphylococcus carnosus* TM300," *Applied and Environmental Microbiology*, vol. 75, no. 3, pp. 811–822, 2009.

[20] A. Grotemeier, S. Alers, S. G. Pfisterer et al., "AMPK-independent induction of autophagy by cytosolic Ca2+ increase," *Cellular Signalling*, vol. 22, no. 6, pp. 914–925, 2010.

[21] R. Rosenstein and F. Götz, "Genomic differences between the food-grade Staphylococcus carnosus and pathogenic staphylococcal species," *International Journal of Medical Microbiology*, vol. 300, no. 2-3, pp. 104–108, 2010.

[22] T. Proikas-Cezanne and S. G. Pfisterer, "Assessing mammalian autophagy by WIPI-1/Atg18 puncta formation," *Methods in Enzymology*, vol. 452, pp. 247–260, 2009.

[23] H. Robenek, M. J. Robenek, I. Buers et al., "Lipid droplets gain PAT family proteins by interaction with specialized plasma membrane domains," *Journal of Biological Chemistry*, vol. 280, no. 28, pp. 26330–26338, 2005.

[24] H. B. J. Jefferies, F. T. Cooke, P. Jat et al., "A selective PIKfyve inhibitor blocks PtdIns(3,5)P(2) production and disrupts endomembrane transport and retroviral budding," *EMBO Reports*, vol. 9, no. 2, pp. 164–170, 2008.

[25] B. Levine, "Eating oneself and uninvited guests: autophagy-related pathways in cellular defense," *Cell*, vol. 120, no. 2, pp. 159–162, 2005.

[26] R. Sumpter Jr. and B. Levine, "Autophagy and innate immunity: triggering, targeting and tuning," *Seminars in Cell and Developmental Biology*, vol. 21, no. 7, pp. 699–711, 2010.

[27] M. C. Lerena, C. L. Vázquez, and M. I. Colombo, "Bacterial pathogens and the autophagic response," *Cellular Microbiology*, vol. 12, no. 1, pp. 10–18, 2010.

[28] S. Shahnazari and J. H. Brumell, "Mechanisms and consequences of bacterial targeting by the autophagy pathway," *Current Opinion in Microbiology*, vol. 14, no. 1, pp. 68–75, 2011.

[29] H. Ham, A. Sreelatha, and K. Orth, "Manipulation of host membranes by bacterial effectors," *Nature Reviews Microbiology*, vol. 9, no. 9, pp. 635–646, 2011.

[30] J. Pizarro-Cerdá and P. Cossart, "Subversion of phosphoinositide metabolism by intracellular bacterial pathogens," *Nature Cell Biology*, vol. 6, no. 11, pp. 1026–1033, 2004.

[31] T. Noda and T. Yoshimori, "Molecular basis of canonical and bactericidal autophagy," *International Immunology*, vol. 21, no. 11, pp. 1199–1204, 2009.

[32] E. Itakura and N. Mizushima, "Characterization of autophagosome formation site by a hierarchical analysis of

mammalian Atg proteins," *Autophagy*, vol. 6, no. 6, pp. 764–776, 2010.

[33] J. M. Gaullier, A. Simonsen, A. D'Arrigo, B. Bremnes, H. Stenmark, and R. Aasland, "FYVE fingers bind PtdIns(3)P," *Nature*, vol. 394, no. 6692, pp. 432–433, 1998.

[34] S. Kageyama, H. Omori, T. Saitoh et al., "The LC3 recruitment mechanism is separate from Atg9L1-dependent membrane formation in the autophagic response against Salmonella," *Molecular Biology of the Cell*, vol. 22, no. 13, pp. 2290–2300, 2011.

11

Reticulophagy and Ribophagy: Regulated Degradation of Protein Production Factories

Eduardo Cebollero,[1, 2] Fulvio Reggiori,[1] and Claudine Kraft[3]

[1] Department of Cell Biology and Institute of Biomembranes, University Medical Centre Utrecht,
  3584 CX Utrecht, The Netherlands
[2] Department of Biochemistry and Cell Biology and Institute of Biomembranes, Utrecht University,
  3508 TD Utrecht, The Netherlands
[3] Max F. Perutz Laboratories, University of Vienna, 1030 Vienna, Austria

Correspondence should be addressed to Fulvio Reggiori, f.reggiori@umcutrecht.nl
and Claudine Kraft, claudine.kraft@univie.ac.at

Academic Editor: Anne Simonsen

During autophagy, cytosol, protein aggregates, and organelles are sequestered into double-membrane vesicles called autophagosomes and delivered to the lysosome/vacuole for breakdown and recycling of their basic components. In all eukaryotes this pathway is important for adaptation to stress conditions such as nutrient deprivation, as well as to regulate intracellular homeostasis by adjusting organelle number and clearing damaged structures. For a long time, starvation-induced autophagy has been viewed as a nonselective transport pathway; however, recent studies have revealed that autophagy is able to selectively engulf specific structures, ranging from proteins to entire organelles. In this paper, we discuss recent findings on the mechanisms and physiological implications of two selective types of autophagy: ribophagy, the specific degradation of ribosomes, and reticulophagy, the selective elimination of portions of the ER.

1. Introduction

Autophagy is a degradative process that allows cells to maintain their homeostasis in numerous physiological situations. It is required, for example, to face prolonged starvation periods and nutritional fluctuations in the environment, developmental tissue remodeling, organelle quality control, and immune responses [1, 2]. In addition, this pathway has been implicated in the physiopathology of multiple diseases [3, 4]. Autophagosomes are the hallmark of autophagy. These double-membrane vesicles are generated in the cytosol and during their formation they engulf the cargo to be delivered into the mammalian lysosomes or yeast and plant vacuoles for degradation [5]. Two types of autophagy have been described: selective and non-selective autophagy. During non-selective autophagy bulk cytosol, including organelles, is randomly sequestered into autophagosomes. On the other hand, during selective autophagy, a specific cargo is exclusively enwrapped by double-membrane vesicles, which contain little cytoplasm with their size corresponding to that of their cargo [6].

Autophagy progression relies on the function of the autophagy-related (Atg) proteins that mediate autophagosome biogenesis, selective cargo recognition, fusion with the lysosome/vacuole, or vesicle breakdown [5, 7, 8]. Upon nutritional stresses, fractions of the cytoplasm are consumed via autophagy and the resulting catabolic products are used as sources of energy or as building blocks for the synthesis of new macromolecules. In these situations autophagy is mainly considered as a non-selective process. Nonetheless an increasing number of selective types of autophagy are being described [6, 9] and these findings challenge the concept whether autophagosomes in fact sequester their cargo randomly.

2. Short Overview of Selective Types of Autophagy

One of the best-studied examples of selective autophagy is the biosynthetic cytoplasm to vacuole targeting (Cvt) pathway in the yeast *Saccharomyces cerevisiae*. During the Cvt pathway a protein oligomer composed of the vacuolar hydrolases aminopeptidase 1 (Ape1), α-mannosidase 1 (Ams1), and aspartyl aminopeptidase (Ape4), is delivered to the vacuolar lumen by small double-membrane vesicles [10–13]. Interestingly, this oligomer is also a specific cargo of autophagosomes under starvation conditions [14]. In higher eukaryotes autophagy also supports the selective destruction of intracellular pathogens, called xenophagy, and protein aggregates, named aggrephagy. In addition, metabolically dispensable or dysfunctional organelles can be specifically degraded by autophagy in both yeast and mammals. Examples of the latter include the exclusive elimination of superfluous or damaged mitochondria, termed mitophagy, and the selective consumption of excessive peroxisomes, called pexophagy [15, 16].

The underlying mechanisms of each of these pathways remain to be characterized in detail but some common principles are emerging. First, a receptor-like recognition of the cargo directing it to the autophagosome or alternatively recruiting the Atg machinery is required for all the selective types of autophagy. Second, the involvement of ubiquitin as a signaling molecule has been described for several selective types of autophagy in higher eukaryotes [17]. Several of the autophagosomal cargos can be degraded in a selective manner under specific conditions or in a random manner during bulk autophagy. It remains to be investigated in more detail how certain autophagy pathways can choose specific cargo in time and space. As the subject of selective autophagy pathways is covered in other reports in this special issue of the *International Journal of Cell Science*, in this review we will discuss the molecular principles and mechanisms underlying two selective types of autophagy that remain less well understood: ribophagy and reticulophagy.

3. Ribophagy: Mechanisms and Physiological Implications

Since the discovery of autophagy, ribosomes have been detected in the interior of autophagosomes by electron microscopy [18, 19]. For a long time these large multiprotein complexes were viewed as a marker for bulk degradation of cytoplasm. However, it has recently been shown that ribosomes are turned over through a selective type of autophagy [20]. Accurate examination of ribosome fate under nutrient starvation conditions in yeast *S. cerevisiae* has revealed that these structures are more rapidly degraded compared to other cytoplasmic components, supporting the notion of a selective degradation process [20]. The involvement of autophagy in this event was demonstrated by uncovering that the transport of ribosomes to the vacuole relies on core autophagy components such as Atg1 and Atg7. A genetic screen in yeast designed to isolate mutant strains with a defect in ribosome turnover revealed that the ubiquitin protease Ubp3 and its cofactor Bre5 are required for this selective type of autophagy, however, not for bulk autophagy [20]. Importantly, a catalytically inactive mutant of Ubp3 also displayed a defect in the autophagy-mediated degradation of ribosomes indicating that ubiquitination plays a key role in this process. This selective autophagic turnover of ribosomes is now termed ribophagy [20] (Figure 1(a)).

4. Ribophagy and Ubiquitination

It remains to be investigated whether ubiquitination is important for either the regulation of signaling pathways triggering ribophagy or in dictating the specificity in the cargo selection. This latter possibility is evoked by the fact that ubiquitin-based modifications are a common theme in the selective elimination of specific structures in higher eukaryotes [17]. As Ubp3 interacts with and influences the ubiquitination status of Atg19 [21], a receptor protein of the Cvt pathway [22], it is plausible that Ubp3 could contribute to other selective types of autophagy in a similar manner. Further evidence for the involvement of ubiquitination in ribophagy comes from the finding that a decrease of the cytoplasmic levels of the ubiquitin ligase Rsp5 together with the deletion of *UBP3* results in a defect in the turnover of ribosomes higher than in the *ubp3Δ* cells [23]. Importantly, cytoplasmic proteins are normally degraded by autophagy in this strain. These findings imply that both ubiquitination and deubiquitination are crucial for the regulation of ribophagy. A reciprocal control mechanism has also been found to be important for the specific removal of midbody rings by autophagy during cytokinesis [24]. To understand the regulation and mechanism of ribophagy, it will be important to identify the targets of Ubp3/Bre5 and Rsp5 during this process.

5. Putative Physiological Roles of Ribophagy

What could be the physiological role of ribophagy? The deletion of *UBP3* results in the inhibition of starvation-induced ribophagy and leads to cell death, without affecting general bulk autophagy [20]. These findings support the notion that not only bulk autophagy, but also ribosomal turnover is important for cell survival during nutrient limiting conditions. This does not come as a surprise as ribosomes constitute half of the cell's protein mass [25], and consequently, represent a major source of amino acids during times of nutrient deprivation. In addition, or alternatively, the importance of ribosomal degradation during starvation might be its contribution to the rapid and simultaneous downregulation of protein translation, a process that consumes large amounts of energy and amino acids.

Interestingly, a ribophagy-like process has also been proposed in plants. The endoribonuclease Rns2, a conserved member of the RNAse T2 protein family, is required for ribosomal RNA decay in plants [26]. Mutant cells lacking Rns2 activity fail to degrade ribosomal RNA. If this results in a failure of disassembling and/or degrading entire ribosomes

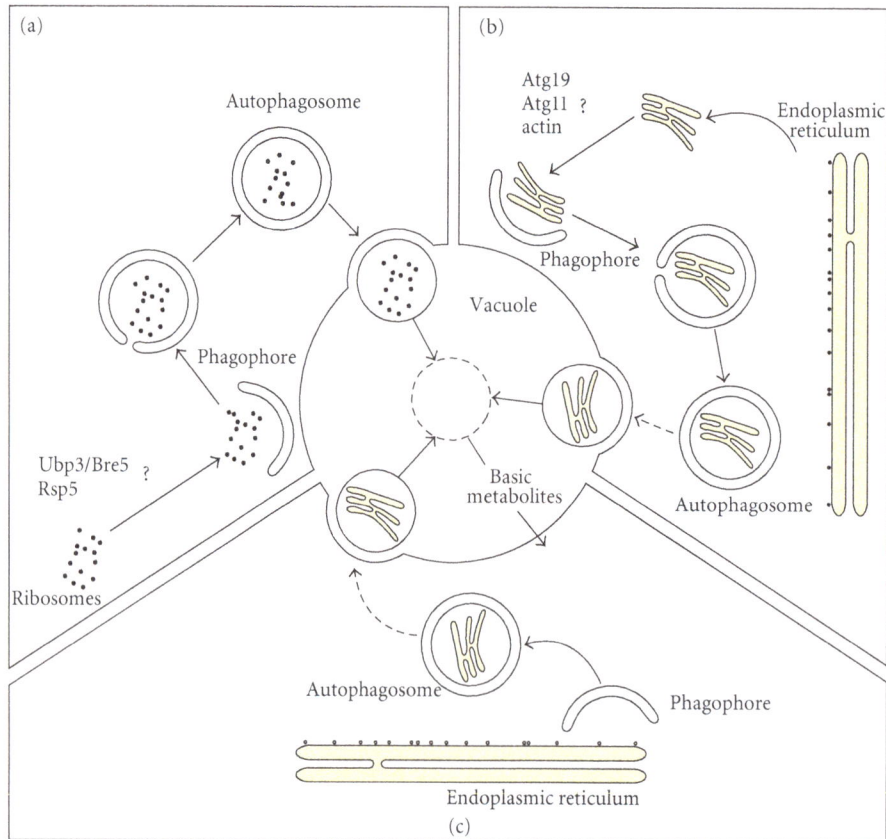

FIGURE 1: Mechanisms of ribophagy and reticulophagy in yeast. (a) A model for ribophagy. Under ribophagy inducing conditions, ribosomes are selectively engulfed into autophagosomes and subsequently degraded in the vacuole. The generated basic metabolites (amino acids, sugars, fatty acids etc.) are then recycled back to the cytoplasm for reuse or as a source of energy. ((b) and (c)) Models for reticulophagy. Under stress conditions, due to an accumulation of unfolded proteins and/or protein aggregates, a partial scission of the ER occurs and the formed fragments are specifically transported to the sites where autophagosome biogenesis takes place (b). ER stress triggers the recruitment of the Atg proteins onto or close to this organelle. There, possibly by utilizing the ER membranes, the Atg proteins mediate the formation of autophagosomes, which expand around the ER sections that have to be removed (c). The dashed arrows indicate that under specific ER stress conditions, autophagosomes do not fuse with the vacuole. Question marks highlight proteins that have been implicated in the transport and selection of the cargo in which the mechanism of action remains to be elucidated.

has not yet been determined. Nevertheless, the defect in the turnover of ribosomal RNA suggests that Rns2 is a component of a ribophagy-like process in plants. The plant *rns2* mutants also exhibit this phenotype in nutrient rich conditions. This suggests that ribophagy might also serve a housekeeping function by recycling some of the ribosomal components such as amino acids and nucleotides. To date, only the degradation of ribosomal RNA has been studied. Consequently, the fate of ribosomal proteins as well as the existence of ribophagy in plants will require more detailed investigations.

6. Ribophagy in Higher Eukaryotes

Ribosomes have also been observed in the interior of autophagosomes in mammalian cells [18]. In particular, the relative abundance of proteins in MCF7 cells during amino acid starvation has been measured using quantitative mass spectrometry [27]. This approach has revealed that in mammalian cells ribosome degradation by autophagy occurs

with different kinetics than that of other cytoplasmic proteins and organelles [27]. It has yet to be explored, however, whether a selective type of autophagy is responsible for the different turnover rates of ribosomes and cytoplasmic proteins. Additional evidence for the possible existence of ribophagy in higher eukaryotes comes from a murine study on neurodegeneration in Purkinje cells, where the disassembly of actively translating polysomes to nontranslational monosomes was observed among other changes [28]. Interestingly, a fraction of the free monosomes was specifically sequestered into autophagosomes suggesting that an autophagy-related pathway is involved in the selective degradation of ribosomes in these cells [28]. Thus, these neuronal cells appear to be an optimal model to study ribophagy and possibly gain additional insight into the involvement of both ubiquitination and the mammalian Ubp3 homologue Usp10 in the turnover of ribosomes in higher eukaryotes.

Autophagy of ribosomal proteins has also been demonstrated to serve an antimicrobial function. Several bacteria are directly captured in the cytosol by the autophagy adapter p62 or NDP52, and subsequently sequestered into autophagosomes to be delivered and degraded in lysosomes [29]. In the case of *Mycobacterium tuberculosis*, autophagy can also be used for its removal from the cell, however, through a different mechanism. Mycobacteria are phagocytosed by macrophages whereupon they delay phagosome maturation, thereby preventing their destruction in the lysosome. In the phagosomes, they persist and replicate often leading to lethal infections. Recently it has been shown that upon autophagy induction the cytosolic ribosomal protein rpS30 precursor FAU and ubiquitin are sequestered into autophagosomes in a p62-dependent manner [30]. In the mature autophagosome, these proteins are processed into peptides possessing antimicrobial properties that direct the killing of *Mycobacterium* [30]. Because of the involvement of p62, this antimicrobial turnover of ribosomal protein precursors appears to have all the characteristics of a selective type of autophagy.

An alternative role for ribophagy in cell homeostasis arises from the possibility that this pathway could also target defective ribosomes under normal growth conditions. In this scenario, by specifically eliminating nonfunctional, incorrectly assembled, and/or damaged ribosomes, ribophagy would have a quality control function. Avoiding the translation of incorrect and potentially harmful proteins might be crucial for cell homeostasis. Along this line, it is important to note that several diseases have been associated with specific mutations in ribosomal subunits [31]. The identification of such a quality control function, as well as the mechanism underlying it will be important directions for future analyses.

7. Protein Folding and ER Stress

While ribosomes located in the cytosol mainly translate cytoplasmic proteins, the synthesis of proteins that are secreted or reside in one of the organelles of the endomembrane system is mediated by ribosomes associated with the ER. As these newly synthesized proteins are cotranslationally translocated into the ER, a conspicuous amount of these molecules remains localized to this compartment. In order to prevent the accumulation of misfolded polypeptides, the ER counts on a specialized group of proteins, the so-called chaperones, which assist the folding of the nascent polypeptides or recognize misfolded proteins and mediate their refolding [32]. Under certain circumstances, this quality control function of the ER can be overcome by the natural occurrence of mutations or peculiar environmental conditions that affect general protein folding. This scenario can also be mimicked by expression of specific mutant proteins or treatment with particular chemical agents [33–37]. These situations may result in the accumulation of unfolded proteins and aggregates in the ER. Two interconnected safeguard mechanisms, the unfolded protein response (UPR) and the ER-associated degradation (ERAD), are in place to cope with misfolded protein buildups [38–40]. The UPR is an intracellular signaling cascade triggered by ER stress. This signal is transduced into cytoplasmic and nuclear

actions aimed at increasing the inherent folding capacity of the ER and eliminating the misfolded proteins accumulated in this organelle. Among the responses initiated by the UPR are inhibition of general translation and upregulation of genes encoding ER chaperones and components of the ERAD machinery. The ERAD in turn, recognizes misfolded proteins and retrotranslocates these proteins into the cytoplasm where they are degraded by the ubiquitin-proteasome system. This elimination is mediated by the retrotranslocon complex, a multiprotein system seated in the ER membrane that facilitates the transport of unfolded proteins across the ER, catalyzes the polyubiquitination of the exported proteins, and mediates their delivery to the proteasome. Autophagy might serve a third cellular mechanism complementing the UPR and ERAD systems in coping with the harmful accumulation of unfolded or aberrant proteins in the ER.

8. Autophagy in ER Stress

Molecular events occurring upon autophagy induction are the association of Atg8/LC3 with autophagosomal membranes through its conjugation to the lipid phosphatidylethanolamine (PE) [41, 42] and the formation of autophagosomes. Yeast and mammalian cells subjected to different ER stresses exhibit levels of lipidated Atg8/LC3 similar to those displayed by starved cells [33, 37, 43, 44]. Additionally, light microscopy studies have revealed that ER stress induces the formation of autophagosomes in all eukaryotes analyzed [33, 44, 45]. Both this lipidation event and the formation of autophagosomes during ER stress can be blocked by chemical inhibitors of autophagy or Atg protein depletion [33, 37, 43, 44]. This, in combination with ultrastructural analyses of both yeast and mammalian cells following ER stress, which showed the induction of autophagosomes and autophagolysosomes, confirms the induction of an autophagy response upon ER stress [36, 43, 45, 46]. A detailed scrutiny of the luminal contents of these carriers has revealed that autophagosomes enclose portions of the ER [45, 46]. The amount of ER sequestered in their interior, however, depends on the nature and strength of the stimuli triggering the reticulophagy response. For example, when yeast are treated with the reducing agent dithiothreitol (DTT), which inhibits disulfide bond formation and thus prevents correct protein folding, autophagosomes are mostly filled with tightly stacked ER membrane cisternae [45]. In contrast, when ER stress is initiated by glucose deprivation, which leads to a defect in the N-glycosylation important for the proper folding of glycoproteins, each autophagosome carries a single ER fragment [46]. The existence of ER-containing autophagosomes is supported by the juxtaposition of Atg8 and the ER marker protein Sec61 in fluorescence microscopy analyses in yeast [45]. Additionally, *in vitro* and *in vivo* studies in mammals on the Z mutant form of α_1-antitrypsin (α_1-ATZ), which aggregates and accumulates in the ER [47], have shown that the cytoplasmic α_1-ATZ aggregates colocalize with GFP-LC3 and ER resident KDEL-containing proteins [48]. Ultrastructural analyses have confirmed that these structures are indeed ER-containing autophagosomes [36].

Several evidences suggest that the sequestration of ER portions by autophagosomes might be a selective process. In yeast, induction of reticulophagy by DTT results in autophagosomes that contain tightly packed ER fragments that are devoid of cytoplasm [45]. Importantly, immuno-electron microscopy analysis in these cells using anti-GFP antibodies directed against GFP-HDEL, an ER marker protein, has demonstrated that the density of the gold particles is higher inside autophagosomes than in the total cell area [46]. This result is in agreement with the concept of a selective type of autophagy, since in a non-selective scenario the label would have been equally distributed outside and inside of the sequestering vesicles [49]. It cannot be excluded, however, that the increased density of the gold particles is the result of a longer half-life of ER components in the interior of autophagosomes. Further support for a selective nature of this pathway emerges from the notion that the actin cytoskeleton and the selectivity adaptor proteins Atg11 and Atg19 are required for the progression of reticulophagy in yeast (see below).

9. Models for the Selective Sorting of ER into Autophagosomes

How is ER targeted for degradation specifically sequestered into autophagosomes? One possibility is that fragments of ER containing unfolded proteins or aggregates are pinching off from the main ER body and are directly transported to the site where autophagosomes arise (Figure 1(b)). During the yeast Cvt pathway, for example, the selective sorting of the cargo oligomer requires the receptor Atg19, the adaptor protein Atg11, and the actin cytoskeleton. Interestingly, these three components have been linked to ER degradation under both stress conditions and nutrient deprivation in yeast [45, 46, 50]. A second possibility is that the selection and enwrapping by autophagosomes occurs in very close proximity to the ER (Figure 1(c)). In contrast to the previous model, this situation does not require a specific machinery to direct the cargo, but rather a system to recruit the Atg proteins to the location where the cargo resides. In both scenarios it remains a mystery how the ER fragments are generated, which factors regulate this scission, and how the ER is selectively sequestered. Interestingly, a recent study in *S. cerevisiae* has shown that Atg8 and Cvt pathway components are recruited onto the ER and negatively regulate the extraction and proteasomal degradation of the misfolded Hmg2 transmembrane protein [51]. Cells could potentially exploit a similar mechanism to recruit the Atg proteins to the ER during reticulophagy. At the ER, the Atg machinery could catalyze the expansion of new membranes destined to sequester an ER fragment or alternatively rearrange a preexisting ER cisterna to constitute the limiting membrane of the sequestering autophagosome. The latter possibility is supported by studies in yeast showing the presence of autophagosomes with ribosomes attached to the membrane surface [45]. In addition, electron tomography analyses in mammalian cells have shown that autophagosomes can be physically connected to the ER, suggesting that these carriers might directly emerge from the ER [52, 53].

10. Regulation of Reticulophagy by ER Quality Control Signaling

The yeast UPR consists of a main signaling pathway initiated by the ER transmembrane kinase inositol requiring enzyme 1 (Ire1). The luminal domain of Ire1 senses the accumulation of unfolded proteins, while the cytoplasmic extension transduces the signal into the nucleus initiating a cellular response at the transcriptional level [54]. Activated Ire1 initiates the nonconventional splicing of *HAC1* mRNA, leading to the production of the transcription factor Hac1, which in turn upregulates the expression of UPR target genes (Figure 2) [54]. Together with the Ire1 counterparts, mammals have two additional ER-stress sensors to induce the UPR: the RNA-dependent protein kinase-like ER kinase (PERK) and the activating transcription factor 6 (ATF6) (Figure 2) [54].

Increasing Atg8 protein levels upon ER stress has been shown to depend on functional Hac1 in yeast (Figure 2) [45]. Additional signaling cascades, however, might be involved in triggering reticulophagy, as the expression of constitutively active spliced Hac1 is not sufficient to stimulate the formation of autophagosomes [45]. Accordingly, cells lacking Hac1 or Ire1 remain capable of inducing the transcription of *ATG8*. This suggests that redundancies or crosstalk among the signaling events regulating autophagy in response to ER stress exist [45].

The lipidation of Atg8/LC3-I also depends on the formation of a large protein complex composed of Atg16 and the conjugate Atg12-Atg5, which is thought to act as an E3-like enzyme conjugating Atg8/LC3-I to PE on autophagosomal membranes [55, 56]. Upregulation of *ATG12*, and the concomitant conversion of Atg8/LC3-I into Atg8-PE/LC3-II, relies on the phosphorylated eIF2α, which itself depends on PERK activation after ER stress in mammalian cells (Figure 2) [37, 44].

Atg1/ULK kinase activity is required to coordinate the action of the Atg proteins during the early events of autophagosome biogenesis [7]. Numerous signaling cascades regulating autophagy such as the mTOR, the AMPK, and the PKA pathways modulate the Atg1/ULK function [7]. Interestingly, Atg1 kinase activity is also enhanced upon ER stress in yeast (Figure 2) [33]. It remains to be established how ER stress acts on this kinase, whether through the above-mentioned cascades or via alternative signaling pathways. For example, depletion of sphingosine-1-phosphate (S1P) phosphatases in mammalian cells leads to an increase of endogenous S1P levels, which cause an ER stress that triggers autophagy [57]. This induction is mTor-independent and PERK-, Ire1-, and ATF6-dependent. Moreover, ER stress causes a release of Ca^{2+} from the ER into the cytosol initiating various signaling cascades, some of which are likely to be involved in autophagy induction [58, 59]. While future research is required to understand the signaling networks regulating autophagy in response to ER stress, it is conceivable that reticulophagy could be induced differently depending on the type and intensity of the ER stress.

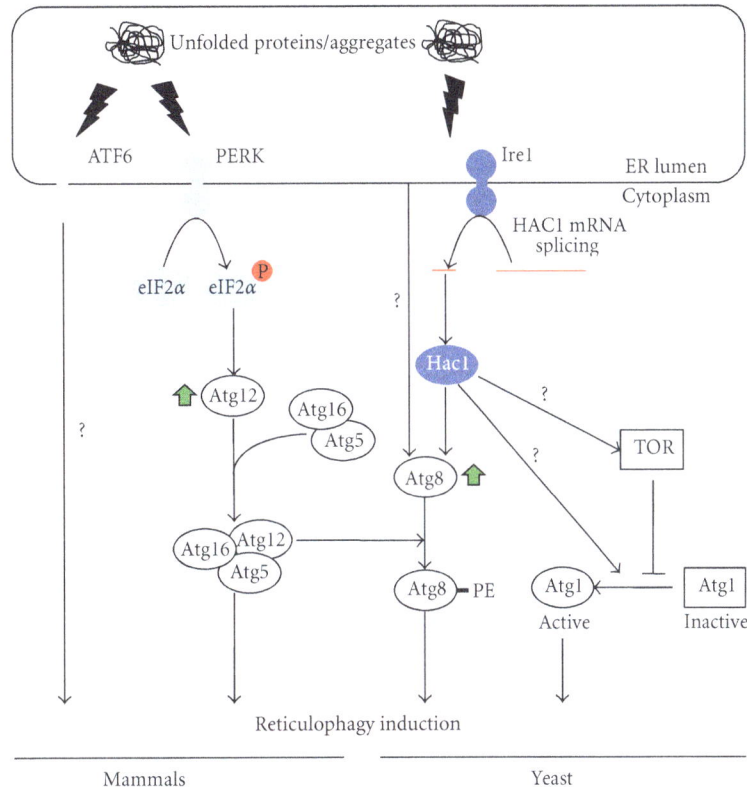

FIGURE 2: Signalling cascades inducing reticulophagy upon ER stress. The transmembrane protein Ire1 (yeast and mammals), ATF6, and PERK (mammals) sense the accumulation of unfolded proteins and/or aggregates, and trigger a general transcriptional response that affect the levels of proteins involved in autophagy. These include Atg8 (signal mediated through Ire1/Hac1 and unidentified alternative pathways in yeast) and Atg12 (mediated by the PERK/eIF2α signalling cascade in mammals). The Atg12-Atg5 (Atg16) complex facilitates the lipidation of Atg8 and autophagy induction. Unknown signalling events in yeast, dependent or independent of the inhibition of the Tor kinase, promote Atg1 activation. Green arrows indicate an increase in protein levels. Question marks indicate signalling cascades that may exist but have not yet been characterized.

11. Putative Physiological Roles of Reticulophagy

Cells subjected to ER stress contain massively expanded ER with increased total length, distance between the lipid bilayers limiting the cisternae and membrane continuity [45]. These morphological changes are not likely caused by the accumulation of unfolded proteins but rather serve as an adaptive response in order to efficiently buffer the ER stress. This might serve to reduce the concentration of unfolded proteins by increasing the space dedicated to protein folding. This idea is supported by the observation that either yeast expressing the constitutively active Hac1, or mammalian cells with the ectopic expression of its metazoan orthologue Xbp1, two proteins capable of inducing a UPR in the absence of unfolded proteins, exhibit an expanded ER [45, 60]. In addition, mammalian cells in which autophagy has been inhibited or genetically ablated display an extended ER upon stress [43]. Conversely, yeast cells accumulating ER-containing autophagosomes do not contain expanded ER [45]. Together, these observations suggest that autophagy could be important to maintain ER homeostasis during UPR by segregating and/or degrading part of the ER. Thus

reticulophagy, through the selective turnover of aggregate-containing and/or damaged ER fragments, could operate in parallel to the ERAD system. This may provide an additional mechanism to dispose unfolded proteins and a way to eliminate damaged membranes. This putative role has been evidenced in yeast expressing pathological mutant versions of human proteins such as the fibrinogen Aguadilla mutation and α_1-ATZ, which accumulate as unfolded aggregates in the ER [34–36]. Knockout strains lacking *ATG* genes expressing these pathological proteins more rapidly amass large amounts of protein aggregates compared to wild-type cells. This suggests that autophagy is important during conditions where the ERAD system is overwhelmed [34, 35]. A similar phenotype was observed in mouse cells lacking Atg5 and expressing expanded polyglutamine repeats [44, 61]. These proteins form cytoplasmic aggregates that trigger ER stress, possibly by impairing ERAD and thus causing an accumulation of unfolded proteins in the ER. Therefore, basal autophagy could serve a similar protective role by preventing the accumulation of misfolded proteins in nonstressed cells. This idea is supported by the observation that an autophagy block caused by the deletion of *ATG6* also induces a UPR in non-stressed cells [35]. The direct implication of autophagy as an ER housekeeping pathway, however, needs to be analyzed in more detail as Atg6 is also required for endosomal trafficking [62, 63].

Paradoxically, autophagy displays a double role in cell viability. It is able to increase the lifespan by protecting against cellular damage; however, in specific pathological situations or when cells have undergone irreversible stress or injuries, autophagy can also contribute to cell death [64]. How reticulophagy contributes to cell fate is not clear and current available data are in part contradictory. Ogata and coworkers concluded that autophagy has a protective role against ER stress-induced cell death as autophagy-deficient cells show higher vulnerability to ER stress and conversely, pretreatment with rapamycin makes cells more resistant to this damage [65]. In contrast, Ding and collaborators proposed a dual role for autophagy according to the status of the cells; autophagy promotes cell survival in cancer cells displaying ER stress, and induces cell death in nononcogenic cells [43]. In yeast, an intact autophagy machinery is essential for cell growth under strong UPR-inducing conditions [45]. Interestingly, it has been proposed that the engulfment of the ER by autophagosomes, without the degradation of the sequestered cargo, is sufficient for autophagy to mitigate ER stress [45]. This hypothesis has been underscored by the finding that in the presence of high concentrations of tunicamycin, an inhibitor of protein glycosylation, Atg proteins are necessary for cell survival while vacuolar proteases are dispensable [45]. Under the same circumstances, ER-containing autophagosomes do not fuse with vacuoles when ER stress is maintained for longer periods [45]. In contrast, when ER stress is initiated by glucose depletion, ER fragments are transported to the lumen of the vacuole indicating that a complete autophagy process occurs [46]. Additional studies are necessary to understand the exact contribution of autophagy as an ER stress response mechanism. A possible scenario is that reticulophagy could have been adapted to differentially modulate its response according to the nature of the stress, and the status of the cell and/or the tissue.

12. Conclusions

Despite their potential relevance in physiological and pathological contexts, the regulation and mechanisms of ribophagy and reticulophagy remain largely unknown. It remains to be determined which of the known or if novel Atg proteins mediate the recognition and selective sequestration of ribosomes and ER fragments into autophagosomes. Moreover, how the cell regulates the segregation of the unwanted parts of the ER and how this breaks away from the organelle need to be further analyzed. A vast field is waiting to be explored.

Acknowledgments

The authors thank Jason Mercer for critical reading of the paper; René Scriwanek for assistance with the preparation of the figures. F. Reggiori is supported by the Netherlands Organization for Health Research and Development (ZonMW-VIDI-917.76.329), by the Netherlands Organization for Scientific Research (Chemical Sciences, ECHO Grant 700.59.003, and Earth and Life Sciences, Open Program Grant 821.02.017). C. Kraft is supported by a WWTF (Wiener-, Wissenschafts-, Forschungs- und Technologie-fonds) "Vienna Research Groups for Young Investigators" Grant. E. Cebollero is supported by an Earth and Life Sciences (ALW-817.02.023) open program grant given to F. Reggiori and Bernd Helms (Department of Biochemistry and Cell Biology, Utrecht University).

References

[1] B. Levine, N. Mizushima, and H. W. Virgin, "Autophagy in immunity and inflammation," *Nature*, vol. 469, no. 7330, pp. 323–335, 2011.

[2] N. Mizushima and B. Levine, "Autophagy in mammalian development and differentiation," *Nature Cell Biology*, vol. 12, no. 9, pp. 823–830, 2010.

[3] B. Levine and G. Kroemer, "Autophagy in the pathogenesis of disease," *Cell*, vol. 132, no. 1, pp. 27–42, 2008.

[4] N. Mizushima, B. Levine, A. M. Cuervo, and D. J. Klionsky, "Autophagy fights disease through cellular self-digestion," *Nature*, vol. 451, no. 7182, pp. 1069–1075, 2008.

[5] H. Nakatogawa, K. Suzuki, Y. Kamada, and Y. Ohsumi, "Dynamics and diversity in autophagy mechanisms: lessons from yeast," *Nature Reviews Molecular Cell Biology*, vol. 10, no. 7, pp. 458–467, 2009.

[6] C. Kraft, F. Reggiori, and M. Peter, "Selective types of autophagy in yeast," *Biochimica et Biophysica Acta*, vol. 1793, no. 9, pp. 1404–1412, 2009.

[7] Z. Yang and D. J. Klionsky, "Mammalian autophagy: core molecular machinery and signaling regulation," *Current Opinion in Cell Biology*, vol. 22, no. 2, pp. 124–131, 2010.

[8] T. Yoshimori and T. Noda, "Toward unraveling membrane biogenesis in mammalian autophagy," *Current Opinion in Cell Biology*, vol. 20, no. 4, pp. 401–407, 2008.

[9] M. Komatsu and Y. Ichimura, "Selective autophagy regulates various cellular functions," *Genes to Cells*, vol. 15, no. 9, pp. 923–933, 2010.

[10] M. U. Hutchins and D. J. Klionsky, "Vacuolar localization of oligomeric α-mannosidase requires the cytoplasm to vacuole targeting and autophagy pathway components in *Saccharomyces cerevisiae*," *Journal of Biological Chemistry*, vol. 276, no. 23, pp. 20491–20498, 2001.

[11] S. V. Scott, M. Baba, Y. Ohsumi, and D. J. Klionsky, "Aminopeptidase I is targeted to the vacuole by a nonclassical vesicular mechanism," *Journal of Cell Biology*, vol. 138, no. 1, pp. 37–44, 1997.

[12] M. A. Lynch-Day and D. J. Klionsky, "The Cvt pathway as a model for selective autophagy," *FEBS Letters*, vol. 584, no. 7, pp. 1359–1366, 2010.

[13] M. Yuga, K. Gomi, D. J. Klionsky, and T. Shintani, "Aspartyl aminopeptidase is imported from the cytoplasm to the vacuole by selective autophagy in Saccharomyces cerevisiae," *Journal of Biological Chemistry*, vol. 286, no. 15, pp. 13704–13713, 2011.

[14] T. Shintani and D. J. Klionsky, "Cargo proteins facilitate the formation of transport vesicles in the cytoplasm to vacuole targeting pathway," *Journal of Biological Chemistry*, vol. 279, no. 29, pp. 29889–29894, 2004.

[15] R. Manjithaya, T. Y. Nazarko, J. C. Farré, and S. S. Suresh, "Molecular mechanism and physiological role of pexophagy," *FEBS Letters*, vol. 584, no. 7, pp. 1367–1373, 2010.

[16] K. Wang and D. J. Klionsky, "Mitochondria removal by autophagy," *Autophagy*, vol. 7, no. 3, pp. 297–300, 2011.

[17] C. Kraft, M. Peter, and K. Hofmann, "Selective autophagy: ubiquitin-mediated recognition and beyond," *Nature Cell Biology*, vol. 12, no. 9, pp. 836–841, 2010.

[18] T. P. Ashford and K. R. Porter, "Cytoplasmic components in hepatic cell lysosomes," *Journal of Cell Biology*, vol. 12, pp. 198–202, 1962.

[19] E. L. Eskelinen, F. Reggiori, M. Baba, A. L. Kovács, and P. O. Seglen, "Seeing is believing: the impact of electron microscopy on autophagy research," *Autophagy*, vol. 7, no. 9, pp. 935–956, 2011.

[20] C. Kraft, A. Deplazes, M. Sohrmann, and M. Peter, "Mature ribosomes are selectively degraded upon starvation by an autophagy pathway requiring the Ubp3p/Bre5p ubiquitin protease," *Nature Cell Biology*, vol. 10, no. 5, pp. 602–610, 2008.

[21] B. K. Baxter, H. Abeliovich, X. Zhang, A. G. Stirling, A. L. Burlingame, and D. S. Goldfarb, "Atg19p ubiquitination and the cytoplasm to vacuole trafficking pathway in yeast," *Journal of Biological Chemistry*, vol. 280, no. 47, pp. 39067–39076, 2005.

[22] S. V. Scott, J. Guan, M. U. Hutchins, J. Kim, and D. J. Klionsky, "Cvt19 is a receptor for the cytoplasm-to-vacuole targeting pathway," *Molecular Cell*, vol. 7, no. 6, pp. 1131–1141, 2001.

[23] C. Kraft and M. Peter, "Is the Rsp5 ubiquitin ligase involved in the regulation of ribophagy?" *Autophagy*, vol. 4, no. 6, pp. 838–840, 2008.

[24] C. Pohl and S. Jentsch, "Midbody ring disposal by autophagy is a post-abscission event of cytokinesis," *Nature Cell Biology*, vol. 11, no. 1, pp. 65–70, 2009.

[25] J. R. Warner, "The economics of ribosome biosynthesis in yeast," *Trends in Biochemical Sciences*, vol. 24, no. 11, pp. 437–440, 1999.

[26] M. S. Hillwig, A. L. Contento, A. Meyer, D. Ebany, D. C. Bassham, and G. C. MacIntosha, "RNS2, a conserved member of the RNase T2 family, is necessary for ribosomal RNA decay in plants," *Proceedings of the National Academy of Sciences of the United States of America*, vol. 108, no. 3, pp. 1093–1098, 2011.

[27] A. R. Kristensen, S. Schandorff, M. Høyer-Hansen et al., "Ordered organelle degradation during starvation-induced autophagy," *Molecular and Cellular Proteomics*, vol. 7, no. 12, pp. 2419–2428, 2008.

[28] F. C. Baltanás, I. Casafont, E. Weruaga, J. R. Alonso, M. T. Berciano, and M. Lafarga, "Nucleolar disruption and cajal body disassembly are nuclear hallmarks of DNA damage-induced neurodegeneration in Purkinje cells," *Brain Pathology*, vol. 21, no. 4, pp. 374–388, 2011.

[29] V. Deretic, "Autophagy in immunity and cell-autonomous defense against intracellular microbes," *Immunological Reviews*, vol. 240, no. 1, pp. 92–104, 2011.

[30] M. Ponpuak, A. S. Davis, E. A. Roberts et al., "Delivery of cytosolic components by autophagic adaptor protein p62 endows autophagosomes with unique antimicrobial properties," *Immunity*, vol. 32, no. 3, pp. 329–341, 2010.

[31] M. F. Campagnoli, U. Ramenghi, M. Armiraglio et al., "RPS19 mutations in patients with Diamond-Blackfan anemia," *Human Mutation*, vol. 29, no. 7, pp. 911–920, 2008.

[32] I. Braakman and N. J. Bulleid, "Protein folding and modification in the mammalian endoplasmic reticulum," *Annual Review of Biochemistry*, vol. 80, pp. 71–99, 2011.

[33] T. Yorimitsu, U. Nair, Z. Yang, and D. J. Klionsky, "Endoplasmic reticulum stress triggers autophagy," *Journal of Biological Chemistry*, vol. 281, no. 40, pp. 30299–30304, 2006.

[34] K. B. Kruse, A. Dear, E. R. Kaltenbrun et al., "Mutant fibrinogen cleared from the endoplasmic reticulum via endoplasmic reticulum-associated protein degradation and autophagy: an explanation for liver disease," *American Journal of Pathology*, vol. 168, no. 4, pp. 1299–1308, 2006.

[35] K. B. Kruse, J. L. Brodsky, and A. A. McCracken, "Characterization of an ERAD gene as *VPS30/ATG6* reveals two alternative and functionally distinct protein quality control pathways: one for soluble Z variant of human α-1 proteinase inhibitor (A1PiZ) and another for aggregates of A1PiZ," *Molecular Biology of the Cell*, vol. 17, no. 1, pp. 203–212, 2006.

[36] J. H. Teckman and D. H. Perlmutter, "Retention of mutant α_1-antitrypsin Z in endoplasmic reticulum is associated with an autophagic response," *American Journal of Physiology*, vol. 279, no. 5, pp. G961–G974, 2000.

[37] E. Fujita, Y. Kouroku, A. Isoai et al., "Two endoplasmic reticulum-associated degradation (ERAD) systems for the novel variant of the mutant dysferlin: ubiquitin/proteasome ERAD(I) and autophagy/lysosome ERAD(II)," *Human Molecular Genetics*, vol. 16, no. 6, pp. 618–629, 2007.

[38] K. Zhang and R. J. Kaufman, "The unfolded protein response: a stress signaling pathway critical for health and disease," *Neurology*, vol. 66, no. 2, pp. S102–S109, 2006.

[39] K. Römisch, "Endoplasmic reticulum-associated degradation," *Annual Review of Cell and Developmental Biology*, vol. 21, pp. 435–456, 2005.

[40] S. Bernales, F. R. Papa, and P. Walter, "Intracellular signaling by the unfolded protein response," *Annual Review of Cell and Developmental Biology*, vol. 22, pp. 487–508, 2006.

[41] Y. Ichimura, T. Kirisako, T. Takao et al., "A ubiquitin-like system mediates protein lipidation," *Nature*, vol. 408, no. 6811, pp. 488–492, 2000.

[42] Y. Kabeya, N. Mizushima, A. Yamamoto, S. Oshitani-Okamoto, Y. Ohsumi, and T. Yoshimori, "LC3, GABARAP and GATE16 localize to autophagosomal membrane depending on form-II formation," *Journal of Cell Science*, vol. 117, no. 13, pp. 2805–2812, 2004.

[43] W. X. Ding, H. M. Ni, W. Gao et al., "Differential effects of endoplasmic reticulum stress-induced autophagy on cell survival," *Journal of Biological Chemistry*, vol. 282, no. 7, pp. 4702–4710, 2007.

[44] Y. Kouroku, E. Fujita, I. Tanida et al., "ER stress (PERK/eIF2α phosphorylation) mediates the polyglutamine-induced LC3 conversion, an essential step for autophagy formation," *Cell Death and Differentiation*, vol. 14, no. 2, pp. 230–239, 2007.

[45] S. Bernales, K. L. McDonald, and P. Walter, "Autophagy counterbalances endoplasmic reticulum expansion during the unfolded protein response.," *PLoS Biology*, vol. 4, no. 12, article e423, 2006.

[46] M. Hamasaki, T. Noda, M. Baba, and Y. Ohsumi, "Starvation triggers the delivery of the endoplasmic reticulum to the vacuole via autophagy in yeast," *Traffic*, vol. 6, no. 1, pp. 56–65, 2005.

[47] P. Salahuddint, "Genetic variants of α1-antitrypsin," *Current Protein and Peptide Science*, vol. 11, no. 2, pp. 107–117, 2010.

[48] T. Kamimoto, S. Shoji, T. Hidvegi et al., "Intracellular inclusions containing mutant α_1-antitrypsin Z are propagated in the absence of autophagic activity," *Journal of Biological Chemistry*, vol. 281, no. 7, pp. 4467–4476, 2006.

[49] M. Baba, K. Takeshige, N. Baba, and Y. Ohsumi, "Ultrastructural analysis of the autophagic process in yeast: detection of autophagosomes and their characterization," *Journal of Cell Biology*, vol. 124, no. 6, pp. 903–913, 1994.

[50] M. J. Mazón, P. Eraso, and F. Portillo, "Efficient degradation of misfolded mutant Pma1 by endoplasmic reticulum-associated degradation requires Atg19 and the Cvt/autophagy pathway," *Molecular Microbiology*, vol. 63, no. 4, pp. 1069–1077, 2007.

[51] E. Kario, N. Amar, Z. Elazar, and A. Navon, "A new autophagy-related checkpoint in the degradation of an ERAD-M target," *Journal of Biological Chemistry*, vol. 286, no. 13, pp. 11479–11491, 2011.

[52] P. Ylä-Anttila, H. Vihinen, E. Jokitalo, and E. L. Eskelinen, "3D tomography reveals connections between the phagophore and endoplasmic reticulum," *Autophagy*, vol. 5, no. 8, pp. 1180–1185, 2009.

[53] M. Hayashi-Nishino, N. Fujita, T. Noda, A. Yamaguchi, T. Yoshimori, and A. Yamamoto, "A subdomain of the endoplasmic reticulum forms a cradle for autophagosome formation," *Nature Cell Biology*, vol. 11, no. 12, pp. 1433–1437, 2009.

[54] K. Kohno, "Stress-sensing mechanisms in the unfolded protein response: similarities and differences between yeast and mammals," *Journal of Biochemistry*, vol. 147, no. 1, pp. 27–33, 2010.

[55] T. Hanada, N. N. Noda, Y. Satomi et al., "The Atg12-Atg5 conjugate has a novel E3-like activity for protein lipidation in autophagy," *Journal of Biological Chemistry*, vol. 282, no. 52, pp. 37298–37302, 2007.

[56] N. Fujita, T. Itoh, H. Omori, M. Fukuda, T. Noda, and T. Yoshimori, "The Atg16L complex specifies the site of LC3 lipidation for membrane biogenesis in autophagy," *Molecular Biology of the Cell*, vol. 19, no. 5, pp. 2092–2100, 2008.

[57] S. Lépine, J. C. Allegood, M. Park, P. Dent, S. Milstien, and S. Spiegel, "Sphingosine-1-phosphate phosphohydrolase-1 regulates ER stress-induced autophagy," *Cell Death and Differentiation*, vol. 18, no. 2, pp. 350–361, 2011.

[58] M. Høyer-Hansen, L. Bastholm, P. Szyniarowski et al., "Control of macroautophagy by calcium, calmodulin-dependent kinase kinase-β, and Bcl-2," *Molecular Cell*, vol. 25, no. 2, pp. 193–205, 2007.

[59] M. Høyer-Hansen and M. Jäättelä, "Connecting endoplasmic reticulum stress to autophagy by unfolded protein response and calcium," *Cell Death and Differentiation*, vol. 14, no. 9, pp. 1576–1582, 2007.

[60] A. L. Shaffer, M. Shapiro-Shelef, N. N. Iwakoshi et al., "XBP1, downstream of Blimp-1, expands the secretory apparatus and other organelles, and increases protein synthesis in plasma cell differentiation," *Immunity*, vol. 21, no. 1, pp. 81–93, 2004.

[61] Y. Kouroku, E. Fujita, A. Jimbo et al., "Polyglutamine aggregates stimulate ER stress signals and caspase-12 activation," *Human Molecular Genetics*, vol. 11, no. 13, pp. 1505–1515, 2002.

[62] A. Kihara, T. Noda, N. Ishihara, and Y. Ohsumi, "Two distinct Vps34 phosphatidylinositol 3-kinase complexes function in autophagy and carboxypeptidase y sorting in Saccharomyces cerevisiae," *Journal of Cell Biology*, vol. 153, no. 3, pp. 519–530, 2001.

[63] M. N. J. Seaman, E. G. Marcusson, J. L. Cereghino, and S. D. Emr, "Endosome to Golgi retrieval of the vacuolar protein sorting receptor, Vps10p, requires the function of the VPS29, VPS30, and VPS35 gene products," *Journal of Cell Biology*, vol. 137, no. 1, pp. 79–92, 1997.

[64] E. H. Baehrecke, "Autophagy: dual roles in life and death?" *Nature Reviews Molecular Cell Biology*, vol. 6, no. 6, pp. 505–510, 2005.

[65] M. Ogata, S. I. Hino, A. Saito et al., "Autophagy is activated for cell survival after endoplasmic reticulum stress," *Molecular and Cellular Biology*, vol. 26, no. 24, pp. 9220–9231, 2006.

12

Plakoglobin: Role in Tumorigenesis and Metastasis

Zackie Aktary and Manijeh Pasdar

Department of Cell Biology, University of Alberta, Edmonton, AB, Canada T6G 2H7

Correspondence should be addressed to Manijeh Pasdar, mpasdar@ualberta.ca

Academic Editor: Eok-Soo Oh

Plakoglobin (γ-catenin) is a member of the Armadillo family of proteins and a homolog of β-catenin. As a component of both the adherens junctions and desmosomes, plakoglobin plays a pivotal role in the regulation of cell-cell adhesion. Furthermore, similar to β-catenin, plakoglobin is capable of participating in cell signaling. However, unlike β-catenin that has well-documented oncogenic potential through its involvement in the Wnt signaling pathway, plakoglobin generally acts as a tumor/metastasis suppressor. The exact roles that plakoglobin plays during tumorigenesis and metastasis are not clear; however, recent evidence suggests that it may regulate gene expression, cell proliferation, apoptosis, invasion, and migration. In this paper, we describe plakoglobin, its discovery and characterization, its role in regulating cell-cell adhesion, and its signaling capabilities in regulation of tumorigenesis and metastasis.

1. Introduction

Plakoglobin (also known as γ-catenin) is a member of the Armadillo family of proteins and a structural and functional homolog of β-catenin. These catenin proteins have two major roles in the cell: the mediation of cell-cell adhesion and cell signaling. As adhesive proteins, both β-catenin and plakoglobin interact with the cytoplasmic domain of cadherins, thereby tethering the cadherin proteins to the cytoskeleton. In addition to their cell-cell adhesive functions, both β-catenin and plakoglobin interact with a number of intracellular partners including signaling proteins and transcription factors, which accounts for their involvement in cellular signaling [1–4]. Despite these similarities, a major difference between β-catenin and plakoglobin emerges when considering their signaling functions. While β-catenin has a well-defined oncogenic potential as the terminal component of the Wnt signaling pathway [5–7], plakoglobin is typically associated with tumor/metastasis suppressor activity [8–10]. However, the mechanisms that underlie this activity remain undefined. In this paper, we have focused on the potential roles of plakoglobin during tumorigenesis and metastasis in an attempt to define how this often overlooked protein contributes to these complex processes.

2. Plakoglobin: Initial Identification and Early Characterization

Plakoglobin was initially identified as an 83 kDa protein component of the desmosomal plaque [11]. Subsequently, using monoclonal antibodies, cDNA cloning, and a combination of biochemical, morphological, and molecular approaches, Cowin et al. [12] demonstrated that this 83 kDa protein was present in both desmosomes and the adherens junction and was given the name plakoglobin.

Although plakoglobin was identified as a junctional protein, the role that it played in these junctional complexes was unclear, and the partners with which plakoglobin interacted were not identified. It was not until several years later that coimmunoprecipitation experiments showed that plakoglobin interacted with the desmosomal cadherin desmoglein, thereby confirming plakoglobin as a constituent of the desmosomes [13]. In addition, several groups showed that E-cadherin (initially known as uvomorulin) immunoprecipitates contained three distinct proteins, which became known as α-, β-, and γ-catenin [14–16]. These studies showed that these three catenin proteins, with molecular weights of approximately 102, 88, and 80 kDa, respectively, interacted with the cytoplasmic domain of E-cadherin.

Further work analyzing the formation and stability of the E-cadherin-catenin complexes suggested that the E-cadherin-β-catenin complex was formed immediately after E-cadherin synthesis and was very stable. Interestingly, these studies also found that α-catenin could not be found in association with E-cadherin independent of β-catenin, suggesting that β-catenin was a physical link between E-cadherin and α-catenin. However, since γ-catenin was found to be only loosely associated with E-cadherin, it was determined that the main adhesive complexes consisted of E-cadherin, β-catenin, and α-catenin, although the existence of a separate E-cadherin-γ-catenin complex could not be ruled out [16].

At this time there was some confusion as to the identity of the catenin proteins and their relationship to plakoglobin. It soon became evident that plakoglobin was a homolog of β-catenin, a 92 kDa E-cadherin-associated protein [17]. However, it was not until the work of Knudsen and Wheelock [18] that it became clear that the 80 kDa protein that was associated with E-cadherin was indeed plakoglobin. In this study, the authors showed that plakoglobin interacted with both E- and N-cadherin and that it was a distinct protein from β-catenin [18]. This finding was confirmed by work from other groups demonstrating that plakoglobin and γ-catenin were indeed the same protein [1, 19].

Subsequent analysis of the kinetics of plakoglobin synthesis and associations with cadherins demonstrated that following synthesis, plakoglobin interacted with both desmoglein and E-cadherin in both the soluble and cytoskeleton-associated pools of cellular proteins. In addition, a distinct, cadherin-independent pool of plakoglobin was also observed, suggesting that plakoglobin may have a role in the cell in addition to cell adhesion. Finally, phosphorylation experiments revealed that whereas the insoluble (cadherin-associated) pool of plakoglobin was serine phosphorylated, the soluble pool was serine, threonine, and tyrosine phosphorylated, suggesting that these different pools of plakoglobin are differentially regulated and perform varying functions [20]. Collectively, these studies demonstrated that plakoglobin is a homolog of β-catenin and a unique protein in that it is the only component common to both E-cadherin and desmosomal cadherin-containing junctions.

3. Plakoglobin Functions: Cell-Cell Adhesion

The most documented role of plakoglobin within the cell is in cell-cell adhesion. As such, plakoglobin is found in both adherens junctions and desmosomes (Figure 1). Adherens junctions are a ubiquitous type of intercellular adhesion structure present in both epithelial and nonepithelial cells, whereas desmosomes are adhesive junctions that confer tensile strength and resilience to cells and are present not only in epithelial cells but also in nonepithelial cells that endure mechanical stress, such as cardiac muscle. Both adherens junctions and desmosomes are cadherin based. Cadherins are single-pass transmembrane glycoproteins that form homotypic interactions with cadherin proteins on neighboring cells. Intracellularly, cadherins interact with proteins of the catenin family. At the adherens junction, the

C-terminal domain of E-cadherin interacts, in a mutually exclusive manner, with β-catenin or plakoglobin, which then interacts with α-catenin, which is an actin-binding protein. A fourth catenin protein, p120-catenin, interacts with the juxtamembrane domain of E-cadherin and is important for E-cadherin dimerization and stability at the membrane (Figure 1; for reviews see [21, 22]). At the desmosome, the desmosomal cadherins (desmocollins and desmogleins) interact intracellularly with plakophilin and plakoglobin, which interact with desmoplakin, an intermediate filament binding protein (Figure 1; for reviews, see [23, 24]).

The identification of plakoglobin as a constituent of both the adherens junction and the desmosomes suggested that it might play an important role in regulating cell-cell adhesion. However, the observation that the adherens junctions could exist as a complex containing E-cadherin, β-catenin, and α-catenin, independent of plakoglobin [16] questioned the necessity of plakoglobin, at least at the adherens junctions. Regardless, it soon became apparent that plakoglobin does have an essential role in regulating cell-cell adhesion.

It had been previously shown that disruption of E-cadherin-based cell-cell adhesion led to a transformed and/or invasive phenotype while reexpression of E-cadherin in cells lacking its expression resulted in a mesenchymal to epithelial phenotypic transition [25–31]. Furthermore, reduced expression of E-cadherin was known to inversely correlate with the differentiation grade of tumors [32–37]. While it was clear that these E-cadherin-based junctions were important for the maintenance of an "epithelial" phenotype, the role of plakoglobin in this phenomenon was not discerned until it was shown that the expression of E- or P-cadherin alone in murine spindle cell carcinomas that lacked endogenous expression of these proteins was not sufficient to modify the morphology or tumorigenicity of these cells [38]. Although these cadherins were expressed in the cells, localized to the cell membrane, and interacted with both α- and β-catenin, they did not interact with plakoglobin. Further analysis showed that the levels of plakoglobin in these cells were very low, thus accounting for the absence of plakoglobin association with E-cadherin. From this work, the authors suggested that the association of the E-cadherin-catenin complex with plakoglobin may be necessary for its tumor suppressing activity.

Another significant role for plakoglobin in the regulation of cell-cell adhesion was discovered when studies showed that A431 epithelial cells treated with dexamethasone (which resulted in the isolation of fibroblastic A431 cells lacking E-cadherin but expressing desmoglein) were unable to form desmosomes upon exogenous expression of E- or P-cadherin, despite the formation of the adherens junction in these cells [39]. Interestingly, the authors observed that although plakoglobin was present at low levels in these cells, it was not coimmunoprecipitated with the exogenously expressed E-cadherin; in fact, the plakoglobin found in these cells coprecipitated with desmoglein. To examine the possibility that plakoglobin plays a regulatory role in desmosome formation, the authors expressed an E-cadherin-plakoglobin chimeric protein capable of forming stable adherens junctions in the cells and observed desmosome

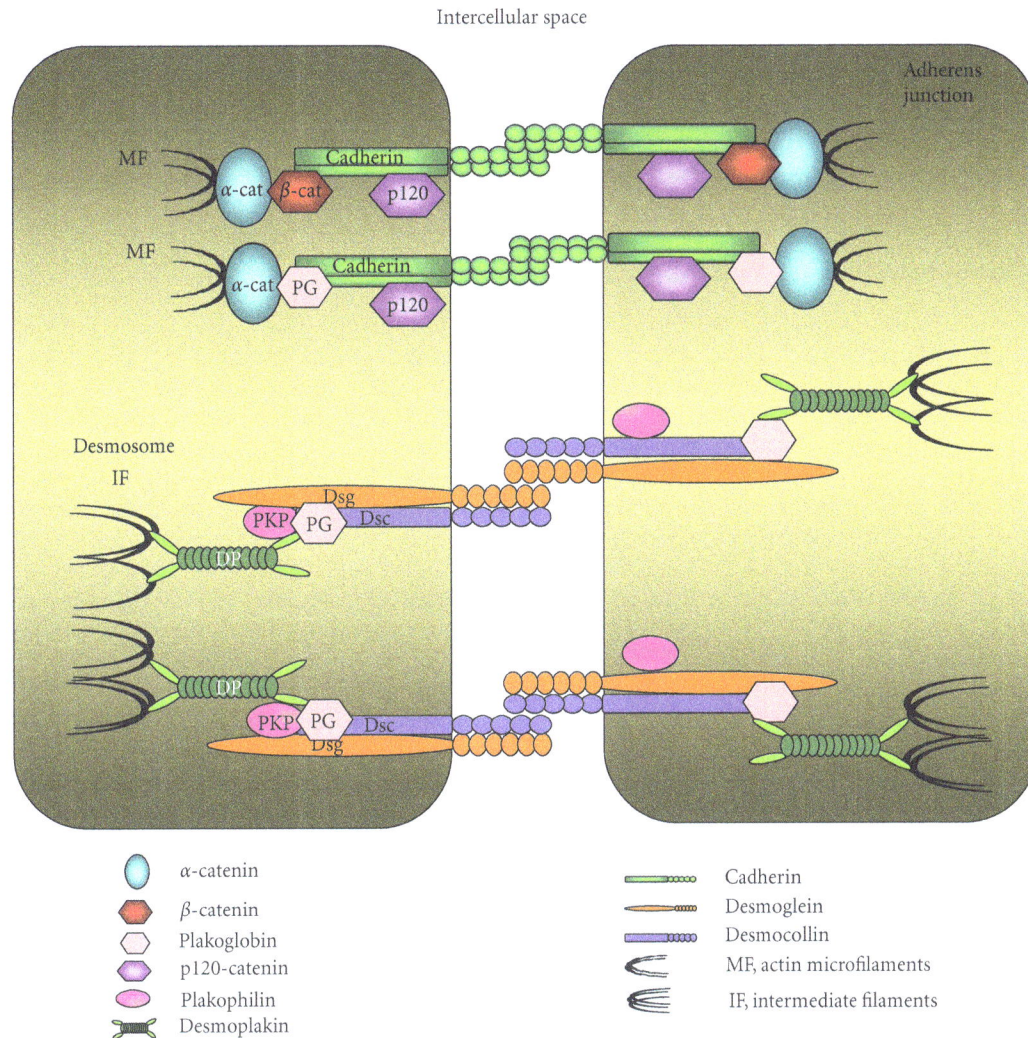

FIGURE 1: Cell adhesion complexes in epithelial cells. Cell-cell adhesion is maintained in epithelial tissues by the adherens junction and desmosomes. At the adherens junctions, E-cadherin forms extracellular interactions with E-cadherin molecules on neighboring cells. Intracellularly, E-cadherin interacts with either β-catenin or plakoglobin, which then interact with α-catenin, an actin-binding protein. A fourth catenin, p120-catenin, also interacts with E-cadherin and regulates its stability at the membrane. At the desmosome, the desmosomal cadherins (desmoglein and desmocollin) interact with plakoglobin and plakophilin, which interact with desmoplakin, which in turn associates with the intermediate filament cytoskeleton. The basic, core protein composition of the desmosomes is represented here: the exact protein constituents of the desmosomes and their interactions vary between different types of cells and tissues.

formation. While it had been previously observed that adherens junction formation not only preceded, but was also a prerequisite for desmosome formation [40–48], this was the first indication that plakoglobin served as a molecule involved in crosstalk between both junctional complexes in epithelia.

Following this study, our laboratory demonstrated the role of plakoglobin in adhesive junction formation by expressing low/physiological levels of plakoglobin in SCC9 cells, a squamous cell carcinoma cell line that lacks the expression of both plakoglobin and E-cadherin [9, 49]. Following exogenous plakoglobin expression, SCC9 cells underwent a mesenchymal to epidermoid phenotypic transition that was concurrent with the stabilization of N-cadherin

and the formation of desmosomes and well-organized N-cadherin-containing adherens junctions [9]. This result confirmed that plakoglobin expression was necessary for desmosome formation and also demonstrated that plakoglobin-N-cadherin interactions could occur prior to desmosome formation. Other studies have further characterized the role of plakoglobin in desmosome assembly and function. Palka and Green [50] demonstrated the role of plakoglobin's C terminus for the proper assembly of the desmosomal plaque, and Acehan et al. showed that plakoglobin is essential for the efficient binding of desmoplakins to the intermediate filaments [51]. Furthermore, plakoglobin was shown to be necessary for the recruitment of plakophilin 3 to the membrane, desmosome formation, efficient cell-cell adhesion, and inhibition of cell migration and invasion

[52, 53]. Finally, work from Birchmeier's laboratory showed that plakoglobin double knockout mice died during embryogenesis as a result of disrupted heart function due to the loss of stable desmosomes in the intercalated discs of cardiac muscle, further confirming the essential role of plakoglobin in desmosome formation and function [48, 54].

4. Plakoglobin Functions: Cell Signaling

4.1. Initial Observations and Controversy. The first clue that plakoglobin might participate in cell signaling came from studies of the exogenous expression of Wnt-1 in PC12 cells. In these cells, plakoglobin levels were increased, and it underwent membrane redistribution, suggesting that, in addition to β-catenin levels, Wnt-1 can modulate plakoglobin levels and localization [55]. Subsequently, Karnovsky and Klymkowsky [56] demonstrated plakoglobin signaling activity by microinjecting mRNAs-encoding plakoglobin into fertilized *Xenopus* embryos, resulting in dorsalized gastrulation and anterior axis duplication. In this study, the exogenously expressed plakoglobin localized both at the plasma membrane and in punctate nuclear aggregates. Furthermore, the coinjection of mRNAs-encoding plakoglobin as well as the cytoplasmic domain of desmoglein suppressed both dorsalized gastrulation and anterior axis duplication. In these embryos, plakoglobin was localized primarily to the plasma membrane with some perinuclear distribution. These results suggested that plakoglobin has signaling ability similar to β-catenin, but when it is sequestered at the plasma membrane (as part of desmosomes), plakoglobin is unable to participate in cell signaling.

This initial finding suggested that plakoglobin may have signaling functions similar to its homologs β-catenin and the *Drosophila* Armadillo protein. However, subsequent studies from various groups have demonstrated that while plakoglobin does indeed have signaling capabilities, it appears to function as a tumor suppressor rather than a tumor promoter. The first demonstration of this phenomenon occurred when Simcha et al. [8] found that plakoglobin expression in SV40-transformed NIH3T3 cells decreased the ability of these cells to form tumors in syngeneic mice. This growth suppressive effect of plakoglobin was augmented by cotransfection with N-cadherin. The authors also expressed plakoglobin in the renal carcinoma cell line KTCTL 60, which lacks endogenous expression of E-cadherin and desmosomal cadherins, α-catenin, β-catenin, plakoglobin, and desmoplakin and induces tumor formation in mice. Plakoglobin expression in KTCTL 60 cells also inhibited the tumorigenicity of these cells in syngeneic mice. Notably, the authors showed that the majority of the plakoglobin in these cells was Triton X-100 soluble, suggesting that it was not junction associated. This result was of significance because it demonstrated that plakoglobin could suppress tumor formation independent of its role in mediating cell-cell adhesion.

These studies made it clear that plakoglobin was capable of cell signaling and able to act as a tumor suppressor. Numerous subsequent studies have described the signaling

function of plakoglobin as primarily one of tumor suppression, although a few reports have suggested that similar to β-catenin, plakoglobin may have oncogenic activity. In the following sections, we will present the experimental evidence for both the tumorigenic and tumor suppressive activities of plakoglobin and propose possible explanations for these observed discrepancies.

4.2. Plakoglobin Oncogenic Activity. Kolligs et al. [57] have shown that the tumor suppressor adenomatous polyposis coli (APC), which was already known to regulate the levels of β-catenin, could also regulate plakoglobin protein levels. In this study, the authors also showed that exogenous expression of plakoglobin in rat RK3E cells, which express considerable amounts of endogenous plakoglobin and β-catenin [57, 58], resulted in a transformed phenotype, which they suggested was dependent on the upregulation of the oncogene c-Myc and activation of Tcf/Lef signaling. More recently, Pan et al. [59] have shown that the exogenous expression of plakoglobin in HCT116 colon carcinoma cells, which express a mutant β-catenin protein that cannot be degraded [60], resulted in genomic instability and increased invasion and migration.

Both of these studies concluded that plakoglobin possessed oncogenic activity. However, it must be noted that several lines of evidence suggest that the oncogenic activity of plakoglobin may be indirect and achieved through modulation of the protein levels and signaling ability of β-catenin [61–67]. Since plakoglobin and β-catenin interact with some of the same proteins and display high sequence homology (Figure 2, [2, 4, 68, 69]), it became evident that plakoglobin may, in fact, be able to promote tumorigenesis by interacting with proteins that would normally sequester β-catenin (e.g., E-cadherin, Axin, APC), which would result in increased levels of cytoplasmic and nuclear β-catenin and in turn enhanced signaling. Indeed, following the observation that plakoglobin expression resulted in *Xenopus* axis duplication [56], the same group showed that this outcome did not depend on the nuclear localization of plakoglobin, since membrane-anchored forms of this protein produced the same axis duplication [70]. This demonstrated that nuclear plakoglobin was inconsequential in inducing a Wnt-like phenotype, since the cytoplasmic plakoglobin induced this same phenotype. At the same time, Salomon et al. [61] showed that overexpression of plakoglobin resulted in the displacement of β-catenin from, and the increased association of plakoglobin with, the N-cadherin-containing adherens junctions. Furthermore, excess cytoplasmic β-catenin was able to translocate into the nucleus. This was supported by other work, which showed that overexpression of plakoglobin in NIH3T3 cells resulted in the nuclear accumulation of β-catenin and that overexpression of the Wnt coactivator Lef-1 in MDCK cells resulted in its preferential interaction with β-catenin (instead of plakoglobin). Subsequently, the β-catenin-Lef complexes were localized to the nucleus [62], suggesting that when both plakoglobin and β-catenin were present within the cell, β-catenin-Lef complexes were more readily formed and transcriptionally active. Further examination of the ability of plakoglobin to

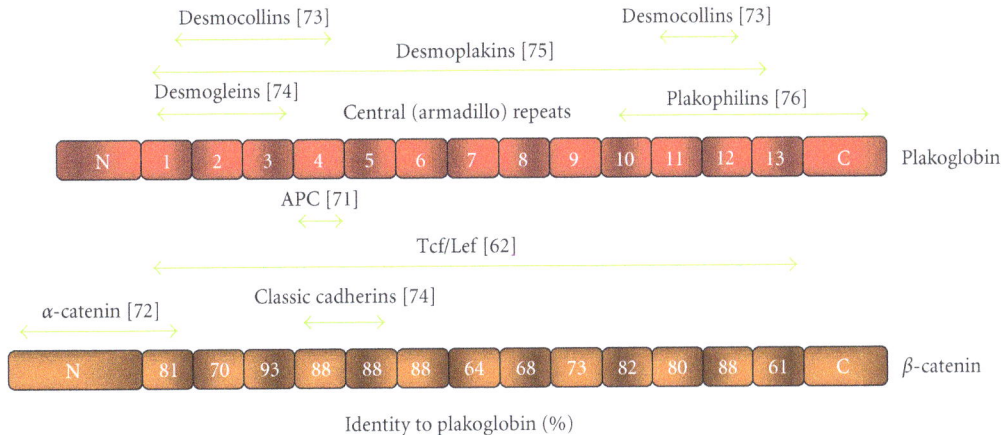

FIGURE 2: Schematic structure of β-catenin and plakoglobin. Both β-catenin and plakoglobin contain 13 Armadillo repeats that are flanked by N- and C-terminal domains, respectively. The degree of homology between β-catenin and plakoglobin for each Armadillo domain is indicated. Protein partners that interact with plakoglobin and the domains involved in these interactions are indicated. The corresponding references are listed in brackets (see [71–76]).

signal via interactions with the Tcf/Lef family of transcription factors showed that although plakoglobin interacted with Lef-1, this complex was inefficient in binding to DNA, whereas β-catenin-Lef-1 complexes more readily bound DNA [63]. This study also demonstrated that overexpression of plakoglobin resulted in increased β-catenin-Lef-1 complex formation and its association with DNA. Further analysis of the transactivation potential of β-catenin and plakoglobin demonstrated that β-catenin was a much stronger activator of Tcf/Lef target genes than plakoglobin [64].

As mentioned earlier, we have previously shown that the expression of low/physiological levels of plakoglobin in plakoglobin-deficient SCC9 cells induced a mesenchymal to epidermoid change in phenotype, whereas its overexpression resulted in foci formation and decreased apoptosis, which was concurrent with the upregulation of the prosurvival protein Bcl-2 [77]. Using cDNAs-encoding plakoglobin fused to nuclear localization or nuclear export signals (NLS and NES), we subsequently showed that Bcl-2 levels were upregulated in plakoglobin overexpressing SCC9 cells regardless of plakoglobin localization. Furthermore, in these cells, β-catenin-N-cadherin interactions were decreased, and β-catenin accumulated in the nucleus, interacted with Tcf, and its signaling was increased [65], confirming that the overexpressed plakoglobin acted indirectly by enhancing the signaling capability of β-catenin.

The above studies describing the oncogenic potential of plakoglobin may also be as a result of β-catenin. In Kolligs's study [57] where plakoglobin was overexpressed in RK3E cells (which express endogenous β-catenin and plakoglobin [58]), it was not determined if plakoglobin could activate c-Myc expression in the absence of β-catenin or whether either of these catenins was detected in the nucleus in association with the c-Myc promoter. In addition, in Pan's study [59] in which HCT116 cells showed increased genomic instability and migration and invasion upon plakoglobin expression, the endogenous β-catenin was a mutant protein that was unable to be phosphorylated and subsequently degraded

[60]. While much of the β-catenin localized to the membrane in these cells [78], plakoglobin expression most likely led to decreased β-catenin-cadherin interactions and increased β-catenin signaling. In support of this prediction, HCT116 cells overexpressing plakoglobin showed increased expression of the oncogenes securin and c-Myc and decreased expression of E-cadherin, all of which are documented β-catenin target genes [79–81]. Taken together, the evidence suggests that although plakoglobin expression may lead to a transformed phenotype, it is likely that this outcome is associated with increased oncogenic β-catenin signaling rather than oncogenic activity due directly to plakoglobin.

4.3. Plakoglobin Signaling in β-Catenin Null Cells. While the oncogenic signaling activity of plakoglobin discussed above can be attributed to the signaling activity of β-catenin rather than plakoglobin itself, this cannot account for all of the observations regarding plakoglobin signaling. Recent studies attempting to discern the signaling activity of plakoglobin independent of β-catenin have used tissue culture cell lines that lack the endogenous expression of β-catenin [82–85]. These studies have shown that in the absence of β-catenin, plakoglobin does indeed have Tcf/Lef-mediated transcriptional activity, although this activity is less than that of β-catenin-Tcf complexes. Interestingly, although these studies have demonstrated that plakoglobin can signal through forming transcriptional complexes with Tcf/Lef transcription factors, they did not assess the tumor-forming properties of these cells, so it remains unclear as to whether these cells possessed transformed or nontransformed properties. To that end, it has been demonstrated that plakoglobin or β-catenin expression in renal carcinomas lacking endogenous β-catenin and plakoglobin resulted in the upregulation of Nr-CAM, a neuronal cell adhesion molecule that can be regulated by both β-catenin and plakoglobin [82]. Furthermore, Nr-CAM expression in NIH3T3 cells conferred a more tumorigenic and invasive phenotype on these cells. Significantly however, although plakoglobin expression resulted in increased Nr-CAM levels in renal carcinomas and although plakoglobin-Tcf/Lef complexes can regulate Nr-CAM expression, the overall phenotype of these cells upon plakoglobin expression was nontumorigenic [8]. This showed that although plakoglobin may regulate β-catenin-target genes in the absence of β-catenin, it still may suppress tumorigenesis in the same cells. The homology between plakoglobin and β-catenin explains the ability of plakoglobin to signal through Tcf/Lef in the absence of β-catenin. Taken together, indeed, it is not surprising that if β-catenin is completely absent from a cell line, plakoglobin cannot only replace it in junctional complexes, but may also be able to regulate some β-catenin target genes (e.g., Survivin [85]). However, as a final note, it is important to consider that β-catenin-null tumors are extremely rare, and in most tumors and cell lines, plakoglobin signaling activity occurs in the presence of β-catenin.

4.4. *Plakoglobin Tumor Suppressor Activity.* Despite the observation that plakoglobin overexpression promotes tumorigenesis mediated by the oncogenic signaling of β-catenin, several studies examining the signaling function of plakoglobin have identified it as a tumor suppressor. We have previously shown that expression of physiological levels of plakoglobin in SCC9 cells, which lack endogenous plakoglobin and E-cadherin, resulted in a mesenchymal to epidermoid phenotypic transition, which was concurrent with the stabilization of N-cadherin, the formation of desmosomes, and the downregulation of β-catenin [9]. Furthermore, we have found that plakoglobin-expressing SCC9 cells showed a decreased growth rate compared to parental SCC9 cells. These results, taken together, demonstrated that not only could plakoglobin act as a tumor suppressor, but that potentially it does so by decreasing the levels of β-catenin.

The ability of plakoglobin to inhibit cell growth and proliferation was next observed when Charpentier et al. [10] expressed plakoglobin (under the control of an epidermal-specific promoter) in the basal cells of the epidermis as well as the hair follicles of transgenic mice. These authors showed that plakoglobin expression resulted in a reduced proliferative potential of the epidermal cells and that plakoglobin-expressing hair follicles had a significantly reduced growth phase, with hairs shorter by roughly 30% after plakoglobin expression.

Further evidence suggesting a growth suppressive activity for plakoglobin was provided in lung cancer, when it was shown that while β-catenin was uniformly expressed in various Nonsmall cell lung cancer (NSCLC) cell lines and lung primary tumors, plakoglobin expression was very low or completely absent [86]. The authors showed that exogenous expression of plakoglobin in the low-plakoglobin-expressing NSCLC cells resulted in decreased β-catenin-Tcf signaling, which was concurrent with decreased cell and anchorage-independent growth. This result further supported the idea that plakoglobin can act as a tumor suppressor by inhibiting the oncogenic activity of β-catenin.

Interestingly, when the authors treated these NSCLC cell lines with the DNA methylation inhibitor 5-aza-2'-deoxycytidine (AZA) or the histone deacetylase inhibitor trichostatin A (TSA), plakoglobin levels were increased. Previous analysis of the plakoglobin promoter had described CpG islands within the promoter [87], and while it had been observed that inhibition of DNA methylation could result in increased plakoglobin protein levels in at least one thyroid carcinoma cell line [88], this was the first indication that both DNA methylation and histone deacetylation played important roles in regulating plakoglobin expression.

The occurrence of methylated CpG islands within the plakoglobin promoter as well as histone deacetylation has not been limited to NSCLC cell lines. Various groups have shown that the plakoglobin promoter is methylated in prostate, bladder, trophoblastic, and mammary carcinomas [89–92], which is concurrent with a transformed phenotype. Canes et al. [90] have shown that treatment of bladder carcinoma cells with TSA resulted in increased plakoglobin expression and a decreased ability of these cells to form tumors in mice, once again suggesting a growth inhibitory activity of plakoglobin. Similarly, when mammary carcinoma cell lines were treated with AZA, increased plakoglobin levels were observed, as well as decreased soft agar colony formation and overall cell growth [92], indicative of decreased tumor-forming ability.

Several lines of evidence suggest that plakoglobin plays a role in regulating apoptosis, in addition to acting as a growth suppressor. In their work describing the effects of plakoglobin on hair growth in transgenic mice, Charpentier et al. [10] showed that plakoglobin expression decreased epithelial proliferation. Moreover, this expression also resulted in premature apoptosis, because TUNEL assays showed that the inner root sheath of the plakoglobin-expressing transgenic follicles underwent apoptosis two days earlier than in normal hair follicles. In agreement with these findings, we have previously shown that SCC9 cells expressing physiological levels of plakoglobin were more prone to undergo staurosporine-induced apoptosis when compared to parental SCC9 cells [77]. We have also observed that SCC9 cells expressing plakoglobin exclusively in the nucleus (SCC9-PG-NLS) showed decreased Bcl-2 levels compared to cells with overexpressed wild-type plakoglobin, which suggests that plakoglobin may play a more direct role in regulating the expression of apoptotic genes. More recently, it has been shown that mouse keratinocytes that lack endogenous plakoglobin expression are protected from etoposide-induced apoptosis, whereas plakoglobin-expressing keratinocytes readily undergo apoptosis upon etoposide treatment [93]. In this study, the authors demonstrated that plakoglobin-null keratinocytes were unable to release cytochrome c from the mitochondria and activate caspase 3, suggesting that plakoglobin plays a role in regulating the apoptotic cascade. Furthermore, the mRNA levels of the antiapoptotic protein Bcl-X_L were higher in the plakoglobin null keratinocytes, which could potentially have prevented the translocation of cytochrome c from the mitochondria. Finally, the expression of plakoglobin in the null keratinocytes resulted in decreased Bcl-X_L levels, caspase 3 activation, and apoptosis induction following etoposide treatment. Taken together, these studies have demonstrated that plakoglobin does have some role in apoptosis signaling and potentially may exert part of its tumor suppressor activity through the modulation of apoptosis.

4.5. Plakoglobin Metastasis Suppressor Activity. As the tumor suppressor activity of plakoglobin began to be revealed, it soon became evident that in addition to inhibiting the growth properties of carcinoma cell lines, plakoglobin also plays a role in regulating the invasive and migratory properties of cancer cells. The initial observation of plakoglobin metastasis suppressor activity was documented in human umbilical vascular endothelial cells (HUVEC), where plakoglobin was typically associated with sites of cell-cell contact [94]. Plakoglobin antisense oligonucleotides increased HUVEC migration, suggesting that the loss of plakoglobin expression led to an increased migratory phenotype. Concurrent with increased migration, the antisense treated HUVEC cells also became more prone to forming tubular structures in Matrigel, suggesting that plakoglobin knock down also promoted angiogenesis.

Mukhina et al. [95] further detailed the metastasis suppressor activity of plakoglobin using MCF-7 cells, which express membrane-localized E-cadherin and plakoglobin, and stable cell junctions. In this study, the authors treated MCF-7 cells with human growth hormone (hGH) and observed a downregulation of plakoglobin, a cytoplasmic distribution of E-cadherin and an increased migratory and invasive phenotype, which was accompanied by an increase in matrix metalloproteinase levels. Furthermore, the authors demonstrated that hGH-mediated invasiveness was dependent on Src kinase and also showed that chemical inhibitors of Src resulted in increased plakoglobin levels and, in turn, decreased invasion and migration. To discern the specific role of plakoglobin in these processes, the authors expressed plakoglobin in the hGH-treated MCF-7 cells, which resulted in both the decreased migration and invasiveness of these cells [95].

The metastasis suppressor activity of plakoglobin has also been described in bladder carcinomas, where the expression of plakoglobin in plakoglobin null cell lines resulted not only in decreased growth and tumorigenicity (as assessed by colony formation in soft agar and tumor formation in nude mice, resp.), but also in decreased invasive and migratory capabilities of the transfectants [96]. Similarly, knock down of plakoglobin using siRNAs resulted in the increased tumorigenic and invasive properties of bladder carcinoma cells relative to their plakoglobin-expressing parental cell lines. This study further demonstrated that plakoglobin expression did not affect Wnt/β-catenin signaling in these bladder carcinomas, which suggested that plakoglobin possessed tumor and metastasis suppressor activities independent of β-catenin.

The ability of plakoglobin to act as a metastasis suppressor independent of its role in cell-cell adhesion has been demonstrated using plakoglobin null keratinocytes [97], which were less adherent to one another and more migratory (as assessed by transwell migration assays). However, when wild-type plakoglobin was expressed in these cells, they became more adherent and less migratory. Using colloidal gold-coated coverslips, the authors were able to

assess the migratory abilities of individual cells and observed that individual plakoglobin null keratinocytes were more migratory than their plakoglobin-expressing counterparts. The authors also showed that plakoglobin may regulate single keratinocyte migration by inhibition of Src signaling, which had been previously shown to promote migration and invasion of mammary carcinomas by downregulation of plakoglobin (see above [95]). These results suggested that plakoglobin could suppress migration through the modulation of cell-cell adhesion, as had been previously suggested. However, to determine whether plakoglobin could have an effect in migration independent of its role in cell-cell adhesion, plakoglobin null keratinocytes were transfected with cDNAs encoding mutant plakoglobin, missing either its N- or C-terminus (α-catenin binding and transactivation domain, resp.). The expression of either of these mutant proteins resulted in increased keratinocyte adhesiveness when compared to the plakoglobin null cells, demonstrating that these domains were dispensable for the adhesive function of plakoglobin. Importantly, the authors showed that whereas individual keratinocytes expressing the N-terminal-deleted plakoglobin were not migratory, those that expressed the C-terminal-deleted plakoglobin were migratory. This showed that plakoglobin could indeed suppress migration independent of its adhesive function (since keratinocytes expressing C-terminal-deleted plakoglobin were as adhesive to one another as wild-type plakoglobin expressing keratinocytes). Subsequent work using these plakoglobin null keratinocytes has suggested that plakoglobin affected individual cell motility by regulating the deposition of the extracellular matrix (ECM) protein fibronectin, actin cytoskeleton organization (which in turn regulates Src signaling), and RhoGTPases [98]. Collectively, these observations clearly demonstrate tumor/metastasis suppressor activity of plakoglobin independent of its role in cell-to-cell adhesion.

5. Plakoglobin Functions: Regulation of Gene Expression

When discussing roles for plakoglobin during tumorigenesis and metastasis, it is important to consider that while plakoglobin may function as both a regulator of cell-cell adhesion and an intracellular signaling molecule, it may also play a more active role in these processes through the regulation of gene expression. Evidence supporting the plakoglobin-mediated regulation of gene expression has started to emerge, and work from several groups, including ours, has suggested that plakoglobin can regulate the expression of genes involved in cell-cycle control, apoptosis, cell proliferation, and invasion.

Williamson et al. [99] have shown that plakoglobin acts as a repressor of the c-Myc gene. Using mouse keratinocytes and reporter assays, the authors of this study showed that plakoglobin suppressed c-Myc expression in a Lef-1-dependent manner, suggesting that when plakoglobin interacted with Lef-1, this complex was unable to promote gene expression. These findings confirmed previous results demonstrating the inefficiency of these complexes in binding

DNA [62–64, 100]. This study further showed that the plakoglobin-mediated suppression was similar in both wild-type and β-catenin null keratinocytes, demonstrating that plakoglobin could regulate gene expression independent of β-catenin. Finally, using chromatin immunoprecipitation with plakoglobin antibodies, the authors demonstrated that plakoglobin and Lef-1 associated with the c-Myc promoter in keratinocytes undergoing growth arrest, which implicated the downregulation of c-Myc gene expression as a possible reason for the suppression of cell growth by plakoglobin.

Plakoglobin-mediated regulation of gene expression has also been shown in renal carcinoma cells. Shtutman et al. [101] found that the exogenous expression of plakoglobin in cells lacking both β-catenin and plakoglobin resulted in the increased expression of the tumor suppressor gene PML, a nuclear protein that forms nuclear bodies and is involved in the regulation of p53 activity. Importantly, the increased PML levels due to plakoglobin expression were independent of β-catenin and Tcf, since β-catenin was not detected in the plakoglobin-expressing cells and the deletion of Tcf/Lef sites in the PML promoter did not affect the ability of plakoglobin to increase PML gene expression. Together, these observations suggest that plakoglobin may regulate gene expression independent of Tcf/Lef.

In β-catenin null mesothelioma and colon carcinoma cells, Wnt3a stimulation led to the nuclear accumulation of plakoglobin and induced the expression of the antiapoptotic gene Survivin [85]. Coimmunoprecipitation and chromatin immunoprecipitation showed that plakoglobin formed a transcriptional complex with both Tcf and the histone acetyltransferase CBP and that this complex was associated with the Survivin promoter [85]. While this study clearly demonstrated that plakoglobin was capable of regulating β-catenin target genes in a β-catenin null background, it is again of importance to emphasize that β-catenin null tumors are very rare and that the plakoglobin-mediated regulation of gene expression occurs mainly in the presence of cellular β-catenin.

As previously discussed, Todorović et al. [98] have shown that plakoglobin can regulate cell motility by regulating Fibronectin and Rho-dependent Src signaling. This study also demonstrated that plakoglobin expression resulted in increased levels of Fibronectin mRNA without increasing expression from the Fibronectin promoter. However, by using Actinomycin D to inhibit transcription, the authors were able to demonstrate that plakoglobin expression led to the increased stability of Fibronectin mRNA, suggesting that in addition to its role in regulating gene expression at the level of transcription, plakoglobin may also regulate gene expression posttranscriptionally. However, the mechanisms underlying this action remain unclear. Overall, these studies suggest that plakoglobin regulates gene expression at the transcriptional and potentially at posttranscriptional levels.

6. Plakoglobin Expression in Human Tumors

The initial characterization of *JUP*, the gene encoding plakoglobin, mapped the gene to chromosome 17q21, proximal to the *BRCA1* gene [102]. In this study, the authors also analyzed RNA isolated from ovarian and breast cancer tumors and showed that loss of heterozygosity in these tumors and low-frequency mutations in the plakoglobin gene predisposed patients to familial breast and ovarian cancer. Since then, several groups have observed the loss of plakoglobin expression in a wide range of tumors, with the majority of these studies examining plakoglobin in conjunction with other adhesive junctional proteins. These studies have demonstrated that loss of plakoglobin expression in conjunction with the lack of expression of other cell-cell adhesion proteins such as E-cadherin, α-catenin, β-catenin, desmoglein, or desmoplakin resulted in increased tumor formation and size and was correlated with increased tumor stage, poor patient survival, and increased metastasis in bladder, pituitary, oral, pharyngeal, skin, prostate, and NSCLC tumors [103–111]. However, several studies have found that decreased levels of plakoglobin alone also occur in various tumors.

The loss of plakoglobin expression has been observed in melanocytic and thyroid tumors [112, 113]. Cerrato et al. [113] found that nearly 90% of papillary and follicular tumors showed decreased or loss of membrane plakoglobin localization. Decreased expression of the plakoglobin gene was also observed in prostate tumors, where methylation of the plakoglobin gene is prevalent in localized prostate cancer when compared to benign prostatic hyperplasia, suggesting that loss of plakoglobin expression was an early step in prostate tumorigenesis [89]. In oropharynx squamous cell carcinomas, decreased plakoglobin expression as well as its abnormal cytoplasmic distribution was correlated with increased tumor size and poor clinical outcome [114].

In colon carcinomas, Lifschitz-Mercer et al. [115] showed that β-catenin accumulated in the nuclei of cells of primary and metastatic adenocarcinoma and adenoma lesions, while the levels of nuclear plakoglobin were decreased in these tumors, suggesting that nuclear plakoglobin did not promote tumorigenesis in the colon. In esophageal cancers, while decreased levels of E-cadherin and plakoglobin were associated with poor differentiation and decreased patient survival, reduced plakoglobin levels alone correlated with lymph node metastasis [116]. The finding that reduced plakoglobin levels alone correlated with increased metastasis was not limited to esophageal tumors. In renal carcinomas, decreased plakoglobin levels have been associated with metastasis, and patients with tumors expressing plakoglobin showed significantly higher survival rates than those that did not [117]. Aberrant or decreased plakoglobin levels have also been reported in Wilms' tumors and soft tissue sarcomas, where the decrease in plakoglobin was associated with increased risk of pulmonary metastasis [118, 119]. In endometrial tumors, the aberrant expression of plakoglobin was correlated with myometrial invasion [120], whereas medulloblastoma tumors expressing plakoglobin were nonmetastatic, with no evidence of subarachnoid or hematogenous metastasis [121]. Finally, reduced plakoglobin expression was also correlated with increased lymph node metastasis in oral squamous cell and bladder tumors [122, 123]. Collectively, these observations suggest that lack or decreased expression of plakoglobin due to genetic or epigenetic causes in tumors of different origins is associated with poor clinical outcome and increased tumor formation and metastasis.

7. Growth/Metastasis Inhibitory Activities of Plakoglobin via Regulation of Gene Expression

We have developed two experimental model systems using squamous and breast carcinoma cell lines with no or very low plakoglobin expression and various degrees of transformation/invasiveness to specifically assess the growth/metastasis inhibitory activities of plakoglobin. Using a combination of molecular and cell biological approaches, including proteomics and transcriptome analysis, we compared the protein and mRNA profiles of plakoglobin-deficient and plakoglobin-expressing cell lines and their *in vitro* migration and invasiveness. These analyses led to the identification of several growth regulatory genes that were differentially expressed in plakoglobin-expressing transfectants compared to their plakoglobin-deficient parental cells.

Comparison of the proteomic profiles of plakoglobin null SCC9 cells and their plakoglobin-expressing transfectants allowed us to identify several tumor/metastasis regulating proteins, which were differentially expressed in plakoglobin-expressing transfectants (SCC9-PG-WT) relative to parental SCC9 cells. We performed RNA microarray experiments to determine whether changes in gene expression upon plakoglobin expression accompanied these changes in protein levels and compared the transcriptome profiles of SCC9 cells and SCC9-PG-WT transfectants. Furthermore, to determine whether the subcellular distribution of plakoglobin had an effect on gene expression, we also compared the RNA profiles of SCC9 and SCC9-PG-WT cells with those of SCC9 cells transfected either with cDNAs-encoding plakoglobin fused with a nuclear localization signal (NLS) to express plakoglobin exclusively in the nucleus (SCC9-PG-NLS), or cDNAs-encoding plakoglobin fused with a nuclear export signal (NES) to express plakoglobin exclusively in the cytoplasm (SCC9-PG-NES). From these experiments, we identified three subsets of genes that were differentially expressed based on plakoglobin expression and its subcellular distribution: those whose differential expression required exclusively cytoplasmic plakoglobin, those whose differential expression required nuclear plakoglobin, and those whose differential expression required the ability of plakoglobin to shuttle between the nucleus and the cytoplasm. Based on the results of these experiments and analysis of the expression patterns of plakoglobin-target genes in relation to plakoglobin subcellular distribution, we propose that plakoglobin can regulate gene expression by three concurrent mechanisms (Figure 3).

The first of these mechanisms involves the action of plakoglobin in the cytoplasm, where it would sequester a protein involved in the regulation of gene expression. In this case, plakoglobin would prevent an inhibitor of a tumor suppressor gene or a promoter of an oncogenic gene from entering the nucleus and affecting gene expression.

Plakoglobin target genes whose expression patterns were similar in SCC9-PG-WT and SCC9-PG-NES cells and were opposite to SCC9-PG-NLS cells would be considered part of this group.

The second mechanism involves nuclear localized plakoglobin, which would directly associate with a nuclear factor and regulate gene expression. In this case, plakoglobin would interact with a transcriptional activator and promote gene expression, or, conversely, it would interact with a transcriptional repressor and silence gene expression. Plakoglobin target genes whose expression patterns were similar in SCC9-PG-WT and SCC9-PG-NLS cells and were opposite to SCC9-PG-NES cells would be considered part of this group.

The vast majority of plakoglobin target genes, however, belonged to the third group of genes: those whose differential expression depended on the ability of plakoglobin to shuttle between the nucleus and the cytoplasm. In this case, plakoglobin would interact with some cytoplasmic cofactor, translocate into the nucleus, and regulate gene expression. Plakoglobin target genes whose expression patterns were similar in SCC9-PG-NES and SCC9-PG-NLS cells and were opposite to SCC9-PG-WT cells would be considered part of this group.

Following these proteomics and microarray analyses, we began our initial characterization of the regulation of potential target genes by plakoglobin. We have recently shown that plakoglobin expression in SCC9 cells resulted in the increased expression of the metastasis suppressors Nm23-H1 and -H2, both at the mRNA and protein levels [124]. Nm23 was the first metastasis suppressor identified, as it is often downregulated in metastatic tumors and its expression in invasive cell lines resulted in decreased migration and invasion (for review, see [125, 126]). We have observed that plakoglobin interacted with Nm23-H1 and -H2 in squamous cell, mammary, renal, and colon epithelial cell lines with the colocalization of these two proteins at sites of cell-cell contact. We have also shown that these interactions occurred in both the cytoskeleton-associated and soluble pool of proteins, suggesting that these interactions have both adhesive and nonadhesive functions. Since plakoglobin was detected in the nucleus of plakoglobin-expressing SCC9 cells and since luciferase reporter assays have shown that β-catenin/Wnt signaling is not activated in SCC9 cells [65], these results together suggested that plakoglobin regulates gene expression in SCC9 cells independent of β-catenin. We are currently characterizing whether plakoglobin directly regulates Nm23 expression. Furthermore, we have also shown that plakoglobin interacts with the transcription factor p53 and regulates the expression of a number of p53 target genes (manuscript in preparation).

8. Concluding Remarks

Recent work has demonstrated that plakoglobin has novel roles in intracellular signaling and the regulation of gene expression, in addition to its previously well-established roles in cell-cell adhesion. Plakoglobin has emerged as a tumor/metastasis suppressor protein based on evidence from

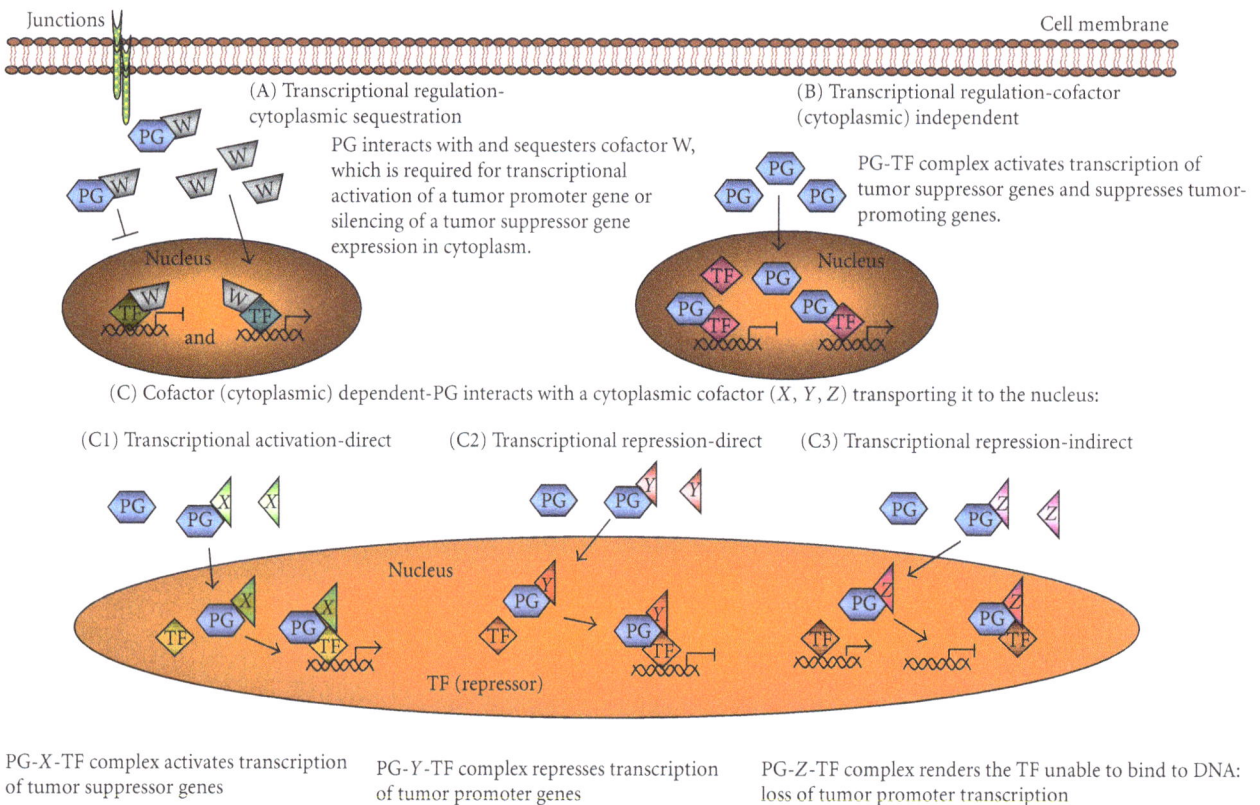

FIGURE 3: A potential model for regulation of gene expression by plakoglobin. Three concurrent mechanisms by which plakoglobin may regulate gene expression are proposed. (A) Cytoplasmic sequestration: plakoglobin sequesters a factor in the cytoplasm which, in the nucleus, suppresses the expression of a tumor suppressor gene or activates the expression of an oncogene. (B) Cytoplasmic cofactor independent: plakoglobin-transcription factor complexes promote the expression of tumor suppressor genes and repress the expression of oncogenes. (C) Cytoplasmic cofactor dependent: plakoglobin interacts with a cytoplasmic cofactor and this complex moves into the nucleus where it activates tumor suppressor gene expression or represses oncogenic gene expression. PG: plakoglobin; TF: transcription factor.

the great majority of the studies that have examined its signaling function. As more work focuses on the role of plakoglobin in tumorigenesis and metastasis, it is becoming clear that plakoglobin is a key, important player in these processes and consequently may be a useful therapeutic target in the treatment of cancer.

Acknowledgments

The authors apologize to any colleagues whose work they may have overlooked during the preparation of this paper. M. Pasdar is supported by grants from the Canadian Breast Cancer Foundation— Prairies/NWT Chapter and Alberta Cancer Foundation. Z. Aktary currently holds a University of Alberta Dissertation Fellowship.

References

[1] M. Peifer, P. D. McCrea, K. J. Green, E. Wieschaus, and B. M. Gumbiner, "The vertebrate adhesive junction proteins β-catenin and plakoglobin and the *Drosophila* segment polarity gene *armadillo* form a multigene family with similar properties," *Journal of Cell Biology*, vol. 118, no. 3, pp. 681–691, 1992.

[2] S. Butz, J. Stappert, H. Weissig et al., "Plakoglobin and β-catenin: distinct but closely related," *Science*, vol. 257, no. 5073, pp. 1142–1144, 1992.

[3] A. Ben-Ze'Ev and B. Geiger, "Differential molecular interactions of β-catenin and plakoglobin in adhesion, signaling and cancer," *Current Opinion in Cell Biology*, vol. 10, no. 5, pp. 629–639, 1998.

[4] J. Zhurinsky, M. Shtutman, and A. Ben-Ze'ev, "Plakoglobin and β-catenin: protein interactions, regulation and biological roles," *Journal of Cell Science*, vol. 113, no. 18, pp. 3127–3139, 2000.

[5] B. T. MacDonald, K. Tamai, and X. He, "Wnt/β-Catenin signaling: components, mechanisms, and diseases," *Developmental Cell*, vol. 17, no. 1, pp. 9–26, 2009.

[6] T. P. Rao and M. Kühl, "An updated overview on wnt signaling pathways: a prelude for more," *Circulation Research*, vol. 106, no. 12, pp. 1798–1806, 2010.

[7] F. Verkaar and G. J. R. Zaman, "New avenues to target Wnt/β-catenin signaling," *Drug Discovery Today*, vol. 16, pp. 35–41, 2011.

[8] I. Simcha, B. Geiger, S. Yehuda-Levenberg, D. Salomon, and A. Ben-Ze'ev, "Suppression of tumorigenicity by plakoglobin: an augmenting effect of N-cadherin," *Journal of Cell Biology*, vol. 133, no. 1, pp. 199–209, 1996.

[9] H. R. Parker, Z. Li, H. Sheinin, G. Lauzon, and M. Pasdar, "Plakoglobin induces desmosome formation and epidermoid phenotype in N- cadherin-expressing squamous

carcinoma cells deficient in plakoglobin and E- cadherin," *Cell Motility and the Cytoskeleton*, vol. 40, no. 1, pp. 87–100, 1998.

[10] E. Charpentier, R. M. Lavker, E. Acquista, and P. Cowin, "Plakoglobin suppresses epithelial proliferation and hair growth in vivo," *Journal of Cell Biology*, vol. 149, no. 2, pp. 503–519, 2000.

[11] W. W. Franke, H. Mueller, S. Mittnacht, H. P. Kapprell, and J. L. Jorcano, "Significance of two desmosome plaque-associated polypeptides of molecular weights 75 000 and 83 000," *The EMBO Journal*, vol. 2, no. 12, pp. 2211–2215, 1983.

[12] P. Cowin, H. P. Kapprell, and W. W. Franke, "Plakoglobin: a protein common to different kinds of intercellular adhering junctions," *Cell*, vol. 46, no. 7, pp. 1063–1073, 1986.

[13] N. J. Korman, R. W. Eyre, V. Klaus-Kovtun, and J. R. Stanley, "Demonstration of an adhering-junction molecule (plakoglobin) in the autoantigens of pemphigus foliaceus and pemphigus vulgaris," *The New England Journal of Medicine*, vol. 321, no. 10, pp. 631–635, 1989.

[14] D. Vestweber and R. Kemler, "Some structural and functional aspects of the cell adhesion molecular uvomorulin," *Cell Differentiation*, vol. 15, no. 2–4, pp. 269–273, 1984.

[15] N. Peyrieras, D. Louvard, and F. Jacob, "Characterization of antigens recognized by monoclonal and polyclonal antibodies directed against uvomorulin," *Proceedings of the National Academy of Sciences of the United States of America*, vol. 82, no. 23, pp. 8067–8071, 1985.

[16] M. Ozawa and R. Kemler, "Molecular organization of the uvomorulin-catenin complex," *Journal of Cell Biology*, vol. 116, no. 4, pp. 989–996, 1992.

[17] P. D. McCrea, C. W. Turck, and B. Gumbiner, "A homolog of the *armadillo* protein in *Drosophila* (plakoglobin) associated with E-cadherin," *Science*, vol. 254, no. 5036, pp. 1359–1361, 1991.

[18] K. A. Knudsen and M. J. Wheelock, "Plakoglobin, or an 83-kD homologue distinct from β-catenin, interacts with E-cadherin and N-cadherin," *Journal of Cell Biology*, vol. 118, no. 3, pp. 671–679, 1992.

[19] P. A. Piepenhagen and W. J. Nelson, "Defining E-cadherin-associated protein complexes in epithelial cells: plakoglobin, β- and γ- catenin are distinct components," *Journal of Cell Science*, vol. 104, no. 3, pp. 751–762, 1993.

[20] M. Pasdar, Z. Li, and V. Chlumecky, "Plakoglobin: kinetics of synthesis, phosphorylation, stability, and interactions with desmoglein and E-cadherin," *Cell Motility and the Cytoskeleton*, vol. 32, no. 4, pp. 258–272, 1995.

[21] W. Meng and M. Takeichi, "Adherens junction: molecular architecture and regulation," *Cold Spring Harbor Perspectives in Biology*, vol. 1, no. 6, pp. 1–13, 2009.

[22] T. J. C. Harris and U. Tepass, "Adherens junctions: from molecules to morphogenesis," *Nature Reviews Molecular Cell Biology*, vol. 11, no. 7, pp. 502–514, 2010.

[23] D. Garrod and M. Chidgey, "Desmosome structure, composition and function," *Biochimica et Biophysica Acta*, vol. 1778, no. 3, pp. 572–587, 2008.

[24] R. L. Dusek and L. D. Attardi, "Desmosomes: new perpetrators in tumour suppression," *Nature Reviews Cancer*, vol. 11, no. 5, pp. 317–323, 2011.

[25] A. Nagafuchi, Y. Shirayoshi, and K. Okazaki, "Transformation of cell adhesion properties by exogenously introduced E-cadherin cDNA," *Nature*, vol. 329, no. 6137, pp. 341–343, 1987.

[26] A. Nose, A. Nagafuchi, and M. Takeichi, "Expressed recombinant cadherins mediate cell sorting in model systems," *Cell*, vol. 54, no. 7, pp. 993–1001, 1988.

[27] R. M. Mege, F. Matsuzaki, W. J. Gallin, J. E. Goldberg, B. A. Cunningham, and G. M. Edelman, "Construction of epithelioid sheets by transfection of mouse sarcoma cells with cDNAs for chicken cell adhesion molecules," *Proceedings of the National Academy of Sciences of the United States of America*, vol. 85, no. 19, pp. 7274–7278, 1988.

[28] J. Behrens, M. M. Mareel, F. M. Van Roy, and W. Birchmeier, "Dissecting tumor cell invasion: epithelial cells acquire invasive properties after the loss of uvomorulin-mediated cell-cell adhesion," *Journal of Cell Biology*, vol. 108, no. 6, pp. 2435–2447, 1989.

[29] U. H. Frixen, J. Behrens, M. Sachs et al., "E-cadherin-mediated cell-cell adhesion prevents invasiveness of human carcinoma cells," *Journal of Cell Biology*, vol. 113, no. 1, pp. 173–185, 1991.

[30] K. Vleminckx, L. Vakaet, M. Mareel, W. Fiers, and F. Van Roy, "Genetic manipulation of E-cadherin expression by epithelial tumor cells reveals an invasion suppressor role," *Cell*, vol. 66, no. 1, pp. 107–119, 1991.

[31] W. Chen and B. Obrink, "Cell-cell contacts mediated by E-cadherin (uvomorulin) restrict invasive behavior of L-cells," *Journal of Cell Biology*, vol. 114, no. 2, pp. 319–327, 1991.

[32] Y. Shimoyama, S. Hirohashi, S. Hirano et al., "Cadherin cell-adhesion molecules in human epithelial tissues and carcinomas," *Cancer Research*, vol. 49, no. 8, pp. 2128–2133, 1989.

[33] Y. Shimoyama and S. Hirohashi, "Expression of E- and P-cadherin in gastric carcinomas," *Cancer Research*, vol. 51, no. 8, pp. 2185–2192, 1991.

[34] J. H. Schipper, U. H. Frixen, J. Behrens, A. Unger, K. Jahnke, and W. Birchmeier, "E-cadherin expression in squamous cell carcinomas of head and neck: inverse correlation with tumor dedifferentiation and lymph node metastasis," *Cancer Research*, vol. 51, no. 23, pp. 6328–6337, 1991.

[35] P. Navarro, M. Gomez, A. Pizarro, C. Gamallo, M. Quintanilla, and A. Cano, "A role for the E-cadherin cell-cell adhesion molecule during tumor progression of mouse epidermal carcinogenesis," *Journal of Cell Biology*, vol. 115, no. 2, pp. 517–533, 1991.

[36] M. M. Mareel, J. Behrens, W. Birchmeier et al., "Down-regulation of E-cadherin expression in Madin Darby canine kidney (MDCK) cells inside tumors of nude mice," *International Journal of Cancer*, vol. 47, no. 6, pp. 922–928, 1991.

[37] C. Gamallo, J. Palacios, A. Suarez et al., "Correlation of E-cadherin expression with differentiation grade and histological type in breast carcinoma," *American Journal of Pathology*, vol. 142, no. 4, pp. 987–993, 1993.

[38] P. Navarro, E. Lozano, and A. Cano, "Expression of E- or P-cadherin is not sufficient to modify the morphology and the tumorigenic behavior of murine spindle carcinoma cells. Possible involvement of plakoglobin," *Journal of Cell Science*, vol. 105, no. 4, pp. 923–934, 1993.

[39] J. E. Lewis, J. K. Wahl, K. M. Sass, P. J. Jensen, K. R. Johnson, and M. J. Wheelock, "Cross-talk between adherens junctions and desmosomes depends on plakoglobin," *Journal of Cell Biology*, vol. 136, no. 4, pp. 919–934, 1997.

[40] R. Kemler, C. Babinet, H. Eisen, and F. Jacob, "Surface antigen in early differentiation," *Proceedings of the National Academy of Sciences of the United States of America*, vol. 74, no. 10, pp. 4449–4452, 1977.

[41] B. Gumbiner and K. Simons, "A functional assay for proteins involved in establishing and epithelial occluding barrier: identification of a uvomorulin-like polypeptide," *Journal of Cell Biology*, vol. 102, no. 2, pp. 457–468, 1986.

[42] H. Hennings and K. A. Holbrook, "Calcium regulation of cell-cell contact and differentiation of epidermal cells in culture. An ultrastructural study," *Experimental Cell Research*, vol. 143, no. 1, pp. 127–142, 1983.

[43] B. Gumbiner, B. Stevenson, and A. Grimaldi, "The role of the cell adhesion molecule uvomorulin in the formation and maintenance of the epithelial junctional complex," *Journal of Cell Biology*, vol. 107, no. 4, pp. 1575–1587, 1988.

[44] M. J. Wheelock and P. J. Jensen, "Regulation of keratinocyte intercellular junction organization and epidermal morphogenesis by E-cadherin," *Journal of Cell Biology*, vol. 117, no. 2, pp. 415–425, 1992.

[45] J. E. Lewis, P. J. Jensen, and M. J. Wheelock, "Cadherin function is required for human keratinocytes to assemble desmosomes and stratify in response to calcium," *Journal of Investigative Dermatology*, vol. 102, no. 6, pp. 870–877, 1994.

[46] P. J. Jensen, B. Telegan, R. M. Lavker, and M. J. Wheelock, "E-cadherin and P-cadherin have partially redundant roles in human epidermal stratification," *Cell and Tissue Research*, vol. 288, no. 2, pp. 307–316, 1997.

[47] M. Amagai, T. Fujimori, T. Masunaga et al., "Delayed assembly of desmosomes in keratinocytes with disrupted classic-cadherin-mediated cell adhesion by a dominant negative mutant," *Journal of Investigative Dermatology*, vol. 104, no. 1, pp. 27–32, 1995.

[48] P. Ruiz, V. Brinkmann, B. Ledermann et al., "Targeted mutation of plakoglobin in mice reveals essential functions of desmosomes in the embryonic heart," *Journal of Cell Biology*, vol. 135, no. 1, pp. 215–225, 1996.

[49] Z. Li, W. J. Gallin, G. Lauzon, and M. Paadar, "L-CAM expression induces fibroblast-epidermoid transition in squamous carcinoma cells and down-regulates the endogenous N-cadherin," *Journal of Cell Science*, vol. 111, no. 7, pp. 1005–1019, 1998.

[50] H. L. Palka and K. J. Green, "Roles of plakoglobin end domains in desmosome assembly," *Journal of Cell Science*, vol. 110, no. 19, pp. 2358–2371, 1997.

[51] D. Acehan, C. Petzold, I. Gumper et al., "Plakoglobin is required for effective intermediate filament anchorage to desmosomes," *Journal of Investigative Dermatology*, vol. 128, no. 11, pp. 2665–2675, 2008.

[52] P. Gosavi, S. T. Kundu, N. Khapare, L. Sehgal, M. S. Karkhanis, and S. N. Dalal, "E-cadherin and plakoglobin recruit plakophilin3 to the cell border to initiate desmosome assembly," *Cellular and Molecular Life Sciences*, vol. 68, no. 8, pp. 1439–1454, 2011.

[53] S. T. Kundu, P. Gosavi, N. Khapare et al., "Plakophilin3 downregulation leads to a decrease in cell adhesion and promotes metastasis," *International Journal of Cancer*, vol. 123, no. 10, pp. 2303–2314, 2008.

[54] P. Ruiz and W. Birchmeier, "The plakoglobin knock-out mouse: a paradigm for the molecular analysis of cardiac cell junction formation," *Trends in Cardiovascular Medicine*, vol. 8, no. 3, pp. 97–101, 1998.

[55] R. S. Bradley, P. Cowin, and A. M. C. Brown, "Expression of Wnt-1 in PC12 cells results in modulation of plakoglobin and E-cadherin and increased cellular adhesion," *Journal of Cell Biology*, vol. 123, no. 6, pp. 1857–1865, 1993.

[56] A. Karnovsky and M. W. Klymkowsky, "Anterior axis duplication in Xenopus induced by the over-expression of the cadherin-binding protein plakoglobin," *Proceedings of the National Academy of Sciences of the United States of America*, vol. 92, no. 10, pp. 4522–4526, 1995.

[57] F. T. Kolligs, B. Kolligs, K. M. Hajra et al., "γ-Catenin is regulated by the APC tumor suppressor and its oncogenic activity is distinct from that of β-catenin," *Genes and Development*, vol. 14, no. 11, pp. 1319–1331, 2000.

[58] G. T. Bommer, C. Jäger, E. M. Dürr et al., "DRO1, a gene down-regulated by oncogenes, mediates growth inhibition in colon and pancreatic cancer cells," *Journal of Biological Chemistry*, vol. 280, no. 9, pp. 7962–7975, 2005.

[59] H. Pan, F. Gao, P. Papageorgis, H. M. Abdolmaleky, D. V. Faller, and S. Thiagalingam, "Aberrant activation of γ-catenin promotes genomic instability and oncogenic effects during tumor progression," *Cancer Biology and Therapy*, vol. 6, no. 10, pp. 1638–1643, 2007.

[60] P. J. Morin, A. B. Sparks, V. Korinek et al., "Activation of β-catenin-Tcf signaling in colon cancer by mutations in β-catenin or APC," *Science*, vol. 275, no. 5307, pp. 1787–1790, 1997.

[61] D. Salomon, P. A. Sacco, S. G. Roy et al., "Regulation of β-catenin levels and localization by overexpression of plakoglobin and inhibition of the ubiquitin-proteasome system," *Journal of Cell Biology*, vol. 139, no. 5, pp. 1325–1335, 1997.

[62] I. Simcha, M. Shtutman, D. Salomon et al., "Differential nuclear translocation and transactivation potential of β-Catenin and plakoglobin," *Journal of Cell Biology*, vol. 141, no. 6, pp. 1433–1448, 1998.

[63] J. Zhurinsky, M. Shtutman, and A. Ben-Ze'ev, "Differential mechanisms of LEF/TCF family-dependent transcriptional activation by β-catenin and plakoglobin," *Molecular and Cellular Biology*, vol. 20, no. 12, pp. 4238–4252, 2000.

[64] B. O. Williams, G. D. Barish, M. W. Klymkowsky, and H. E. Varmus, "A comparative evaluation of β-catenin and plakoglobin signaling activity," *Oncogene*, vol. 19, no. 50, pp. 5720–5728, 2000.

[65] L. Li, K. Chapman, X. Hu, A. Wong, and M. Pasdar, "Modulation of the oncogenic potential of β-catenin by the subcellular distribution of plakoglobin," *Molecular Carcinogenesis*, vol. 46, no. 10, pp. 824–838, 2007.

[66] J. Teulière, M. M. Faraldo, M. Shtutman et al., "β-catenin-dependent and -independent effects of ΔN-plakoglobin on epidermal growth and differentiation," *Molecular and Cellular Biology*, vol. 24, no. 19, pp. 8649–8661, 2004.

[67] M. W. Klymkowsky, B. O. Williams, G. D. Barish, H. E. Varmus, and Y. E. Vourgourakis, "Membrane-anchored plakoglobins have multiple mechanisms of action in Wnt signaling," *Molecular Biology of the Cell*, vol. 10, no. 10, pp. 3151–3169, 1999.

[68] T. Shibata, M. Gotoh, A. Ochiai, and S. Hirohashi, "Association of plakoglobin with APC, a tumor suppressor gene product, and its regulation by tyrosine phosphorylation," *Biochemical and Biophysical Research Communications*, vol. 203, no. 1, pp. 519–522, 1994.

[69] S. Kodama, S. Ikeda, T. Asahara, M. Kishida, and A. Kikuchi, "Axin directly interacts with plakoglobin and regulates its stability," *Journal of Biological Chemistry*, vol. 274, no. 39, pp. 27682–27688, 1999.

[70] J. M. Merriam, A. B. Rubenstein, and M. W. Klymkowsky, "Cytoplasmically anchored plakoglobin induces a WNT-like

phenotype in Xenopus," *Developmental Biology*, vol. 185, no. 1, pp. 67–81, 1997.

[71] M. Ozawa, H. Terada, and C. Pedraza, "The fourth *armadillo* repeat of plakoglobin (γ-catenin) is required for its high affinity binding to the cytoplasmic domains of E-cadherin and desmosomal cadherin Dsg2, and the tumor suppressor APC protein," *Journal of Biochemistry*, vol. 118, no. 5, pp. 1077–1082, 1995.

[72] H. Aberle, H. Schwartz, H. Hoschuetzky, and R. Kemler, "Single amino acid substitutions in proteins of the *armadillo* gene family abolish their binding to α-catenin," *Journal of Biological Chemistry*, vol. 271, no. 3, pp. 1520–1526, 1996.

[73] L. L. Witcher, R. Collins, S. Puttagunta et al., "Desmosomal cadherin binding domains of plakoglobin," *Journal of Biological Chemistry*, vol. 271, no. 18, pp. 10904–10909, 1996.

[74] R. B. Troyanovsky, N. A. Chitaev, and S. M. Troyanovsky, "Cadherin binding sites of plakoglobin: localization, specificity and role in targeting to adhering junctions," *Journal of Cell Science*, vol. 109, no. 13, pp. 3069–3078, 1996.

[75] A. P. Kowalczyk, E. A. Bornslaeger, J. E. Borgwardt et al., "The amino-terminal domain of desmoplakin binds to plakoglobin and clusters desmosomal cadherin-plakoglobin complexes," *Journal of Cell Biology*, vol. 139, no. 3, pp. 773–784, 1997.

[76] M. Hatzfeld, K. J. Green, and H. Sauter, "Targeting of p0071 to desmosomes and adherens junctions is mediated by different protein domains," *Journal of Cell Science*, vol. 116, no. 7, pp. 1219–1233, 2003.

[77] S. Hakimelahi, H. R. Parker, A. J. Gilchrist et al., "Plakoglobin regulates the expression of the anti-apoptotic protein BCL-2," *Journal of Biological Chemistry*, vol. 275, no. 15, pp. 10905–10911, 2000.

[78] R. Rosin-Arbesfeld, A. Cliffe, T. Brabletz, and M. Bienz, "Nuclear export of the APC tumour suppressor controls β-catenin function in transcription," *The EMBO Journal*, vol. 22, no. 5, pp. 1101–1113, 2003.

[79] T. C. He, A. B. Sparks, C. Rago et al., "Identification of c-MYC as a target of the APC pathway," *Science*, vol. 281, no. 5382, pp. 1509–1512, 1998.

[80] C. Zhou, S. Liu, X. Zhou et al., "Overexpression of human pituitary tumor transforming gene (hPTTG), is regulated by β-catenin /TCF pathway in human esophageal squamous cell carcinoma," *International Journal of Cancer*, vol. 113, no. 6, pp. 891–898, 2005.

[81] D. ten Berge, W. Koole, C. Fuerer, M. Fish, E. Eroglu, and R. Nusse, "Wnt signaling mediates self-organization and axis formation in embryoid bodies," *Cell Stem Cell*, vol. 3, no. 5, pp. 508–518, 2008.

[82] M. E. Conacci-Sorrell, T. Ben-Yedidia, M. Shtutman, E. Feinstein, P. Einat, and A. Ben-Ze'ev, "Nr-CAM is a target gene of the β-catenin/LEF-1 pathway in melanoma and colon cancer and its expression enhances motility and confers tumorigenesis," *Genes and Development*, vol. 16, no. 16, pp. 2058–2072, 2002.

[83] O. Maeda, N. Usami, M. Kondo et al., "Plakoglobin (γ-catenin) has TCF/LEF family-dependent transcriptional activity in β-catenin-deficient cell line," *Oncogene*, vol. 23, no. 4, pp. 964–972, 2004.

[84] M. Shimizu, Y. Fukunaga, J. Ikenouchi, and A. Nagafuchi, "Defining the roles of β-catenin and plakoglobin in LEF/T-cell factor-dependent transcription using β-catenin/plakoglobin-null F9 cells," *Molecular and Cellular Biology*, vol. 28, no. 2, pp. 825–835, 2008.

[85] Y. M. Kim, H. Ma, V. G. Oehler et al., "The Gamma catenin/CBP complex maintains survivin transcription in β-catenin deficient/depleted cancer cells," *Current Cancer Drug Targets*, vol. 11, no. 2, pp. 213–225, 2011.

[86] R. A. Winn, R. M. Bremnes, L. Bemis et al., "γ-catenin expression is reduced or absent in a subset of human lung cancers and re-expression inhibits transformed cell growth," *Oncogene*, vol. 21, no. 49, pp. 7497–7506, 2002.

[87] E. Pötter, S. Braun, U. Lehmann, and G. Brabant, "Molecular cloning of a functional promoter of the human plakoglobin gene," *European Journal of Endocrinology*, vol. 145, no. 5, pp. 625–633, 2001.

[88] J. Husmark, N. E. Heldin, and M. Nilsson, "N-cadherin-mediated adhesion and aberrant catenin expression in anaplastic thyroid-carcinoma cell lines," *International Journal of Cancer*, vol. 83, no. 5, pp. 692–699, 1999.

[89] H. Shiina, J. E. Breault, W. W. Basset et al., "Functional loss of the γ-catenin gene through epigenetic and genetic pathways in human prosthate cancer," *Cancer Research*, vol. 65, no. 6, pp. 2130–2138, 2005.

[90] D. Canes, G. J. Chiang, B. R. Billmeyer et al., "Histone deacetylase inhibitors upregulate plakoglobin expression in bladder carcinoma cells and display antineoplastic activity in vitro and in vivo," *International Journal of Cancer*, vol. 113, no. 5, pp. 841–848, 2005.

[91] F. Rahnama, F. Shafiei, P. D. Gluckman, M. D. Mitchell, and P. E. Lobie, "Epigenetic regulation of human trophoblastic cell migration and invasion," *Endocrinology*, vol. 147, no. 11, pp. 5275–5283, 2006.

[92] F. Shafiei, F. Rahnama, L. Pawella, M. D. Mitchell, P. D. Gluckman, and P. E. Lobie, "DNMT3A and DNMT3B mediate autocrine hGH repression of plakoglobin gene transcription and consequent phenotypic conversion of mammary carcinoma cells," *Oncogene*, vol. 27, no. 18, pp. 2602–2612, 2008.

[93] R. L. Dusek, L. M. Godsel, F. Chen et al., "Plakoglobin deficiency protects keratinocytes from apoptosis," *Journal of Investigative Dermatology*, vol. 127, no. 4, pp. 792–801, 2007.

[94] H. Nagashima, M. Okada, C. Hidai, S. Hosoda, H. Kasanuki, and M. Kawana, "The role of cadherin-catenin-cytoskeleton complex in angiogenesis: antisense oligonucleotide of plakoglobin promotes angiogenesis in vitro, and protein kinase C (PKC) enhances angiogenesis through the plakoglobin signaling pathway," *Heart and Vessels*, vol. 12, no. 12, pp. 110–112, 1997.

[95] S. Mukhina, H. C. Mertani, K. Guo, K. O. Lee, P. D. Gluckman, and P. E. Lobie, "Phenotypic conversion of human mammary carcinoma cells by autocrine human growth hormone," *Proceedings of the National Academy of Sciences of the United States of America*, vol. 101, no. 42, pp. 15166–15171, 2004.

[96] K. M. Rieger-Christ, L. Ng, R. S. Hanley et al., "Restoration of plakoglobin expression in bladder carcinoma cell lines suppresses cell migration and tumorigenic potential," *British Journal of Cancer*, vol. 92, no. 12, pp. 2153–2159, 2005.

[97] T. Yin, S. Getsios, R. Caldelari et al., "Plakoglobin supresses keratinocyte motility through both cell-cell adhesion-dependent and -independent mechanisms," *Proceedings of the National Academy of Sciences of the United States of America*, vol. 102, no. 15, pp. 5420–5425, 2005.

[98] V. Todorović, B. V. Desai, M. J. S. Patterson et al., "Plakoglobin regulates cell motility through Rho- and fibronectin-dependent Src signaling," *Journal of Cell Science*, vol. 123, no. 20, pp. 3576–3586, 2010.

[99] L. Williamson, N. A. Raess, R. Caldelari et al., "Pemphigus vulgaris identifies plakoglobin as key suppressor of c-Myc in the skin," *The EMBO Journal*, vol. 25, no. 14, pp. 3298–3309, 2006.

[100] S. Miravet, J. Piedra, F. Miró, E. Itarte, A. G. De Herreros, and M. Duñach, "The transcriptional factor Tcf-4 contains different binding sites for β-catenin and plakoglobin," *Journal of Biological Chemistry*, vol. 277, no. 3, pp. 1884–1891, 2002.

[101] M. Shtutman, J. Zhurinsky, M. Oren, E. Levina, and A. Ben-Ze'ev, "PML is a target gene of β-catenin and plakoglobin, and coactivates β-catenin-mediated transcription," *Cancer Research*, vol. 62, no. 20, pp. 5947–5954, 2002.

[102] H. Aberle, C. Bierkamp, D. Torchard et al., "The human plakoglobin gene localizes on chromosome 17q21 and is subjected to loss of heterozygosity in breast and ovarian cancers," *Proceedings of the National Academy of Sciences of the United States of America*, vol. 92, no. 14, pp. 6384–6388, 1995.

[103] K. N. Syrigos, K. Harrington, J. Waxman, T. Krausz, and M. Pignatelli, "Altered γ-catenin expression correlates with poor survival in patients with bladder cancer," *Journal of Urology*, vol. 160, no. 5, pp. 1889–1893, 1998.

[104] A. Clairotte, I. Lascombe, S. Fauconnet et al., "Expression of E-cadherin and α-, β-, γ-catenins in patients with bladder cancer: identification of γ-catenin as a new prognostic marker of neoplastic progression in T1 superficial urothelial tumors," *American Journal of Clinical Pathology*, vol. 125, no. 1, pp. 119–126, 2006.

[105] V. Tziortzioti, K. H. Ruebel, T. Kuroki, L. Jin, B. W. Scheithauer, and R. V. Lloyd, "Analysis of β-catenin mutations and α-, β-, and γ-catenin expression in normal and neoplastic human pituitary tissues," *Endocrine Pathology*, vol. 12, no. 2, pp. 125–136, 2001.

[106] H. Tada, M. Hatoko, A. Tanaka, M. Kuwahara, and T. Muramatsu, "Expression of desmoglein I and plakoglobin in skin carcinomas," *Journal of Cutaneous Pathology*, vol. 27, no. 1, pp. 24–29, 2000.

[107] L. L. Muzio, S. Staibano, G. Pannone et al., "Beta- and gamma-catenin expression in oral squamous cell carcinomas," *Anticancer Research*, vol. 19, no. 5 B, pp. 3817–3826, 1999.

[108] N. Morita, H. Uemura, K. Tsumatani et al., "E-cadherin and α-, β- and γ-catenin expression in prostate cancers: correlation with tumour invasion," *British Journal of Cancer*, vol. 79, no. 11-12, pp. 1879–1883, 1999.

[109] G. Ueda, H. Sunakawa, K. Nakamori et al., "Aberrant expression of β- and γ-catenin is an independent prognostic marker in oral squamous cell carcinoma," *International Journal of Oral and Maxillofacial Surgery*, vol. 35, no. 4, pp. 356–361, 2006.

[110] J. Depondt, A. H. Shabana, S. Florescu-Zorila, P. Gehanno, and N. Forest, "Down-regulation of desmosomal molecules in oral and pharyngeal squamous cell carcinomas as a marker for tumour growth and distant metastasis," *European Journal of Oral Sciences*, vol. 107, no. 3, pp. 183–193, 1999.

[111] R. M. Bremnes, R. Veve, E. Gabrielson et al., "High-throughput tissue microarray analysis used to evaluate biology and prognostic significance of the E-cadherin pathway in non-small-cell lung cancer," *Journal of Clinical Oncology*, vol. 20, no. 10, pp. 2417–2428, 2002.

[112] D. S. A. Sanders, K. Blessing, G. A. R. Hassan, R. Bruton, J. R. Marsden, and J. Jankowski, "Alterations in cadherin and catenin expression during the biological progression of melanocytic tumours," *Journal of Clinical Pathology*, vol. 52, no. 3, pp. 151–157, 1999.

[113] A. Cerrato, F. Fulciniti, A. Avallone, G. Benincasa, L. Palombini, and M. Grieco, "Beta- and gamma-catenin expression in thyroid carcinomas," *Journal of Pathology*, vol. 185, no. 3, pp. 267–272, 1998.

[114] S. Papagerakis, A. H. Shabana, J. Depondt, L. Pibouin, C. Blin-Wakkach, and A. Berdal, "Altered Plakoglobin Expression at mRNA and Protein Levels Correlates with Clinical Outcome in Patients with Oropharynx Squamous Carcinomas," *Human Pathology*, vol. 35, no. 1, pp. 75–85, 2004.

[115] B. Lifschitz-Mercer, R. Amitai, B. B. S. Maymon et al., "Nuclear localization of β-catenin and plakoglobin in primary and metastatic human colonic carcinomas, colonic adenomas, and normal colon," *International Journal of Surgical Pathology*, vol. 9, no. 4, pp. 273–279, 2001.

[116] Y. C. Lin, M. Y. Wu, D. R. Li, X. Y. Wu, and R. M. Zheng, "Prognostic and clinicopathological features of E-cadherin, α -catenin, β-catenin, γ-catenin and cyclin D1 expression in human esophageal squamous cell carcinoma," *World Journal of Gastroenterology*, vol. 10, no. 22, pp. 3235–3239, 2004.

[117] A. Buchner, R. Oberneder, R. Riesenberg, E. Keiditsch, and A. Hofstetter, "Expression of plakoglobin in renal cell carcinoma," *Anticancer Research*, vol. 18, no. 6 A, pp. 4231–4235, 1998.

[118] G. Basta-Jovanovic, E. Gvozdenovic, I. Dimitrijević et al., "Immunohistochemical analysis of gamma catenin in Wilms' tumors," *Fetal and Pediatric Pathology*, vol. 27, no. 2, pp. 63–70, 2008.

[119] Y. Kanazawa, Y. Ueda, M. Shimasaki et al., "Down-regulation of Plakoglobin in soft tissue sarcoma is associated with a higher risk of pulmonary metastasis," *Anticancer Research*, vol. 28, no. 2 A, pp. 655–664, 2008.

[120] Y. T. Kim, E. K. Choi, J. W. Kim, D. K. Kim, S. H. Kim, and W. I. Yang, "Expression of E-cadherin and α-, β-, γ-catenin proteins in endometrial carcinoma," *Yonsei Medical Journal*, vol. 43, no. 6, pp. 701–711, 2002.

[121] K. Misaki, K. Marukawa, Y. Hayashi et al., "Correlation of γ-catenin expression with good prognosis in medulloblastomas," *Journal of Neurosurgery*, vol. 102, no. 2, pp. 197–206, 2005.

[122] M. Närkiö-Mäkelä, M. Pukkila, E. Lagerstedt et al., "Reduced γ-catenin expression and poor survival in oral squamous cell carcinoma," *Archives of Otolaryngology—Head and Neck Surgery*, vol. 135, no. 10, pp. 1035–1040, 2009.

[123] E. Baumgart, M. S. Cohen, B. S. Neto et al., "Identification and prognostic significance of an epithelial-mesenchymal transition expression profile in human bladder tumors," *Clinical Cancer Research*, vol. 13, no. 6, pp. 1685–1694, 2007.

[124] Z. Aktary, K. Chapman, L. Lam et al., "Plakoglobin interacts with and increases the protein levels of metastasis suppressor Nm23-H2 and regulates the expression of Nm23-H1," *Oncogene*, vol. 29, no. 14, pp. 2118–2129, 2010.

[125] P. S. Steeg, C. E. Horak, and K. D. Miller, "Clinical-translational approaches to the Nm23-H1 metastasis Suppressor," *Clinical Cancer Research*, vol. 14, no. 16, pp. 5006–5012, 2008.

[126] J. C. Marshall, J. Collins, N. Marino, and P. Steeg, "The Nm23-H1 metastasis suppressor as a translational target," *European Journal of Cancer*, vol. 46, no. 7, pp. 1278–1282, 2010.

13

Alternative Macroautophagic Pathways

Katrin Juenemann and Eric A. Reits

Department of Cell Biology and Histology, Academic Medical Center, Meibergdreef 9, 1105 AZ Amsterdam, The Netherlands

Correspondence should be addressed to Katrin Juenemann, k.junemann@amc.uva.nl and Eric A. Reits, e.a.reits@amc.uva.nl

Academic Editor: Kim Finley

Macroautophagy is a bulk degradation process that mediates the clearance of long-lived proteins, aggregates, or even whole organelles. This process includes the formation of autophagosomes, double-membrane structures responsible for delivering cargo to lysosomes for degradation. Currently, other alternative autophagy pathways have been described, which are independent of macroautophagic key players like Atg5 and Beclin 1 or the lipidation of LC3. In this review, we highlight recent insights in indentifying and understanding the molecular mechanism responsible for alternative autophagic pathways.

1. Introduction

Autophagy, which is highly conserved from yeast to human, is a cellular degradation pathway that delivers cytoplasmic substrates to lysosomes for subsequent degradation. In contrast to the Ubiquitin-Proteasome System (UPS), which directly degrades monomeric proteins in the cytoplasm or nucleus, autophagy targets a wide spectrum of substrates including long-lived proteins, protein aggregates, and organelles towards lysosomes for subsequent degradation. In mammalian cells, autophagy occurs under basal conditions but can be stimulated by various stress conditions including starvation, hypoxia, and treatment with apoptosis-inducing compounds like rapamycin. In addition to its role in maintaining cellular homeostasis, autophagy is implicated in a wide range of physiological and pathological conditions, including early embryological development, clearance of pathogens, tumor suppression, and antigen processing and presentation [1]. In order to target cytoplasmic proteins to the lysosomes, several autophagic pathways exist, including microautophagy, chaperone-mediated autophagy (CMA), and macroautophagy. While micro- and macroautophagy can occur both in eukaryotes, plants, and fungi, CMA has only been observed in mammals. Microautophagy is the direct engulfment of cytoplasm or whole organelles by invagination or protrusion of arm-like structures of the lysosomal membrane. Here, the sequestration of cytoplasmic cargo occurs directly at the vacuole surface [2–5]. The second type of autophagy is CMA, which selectively degrades specific cytosolic proteins containing a pentapeptide motif (KFERQ) that is recognized by the heat shock cognate protein 70 (Hsc70) [6, 7]. The chaperone-substrate complex subsequently binds the lysosome through interaction with the receptor Lamp-2a on the lysosomal membrane [8]. Upon delivery by Hsc70, the substrate protein is unfolded before crossing the lysosomal membrane and lysosomal Hsc70 pulls the substrate into the lysosomal matrix where it is degraded by proteases [9]. The last but main type of autophagy is macroautophagy. Here, double-membrane vesicles, termed autophagosomes, are formed and sequester portions of cytosolic content or intact organelles (such as mitochondria) [10]. These autophagosomes are subsequently transported in a dynein-dependent manner along microtubules and fuse with endosomes or directly with lysosomes to form autolysosomes, resulting in breakdown of their contents by hydrolytic enzymes [11]. Macroautophagy is the major cellular pathway to recycle cell components including long-lived proteins and organelles, thereby providing nutrients for the eukaryotic cell, and it is activated under nutrient starvation. Additionally, macroautophagy is essential for development, cell survival, and tissue-specific processes [12, 13]. The initiation of autophagosome formation starts with the phagophore (autophagosome precursor), and recent studies indicate that the source of the membrane is the endoplasmatic reticulum (ER) [14, 15]. However, alternative sources for the autophagosomal membrane have been proposed, including the Golgi apparatus, and therefore the origin of the phagophore membrane still remains unresolved [16, 17].

2. Macroautophagy

Macroautophagy is a multistep process controlled by proteins termed autophagy-related (Atg) proteins [18]. The formation of the phagophore requires the class-III-phosphatidylinositol 3-kinase (PI3K) Vps34 that forms a complex with Beclin 1 (the mammalian orthologue of yeast Atg6). Inhibitors of Vps34 such as methyladenine (3-MA) or wortmannin can be used to inhibit macroautophagy since they prevent autophagosome nucleation [19–22]. The elongation of the autophagosomal membrane is dependent on two ubiquitin-like conjugation systems [23]. Atg5-Atg12 controls autophagy, where Atg12 is conjugated to Atg5 in a step that requires Atg7 (ubiquitin-activating-enzyme (E1)-like) and Atg10 (ubiquitin-conjugating-enzyme (E2)-like). The Atg5-Atg12 conjugation depends on Vps34 activity and is localized onto the phagophore where it dissociates upon formation of the autophagosome. Atg5-Atg12 forms a complex with Atg16L that modulates the next process, the ubiquitin-like conjugation of LC3-I (mammalian orthologue of Atg8). The protein LC3 is proteolytic activated by Atg4, which cleaves the C-terminus of LC3, thereby generating a cytosolic LC3-I, which subsequently conjugates with phosphatidylethanolamine (PE) to form membrane-associated LC3-II [24]. This process requires Atg7 and Atg3, and the Atg16L complex modulates the LC3-I lipidation by acting like an E3-like enzyme [25]. Although the Atg5-Atg12 conjugation dissociates upon completion of the autophagosome formation, LC3-II persists with the autophagosomal membrane even after fusion with a lysosomes and is regarded as a key marker for autophagosomes. Atg4 is also involved in the deconjugation reaction of LC3-II, as Atg4 delipidates LC3-II and removes it from the autophagosomal membrane [24, 26]. A pathway that negatively regulates macroautophagy is controlled by mTOR (mammalian target of rapamycin). mTOR activity is inhibited under starvation conditions, which activates starvation-induced macroautophagy. Recently, two new key regulators of macroautophagy, named NIX and DOR, which directly interact with the autophagosome-membrane-associated protein LC3, were identified [27]. Nix, a Bcl2-related protein localized the outer mitochondrial membrane, has a function as an adaptor protein and recruits autophagic components to mitochondria via its WXXL-like domain facing the cytoplasm [28–30]. NIX is upregulated during erythroid differentiation where a lack of mitochondria is achieved by mitophagy [27, 31, 32]. Interestingly, NIX-deficient mice show remaining mitochondria in matured red blood cells suggesting that NIX is a selective autophagy receptor that mediates mitochondrial clearance, as it directly binds LC3, but it may also target mitochondria for degradation in an LC3-independent manner [27, 33, 34]. Intriguingly, in the same issue of EMBO reports, another new autophagy-related protein was reported. Mauvezin et al. identified the nuclear cofactor of thyroid hormone receptors, termed DOR (diabetes- and obesity-regulated gene), as a new player of macroautophagy [35]. Stress-induced macroautophagy by starvation or rapamycin leads to release of DOR from the nucleus in DOR-transfected HeLa cells. This relocalization was not observed in the absence of cellular stress, indicating that cellular stress is essential to trigger DOR recruitment to the cytoplasm. DOR is associated with early autophagosomes via interaction with LC3 and GATE16 but does not colocalize with autolysosomes suggesting that DOR has a regulatory role in recruiting substrates for autophagic clearance. In addition, DOR-transfected HeLa cells show increased turnover of proteins and elevated numbers of autophagosomes compared to untreated cells. It has yet to be discovered which role DOR is playing, as it may be involved in targeting proteins to autophagy or in the formation and nucleation of the autophagosome. Whether DOR activation affects autophagy-induced alterations in cell survival remains to be established.

Macroautophagy was originally described to target intracellular organelles such as mitochondria and big protein complexes, but over the years it became clear that also most long-lived proteins are degraded via autophagic pathways. In contrast, the other main degradation machinery in the cell, the UPS, degrades mainly soluble short-lived and misfolded proteins that are targeted to the proteasome following ubiquitination (using a series of E1-E2-E3 enzymes to specifically target proteins for destruction). The proteasome is present in both the cytoplasm and the nucleus and can unfold and degrade single proteins into small peptide fragments that are subsequently recycled by peptidases. Interestingly, impairment of the proteasome leads to an increase in macroautophagy, indicating that macroautophagy can target accumulating ubiquitinated proteasomal clients when required [36–39]. In contrast, impairment of macroautophagy does not lead to increased proteasome activity. Inhibition of macroautophagy does not affect the catalytic activity of the proteasome but results in the accumulation of the macroautophagy cargo receptor p62 (also termed SQSTM1) which competes with the proteasome for ubiquitinated substrates. Indeed, silencing of p62 increases the amount of UPS clients, whereas overexpression of p62 inhibits degradation of the proteasomal substrates p53 and Ub^{G76V}-GFP [40, 41]. As p62 links ubiquitinated proteins via its ubiquitin-associated (UBA) domain to the autophagic protein LC3-II and is itself degraded in the process, inhibition of macroautophagy leads to p62 accumulation which will compete and frustrate other ubiquitin-binding proteins that participate in proteasome-mediated degradation.

3. Alternative Autophagic Pathways

Failure of the UPS or autophagic pathways to efficiently clear proteins leads to the accumulation and subsequent aggregation of these proteins, which is a hallmark of various neurodegenerative disorders including polyglutamine (polyQ) disorders such as Huntington's disease. Here, fragments of the disease-related protein containing the polyQ tract initiate aggregation and toxicity, which can be mimicked by expressing the expanded polyQ sequence as a peptide [42]. Apparently, not all peptides are efficiently degraded by peptidases, which led to our recently published study where we examined potential alternative degradation machineries when peptidases would fail in degrading protein fragments [43]. In this study, we introduced peptidase-resistant peptides into living cells and observed a perinuclear accumulation of these peptides in time. Surprisingly, these structures did not represent aggregates or inclusion bodies as observed previously for aggregation-prone protein fragments, as no UPS components or chaperones were recruited. Although initially present in the nucleus and cytoplasm, the peptides were efficiently targeted to lysosomes within a few hours upon introduction into cells, and subsequently degraded. Our results indicate, therefore, that similar to the described increase in autophagy upon proteasome impairment, a backup mechanism exists for small protein fragments that show peptidase resistance. Intriguingly, this mechanism was very efficient for peptides of the average size of proteasomal products (6–9 amino acids), but far less for extended peptides over 25–30 amino acids which remained cytoplasmic for prolonged periods [43]. Similar to expanded polyQ peptides of disease-related lengths, these expanded peptidase-resistant peptides were more resistant to clearance by lysosomes suggesting that this pathway is particularly efficient for small peptides generated by the proteasome. It is tempting to speculate that this mechanism evolved as a backup to peptidases in the clearance of proteasome-derived peptides and emphasizes the need to identify the involved proteins. Using correlative microscopy, we mainly observed double-membrane vesicles that contained peptides and that colocalized with LC3. The colocalization increased when we used Bafilomycin A1 to impair maturation into autolysosomes. In contrast, we could prevent colocalization of LC3 with the macroautophagy inhibitor 3-MA, suggesting that the macroautophagic pathway took over the clearance of these peptides. Unexpectedly, inhibition of macroautophagy by inhibitors such as 3-MA or knockdown of Atg5 prevented recruitment of LC3 but did not affect the trafficking of these peptides into lysosomes or their subsequent degradation. Apparently, LC3 was recruited during the trafficking of peptides towards lysosomes yet was not essential. Similar to the knockdown of the various LC3 isoforms (LC3A-C), knockdown of the Atg8-related GABARAP proteins, that can interact with autophagosomes, did not affect the targeting of peptides towards lysosomes [44, 45]. As knockdown of Atg5 or WIPI-1 did not affect the trafficking and subsequent degradation of peptides in lysosomes, we concluded that these peptides entered lysosomes via a pathway different from macroautophagy. CMA is also unlikely to contribute to this pathway as the peptides lack a CMA motif and peptides composed of D-amino acids, which are unable to bind chaperones like Hsc70, were also trafficking via this pathway. Finally, we also examined endosomal microautophagy, a process that delivers soluble cytosolic material to vesicles of late endosomes or multivesicular bodies (MVBs) [46, 47].

Although accumulated peptides colocalized with internalized MHC class II molecules which may lead to so-called cross-presentation to the immune system (unpublished observation), knockdown of the sorting complexes required for transport (ESCRTs) I and III showed no effect on peptide accumulation in lysosomes. As no recruitment of ESCRT regulators towards accumulated peptides was observed, this indicates that the endosomal microautophagy pathway is not involved in the trafficking and clearance of the peptidase-resistant peptides.

The accumulation and subsequent lysosomal degradation of cytoplasmic proteins independent of known autophagy pathways have been previously observed in several studies (as described below), although in each case differences in sensitivity to autophagy inhibitors and the involvement of various Atg proteins were reported. Interestingly, in a study using Green Fluorescent Protein (GFP) like fluorophores, a pathway reminiscent of that we observed for the peptidase-resistant peptides was observed [40]. Various GFP-like fluorophores have been shown to form dimers, tetramers, or even larger complexes. Upon expression, these fluorescent proteins formed cytoplasmic fluorescent puncta that resembled lysosomes, similar as observed for the peptidase-resistant peptides [48]. However, the accumulating fluorophore proteins including monomeric RFP1 (mRFP1) showed resistance to lysosomal degradation and retain fluorescence, in contrast to the peptides. Trafficking of the GFP-like proteins and the peptidase-resistant peptides was not affected in Atg5-deficient mouse embryonic fibroblasts, suggesting that they may be targeted to lysosomes by a similar pathway (although no other macroautophagy markers were examined for the fluorescent proteins). So is the constitutive macroautophagy-independent targeting of cytoplasmic proteins and peptides to autolysosomes restricted to introduced peptides and GFP-like fluorophores?

At least two alternative autophagy pathways have been described: an Atg5/Atg7-independent pathway and the so-called noncanonical autophagy pathway, which is independent of Beclin 1 (Table 1). The Atg5/Atg7-independent autophagic pathway was recently discovered in mouse embryonic fibroblasts (MEF) lacking Atg5 and Atg7 that were treated with the cytotoxic stressor etoposide, which caused an equivalent appearance of autophagic vacuoles when compared to wild-type cells [49]. Moreover, autophagic vacuoles were also found in starved Atg5-/- cells. The Atg5/Atg7-independent form of autophagy does not involve the lipidated conjugate LC3-II, which is membrane associated. Interestingly, equivalent numbers of LC3-positive and LC3-negative autophagosomes were observed in etoposide-treated wild-type cells, suggesting that conventional and alternative autophagic pathway occur at the same time. The proteins Atg5, Atg7, and LC3, which are important in the ubiquitin-like conjugation system for the autophagosome elongation, are not involved in this alternative form of autophagy. However, silencing of Beclin 1 and Vps34 decreased the amount of autophagosomes, indicating that the PI3K complex, which acts upstream of initiation of autophagosome formation, is still required in etoposide- or starvation-induced autophagy in Atg5-/- cells. Accordingly,

TABLE 1: Types of alternative macroautophagic pathways.

Alternative macroautophagic pathways	Macroautopagic molecules involved	Macroautopagic molecules not involved	Induction	Cell type	Reference
Beclin 1-independent	Atg5 Atg7 Ulk1/2 LC3	Beclin 1 (Vps34)	Resveratrol	MCF-7 (breast cancer cells)	[50]
			Staurosporine Etoposide MK801	primary cortical neurons	[51]
			H_2O_2	RAW 264.7 (macrophage cells)	[52]
			MPP+	SH-SY5Y (neuroblastoma cells)	[53]
				primary dopaminergic neurons	
			As_2O_3	ovarian cells	[55]
Atg5/Atg7-independent	Beclin 1 Vps34 Ulk1 Fip200	Atg5 Atg7 Atg9 Atg12 Atg16 LC3	Etoposide Staurosporine Starvation	Atg5-/- MEF Atg7-/- MEF wt MEF	[49]
Degradation of peptidase-resistant peptides	LC3 (but not essential)	Atg5 WIPI-1 p62 Tsg101 Vps24	Resistance against cytoplasmic peptidases	HeLa Atg5-/- MEF wt MEF	[43]

protein degradation via this pathway was inhibited by the PI3K inhibitor 3-MA. Furthermore, silencing of components of the Ulk1 complex, a mammalian serine/threonine protein kinase that plays a key role in the initial stages of autophagy, decreased autophagic vacuoles, suggesting that the Ulk1 complex is needed for Atg5/Atg7-independent autophagy [49].

Apoptosis-induced stress, for example, by staurosporine, resveratrol, or H_2O_2 can also induce the so-called non-canonical autophagy pathway, where autophagosomes can be formed independent of Beclin 1 or Vps34 and with an insensitivity to 3-MA [50–52]. However, this specific pathway still requires Atg7-activity for LC3-I lipidation and is, therefore, different from the Atg5/Atg7-independent pathway described above [49]. Furthermore, Scarlatti et al. have shown that resveratrol inhibits the mTOR activation by a direct inhibitory effect on the upstream class 1A PI3K [50]. Similarly, a Beclin 1-independent pathway has been reported in neuronal cells treated with the neurotoxin 1-methyl-4-phenylpyridinium (MPP+) [53] and in other cellular systems in response to various drugs [54, 55]. These studies have shown that several agents stimulate autophagic cell death through Beclin 1 in canonical autophagy pathways [56]. Recently, evidence emerged that autophagy and cell death are induced independent of Beclin 1 and Vps34. In breast cancer cells, resveratrol induces autophagic cell death in a Beclin 1-independent manner [50]. Silencing of Atg7 impairs the cellular death elicited by resveratrol. In dopaminergic neuronal cells, the neutotoxin MPP+ induces Beclin 1-independent autophagy and cell death [53]. As most studies on the noncanonical pathway used compounds to induce

cell death, it is tempting to link the noncanonical autophagy pathway to a death execution mechanism or cell survival. However, it has also been suggested that the independency of the noncanonical autophagy pathway may provide an adaptation to loss of Beclin 1, for example, in various tumors where Beclin 1 is deleted, in immune cell development, and may even be an evolutionary way to circumvent inhibition of Beclin 1 by various viruses in order to prevent autophagy [57–59].

None of these alternative autophagy pathways seem to correspond to the trafficking we observed for the peptidase-resistant peptides, as the Atg5/Atg7-independent pathway is still 3-MA sensitive (in contrast to the peptide targeted to lysosomes), while the noncanonical pathway (Beclin 1-independent) is 3-MA insensitive but still depends on LC3. Thus, lysosomal degradation of peptidase-resistant peptides and proteins, as we and others have demonstrated [35, 41–45], defines a novel authophagy route independent of known regulators of the constitutive macroautophagic pathway like Beclin 1, Atg5 or LC3. A better understanding of the role of these alternative autophagic pathways and their molecular regulators raise to two crucial questions: (1) What is the origin of the autophagic membrane in the different autophagic routes, and (2) Which stimuli trigger the different autophagic pathways?

In mammalian macroautophagy, various sources for the origin of the autophagosome membrane have been proposed including the ER, the Golgi complex, the plasma membrane, and the mitochondria [17, 60–66]. Alternatively, *de novo* synthesis of a nucleating structure, the phagophore, is proposed to elongate by the addition of lipids via the integral

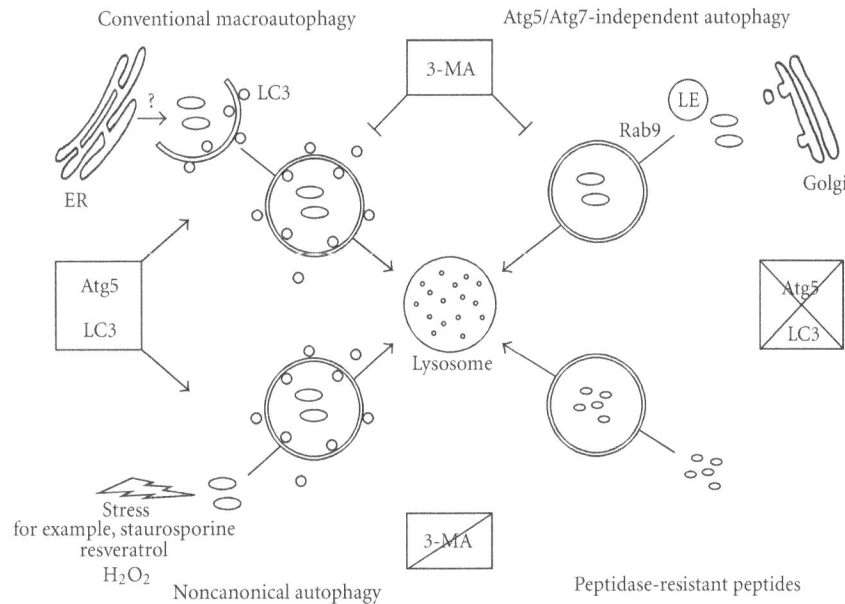

FIGURE 1: Alternative macroautophagic pathways lead to lysosomal degradation. At least four autophagic pathways can be distinguished that all show double-membrane autophagic structures and end in lysosomal degradation of cytoplasmic cargo. Conventional macroautophagy is hallmarked by the recruitment of lipidated LC3 to autophagosomal membranes that may origin from the endoplasmatic reticulum (ER). This process is dependent on Atg5 and Beclin 1 and can be inhibited by 3-methyladenine (3-MA). In contrast, the observed Atg5/Atg7-independent autophagy pathway forms Rab9-positive double-membrane vesicles derived from the *trans*-Golgi network and late endosomes (LE), and while it can be inhibited by 3-MA and is dependent on Beclin 1, the process is independent of Atg5 and LC3. Almost similar, the degradation of accumulated peptidase-resistant peptides is independent of Atg5 and LC3 and is also insensitive to 3-MA treatment. Finally, the noncanonical autophagy pathway induced by different stress factors is dependent on Atg5 and LC3 and independent of Beclin 1 but cannot be impaired by 3-MA.

membrane protein Atg9 [67–70]. Atg9 seems to be a key regulator in regulating the formation and expansion of nascent autophagosomes. Unfortunately, the identity of proteins that partition to the autophagosomal membrane remains largely unknown. Therefore, attempting to determine the origin of the autophagosomal membrane based on the associated proteins remains a challenge [71]. Alternatively, others attempted to determine the source of the autophagosomal membrane by inspecting its thickness and lipid composition [15]. Several studies reported that the autophagosomal membrane can be classified as of a thin type (6–8 nm), similar to membranes of the ER and mitochondria [60, 72–75]. Furthermore, lipid structures enriched in PI3P (known as omegasomes) were formed in the vicinity of ER membranes after amino acid starvation, suggesting that these omegasomes originate from the ER [76–79]. As the omegasomes carry autophagosomal proteins like Atg5 and LC3, they may represent the source of isolated membranes required for autophagosome expansion. In contrast, in the Atg5/Atg7-independent autophagic pathway, autophagosomes with membranes of the thick type (9-10 nm) were observed, similar to membranes of lysosomes and the *trans*-Golgi network [49]. Intriguingly, unlike the conventional pathway the alternative Atg5/Atg7-independent form of autophagy is blocked by brefeldin A, indicating that autophagosomes are derived from the Golgi-apparatus. Etoposide-induced Atg5/Atg7-independent autophagy is accompanied by colocalization of markers of the *trans*-Golgi and late endosomes (such

as the mannose 6-phosphate receptor, TGN38, and Rab9) with Lamp-2-positive autolysosomes, further pointing to the requirement of the *trans*-Golgi or late endosomes in this alternative form of autophagy. Indeed, silencing of Rab9 or expression of a Rab9 dominant negative mutant established an essential role for Rab9 in membrane expansion from isolated membranes and led to an accumulation of isolated membranes after silencing of Rab9 but not upon inhibition of Ulk1 or Beclin 1. Since the Atg5/Atg7-independent type of alternative autophagy is activated by starvation and the stress-inducing reagent etoposide, but not by rapamycin, this suggests that a specific stimulus for induction of autophagy activates nonconventional macroautophagy with different lipid structures compared to conventional macroautophagy. To the best of our knowledge, there is no clear data on the source of membrane for the Beclin 1-independent noncanonical autophagy pathway.

So far, several sources have been proposed to provide the putative moiety of autophagosomal membranes. However, autophagosomal membranes could derive from multiple membrane sources and the origin of lipids may vary dependent on the cell type, the stimulus that triggers the degradation, and the type of cargo for autophagic destruction (proteins, aggregates or even whole organelles). As shown in Figure 1, there are now at least three alternative pathways that target cytosolic content to lysosomes, which can be discriminated by their dependence on Atg5 and 3-MA (Figure 1). The identification of key players and the origin of membrane structures involved in alternative autophagic pathways will be important for the understanding of molecular mechanism regulating these various types of autophagy.

References

[1] B. Ravikumar, S. Sarkar, J. E. Davies et al., "Regulation of mammalian autophagy in physiology and pathophysiology," *Physiological Reviews*, vol. 90, no. 4, pp. 1383–1435, 2010.

[2] G. E. Mortimore, N. J. Hutson, and C. A. Surmacz, "Quantitative correlation between proteolysis and macro- and microautophagy in mouse hepatocytes during starvation and refeeding," *Proceedings of the National Academy of Sciences of the United States of America*, vol. 80, no. 8, pp. 2179–2183, 1983.

[3] E. J. De Waal, H. Vreeling-Sindelarova, and J. P. M. Schellens, "Quantitative changes in the lysosomal vacuolar system of rat hepatocytes during short-term starvation. A morphometric analysis with special reference to macro- and microautophagy," *Cell and Tissue Research*, vol. 243, no. 3, pp. 641–648, 1986.

[4] O. Müller, T. Sattler, M. Flötenmeyer, H. Schwarz, H. Plattner, and A. Mayer, "Autophagic tubes: vacuolar invaginations involved in lateral membrane sorting and inverse vesicle budding," *Journal of Cell Biology*, vol. 151, no. 3, pp. 519–528, 2000.

[5] T. Sattler and A. Mayer, "Cell-free reconstitution of microautophagic vacuole invagination and vesicle formation," *Journal of Cell Biology*, vol. 151, no. 3, pp. 529–538, 2000.

[6] J. F. Dice, H. L. Chiang, E. P. Spencer, and J. M. Backer, "Regulation of catabolism of microinjected ribonuclease A. Identification of residues 7–11 as the essential pentapeptide," *Journal of Biological Chemistry*, vol. 261, no. 15, pp. 6853–6859, 1986.

[7] J. F. Dice, "Molecular determinants of protein half-lives in eukaryotic cells," *The FASEB Journal*, vol. 1, no. 5, pp. 349–357, 1987.

[8] A. M. Cuervo and J. F. Dice, "A receptor for the selective uptake and degradation of proteins by lysosomes," *Science*, vol. 273, no. 5274, pp. 501–503, 1996.

[9] F. A. Agarraberes, S. R. Terlecky, and J. F. Dice, "An intralysosomal hsp70 is required for a selective pathway of lysosomal protein degradation," *Journal of Cell Biology*, vol. 137, no. 4, pp. 825–834, 1997.

[10] P. O. Seglen, P. B. Gordon, and I. Holen, "Non-selective autophagy," *Seminars in Cell Biology*, vol. 1, no. 6, pp. 441–448, 1990.

[11] Z. Xie and D. J. Klionsky, "Autophagosome formation: core machinery and adaptations," *Nature Cell Biology*, vol. 9, no. 10, pp. 1102–1109, 2007.

[12] N. Mizushima and B. Levine, "Autophagy in mammalian development and differentiation," *Nature Cell Biology*, vol. 12, no. 9, pp. 823–830, 2010.

[13] B. Levine and D. J. Klionsky, "Development by self-digestion: molecular mechanisms and biological functions of autophagy," *Developmental Cell*, vol. 6, no. 4, pp. 463–477, 2004.

[14] E. L. Axe, S. A. Walker, M. Manifava et al., "Autophagosome formation from membrane compartments enriched in phosphatidylinositol 3-phosphate and dynamically connected to the endoplasmic reticulum," *Journal of Cell Biology*, vol. 182, no. 4, pp. 685–701, 2008.

[15] G. Juhasz and T. P. Neufeld, "Autophagy: a forty-year search for a missing membrane source," *Plos Biology*, vol. 4, no. 2, article e36, 2006.

[16] T. Yoshimori and T. Noda, "Toward unraveling membrane biogenesis in mammalian autophagy," *Current Opinion in Cell Biology*, vol. 20, no. 4, pp. 401–407, 2008.

[17] F. Reggiori, "Membrane origin for autophagy," *Current Topics in Developmental Biology*, vol. 74, pp. 1–30, 2006.

[18] D. J. Klionsky, J. M. Cregg, W. A. Dunn et al., "A unified nomenclature for yeast autophagy-related genes," *Developmental Cell*, vol. 5, no. 4, pp. 539–545, 2003.

[19] K. B. Hendil, A. M. Lauridsen, and P. O. Seglen, "Both endocytic and endogenous protein degradation in fibroblasts is stimulated by serum/amino acid deprivation and inhibited by 3-methyladenine," *Biochemical Journal*, vol. 272, no. 3, pp. 577–581, 1990.

[20] P. O. Seglen and P. B. Gordon, "3-Methyladenine: specific inhibitor of autophagic/lysosomal protein degradation in isolated rat hepatocytes," *Proceedings of the National Academy of Sciences of the United States of America*, vol. 79, no. 6, pp. 1889–1892, 1982.

[21] Y. T. Wu, H. L. Tan, G. Shui et al., "Dual role of 3-methyladenine in modulation of autophagy via different temporal patterns of inhibition on class I and III phosphoinositide 3-kinase," *Journal of Biological Chemistry*, vol. 285, no. 14, pp. 10850–10861, 2010.

[22] E. F. Blommaart, U. Krause, J. P. M. Schellens, H. Vreeling-Sindelárová, and A. J. Meijer, "The phosphatidylinositol 3-kinase inhibitors wortmannin and LY294002 inhibit in isolated rat hepatocytes," *European Journal of Biochemistry*, vol. 243, no. 1-2, pp. 240–246, 1997.

[23] Y. Ohsumi and N. Mizushima, "Two ubiquitin-like conjugation systems essential for autophagy," *Seminars in Cell and Developmental Biology*, vol. 15, no. 2, pp. 231–236, 2004.

[24] I. Tanida, Y. S. Sou, J. Ezaki, N. Minematsu-Ikeguchi, T. Ueno, and E. Kominami, "HsAtg4B/HsApg4B/autophagin-1 cleaves the carboxyl termini of three human Atg8 homologues and delipidates microtubule-associated protein light chain 3- and GABAA receptor-associated protein-phospholipid conjugates," *Journal of Biological Chemistry*, vol. 279, no. 35, pp. 36268–36276, 2004.

[25] T. Hanada, N. N. Noda, Y. Satomi et al., "The Atg12-Atg5 conjugate has a novel E3-like activity for protein lipidation in autophagy," *Journal of Biological Chemistry*, vol. 282, no. 52, pp. 37298–37302, 2007.

[26] K. Satoo, N. N. Noda, H. Kumeta et al., "The structure of Atg4B-LC3 complex reveals the mechanism of LC3 processing and delipidation during autophagy," *The EMBO Journal*, vol. 28, no. 9, pp. 1341–1350, 2009.

[27] R. L. Schweers, J. Zhang, M. S. Randall et al., "NIX is required for programmed mitochondrial clearance during reticulocyte maturation," *Proceedings of the National Academy of Sciences of the United States of America*, vol. 104, no. 49, pp. 19500–19505, 2007.

[28] M. Schwarten, J. Mohrlüder, P. Ma et al., "Nix directly binds to GABARAP: a possible crosstalk between apoptosis and autophagy," *Autophagy*, vol. 5, no. 5, pp. 690–698, 2009.

[29] I. Novak and I. Dikic, "Autophagy receptors in developmental clearance of mitochondria," *Autophagy*, vol. 7, no. 3, pp. 301–303, 2011.

[30] I. Novak, V. Kirkin, D. G. McEwan et al., "Nix is a selective autophagy receptor for mitochondrial clearance," *EMBO Reports*, vol. 11, no. 1, pp. 45–51, 2010.

[31] M. Kundu, T. Lindsten, C. Y. Yang et al., "Ulk1 plays a critical role in the autophagic clearance of mitochondria and ribosomes during reticulocyte maturation," *Blood*, vol. 112, no. 4, pp. 1493–1502, 2008.

[32] W. Aerbajinai, M. Giattina, Y. T. Lee, M. Raffeld, and J. L. Miller, "The proapoptotic factor Nix is coexpressed with Bcl-xL during terminal erythroid differentiation," *Blood*, vol. 102, no. 2, pp. 712–717, 2003.

[33] A. Diwan, A. G. Koesters, A. M. Odley et al., "Unrestrained erythroblast development in Nix-/- mice reveals a mechanism for apoptotic modulation of erythropoiesis," *Proceedings of the National Academy of Sciences of the United States of America*, vol. 104, no. 16, pp. 6794–6799, 2007.

[34] H. Sandoval, P. Thiagarajan, S. K. Dasgupta et al., "Essential role for Nix in autophagic maturation of erythroid cells," *Nature*, vol. 454, no. 7201, pp. 232–235, 2008.

[35] C. Mauvezin, M. Orpinell, V. A. Francis et al., "The nuclear cofactor DOR regulates autophagy in mammalian and Drosophila cells," *EMBO Reports*, vol. 11, no. 1, pp. 37–44, 2010.

[36] U. B. Pandey, Z. Nie, Y. Batlevi et al., "HDAC6 rescues neurodegeneration and provides an essential link between autophagy and the UPS," *Nature*, vol. 447, no. 7146, pp. 859–863, 2007.

[37] W. X. Ding, H. M. Ni, W. Gao et al., "Linking of autophagy to ubiquitin-proteasome system is important for the regulation of endoplasmic reticulum stress and cell viability," *American Journal of Pathology*, vol. 171, no. 2, pp. 513–524, 2007.

[38] A. Iwata, B. E. Riley, J. A. Johnston, and R. R. Kopito, "HDAC6 and microtubules are required for autophagic degradation of aggregated huntingtin," *Journal of Biological Chemistry*, vol. 280, no. 48, pp. 40282–40292, 2005.

[39] Y. Du, D. Yang, L. Li et al., "An insight into the mechanistic role of p53-mediated autophagy induction in response to proteasomal inhibition-induced neurotoxicity," *Autophagy*, vol. 5, no. 5, pp. 663–675, 2009.

[40] V. I. Korolchuk, F. M. Menzies, and D. C. Rubinsztein, "A novel link between autophagy and the ubiquitin-proteasome system," *Autophagy*, vol. 5, no. 6, pp. 862–863, 2009.

[41] V. I. Korolchuk, A. Mansilla, F. M. Menzies, and D. C. Rubinsztein, "Autophagy inhibition compromises degradation of ubiquitin-proteasome pathway substrates," *Molecular Cell*, vol. 33, no. 4, pp. 517–527, 2009.

[42] M. Raspe, J. Gillis, H. Krol et al., "Mimicking proteasomal release of polyglutamine peptides initiates aggregation and toxicity," *Journal of Cell Science*, vol. 122, no. 18, pp. 3262–3271, 2009.

[43] J. M. Gillis, W. Benckhuijsen, H. van Veen, A. S. Sanz, J. W. Drijfhout, and E. A. Reits, "Aminopeptidase-resistant peptides are targeted to lysosomes and subsequently degraded," *Traffic*, vol. 12, no. 12, pp. 1897–1910, 2011.

[44] Y. Kabeya, N. Mizushima, A. Yamamoto, S. Oshitani-Okamoto, Y. Ohsumi, and T. Yoshimori, "LC3, GABARAP and GATE16 localize to autophagosomal membrane depending on form-II formation," *Journal of Cell Science*, vol. 117, no. 13, pp. 2805–2812, 2004.

[45] H. Weidberg, T. Shpilka, E. Shvets, A. Abada, F. Shimron, and Z. Elazar, "LC3 and GATE-16 N termini mediate membrane fusion processes required for autophagosome biogenesis," *Developmental Cell*, vol. 20, no. 4, pp. 444–454, 2011.

[46] R. Sahu, S. Kaushik, C. C. Clement et al., "Microautophagy of cytosolic proteins by late endosomes," *Developmental Cell*, vol. 20, no. 1, pp. 131–139, 2011.

[47] L. Marzella, J. Ahlberg, and H. Glaumann, "Autophagy, heterophagy, microautophagy and crinophagy as the means for intracellular degradation," *Virchows Archiv Abteilung B Cell Pathology*, vol. 36, no. 2-3, pp. 219–234, 1981.

[48] H. Katayama, A. Yamamoto, N. Mizushima, T. Yoshimori, and A. Miyawaki, "GFP-like proteins stably accumulate in lysosomes," *Cell Structure and Function*, vol. 33, no. 1, pp. 1–12, 2008.

[49] Y. Nishida, S. Arakawa, K. Fujitani et al., "Discovery of Atg5/Atg7-independent alternative macroautophagy," *Nature*, vol. 461, no. 7264, pp. 654–658, 2009.

[50] F. Scarlatti, R. Maffei, I. Beau, P. Codogno, and R. Ghidoni, "Role of non-canonical Beclin 1-independent autophagy in cell death induced by resveratrol in human breast cancer cells," *Cell Death and Differentiation*, vol. 15, no. 8, pp. 1318–1329, 2008.

[51] Y. Grishchuk, V. Ginet, A. C. Truttmann, P. G. Clarke, and J. Puyal, "Beclin 1-independent autophagy contributes to apoptosis in cortical neurons," *Autophagy*, vol. 7, no. 10, pp. 1115–1131, 2011.

[52] G. Seo, S. K. Kim, Y. J. Byun et al., "Hydrogen peroxide induces Beclin 1-independent autophagic cell death by suppressing the mTOR pathway via promoting the ubiquitination and degradation of Rheb in GSH-depleted RAW 264.7 cells," *Free Radical Research*, vol. 45, no. 4, pp. 389–399, 2011.

[53] J. H. Zhu, C. Horbinski, F. Guo, S. Watkins, Y. Uchiyama, and C. T. Chu, "Regulation of autophagy by extracellular signal-regulated protein kinases during 1-methyl-4-phenylpyridinium-induced cell death," *American Journal of Pathology*, vol. 170, no. 1, pp. 75–86, 2007.

[54] M. Mauthe, A. Jacob, S. Freiberger et al., "Resveratrol-mediated autophagy requires WIPI-1-regulated LC3 lipidation in the absence of induced phagophore formation," *Autophagy*, vol. 7, no. 12, pp. 1448–1461, 2011.

[55] D. M. Smith, S. Patel, F. Raffoul, E. Haller, G. B. Mills, and M. Nanjundan, "Arsenic trioxide induces a beclin-1-independent autophagic pathway via modulation of SnoN/SkiL expression in ovarian carcinoma cells," *Cell Death and Differentiation*, vol. 17, no. 12, pp. 1867–1881, 2010.

[56] Y. Kondo, T. Kanzawa, R. Sawaya, and S. Kondo, "The role of autophagy in cancer development and response to therapy," *Nature Reviews Cancer*, vol. 5, no. 9, pp. 726–734, 2005.

[57] X. H. Liang, S. Jackson, M. Seaman et al., "Induction of autophagy and inhibition of tumorigenesis by beclin 1," *Nature*, vol. 402, no. 6762, pp. 672–676, 1999.

[58] I. Arsov, A. Adebayo, M. Kucerova-Levisohn et al., "A role for autophagic protein Beclin 1 early in lymphocyte development," *Journal of Immunology*, vol. 186, no. 4, pp. 2201–2209, 2011.

[59] P. Codogno, M. Mehrpour, and T. Proikas-Cezanne, "Canonical and non-canonical autophagy: variations on a common theme of self-eating?" *Nature Reviews Molecular Cell Biology*, vol. 13, pp. 7–12, 2011.

[60] D. W. Hailey, A. S. Rambold, P. Satpute-Krishnan et al., "Mitochondria supply membranes for autophagosome biogenesis during starvation," *Cell*, vol. 141, no. 4, pp. 656–667, 2010.

[61] B. Ravikumar, K. Moreau, L. Jahreiss, C. Puri, and D. C. Rubinsztein, "Plasma membrane contributes to the formation of pre-autophagosomal structures," *Nature Cell Biology*, vol. 12, no. 8, pp. 747–757, 2010.

[62] N. Araki, Y. Takashima, and T. Makita, "Redistribution and fate of colchicine-induced alkaline phosphatase in rat hepatocytes: possible formation of autophagosomes whose membrane is derived from excess plasma membrane," *Histochemistry and Cell Biology*, vol. 104, no. 4, pp. 257–265, 1995.

[63] T. Ueno, D. Muno, and E. Kominami, "Membrane markers of endoplasmic reticulum preserved in autophagic vacuolar membranes isolated from leupeptin-administered rat liver,"

Journal of Biological Chemistry, vol. 266, no. 28, pp. 18995–18999, 1991.

[64] A. Yamamoto, R. Masaki, and Y. Tashiro, "Characterization of the isolation membranes and the limiting membranes of autophagosomes in rat hepatocytes by lectin cytochemistry," *Journal of Histochemistry and Cytochemistry*, vol. 38, no. 4, pp. 573–580, 1990.

[65] R. Scherz-Shouval, E. Shvets, E. Fass, H. Shorer, L. Gil, and Z. Elazar, "Reactive oxygen species are essential for autophagy and specifically regulate the activity of Atg4," *The EMBO Journal*, vol. 26, no. 7, pp. 1749–1760, 2007.

[66] E. L. Eskelinen, F. Reggiori, M. Baba, A. L. Kovacs, and P. O. Seglen, "Seeing is believing: the impact of electron microscopy on autophagy research," *Autophagy*, vol. 7, no. 9, pp. 935–956, 2011.

[67] A. Simonsen and S. A. Tooze, "Coordination of membrane events during autophagy by multiple class III PI3-kinase complexes," *Journal of Cell Biology*, vol. 186, no. 6, pp. 773–782, 2009.

[68] A. L. Kovács, Z. Pálfia, G. Réz, T. Vellai, and J. Kovács, "Sequestration revisited: integrating traditional electron microscopy, de novo assembly and new results," *Autophagy*, vol. 3, no. 6, pp. 655–662, 2007.

[69] M. Mari and F. Reggiori, "Atg9 reservoirs, a new organelle of the yeast endomembrane system?" *Autophagy*, vol. 6, no. 8, pp. 1221–1223, 2010.

[70] M. Mari, J. Griffith, E. Rieter, L. Krishnappa, D. J. Klionsky, and F. Reggiori, "An Atg9-containing compartment that functions in the early steps of autophagosome biogenesis," *Journal of Cell Biology*, vol. 190, no. 6, pp. 1005–1022, 2010.

[71] D. Mijaljica, M. Prescott, and R. J. Devenish, "Endoplasmic reticulum and golgi complex: contributions to, and turnover by, autophagy," *Traffic*, vol. 7, no. 12, pp. 1590–1595, 2006.

[72] M. Mari, S. A. Tooze, and F. Reggiori, "The puzzling origin of the autophagosomal membrane," *F1000 Biology Reports*, vol. 3, no. 1, article 25, 2011.

[73] W. A. Dunn Jr., "Studies on the mechanisms of autophagy: formation of the autophagic vacuole," *Journal of Cell Biology*, vol. 110, no. 6, pp. 1923–1933, 1990.

[74] M. Hayashi-Nishino, N. Fujita, T. Noda, A. Yamaguchi, T. Yoshimori, and A. Yamamoto, "A subdomain of the endoplasmic reticulum forms a cradle for autophagosome formation," *Nature Cell Biology*, vol. 11, no. 12, pp. 1433–1437, 2009.

[75] P. Ylä-Anttila, H. Vihinen, E. Jokitalo, and E. L. Eskelinen, "3D tomography reveals connections between the phagophore and endoplasmic reticulum," *Autophagy*, vol. 5, no. 8, pp. 1180–1185, 2009.

[76] E. L. Axe, S. A. Walker, M. Manifava et al., "Autophagosome formation from membrane compartments enriched in phosphatidylinositol 3-phosphate and dynamically connected to the endoplasmic reticulum," *Journal of Cell Biology*, vol. 182, no. 4, pp. 685–701, 2008.

[77] Q. Lu, P. Yang, X. Huang et al., "The WD40 repeat PtdIns(3)P-binding protein EPG-6 regulates progression of omegasomes to autophagosomes," *Developmental Cell*, vol. 21, no. 2, pp. 343–357, 2011.

[78] K. Matsunaga, E. Morita, T. Saitoh et al., "Autophagy requires endoplasmic reticulum targeting of the PI3-kinase complex via Atg14L," *Journal of Cell Biology*, vol. 190, no. 4, pp. 511–521, 2010.

[79] H. E. J. Polson, J. De Lartigue, D. J. Rigden et al., "Mammalian Atg18 (WIPI2) localizes to omegasome-anchored phagophores and positively regulates LC3 lipidation," *Autophagy*, vol. 6, no. 4, pp. 506–522, 2010.

14

MAP1B Interaction with the FW Domain of the Autophagic Receptor Nbr1 Facilitates Its Association to the Microtubule Network

Katie Marchbank,[1] **Sarah Waters,**[2] **Roland G. Roberts,**[1] **Ellen Solomon,**[1] **and Caroline A. Whitehouse**[1]

[1] *Department of Medical and Molecular Genetics, Kings College London, London SE1 9RT, UK*
[2] *The Randall Division for Cell and Molecular Biophysics and Cardiovascular Division, British Heart Foundation Centre of Research Excellence, King's College London, London SE1 1UL, UK*

Correspondence should be addressed to Caroline A. Whitehouse, caroline.whitehouse@kcl.ac.uk

Academic Editor: Anne Simonsen

Selective autophagy is a process whereby specific targeted cargo proteins, aggregates, or organelles are sequestered into double-membrane-bound phagophores before fusion with the lysosome for protein degradation. It has been demonstrated that the microtubule network is important for the formation and movement of autophagosomes. Nbr1 is a selective cargo receptor that through its interaction with LC3 recruits ubiquitinated proteins for autophagic degradation. This study demonstrates an interaction between the evolutionarily conserved FW domain of Nbr1 with the microtubule-associated protein MAP1B. Upon autophagy induction, MAP1B localisation is focused into discrete vesicles with Nbr1. This colocalisation is dependent upon an intact microtubule network as depolymerisation by nocodazole treatment abolishes starvation-induced MAP1B recruitment to these vesicles. MAP1B is not recruited to autophagosomes for protein degradation as blockage of lysosomal acidification does not result in significant increased MAP1B protein levels. However, the protein levels of phosphorylated MAP1B are significantly increased upon blockage of autophagic degradation. This is the first evidence that links the ubiquitin receptor Nbr1, which shuttles ubiquitinated proteins to be degraded by autophagy, to the microtubule network.

1. Introduction

Cellular turnover of damaged and misfolded proteins is mediated by two main degradation pathways; macroautophagy (hereafter referred to as autophagy) and the ubiquitin proteasome system (UPS). The UPS targets soluble, cytosolic proteins to the proteasome where they are degraded. Proteins targeted for degradation are covalently modified by the small, highly conserved, ubiquitously expressed protein ubiquitin. Ubiquitin can form chains at all seven lysine residues and typically, chains of four or more ubiquitin molecules are required for the targeting of proteins to the proteasome [1]. However, misfolded proteins can form large aggregates which render them resistant to proteasomal degradation [2]. Autophagy is an evolutionary conserved catabolic process that serves to deliver large polyubiquitinated protein aggregates and whole organelles to the lysosome for degradation [3]. A block in this process can cause the accumulation of ubiquitinated protein aggregates and ultimately cell death [4].

Autophagy requires the coordinated action of 35 to date autophagy-related genes (ATG) that mediate the formation of the double-membrane bound autophagosome which encloses a portion of the cytoplasm and delivers it to the lysosome [5, 6]. There are two ubiquitin-like conjugation systems that are required for autophagosomal formation. The Atg12-Atg5-Atg16L complex is important for elongation of the isolation membrane [7] whilst Atg8/LC3, covalently attached to phosphatidylethanolamine (PE) is essential for autophagosome biogenesis [8]. LC3 is often used as a marker for autophagosomes and has been shown to bind and stabilise microtubules [9, 10]. The microtubule network is important for autophagosomal formation [11, 12]; however, its requirement for fusion of autophagosomes with lysosomes is still unclear [11–13]. Roles for distinct populations of microtubules have also been proposed whereby labile microtubules specifically recruit markers of the isolation membrane such as Atg5, Atg12, and LC3 to sites of autophagosomal formation whereas stable microtubules facilitate the movement of mature autophagosomes [14].

Recent evidence demonstrates that autophagy can be a selective process, whereby single proteins and cellular structures such as aggregates and organelles can be specifically targeted to autophagosomes [15, 16], but the molecular mechanism of cargo recognition is poorly understood. Recently autophagic receptors have been described which include the structurally similar proteins p62 and NBR1, as well as the TBK1 adaptor NDP52 [17–19]. These receptors are thought to bind to polyubiquitinated proteins via their C-terminal-ubiquitin-associated (UBA/UBZ) domains and sort them to sites of autophagosomal formation via their interaction with LC3 [20, 21]. Both NBR1 and p62 colocalise with ubiquitin in Mallory bodies in the liver of patients with alcoholic steatohepatitis [18] and accumulate with ubiquitin in muscle fibres of sporadic inclusion-body myositis [22]. In contrast to p62, NBR1 has not been extensively studied, however growing evidence has implicated it in a diverse range of biological functions. NBR1 interacts with the giant sarcomeric protein titin and is part of a signalling complex that regulates muscle gene expression [23]. A genetically modified mouse model expressing a C-terminally truncated form of Nbr1 identified a role for Nbr1 in bone remodelling whilst a T-cell-specific knock-out of full length Nbr1 has implicated NBR1 as a mediator of T-cell differentiation and allergic inflammation [24, 25]. NBR1 has also recently been shown to direct autophagic degradation of midbody derivatives, independent of p62 [26]. Additionally, NBR1 inhibits receptor tyrosine kinase (RTK) degradation by trapping the receptor at the cell surface [27] and via its interaction with SPRED2, mediates the lysosomal degradation of activated receptors and the attenuation of fibroblast growth factor (FGF) signalling [28]. Identification of other protein interactors of NBR1 such as calcium- and integrin-binding protein (CIB) and fasciculation and elongation protein zeta-1 (FEZ1) [29] have suggested additional roles for NBR1 in cardiac dysfunction [30] and neuronal development, respectively [31]. It has been shown that both NBR1 and p62 are recruited to autophagosomal formation sites independent of LC3; however, the mechanism is unclear [32].

In this paper, we identify NBR1 as an interaction partner of the microtubule-associated protein MAP1B. This occurs via the evolutionarily conserved FW domain. We show that whilst MAP1B is not itself a substrate for autophagosomal protein degradation, the phosphorylated form of MAP1B is stabilised by lysosomal inhibition. We propose that this interaction provides a mechanism by which NBR1 is targeted to the microtubule network to promote degradation of proteins via the autophagosome.

2. Materials and Methods

2.1. Bioinformatics. BioEdit was used to curate sequences and compile alignments. BLAST was used on various databases to identify FW-like sequences from animal, plant, fungal, protist, bacterial, and metagenome sequences. Phyre was used for structural predictions.

2.2. Primary Antibodies and Constructs. For western blot analysis and immunofluorescence the following antibodies were used: polyclonal anti-myc (A14, Santa Cruz), monoclonal anti-HA (Roche), monoclonal anti-myc (9E10, Santa Cruz), and polyclonal anti-MAP1B-HC (kindly provided by Prof. Gordon-Weeks, King's College London [33, 34], polyclonal anti-MAP1B (N19, Santa Cruz), polyclonal anti-MAP1B (C20, Santa Cruz), polyclonal anti-pThr1265-MAP1B (Novus Biologicals), monoclonal anti-Nbr1 (Abcam), monoclonal anti-p62 (Abnova), and polyclonal anti-p62 (kindly provided by Prof. Gautel, King's College London), polyclonal anti-ULK1 (Sigma), polyclonal anti-ubiquitin (Dako), polyclonal anti-EEA1 (Cell Signaling), polyclonal anti β-actin (Abcam), and monoclonal anti-His (Novagen).

Yeast two-hybrid bait for Nbr1 was amplified by PCR and cloned into pGBKT7 (Nbr1 aa346-498) (Clontech). Full length Nbr1 was cloned into pHM6 (Roche) and MAP1B aa2216- 2464 was cloned into pcDNA3.1 (Invitrogen) for the coimmunoprecipitation assay. Nbr1 aa346-498 was cloned into pGEX2T (GE Healthcare) for the GST-binding assay and MAP1B aa2227-2464 was cloned into PET6H (a modified version of pET11d-Novagen) for the recombinant binding assay.

2.3. Yeast-2-Hybrid. Yeast strain Y187 was transformed with the Nbr1 bait construct and mated with a pretransformed (yeast strain AH109) mouse neonatal calvarial cDNA library kindly supplied by Prof. Ikramuddin Aukhil, University of Florida. Resulting colonies were screened by HIS3 reporter gene activity, replated three times and inserts were sequenced. Y187 transformed with pGBKT7 Nbr1 aa346-498 was mated with yeast strain AH109 expressing the library MAP1B clone pGADT7 MAP1B aa2238-2465 and plated onto SD medium lacking leucine, tryptophan, histidine and adenine and were cultured at 30°C to verify the interaction.

2.4. Coimmunoprecipitation. COS7 cells were cotransfected with HA-Nbr1 and MAP1B-myc and after 48 hours, lysed in IP buffer (50 mM Tris pH 7.5, 150 mM NaCl, 0.5% NP-40, supplemented with protease and phosphatase inhibitors (Roche)), and cell lysates incubated with rabbit polyclonal anti-myc antibody overnight at 4°C. Protein A beads (Millipore) were then added to the lysates for a further 2 hours, beads were then washed three times in IP buffer. Proteins retained on the beads were separated by SDS-PAGE and transferred onto a nitrocellulose membrane following standard procedures. Blots were probed with mouse monoclonal anti-myc and rat monoclonal anti-HA antibodies and subsequently with a secondary antibody

(HRP-conjugated anti-mouse or anti-rat, Dako, Abcam). Detection was performed by ECL (GE Healthcare).

2.5. Bacterial Expression of Fusion Proteins. Nbr1 aa346-498 fused to GST and GST alone were expressed in Bl21(DE3) bacterial cells and proteins purified by glutathione affinity chromatography as previously described [35]. MAP1B aa2227-2464 fused to His6 was also expressed in Bl21(DE3) bacterial cells and purified in the presence of urea. Briefly, bacterial cells expressing His_6-MAP1B aa2227-2464 were lysed in lysis buffer (100 mM NaH_2PO_4, 10 mM Tris pH 8, 6 M Urea, 5 mM Imidazole pH 8, supplemented with EDTA-free protease inhibitors (Roche)). The sample was sonicated, centrifuged, and the supernatant was incubated with Ni Sepharose 6 fast flow beads (Amersham Biosciences) for 2 hours at 4°C. Beads were then washed in low Imidazole elution buffer (100 mM NaH_2PO_4, 10 mM Tris pH 8, 6 M Urea, 20 mM Imidazole pH 8, supplemented with EDTA free protease inhibitors (Roche)) and bound proteins eluted from the beads using high Imidazole elution buffer (100 mM NaH_2PO_4, 10 mM Tris pH 8, 6 M Urea, 250 mM Imidazole pH 8, supplemented with EDTA-free protease inhibitors (Roche)). The resulting purified His-tagged protein was dialysed into 50 mM Tris pH 7.5, 150 mM NaCl and used in the GST pull-down assay.

2.6. GST Pull-Down Assays. COS7 cells were transfected with a MAP1B-myc construct and after 48 hours expression, lysed in IP buffer (as above), and lysates incubated with beads coupled with either GST-Nbr1 aa346-498 or GST alone for 2 hours at 4°C. Following incubation, beads were washed three times in IP wash buffer (50 mM Tris pH 7.5, 200 mM NaCl, 0.5% NP-40, supplemented with protease and phosphatase inhibitors (Roche)) and proteins retained on the beads were analysed by western blotting as described above, using the monoclonal anti-myc antibody, 9E10. Alternatively purified GST or GST-Nbr1 aa346-498 attached to glutathione agarose beads (Sigma) was incubated in IP buffer (50 mM Tris pH 7.5, 150 mM NaCl, 0.5% NP-40, supplemented with protease and phosphatase inhibitors (Roche)) with purified His_6-MAP1B aa2227-2464 for 2 hours at 4°C. Following incubation, beads were washed three times in IP wash buffer (50 mM Tris pH 7.5, 200 mM NaCl, 0.5% NP-40), and proteins retained on the beads were separated by SDS-PAGE and analysed by western blotting as described above, using the anti-His tag monoclonal antibody.

2.7. Cell Culture, Treatments, Transfection and Immunostaining. COS7 cells were cultured in DMEM/10% FCS by standard protocols and transfected using Fugene 6 (Roche). Cells were lysed 48 hours later in 200 μL IP buffer for pull-down and coimmunoprecipitation assays. For immunostaining PC12 cells were cultured on coverslips in DMEM/10% FCS, treated with DMSO or Bafilomycin A1 (Sigma) for 4 or 8 hours, or starved in Hanks-Balanced Salt Solution (Sigma) for 4 hours or treated with 5 μg/mL nocodazole (Sigma) for 30 minutes before or after 2 hours starvation and fixed in 4% paraformaldehyde/PBS for 10 minutes. Cells were then permeabilised in 0.1% Triton X100/PBS and incubated consecutively with primary and secondary antibodies (Dako) for one hour each prior to mounting. Cells were imaged using a Zeiss LSM 510 confocal microscope in sequential scanning mode with a Plan-Apochromat 63 x/1.4 Oil DIC objective. Quantification of MAP1B/Nbr1 colocalisation was performed using Zeiss ZEN2010 software, data represent mean ± SEM of 22 images.

3. Results

3.1. The Predicted Structure and Evolution of Nbr1 FW Domain. We used BLAST-based searches to acquire NBR1-related sequences from multiple available genomic and transcriptomic sources across a broad range of eukaryotes. These identified a region of pronounced conservation of 105 amino acids (residues 374–478 of human NBR1) which is recognisable in the single NBR1 orthologue found in most eukaryotes but is absent from p62. This novel domain has been named the NBR1 domain [36] and FW domain by Terje Johansen's group [37] after its four strikingly conserved tryptophan residues, and we will use FW nomenclature here for clarity.

Single NBR1 orthologues were found in all animals, most plants, most fungi (though notably not *Saccharomyces cerevisiae*) and some single-celled eukaryotes (such as *Dictyostelium discoideum*). In each case the NBR1-like molecule possessed an N-terminal PB1 domain, one (animals, plants) or more (fungi) ZZ domains, an FW domain, and a C-terminal UBA domain (Figure 1).

The FW domain was also found in a second, otherwise unrelated animal protein. As the human version has been named c6ORF106, we will use this name. Single c6ORF106 orthologues are found in all animal species examined, plus the single-celled metazoan sister-group choanoflagellates. No c6ORF106 orthologues were found in any other organisms. The proteins tend to be small (the human c6ORF106 is 298 amino acids long), comprising a universally conserved N-terminal α-helical domain of ~70–80 amino acids, then the FW domain and finally a poorly structured and variable length C-terminal region (Figure 1).

Intriguingly, FW domains were also found in a wide range of eubacteria. The eubacterial FW-containing proteins are strikingly diverse in domain structure, with the only common theme being that the FW domain tends to be very close to the C-terminus. Although most eubacterial genomes do not encode an FW domain-containing protein, we find that it is broadly distributed across eubacterial clades (γ-proteobacteria, chloroflexi, actinobacteria, and several unclassified metagenomes). In several of the bacterial proteins (*Halorhodospira halophila, Methylomonas methanica, Kribbella flavida, Variovorax paradoxus,* and one from a fresh water environmental metagenome), the FW domain appears immediately C-terminal to a robustly predicted "helix-turn-helix" DNA-binding motif of the XRE family. We call these XRE-FW proteins. XRE domains tend to appear either alone or with multimerisation domains (as in the *Bacillus subtilis* repressor of sporulation and biofilm formation, SinR and

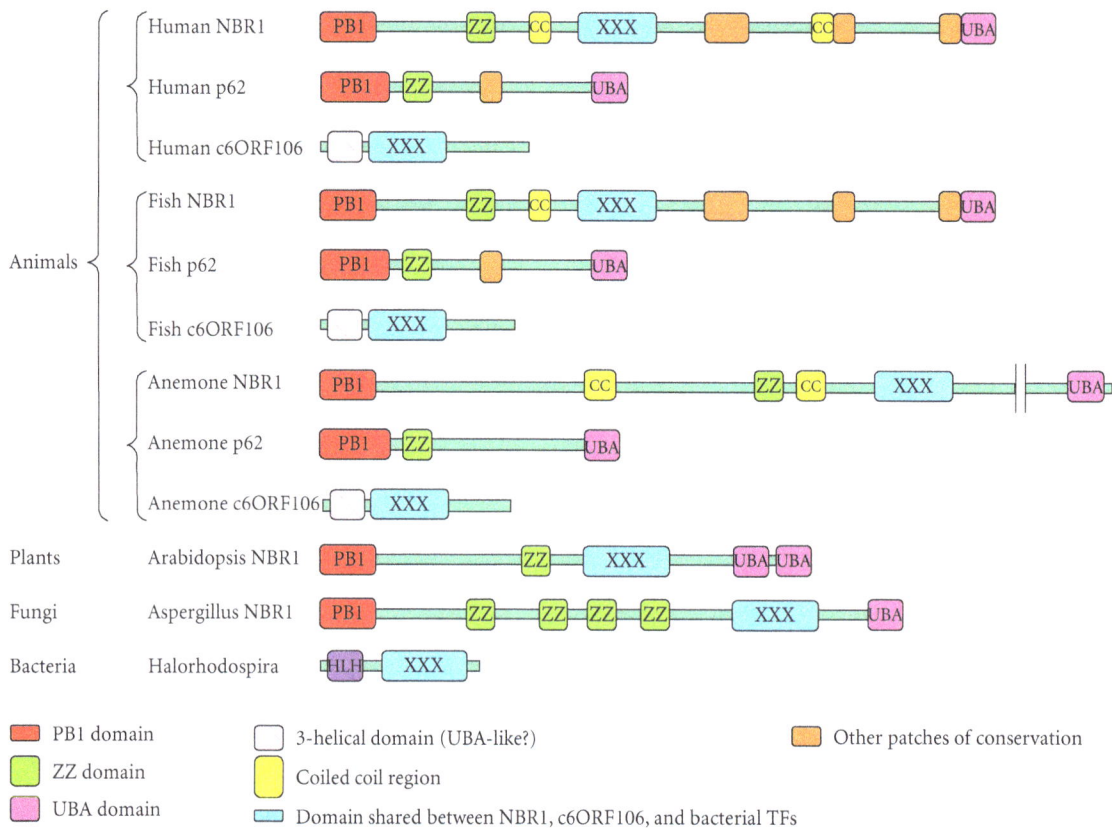

FIGURE 1: Schematic representation of the commonest members of the FW-containing protein families- the NBR1 proteins found in almost all eukaryotes, the c6ORF106 proteins found in almost all metazoans and the XRE-FW proteins found in some bacteria. The metazoan p62 family is also included to show its relationship to NBR1. Proteins are drawn to scale. A key to the domains appears at the bottom of the figure.

the bacteriophage repressors CI and Cro). This juxtaposition raises the possibility that the FW domain might mediate homo- and/or heterodimerisation or could bind a small signalling molecule. Some other eubacterial FW proteins consist solely of two tandem FW domains and little else (e.g., that from *Coprococcus*), while others contain transmembrane domains (e.g., that from *Streptomyces* sp.). This structural diversity suggests that the FW domain has a generically useful function that has been exploited in many ways.

We used Phyre to predict the secondary structure of all FW domains separately. This robustly predicted the same alignable structural features in every sequence, regardless of sequence divergence (Figure 2). Thus we feel that we are able to say with some confidence that the FW domain consists of two sets of three β-strands separated by a central unstructured region of more variable length. Striking sequence features include four almost invariant tryptophan residues, which lie in the middle of strand β2, in the linker between β2 and β3, and in the middle of strands β5 and β6. These give the domain its name and are only rarely replaced by other aromatic residues. There are also several invariant glycines and prolines in some of the unstructured linkers. It is conceivable that the domain folds into a sandwich of two three-strand β-sheets with the tryptophans projecting into the hydrophobic core.

Phylogenetic analysis showed that all NBR1 FW domains clustered together, as did all c6ORF106 FW domains. Two FW sequences from metagenomic sources clustered with NBR1 sequences; one of these had a C-terminal UBA domain, and we assume that these are from eukaryotic species in the environmental metagenome sources. The reproducible monophyletic clustering of bacterial FW domains (to the exclusion of eukaryotic NBR1 and c6ORF106 sequences) argues against multiple eukaryote-to-prokaryote horizontal gene transfer events and suggests that the FW domain may be ancient, predating the split between eukaryotic and eubacterial domains.

3.2. Identification of the FW Domain of Nbr1 as an Interaction Partner of Microtubule-Associated Protein MAP1B. To identify novel protein interactors of the highly conserved FW domain of Nbr1 and therefore elucidate a function, we performed a yeast-2-hybrid screen with the FW domain of Nbr1 as bait. A neonatal calvarial cDNA library was screened and the light chain of the microtubule-associated protein 1B (MAP1B-LC1) was identified as an interaction partner of Nbr1. This interaction was verified by a directed yeast two-hybrid assay by retransforming the isolated prey vector encoding the partial MAP1B-LC1 sequence (aa2238-2465)

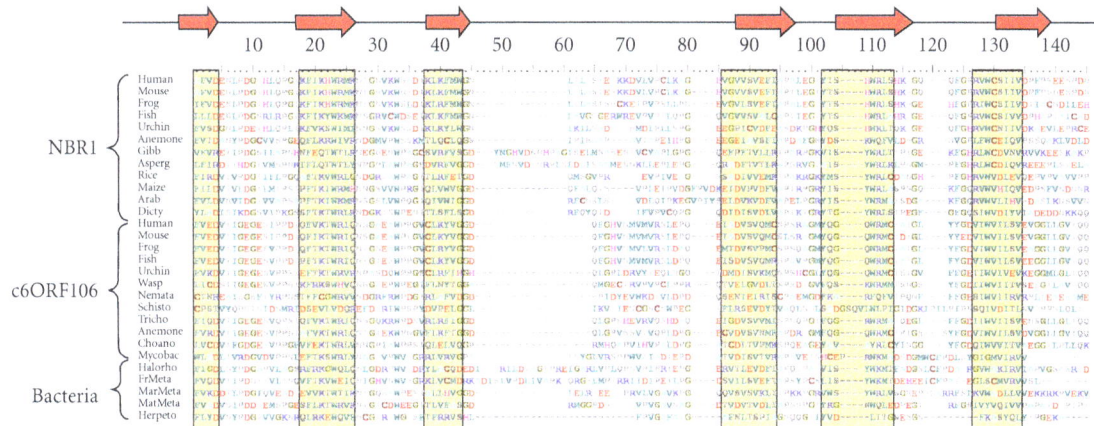

FIGURE 2: Alignment of NBR1, c60RFI06 and eubacterial FW domain sequences. Yellow boxes and red arrows indicate β-strands predicted by Phyre. The amino acids are colour coded as follows; red-positively charged, dark blue-negatively charged, grey-non-charged polar, dark green-aliphatic and aromatic, cyan-alanine, brown-cysteine, magenta-histidine, gold-glycine. Brackets at left indicate broad origin of FW domains (according to gross structure of host protein or phylogenetic affinity). Species are indicated as follows: Human—*Homo sapiens*; Mouse—*Mus musculus*; Frog—*Xenopus tropicalis*; Fish—*Danio rerio*; Urchin—*Strongylocentrotus purpuratus*; Anemone—*Nematostella vectensis*; Gibb—*Gibberella zeae*; Asperg—*Aspergillus nidulans*; Rice—*Oryza sativa*; Maize—*Zea mays*; Arab—*Arabidopsis thaliana*; Dicty—*Dictyostelium discoideum*; Wasp—*Nasonia vitripennis*; Nemato—*Caenorhabditis elegans*; Schisto—*Schistosoma mansoni*; Tricho—*Trichoplax adhaerens*; Choano—*Monosiga brevicollis*; Mycobac—*Mycobacterium* sp. MCS; Halorho—*Halorhodospira halophila*; FrMeta—Fresh water metagenome; MarMeta—Marine metagenome; MatMeta—Mat metagenome; Herpeto—*Herpetosiphon* sp.

into yeast strain AH109 and mating it with yeast strain Y187 that was expressing the FW domain of Nbr1 (Figure 3(a)). MAP1B is transcribed as a single mRNA, translated into a polypeptide, and subsequently cleaved producing a heavy chain (2214aa) and a light chain (250aa) [38]. Both the heavy chain (MAP1B HC) and the light chain (MAP1B-LC1) can bind to microtubules [39, 40] and to each other [41]. MAP1B has been implicated in the regulation of autophagy, as it interacts with LC3 and targets autophagosomes to axon terminals during neurodegeneration [42].

3.3. Nbr1 is Found in a Complex with MAP1B-LC1 In Vivo. To determine whether Nbr1 forms a complex with MAP1B-LC1 *in vivo*, we performed a coimmunoprecipitation experiment using COS7 cells transiently transfected with HA-Nbr1 and MAP1B-LC1-myc constructs. Using an anti-myc antibody for immunoprecipitation of MAP1B-LC1-myc, we found that HA-Nbr1 did coimmunoprecipitate with MAP1B-LC1-myc (Figure 3(b)) confirming that they are found in a complex *in vivo*. We were unable to show coimmunoprecipitation of endogenous Nbr1 and MAP1B-LC1 in PC12 cells. This is likely to be due to the levels of interacting protein being below the detection level possible by western blot analysis with the available antibodies (data not shown).

3.4. Nbr1 Interacts with MAP1B-LC1 In Vitro. The yeast-2-hybrid data suggested that the FW domain of Nbr1 interacts directly with MAP1B-LC1. To more rigorously test this hypothesis, we performed GST pull-down assays using extracts from COS7 cells overexpressing MAP1B-LC1-myc and a GST fusion of the FW domain of Nbr1 and GST alone.

Indeed, the FW domain of Nbr1 interacted with MAP1B-LC1 whilst GST alone did not (Figure 4(a)).

To verify the interaction between the FW domain of Nbr1 and MAP1B-LC1 in a cell-free environment, His$_6$-MAP1B-LC1 was purified and incubated with the GST fusions of the FW domain of Nbr1 or GST alone. This demonstrated that the FW domain of Nbr1 interacts directly with the light chain of MAP1B (Figure 4(b)).

3.5. MAP1B Is Not Degraded by Autophagy. It has previously been observed that MAP1B-HC is not degraded by autophagy [42] however, it has not been reported whether the same is true for MAP1B-LC1. To establish if the function of the interaction between Nbr1 and MAP1B-LC1 is to facilitate the degradation of MAP1B-LC1 via autophagy, MAP1B-LC1 protein levels were analysed under conditions where autophagic protein degradation was blocked. PC12 cells, a neuronal cell line that expresses elevated levels of endogenous MAP1B, were treated with Bafilomycin A1 or DMSO for 8 hours before protein extracts were resolved by SDS PAGE and detected using antibodies that recognise p62, Nbr1, MAP1B-LC1, MAP1B-HC and β-actin. Upon blockage of autophagic protein turnover, the levels of p62 and Nbr1 were increased by 60% and 130%, respectively, demonstrating that autophagic protein degradation was blocked by Bafilomycin A1 treatment (Figures 5(a) and 5(b)). MAP1B-HC, and MAP1B-LC1 protein levels showed a negligible increase upon the blockage of autophagic protein degradation suggesting that they are not degraded by autophagy and that the function of the Nbr1-MAP1B-LC1 interaction is not to target MAP1B-LC1 for autophagic protein turnover. Surprisingly, although total MAP1B levels

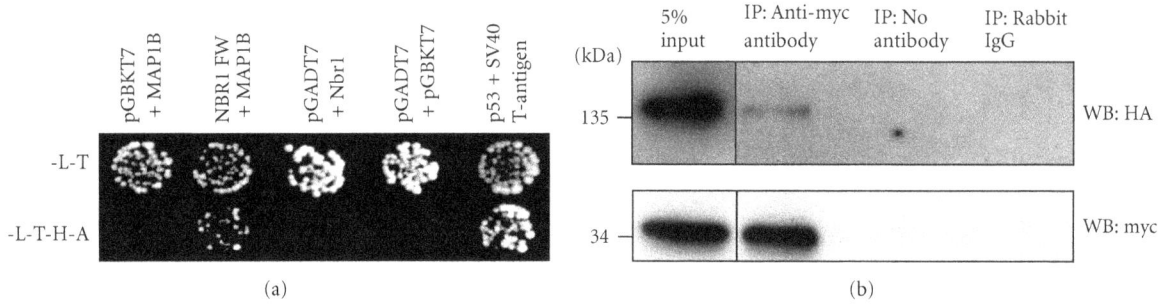

(a)   (b)

FIGURE 3: Nbr1 interacts with MAP1B *in vivo*. (a) Identification of Nbr1 as an interaction partner of MAP1B-LC1. Yeast-2-hybrid retransformation assay confirming the interaction between the FW domain of Nbr1 (aa346-498) and the light chain of MAP1B (aa2238-2465). Interaction was assessed by yeast growth on SD-L/-T/-H/-A medium. Empty vectors were used as negative controls, SV40 large T antigen and p53 were used as positive controls. (b) Nbr1 is found in a complex with MAP1B-LC1. Coimmunoprecipitation of HA-Nbr1 and MAP1B-LC1-myc from COS7 cells transfected with HA-Nbr1 and MAP1B-LC1-myc constructs. Extracts and precipitates were analysed by western blot using the indicated antibodies.

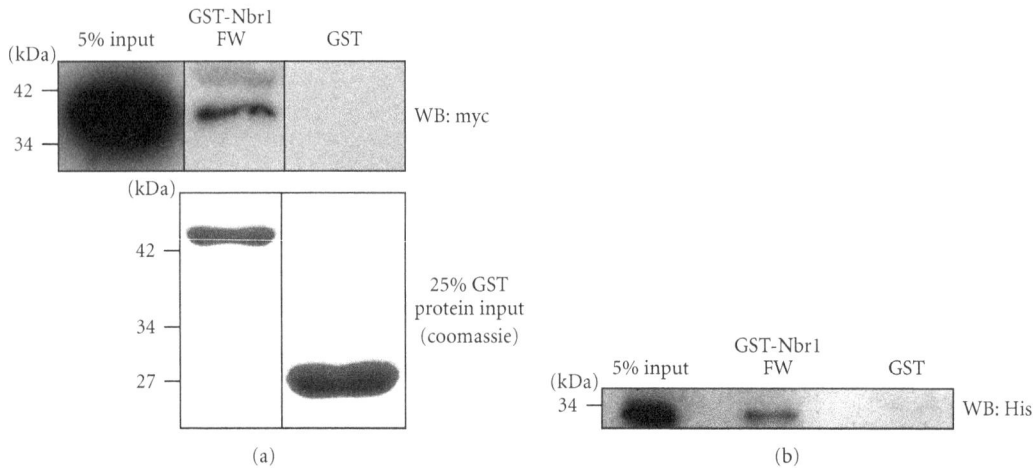

(a)   (b)

FIGURE 4: Nbr1 interacts with MAP1B-LC1. (a) GST pulldown assay using cell extracts from COS7 cells transfected with MAP1B-LC1-myc and immobilised GST or GST-Nbr1 FW domain. Upper panel: coprecipitated proteins were detected with an anti-myc antibody. The FW domain of Nbr1 interacts with MAP1B-LC1. Lower panel: coomassie stained SDS PAGE gel showing 25% of GST-tagged protein input. (b) GST pulldown assay using purified His-MAP1B-LC1 and immobilised GST or GST-Nbr1 FW domain. Coprecipitated proteins were detected using an anti-His antibody and demonstrated that the FW domain of Nbr1 interacts with MAP1B-LC1. The same amount of GST or GST-Nbr1 FW domain fusion protein was used as shown in (a) (lower panel).

are largely unaffected by blocking autophagic protein degradation, levels of phospho-pThr1265-MAP1B are increased following Bafilomycin A1 treatment (Figure 5(c)). This phosphorylated form of MAP1B is expressed in differentiating neurons and is a major substrate for glycogen synthase kinase-3beta (GSK-3beta) and is thought to be involved in regulating microtubule dynamics by MAP1B [43].

3.6. Nbr1 Colocalises with MAP1B upon Induction of Autophagy.
Next, we analysed the subcellular localisation of endogenous Nbr1 and MAP1B by confocal microscopy. To establish if Nbr1 and MAP1B colocalise *in vivo*, PC12 cells were treated with DMSO, Bafilomycin A1 to block autophagic protein degradation or starved to induce autophagy and analysed by immunofluorescence. Under basal conditions, when levels of Nbr1 are low, there was little

colocalisation between Nbr1 and MAP1B (Figure 6(A)). Upon blockage of autolysosomal protein degradation by Bafilomycin A1 treatment, Nbr1 is no longer turned over by autophagy and accumulates (Figure 6(B)) but total MAP1B is unaffected and appears excluded from Nbr1-positive vesicles. This confirms that MAP1B is not itself degraded by the autolysosomal pathway. Upon starvation and induction of autophagy, Nbr1 and MAP1B colocalise to distinct perinuclear vesicular structures (Figure 6(C)). Although this does not occur in all cells, only under starvation conditions were MAP1B-/Nbr1-positive vesicles observed. Under starvation conditions where MAP1B/Nbr1 positive punctate structures were observed, quantification of colocalisation showed a Mander's colocalisation coefficient of 64 ± 10%. These MAP1B-/Nbr1-positive vesicles also colocalise with the autophagic protein p62 (Figure 6(D)) but few colocalise with ubiquitin, suggesting that these vesicles

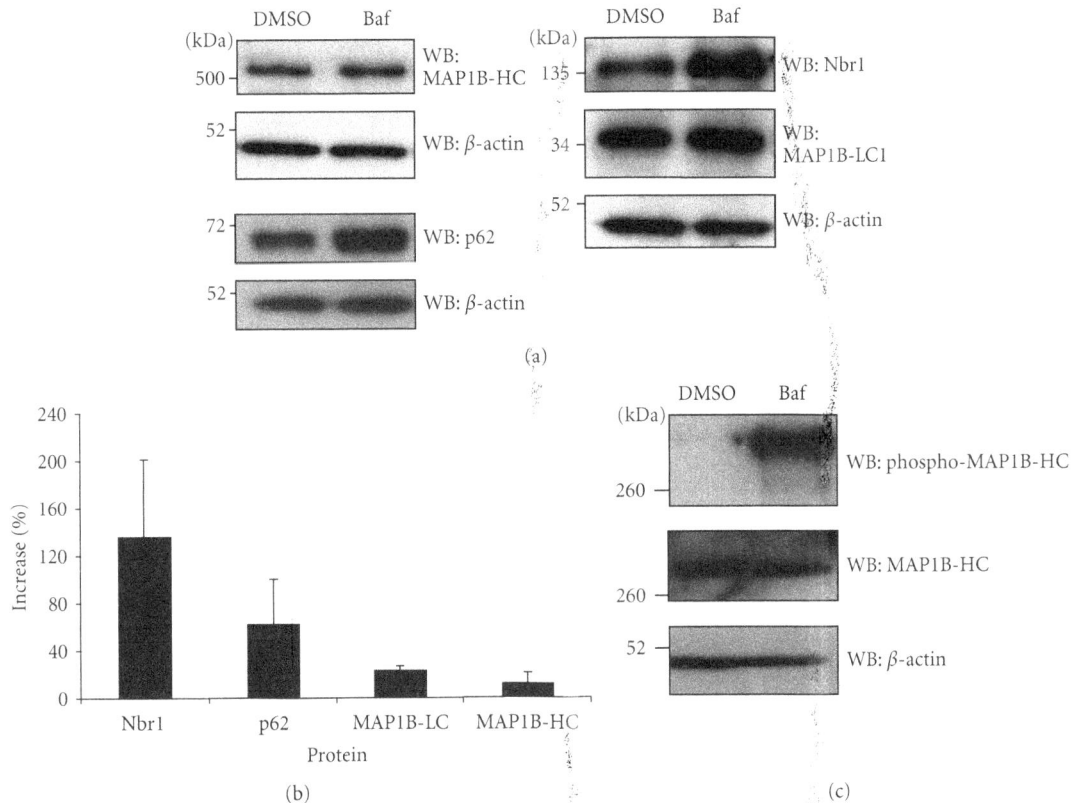

FIGURE 5: Western blot analysis of p62, Nbr1, MAP1B-HC, and MAP1B-LC1 protein levels following blockage of autophagic protein degradation. (a) Western blots showing protein levels in cells treated with DMSO (control) or Bafilomycin A1 (Baf). (b) Quantification of protein band intensity. MAP1B-LC1 and MAP1B-HC levels are increased by a negligable amount compared with Nbr1 and p62 upon treatment with Bafilomycin A1; Error bars represent SD, $n = 3$. (c) Phospho-MAP1B-HC is degraded by autophagy. Upon blockage of autophagic degradation with Bafilomycin A1: (Baf), levels of phospho-MAP1B-HC increase compared with levels of total MAP1B-HC.

are not aggresomes or mature autophagosomes loaded with ubiquitinated cargo (Figure 6(F)). We found little overlapping distribution with ULK1 and Nbr1/MAP1B vesicles under starvation conditions, in comparison with previous analysis of Nbr1/ULK1 colocalisation under these conditions [32] (Figure 6(E)) or with the early endosomal marker EEA1 (Figure 6(G)). This demonstrates that upon induction of autophagy, Nbr1 is recruited to MAP1B positive structures which are colocalising with p62, suggesting these may be early autophagosomes but downstream of autophagosomal formation sites.

To determine if colocalisation of Nbr1 and MAP1B in response to starvation-induced autophagy was dependent upon an intact microtubule network, PC12 cells were treated with the depolymerisation agent nocodazole under starvation conditions and examined for colocalisation. Depolymerisation of the microtubule network was confirmed by α-tubulin staining (data not shown) and resulted in loss of the punctate colocalisation of MAP1B and Nbr1 but intact Nbr1 vesicles were retained (Figure 6(H)). This suggests that MAP1B is not essential for the formation of Nbr1-positive vesicles but that an intact microtubule network is essential for colocalisation of Nbr1 and MAP1B under starvation conditions.

4. Discussion

The FW domain of Nbr1 is highly conserved throughout the eukaryotic kingdom and is also present in a number of bacterial proteins. It contains two internal repeats of ~55 residues and has a predicted secondary structure consisting of two, three β-stranded sheets. The high conservation of this region and its absence in p62 [37] suggests that it has a function that is distinct from p62. We therefore performed a yeast-2-hybrid screen with the FW domain of Nbr1 in order to determine a specific function for this region. The light chain of MAP1B (MAP1B-LC1) was identified as an interaction partner of the FW domain. As Nbr1 has previously been identified as an autophagic receptor that targets ubiquitinated proteins for degradation via its interaction with LC3 [18, 19], it was reasonable to hypothesise that the function of the interaction between Nbr1 and MAP1B-LC1 is to facilitate the autophagic degradation of MAP1B-LC1. Analysis of protein levels after autophagy blockage demonstrated that the levels of MAP1B-LC1 increased by a negligable amount suggesting that it is not degraded by autophagy (Figure 5). Blockage of autophagosomal protein degradation can also result in a reduction of protein turnover by the UPS [44] therefore, as MAP1B-LC1 is known to be degraded by the

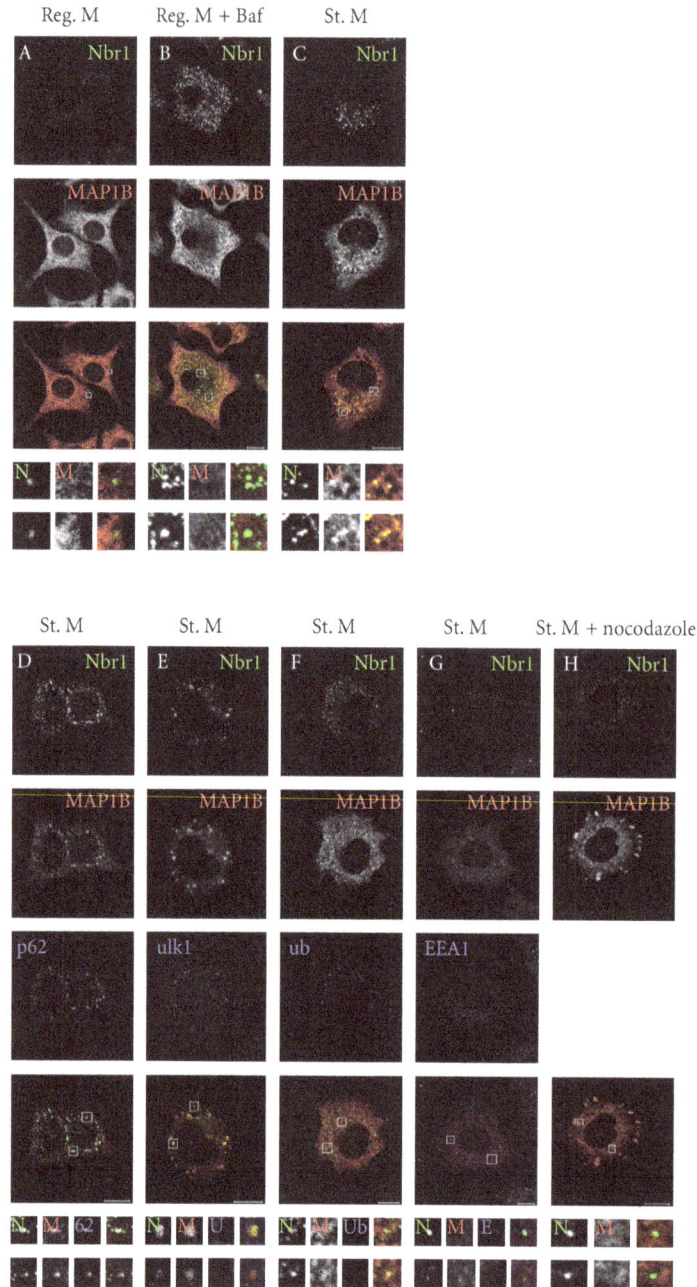

FIGURE 6: Nbr1 and MAP1B colocalise in discrete perinuclear vesicles upon induction of autophagy. PC12 cells were treated with DMSO, Bafilomycin or starved then fixed and stained with antibodies against the indicated proteins. Under basal conditions (A) or when autophagic degradation is blocked by Bafilomycin A1 treatment (B), very little or no colocalisation was observed between Nbr1 and MAP1B. When cells were starved to induce autophagy (C) MAP1B and Nbr1 colocalise in distinct perinuclear vesicles which are also positive for p62 (D) but are largely negative for ULK1 (E), ubiquitin (F), and EEA1 (G). Upon depolymerisation of the microtubule network and subsequent induction of autophagy by starvation, MAP1B no longer colocalised in distinct perinuclear vesicles with Nbr1 (H). Antibodies used: anti-Nbr1 (abcam), anti-MAP1B (N19, Santa Cruz), anti-p62 (M. Gautel, KCL), anti-ULK1 (Sigma), anti-ubiquitin (Ub) (Sigma), and anti-EEA1 (Cell Signaling). Scale bar; 10 μm.

UPS [45], this could suggest that Bafilomycin A1 treatment results in the inhibition of MAP1B-LC1 degradation via the proteasome rather than by autophagy. Interestingly, we observed that inhibition of autophagic degradation resulted in an increase in phospho-Thr1265 MAP1B, perhaps also reflected in the small increase in total MAP1B levels observed. Expression of this phosphorylated form of MAP1B is spatially regulated in differentiating neurons, and the kinase responsible for phosphorylation at this site has been identified as glycogen synthase kinase-3 beta. GSK-3 beta inhibition has been linked to Bif-1-dependent autophagic induction under serum starvation to modulate cell survival [46].

Further biochemical analysis confirmed that the interaction between Nbr1 and MAP1B-LC1 is direct and that these proteins can be found in a complex together *in vivo*. As both Nbr1 and the microtubule network have been identified as key players in the facilitation of protein degradation via autophagy [11, 12, 18, 19], this could suggest that the Nbr1-MAP1B-LC1 interaction is important for this process. MAP1B interacts with LC3 and through this interaction it has been proposed that autophagosomes are targeted to axon terminals during neurodegeneration [47]. Additionally MAP1B has been predicted to interact with Atg12 and Atg3 suggesting that in addition to LC3, MAP1B is important for targeting other components of the autophagosomal machinery to sites of autophagosomal formation [48]. The interaction data presented here and the colocalisation of Nbr1 and MAP1B to perinuclear vesicles suggest that via its interaction with MAP1B, Nbr1 is targeted to the microtubule network, thus providing a mechanism by which proteins can be targeted to autophagosomes. The MAP1B-/Nbr1-positive vesicles do not however colocalise with ubiquitin, suggesting that these vesicles are not yet loaded with ubiquitinated cargo. Alternatively, they could represent vesicles loaded with other nonubiquitinated proteins that have been targeted for degradation. Whilst there are currently no known proteins that are targeted for autophagy by Nbr1 in a ubiquitin-independent manner, STAT5A-ΔE18 can be targeted for autophagic degradation by the PB1 domain of p62 independent of ubiquitin [49]. This suggests that Nbr1 could also be acting by a similar mechanism to target proteins for degradation independent of ubiquitin. Nbr1/MAP1B vesicles did not colocalise with EEA1, showing that these are not early endosomes. Likewise, we saw largely no colocalisation of MAP1B/Nbr1 vesicles with ULK1, suggesting that MAP1B- and Nbr1-positive structures are not present at sites of autophagosomal formation but do perhaps represent early autophagosomes that are positive for p62 and nonubiquitinated protein cargo.

This is the first evidence linking Nbr1 to the microtubule network and also demonstrates a distinct function for the FW domain of Nbr1. A similar mechanism has previously been demonstrated whereby HDAC6 is able to interact with polyubiquitinated protein aggregates and to dynein motors thereby coupling protein aggregates to the microtubule network where they can be transported to sites of autophagosomal formation [50]. Furthermore, adaptor proteins such as FYCO can interact with LC3 and microtubule motor proteins and through these interactions it has been suggested that preautophagosomal membranes are targeted to sites of autophagosomal formation [51]. Roles for MAP1S (a MAP1B homologue) in autophagic degradation of mitochondria have also been demonstrated. MAP1S interacts with LC3 and this interaction functions to target LC3, to the microtubule network. Genetic ablation of MAP1S causes the accumulation of defective mitochondria and severe defects in response to nutritive stress suggesting defects in autophagosomal biogenesis and clearance [52].

It has been suggested that recruitment of autophagosomal cargo receptors like Nbr1 and p62 to the autophagosomal formation site may be a general feature of this type of receptor, but that it is independent of Atg factors downstream of the PI3-kinase complex [32]. This study further highlights the role for microtubule associated proteins in the targeting of autophagosome machinery to the microtubule network and complements the work presented here that suggests a link between microtubule-associated proteins and autophagic receptors.

The high evolutionary conservation of the FW domain within Nbr1 homologues implicates it to have a critical role in Nbr1 function. The predicted secondary structure of the FW domain that consists of two three β-stranded sheets that form a compact "sandwich" is also present in the cholesterol-binding protein Niemann-Pick C2 (NPC2) [53, 54] suggesting additional roles for the FW domain in lipid binding.

In summary, we present the first evidence linking the autophagic receptor protein Nbr1 and the microtubule network via a direct interaction of the evolutionary-conserved FW domain of Nbr1 with MAP1B. Nbr1 is a ubiquitously expressed protein that has been implicated in several diseases [18, 22, 23], and it therefore will be of significant value to assess this interaction in tissue-specific physiological studies.

Acknowledgments

The authors would like to acknowledge Professor Gordon-Weeks (King's College London) for helpful discussion and a polyclonal MAP1B antibody and Prof. Ikramuddin Aukhil, (University of Florida) for the yeast two-hybrid calvarial cDNA library. K. Marchbank was supported by a BBSRC PhD studentship, S. Waters is supported by the Wellcome Trust, R. G. Roberts was funded by the Muscular Dystrophy Campaign, and C. A. Whitehouse is supported by an Arthritis Research UK Fellowship.

References

[1] J. S. Thrower, L. Hoffman, M. Rechsteiner, and C. M. Pickart, "Recognition of the polyubiquitin proteolytic signal," *EMBO Journal*, vol. 19, no. 1, pp. 94–102, 2000.

[2] I. Weinhofer, S. Forss-Petter, M. Žigman, and J. Berger, "Aggregate formation inhibits proteasomal degradation of polyglutamine proteins," *Human Molecular Genetics*, vol. 11, no. 22, pp. 2689–2700, 2002.

[3] T. Yoshimori, "Autophagy: a regulated bulk degradation process inside cells," *Biochemical and Biophysical Research Communications*, vol. 313, no. 2, pp. 453–458, 2004.

[4] J. S. Carew, E. C. Medina, J. A. Esquivel et al., "Autophagy inhibition enhances vorinostat-induced apoptosis via ubiquitinated protein accumulation," *Journal of Cellular and Molecular Medicine*, vol. 14, no. 10, pp. 2448–2459, 2010.

[5] Z. Yang and D. J. Klionsky, "Eaten alive: a history of macroautophagy," *Nature Cell Biology*, vol. 12, no. 9, pp. 814–822, 2010.

[6] N. Mizushima, T. Yoshimori, and Y. Ohsumi, "The role of Atg proteins in autophagosome formation," *Annual Review of Cell and Developmental Biology*, vol. 27, no. 1, pp. 107–132, 2011.

[7] N. Mizushima, A. Yamamoto, M. Hatano et al., "Dissection of autophagosome formation using apg5-deficient mouse embryonic stem cells," *Journal of Cell Biology*, vol. 152, no. 4, pp. 657–667, 2001.

[8] H. Weidberg, E. Shvets, T. Shpilka, F. Shimron, V. Shinder, and Z. Elazar, "Lc3 and gate-16/gabarap subfamilies are both essential yet act differently in autophagosome biogenesis," *EMBO Journal*, vol. 29, no. 11, pp. 1792–1802, 2010.

[9] E. M. Faller, T. S. Villeneuve, and D. L. Brown, "Map1a associated light chain 3 increases microtubule stability by suppressing microtubule dynamics," *Molecular and Cellular Neuroscience*, vol. 41, no. 1, pp. 85–93, 2009.

[10] S. S. Mann and J. A. Hammarback, "Molecular characterization of light chain 3. a microtubule binding subunit of map1a and map1b," *Journal of Biological Chemistry*, vol. 269, no. 15, pp. 11492–11497, 1994.

[11] E. Fass, E. Shvets, I. Degani, K. Hirschberg, and Z. Elazar, "Microtubules support production of starvation-induced autophagosomes but not their targeting and fusion with lysosomes," *Journal of Biological Chemistry*, vol. 281, no. 47, pp. 36303–36316, 2006.

[12] R. Köchl, X. W. Hu, E. Y. W. Chan, and S. A. Tooze, "Microtubules facilitate autophagosome formation and fusion of autophagosomes with endosomes," *Traffic*, vol. 7, no. 2, pp. 129–145, 2006.

[13] R. Xie, S. Nguyen, W. L. McKeehan, and L. Liu, "Acetylated microtubules are required for fusion of autophagosomes with lysosomes," *BMC Cell Biology*, vol. 11, article no. 89, p. 897, 2010.

[14] C. Geeraert, A. Ratier, S. G. Pfisterer et al., "Starvation-induced hyperacetylation of tubulin is required for the stimulation of autophagy by nutrient deprivation," *Journal of Biological Chemistry*, vol. 285, no. 31, pp. 24184–24194, 2010.

[15] M. Komatsu, H. Kurokawa, S. Waguri et al., "The selective autophagy substrate p62 activates the stress responsive transcription factor nrf2 through inactivation of keap1," *Nature Cell Biology*, vol. 12, no. 3, pp. 213–223, 2010.

[16] T. Johansen and T. Lamark, "Selective autophagy mediated by autophagic adapter proteins," *Autophagy*, vol. 7, no. 3, pp. 279–296, 2011.

[17] T. L. Thurston, G. Ryzhakov, S. Bloor, N. von Muhlinen, and F. Randow, "The tbk1 adaptor and autophagy receptor ndp52 restricts the proliferation of ubiquitin-coated bacteria," *Nature Immunology*, vol. 10, no. 11, pp. 1215–1221, 2009.

[18] V. Kirkin, T. Lamark, Y. S. Sou et al., "A role for nbr1 in autophagosomal degradation of ubiquitinated substrates," *Molecular Cell*, vol. 33, no. 4, pp. 505–516, 2009.

[19] S. Waters, K. Marchbank, E. Solomon, C. Whitehouse, and M. Gautel, "Interactions with lc3 and polyubiquitin chains link nbr1 to autophagic protein turnover," *FEBS Letters*, vol. 583, no. 12, pp. 1846–1852, 2009.

[20] S. Pankiv, T. H. Clausen, T. Lamark et al., "P62/sqstm1 binds directly to atg8/lc3 to facilitate degradation of ubiquitinated protein aggregates by autophagy," *Journal of Biological Chemistry*, vol. 282, no. 33, pp. 24131–24145, 2007.

[21] M. L. Seibenhener, J. R. Babu, T. Geetha, H. C. Wong, N. R. Krishna, and M. W. Wooten, "Sequestosome 1/p62 is a polyubiquitin chain binding protein involved in ubiquitin proteasome degradation," *Molecular and Cellular Biology*, vol. 24, no. 18, pp. 8055–8068, 2004.

[22] C. D'Agostino et al., "Abnormalities of NBR1, a novel autophagy-associated protein, in muscle fibers of sporadic inclusion-body myositis," *Acta Neuropathologica*, vol. 122, no. 5, pp. 627–636, 2011.

[23] S. Lange, F. Xiang, A. Yakovenko et al., "Cell biology: the kinase domain of titin controls muscle gene expression and protein turnover," *Science*, vol. 308, no. 5728, pp. 1599–1603, 2005.

[24] C. A. Whitehouse, S. Waters, K. Marchbank et al., "Neighbor of brca1 gene (nbr1) functions as a negative regulator of postnatal osteoblastic bone formation and p38 mapk activity," *Proceedings of the National Academy of Sciences of the United States of America*, vol. 107, no. 29, pp. 12913–12918, 2010.

[25] J. Q. Yang, H. Liu, M. T. Diaz-Meco, and J. Moscat, "Nbr1 is a new pb1 signalling adapter in th2 differentiation and allergic airway inflammation *in vivo*," *EMBO Journal*, vol. 29, no. 19, pp. 3421–3433, 2010.

[26] T. C. Kuo et al., "Midbody accumulation through evasion of autophagy contributes to cellular reprogramming and tumorigenicity," *Nature Cell Biology*, vol. 13, no. 10, pp. 1214–1223, 2011.

[27] F. K. Mardakheh et al., "Nbr1 is a novel inhibitor of ligand-mediated receptor tyrosine kinase degradation," *Molecular and Cellular Biology*, vol. 30, no. 24, pp. 5672–5685, 2010.

[28] F. K. Mardakheh, M. Yekezare, L. M. Machesky, and J. K. Heath, "Spred2 interaction with the late endosomal protein nbr1 down-regulates fibroblast growth factor receptor signaling," *Journal of Cell Biology*, vol. 187, no. 2, pp. 265–277, 2009.

[29] C. Whitehouse, J. Chambers, K. Howe, M. Cobourne, P. Sharpe, and E. Solomon, "Nbr1 interacts with fasciculation and elongation protein zeta-1 (fez1) and calcium and integrin binding protein (cib) and shows developmentally restricted expression in the neural tube," *European Journal of Biochemistry*, vol. 269, no. 2, pp. 538–545, 2002.

[30] J. Heineke, M. Auger-Messier, R. N. Correll et al., "Cib1 is a regulator of pathological cardiac hypertrophy," *Nature Medicine*, vol. 16, no. 8, pp. 872–879, 2010.

[31] N. Sakae, N. Yamasaki, K. Kitaichi et al., "Mice lacking the schizophrenia-associated protein fez1 manifest hyperactivity and enhanced responsiveness to psychostimulants," *Human Molecular Genetics*, vol. 17, no. 20, pp. 3191–3203, 2008.

[32] E. Itakura and N. Mizushima, "P62 targeting to the autophagosome formation site requires self-oligomerization but not lc3 binding," *Journal of Cell Biology*, vol. 192, no. 1, pp. 17–27, 2011.

[33] S. R. Tymanskyj, T. M. Scales, and P. R. Gordon-Weeks, "MAP1B enhances microtubule assembly rates and axon extension rates in developing neurons," *Mol Cell Neurosci*, vol. 49, no. 2, pp. 110–119, 2011.

[34] M. Johnstone, R. G. Goold, I. Fischer, and P. R. Gordon-Weeks, "The neurofilament antibody rt97 recognises a developmentally regulated phosphorylation epitope on microtubule-associated protein 1b," *Journal of Anatomy*, vol. 191, no. 2, pp. 229–244, 1997.

[35] S. Lange, D. Auerbach, P. McLoughlin et al., "Subcellular targeting of metabolic enzymes to titin in heart muscle may be mediated by dral/fhl-2," *Journal of Cell Science*, vol. 115, no. 24, pp. 4925–4936, 2002.

[36] C. Kraft, M. Peter, and K. Hofmann, "Selective autophagy: ubiquitin-mediated recognition and beyond," *Nature Cell Biology*, vol. 12, no. 9, pp. 836–841, 2010.

[37] S. Svenning et al., "Plant NBR1 is a selective autophagy substrate and a functional hybrid of the mammalian autophagic adapters NBR1 and p62/SQSTM1," *Autophagy*, vol. 7, no. 9, pp. 993–1010, 2011.

[38] M. Tögel, R. Eichinger, G. Wiche, and F. Propst, "A 45 amino acid residue domain necessary and sufficient for proteolytic cleavage of the map1b polyprotein precursor," *FEBS Letters*, vol. 451, no. 1, pp. 15–18, 1999.

[39] J. A. Hammarback, R. A. Obar, S. M. Hughes, and R. B. Vallee, "Map1b is encoded as a polyprotein that is processed to form a complex N-terminal microtubule-binding domain," *Neuron*, vol. 7, no. 1, pp. 129–139, 1991.

[40] R. Noiges, R. Eichinger, W. Kutschera et al., "Microtubule-associated protein 1a (map1a) and map1b: light chains determine distinct functional properties," *Journal of Neuroscience*, vol. 22, no. 6, pp. 2106–2114, 2002.

[41] M. Tögel, G. Wiche, and F. Propst, "Novel features of the light chain of microtubule-associated protein map1b: microtubule stabilization, self interaction, actin filament binding, and regulation by the heavy chain," *Journal of Cell Biology*, vol. 143, no. 3, pp. 695–707, 1998.

[42] Q. J. Wang, Y. Ding, S. Kohtz et al., "Induction of autophagy in axonal dystrophy and degeneration," *Journal of Neuroscience*, vol. 26, no. 31, pp. 8057–8068, 2006.

[43] N. Trivedi, P. Marsh, R. G. Goold, A. Wood-Kaczmar, and P. R. Gordon-Weeks, "Glycogen synthase kinase-3β phosphorylation of map1b at ser1260 and thr1265 is spatially restricted to growing axons," *Journal of Cell Science*, vol. 118, no. 5, pp. 993–1005, 2005.

[44] L. Qiao and J. Zhang, "Inhibition of lysosomal functions reduces proteasomal activity," *Neuroscience Letters*, vol. 456, no. 1, pp. 15–19, 2009.

[45] E. Allen, J. Ding, W. Wang et al., "Gigaxonin-controlled degradation of map1b light chain is critical to neuronal survival," *Nature*, vol. 438, no. 7065, pp. 224–228, 2005.

[46] J. Yang, Y. Takahashi, E. Cheng et al., "Gsk-3β promotes cell survival by modulating bif-1-dependent autophagy and cell death," *Journal of Cell Science*, vol. 123, no. 6, pp. 861–870, 2010.

[47] Q. J. Wang, Y. Ding, S. Kohtz et al., "Induction of autophagy in axonal dystrophy and degeneration," *Journal of Neuroscience*, vol. 26, no. 31, pp. 8057–8068, 2006.

[48] C. Behrends, M. E. Sowa, S. P. Gygi, and J. W. Harper, "Network organization of the human autophagy system," *Nature*, vol. 466, no. 7302, pp. 68–76, 2010.

[49] Y. Watanabe and M. Tanaka, "P62/sqstm1 in autophagic clearance of a non-ubiquitylated substrate," *Journal of Cell Science*, vol. 124, no. 16, pp. 2692–2701, 2011.

[50] Y. Kawaguchi, J. J. Kovacs, A. McLaurin, J. M. Vance, A. Ito, and T. P. Yao, "The deacetylase hdac6 regulates aggresome formation and cell viability in response to misfolded protein stress," *Cell*, vol. 115, no. 6, pp. 727–738, 2003.

[51] S. Pankiv, E. A. Alemu, A. Brech et al., "Fyco1 is a rab7 effector that binds to lc3 and pi3p to mediate microtubule plus end - directed vesicle transport," *Journal of Cell Biology*, vol. 188, no. 2, pp. 253–269, 2010.

[52] R. Xie, S. Nguyen, K. McKeehan, F. Wang, W. L. McKeehan, and L. Liu, "Microtubule-associated protein 1s (map1s) bridges autophagic components with microtubules and mitochondria to affect autophagosomal biogenesis and degradation," *Journal of Biological Chemistry*, vol. 286, no. 12, pp. 10367–10377, 2011.

[53] D. C. Ko, J. Binkley, A. Sidow, and M. P. Scott, "The integrity of a cholesterol-binding pocket in niemann-pick c2 protein is necessary to control lysosome cholesterol levels," *Proceedings of the National Academy of Sciences of the United States of America*, vol. 100, no. 5, pp. 2518–2525, 2003.

[54] N. Friedland, H. L. Liou, P. Lobel, and A. M. Stock, "Structure of a cholesterol-binding protein deficient in niemann—pick type C2 disease," *Proceedings of the National Academy of Sciences of the United States of America*, vol. 100, no. 5, pp. 2512–2517, 2003.

15

Plastin Family of Actin-Bundling Proteins: Its Functions in Leukocytes, Neurons, Intestines, and Cancer

Hiroto Shinomiya

Department of Immunology and Host Defenses, Ehime University Graduate School of Medicine, Toon, Ehime 791-0295, Japan

Correspondence should be addressed to Hiroto Shinomiya, hiroto@m.ehime-u.ac.jp

Academic Editor: Liza Pon

Sophisticated regulation of the actin cytoskeleton by a variety of actin-binding proteins is essential for eukaryotic cells to perform their diverse functions. The plastin (also know, as fimbrin) protein family belongs to actin-bundling proteins, and the protein family is evolutionarily conserved and expressed in yeast, plant, and animal cells. Plastins are characterized by EF-hand Ca^{2+}-binding domains and actin-binding domains and can cross-link actin filaments into higher-order assemblies like bundles. Three isoforms have been identified in mammals. T-plastin is expressed in cells from solid tissues, such as neurons in the brain. I-plastin expression is restricted to intestine and kidney; the isoform plays a vital role in the function of absorptive epithelia in these organs. L-plastin is expressed in hematopoietic cell lineages and in many types of cancer cells; the isoform is thus considered to be a useful biomarker for cancer.

1. Introduction

Dynamics of the actin cytoskeleton is one of the cardinal features of eukaryotic cells, which is essential for fundamental cellular functions such as cell division, intracellular traffic of organelles, cell morphology, and cell motility [1]. The architecture of the actin cytoskeleton is regulated by a variety of proteins termed actin-binding proteins [1]. Actin filaments are organized into two types of arrays: bundles and weblike networks. Likewise, the actin filaments cross-linking proteins that help to stabilize and maintain these distinct structures are divided into two classes: bundling proteins and web-forming proteins. The plastin (also known as fimbrin) protein family belongs to bundling proteins and is evolutionarily conserved from yeast to mammalian cells. In mammals, three isoforms are known to be expressed in a cell-type-specific manner and exhibit distinct properties (Table 1 and Figure 1). In this paper, studies regarding the structure and biological functions of plastins are reviewed.

2. Structure of Plastins

The structure of plastins is well conserved from lower eukaryotes to humans and is characterized by actin binding domains (ABD). An ABD consists of a pair of ~ 125 residue calponin-homology (CH) domains (Figure 2(a)). ABD-containing proteins include proteins such as spectrin, α-actinin, dystrophin, cortexillin, and plastin/fimbrin [2]. The plastins are unique among these, as they possess two tandem repeats of ABD (ABD1 and ABD2) within a single polypeptide chain and cross-link actin filaments into higher order assemblies like bundles through this tandem pair of ABDs [3]. N-terminal EF-hand Ca^{2+}-binding domains (Figure 2(a)) are also important, since the actin bundling activity of plastins is regulated by Ca^{2+} [4].

Though ABD1 was previously solved by X-ray crystallography [5], the complete crystal structure of plastins has not yet been resolved. We have recently performed conformational analyses of murine L-plastin by X-ray scattering in solution and for the first time shown the overall structure of full-length plastin protein by reconstructing from data using the DAMMIN program (Figure 2(b)) [6]. The program DAMMIN is an advanced modeling procedure designed to reconstruct the shape of a molecule from the scattering intensity of small angle X-ray scattering data. Our results, taken together with those by Klein et al. on the plastin core region [7], demonstrated that plastin has a compact globular structure rather than a dumbbell-like shape and that the two

TABLE 1: Plastin isoforms expressed in distinct cell types.

Isoforms	Cell/tissue types
I-plastin (Plastin 1)	Intestine and kidney
L-plastin (Plastin 2)	Leukocytes and cancer
T-plastin (Plastin 3)	Solid tissues

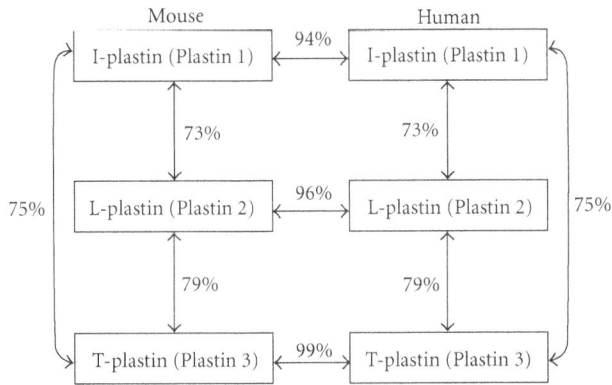

FIGURE 1: Human and mouse plastin isoforms. The homology between the amino acid sequences of these isoforms is showed.

ABDs are packed together in an approximately antiparallel arrangement with the N- and C-terminal CH domains (CH1 and CH4) making direct contact (Figure 2(c)). We also demonstrated that significant conformational changes of the protein were induced in the presence of Ca^{2+} [6]. These findings should shed light on the molecular mechanisms of how plastins regulate the architecture of actin filaments.

3. Functions of Plastin Isoforms

Since plastins are expressed in a cell-type-specific manner, it is conceivable that the most suitable isoform is expressed to regulate the actin cytoskeleton in a particular type of cell. Functions of plastin isoforms in distinct types of cells or tissues in mammals are reviewed.

3.1. Plastin Functions in Leukocytes. We isolated a 65-kDa cytosolic protein that was phosphorylated in murine macrophages by stimulation with bacterial lipopolysaccharide (LPS) and determined its complete primary structure as a novel protein [8–10]. The sequence of the 65-kDa protein revealed that it was a murine homolog of human L-plastin that had been identified as a transformation-induced polypeptide of neoplastic human fibroblasts [11]. We further demonstrated that L-plastin plays an important role in macrophage functions such as host defense against bacterial infections [12–15]. We, and others, have clarified that L-plastin is exclusively expressed in leukocytes such as lymphocytes, macrophages, and granulocytes under physiological conditions. Representative studies addressing the role of L-plastin in leukocyte functions are shown in Table 2 [8–10, 12–39]; the isoform serves important functions not only in cells of innate immunity such as macrophages and

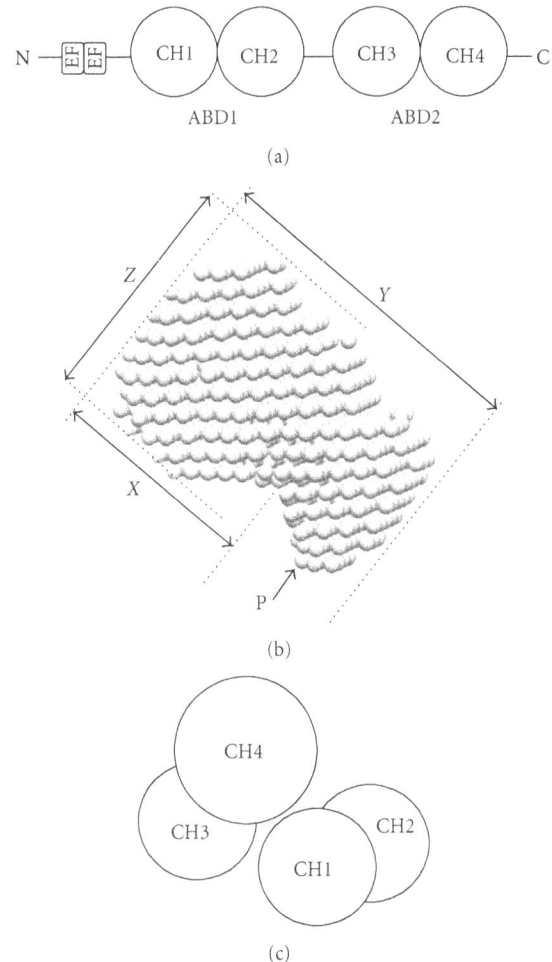

FIGURE 2: Schematic diagram of L-plastin structure. (a) Domain organization of L-plastin. The protein possesses an N-terminal headpiece of ~100 amino acids containing two EF-hand Ca^{2+}-binding motifs and two actin-binding domains (ABDs) consisting of ABD1 (residues 120–379) and ABD2 (residues 394–623), and each ABD contains two calponin-homology (CH) domains. (b) Reconstructed molecular shape of L-plastin. Conformational analyses of L-plastin by X-ray scattering in solution revealed that plastin has a compact globular structure rather than a dumbbell-like shape. It is conceivable that the two ABDs are packed together in an approximately antiparallel arrangement with the N- and C-terminal CH domains (CH1 and CH4) making direct contact as shown in Figure 2(c); that is, X, Y, and Z correspond to CH1–CH3, CH2–CH4, and CH3-CH4 of the plastin protein, respectively. P indicates a putative N-terminal headpiece. See more details in [6, 7]. (c) Possible arrangement of the ABDs (CH1–CH4) of L-plastin in solution without Ca^{2+}.

granulocytes (neutrophils and eosinophils) [8–10, 12–29], but also in those of adaptive immunity such as T and B lymphocytes [30–37].

Many of the studies have addressed the phosphorylation of L-plastin during leukocyte activation by various stimuli (Table 2). Only L-plastin has, so far, been known to be phosphorylated in cells among three isoforms. We demonstrated that L-plastin was phosphorylated exclusively on Ser5 in

TABLE 2: Studies on L-plastin in leukocytes.

Leukocyte types	Key words of the study	References
Innate immunity		
Macrophages	Bacterial lipopolysaccharide (LPS)-induced activation	[8–10, 12]*, [13, 14]
Macrophages	Grancalcin and host defence	[15]
Macrophages	IL-1/TNF-induced activation	[16]*
Macrophages	Podosome formation	[17]
Macrophages	Zebrafish and lineage marker	[18–20]
Macrophages	*Toxoplasma gondii*-infection	[21]
Neutrophils	IL-8-induced activation	[22]*
Neutrophils	Fc receptor and phagocytosis	[23, 24]*
Neutrophils	L-plastin KO# and integrin	[25, 26]*
Eosinophils	GM-CSF-induced priming	[27]*
Osteoclasts	Podosome formation	[28]
Osteoclasts	Sealing ring formation	[29]
Adaptive immunity		
T-cells	IL-2-induced activation	[30]*
T-cells	Accessory receptors	[31]*
T-cells	Lymphokine-activated killer cells	[32]*
T-cells	Costimulation	[33]*
T-cells	CCR7 and thymus	[34]
T-cells	LFA-1 and immune synapse	[35]
T-cells	L-plastin KO# and impaired T cell responses	[36]
B-cells	Marginal zone B cell development	[37]
Suppression immunotherapies targeting L-plastin		
Leukocytes	Cannabinoid receptor agonists	[38]
Leukocytes	Glucocorticoid dexamethasone	[39]

* These studies addressed the L-plastin phosphorylation in leukocytes.
# Disruption of the L-plastin gene in mice.

macrophages by stimulation with LPS [10]. In connection with this, Jones et al. showed that the Ser5-phosphorylated L-plastin peptide induced adhesion in neutrophils [25]. It was also demonstrated that phosphorylation on Ser5 increased the F-actin-binding activity of L-plastin and promoted its targeting to sites of actin assembly in cells [40]. In addition to phosphorylation, the actin-bundling action of L-plastin is also regulated by intracellular Ca^{2+} [4]. The above findings indicate that L-plastin regulates the actin cytoskeleton in a phosphorylation- and/or Ca^{2+}-dependent manner. These features of L-plastin seem to help leukocytes rapidly rearrange their actin cytoskeleton when they need to quickly respond to a variety of extracellular stimuli. Indeed, it has recently been demonstrated by using L-plastin gene-disrupted mice that neutrophils lacking L-plastin are deficient in killing bacterial pathogens [26] and that T and B cell functions are also impaired in the mice; T cell responses to antigens are impaired [36], and splenic maturation of B cells and an antibody response to *Streptococcus pneumoniae* are largely diminished [37].

These studies regarding the key roles of L-plastin in leukocytes functions may provide the basis for new therapy. Several suppression immunotherapies targeting L-plastin

have been proposed (Table 2) [38, 39]. Cannabinoid CB2 receptor-specific compounds were found to inhibit the L-plastin phosphorylation in human monocytes and blocked experimental autoimmune encephalomyelitis in the rat [38]. In addition, it has recently been shown that glucocorticoid dexamethasone inhibits the L-plastin phosphorylation in human T cells, which prevents the immune synapse formation and subsequent T cell activation [39].

3.2. Plastin Functions in Neurons. The nervous system chiefly expresses T-plastin among the three isoforms. Though microglia in the brain are a kind of resident macrophage, their expression of L-plastin is very low under physiological conditions (unpublished observations). With regard to T-plastin in neurons, an interesting article has recently been published; the authors showed that high expression of T-plastin acts as a protective modifier of spinal muscular atrophy (SMA), the most frequent genetic cause of early childhood lethality, and that T-plastin is important for axonogenesis, as its overexpression rescues the axon length and outgrowth defects in neurons of the SMA mouse [41]. Therefore, T-plastin seems to play a vital role in neuronal differentiation. Another interesting study addressed the preventive role of T-plastin in neurodegenerative diseases, demonstrating that T-plastin lessens the toxicity of polyglutamine proteins in neurons, such as ataxin and huntingtin, that cause neurodegenerative disorder [42]. These, and related studies, are summarized in Table 3 [41–43].

Since sensory cells are closely related to the nervous system, studies regarding the plastin expression in sensory cells are reviewed here as shown in Table 3. Several studies using specific antibodies against plastins/fimbrin revealed that the proteins are expressed in the stereocilia of auditory hair cells in the chicken and mammals [44–47], suggesting that plastins may contribute to the physiological auditory sense. In the rat cochlear hair cells, T- and I-plastin, but not L-plastin, were found to be expressed [47]. T-plastin is expressed only in the early stage of hair cell differentiation, while I-plastin is constantly expressed from the early to the adult stages, indicating that I-plastin is the major isoform expressed in the adult cochlear hair cells. The architecture of actin filaments cross-linked by plastins may be essential for the transduction of auditory signals by cochlear hair cells.

3.3. Plastin Functions in the Intestine.
The first protein of the plastin/fimbrin family was discovered in microvilli of the chicken intestinal brush border and named fimbrin [48]. Chicken fimbrin was characterized as a cytoskeletal protein that binds and cross-links F-actin filaments [49, 50]. Meanwhile, a third plastin isoform, named I-plastin, was identified in humans and was found to be expressed in the small intestine, colon, and kidney [51]. Thus, I-plastin is considered to be the human homolog of chicken fimbrin. Using I-plastin gene-disrupted mice, it has recently been reported that a lack of I-plastin results in increased fragility of the intestinal epithelium and decreased transepithelial resistance [52], suggesting that I-plastin is an important regulator of brush border morphology and stability. Recently, a model of the microvillar cytoskeleton of the brush border that includes plastin/fimbrin cross-linking actin filaments has been proposed [53]. These studies are summarized in Table 4 [48–53].

3.4. Plastin Functions in Cancer.
The cell-type-specific expression of plastin isoforms is strictly regulated under physiological conditions. However, ectopic expression of plastins in malignant cells has been observed in many studies as shown in Table 5 [11, 54–69]. L-plastin that is normally expressed only in hematopoietic cells is especially expressed in a variety of cancer cells of nonhematopoietic origin. Lin et al. demonstrated that 68% of cancers derived from epithelia express L-plastin [54]. It was further clarified by using sensitive RT-PCR that the L-plastin gene is activated in most human cancer cells [56]. Thus, L-plastin has been considered to be a common marker of many types of human cancer. In particular, a high percentage of cancer cells arising from female reproductive tissues express L-plastin constitutively and abundantly though its expression is ovarian steroid hormone-independent in cancer cells [58]. In addition, the expression of L-plastin in prostatic epithelial cells is linked to the malignant state, and once expressed in

TABLE 3: Studies on plastins in the nervous system and sensory cells.

Key words of the study	References
T-plastin in neurons	
Spinal muscular atrophy and axonogenesis	[41, 43]
Spinocerebellar ataxia and polyglutamine protein	[42]
Plastin/fimbrin in sensory cells	
Fimbrin, chicken, stereocilia, auditory hair cells	[44, 45]
I-plastin, mouse/rat, auditory hair cells	[46, 47]

TABLE 4: Studies on fimbrin/I-plastin in the intestine.

Key words of the study	References
Fimbrin, Microvilli	[48]
Intestinal brush border	[49, 50]
I-plastin, Intestine, Kidney	[51]
I-plastin KO[#], Intestinal epithelium	[52]
Model of microvillar cytoskeleton	[53]

[#] Disruption of the I-plastin gene in mice.

carcinoma, its expression is regulated by steroid hormone receptors [59]. The L-plastin gene promoter was found to include several hormone receptor-responsive elements [55, 57]. These evidences support a *trans*-activation mechanism for the activation of L-plastin synthesis accompanying tumorigenesis. In contrast, T-plastin gene expression was found to be suppressed in human colorectal cancer cells, suggesting that downregulation of T-plastin is involved in cancer development [65].

Since L-plastin is normally expressed in leukocytes that are able to move rapidly to infectious and inflammatory sites, cancer cells may gain the ability to metastasize to other parts of the body by expressing L-plastin. In other words, L-plastin tends to be expressed in freely movable cells such as leukocytes and cancer cells. On the other hand, it was found that T-plastin is expressed in certain types of malignant cells of leukocyte origin, including cutaneous T cell lymphoma and Sezary syndrome (Table 5) [66–69]. In cutaneous T cell lymphoma, dense clusters or nodules of malignant cells are observed [70]; these cells appear to lose their motility. This seems to contrast sharply with the case of cancer cells that express L-plastin. Considering the above, the expression of plastin isoforms could be dysregulated when cells, regardless of their origin, become malignant, which may endow tumor cells with properties distinct from those of their normal counterparts.

On the basis of these observations regarding abnormal expression of plastins, clinical applications have been developed. Cancer screening and diagnosis methods assessing the expression of plastins as a biomarker have been described. These include choroids plexus tumors, urinary bladder cancer, ovarian cancer, and colorectal cancer (Table 5) [71–74]. Furthermore, gene therapy experiments targeting the L-plastin gene or L-plastin promoters in cancer cells have been started, as shown in Table 5 [75–78]. These approaches have shown promising results. For example, Peng et al.

TABLE 5: Studies on plastins in cancer cells.

Key words of the study	References
Ectopic expression of L-plastin in cancer	
Transformed human fibroblasts	[11]
Many types of human cancer	[54]
L-plastin gene promoter in cancer	[55–57]
Ovarian steroid hormones	[58]
Prostate cancer and steroid hormone	[59]
Chromosome translocation	[60]
Breast cancer and expression pattern	[61]
Colorectal cancer and metastasis	[62]
Colon cancer, invasion, and loss of E-cadherin	[63]
Melanoma tumor invasion	[64]
T-plastin downregulation and CpG methylation	[65]
Ectopic expression of T-plastin in lymphoma	
Cutaneous T cell lymphoma	[66, 67]
Sezary cells	[68, 69]
As a biomarker for cancer screening and diagnosis	
Choroid plexus tumors and diagnostic marker	[71]
Bladder cancer and biomarker	[72]
Proteomics imaging and mass spectrometry	[73]
Colorectal cancer and human feces	[74]
Gene therapies targeting L-plastin gene	
L-plastin promoter and gene therapy	[75–77]
Antisense L-plastin gene and tumor suppression	[78]

prepared adenoviral vectors in which a truncated human L-plastin promoter and the *cytosine deaminase* (*CD*) gene were coinserted [76]. *CD* is a bacterial gene which converts 5-fluorocytosine (5FC: nontoxic to cells) to 5-fluorouracil (5FU: toxic to most cells). When the vector is transfected into human ovarian or bladder cancer cells *in vitro*, CD transcription is increased through the L-plastin promoter activation, leading to tumor cell death via the conversion of 5FC to 5FU in the cells. The authors also performed *in vivo* study and demonstrated that human tumor masses grown in nude mice were reduced by this method [76]. In another study by Zheng et al., the authors constructed retroviral vectors to express regions of the human L-plastin gene in antisense orientation and found that introduction of the vectors into prostate carcinoma cells reduced the growth rates of the cells and suppressed their invasion and motility *in vitro* [78]. This suggests that overexpression of L-plastin is involved in cancer invasion and metastasis and that downregulation of L-plastin by antisense delivery is potentially a useful approach to interfere with prostate cancer progression.

4. Perspectives

Since plastins are evolutionarily conserved and expressed in yeast, plant, and animal cells, they should play a fundamental role in cellular activities. Sophisticated regulation of the actin cytoskeleton seems to be a mandatory event for the functions of eukaryotic cells. It has been investigated in detail how plastins interact with actin filaments *in vitro*. Extensive studies have clarified that three plastin isoforms are expressed in distinct types of cells in mammals and that dysregulated expression of plastins appears to be important in the progression of various cancers. As described in this paper, plastins can serve as versatile regulators of the actin cytoskeleton in many aspects of cellular functions. Further studies will be expected to reveal how plastins, together with other actin-binding proteins, dynamically regulate the remodeling of cytoskeletal architecture during diverse cellular activities.

Abbreviations

ABD: Actin binding domain
CH: Calponin-homology
IL: Interleukin
LPS: Bacterial lipopolysaccharide
SMA: Spinal muscular atrophy
TNF: Tumor necrosis factor.

Acknowledgment

This work was supported by Grant-in-Aid for Scientific Research (C). No conflict of interests is present.

References

[1] T. D. Pollard and J. A. Cooper, "Actin, a central player in cell shape and movement," *Science*, vol. 326, no. 5957, pp. 1208–1212, 2009.

[2] S. Bañuelos, M. Saraste, and K. D. Carugo, "Structural comparisons of calponin homology domains: implications for actin binding," *Structure*, vol. 6, no. 11, pp. 1419–1431, 1998.

[3] M. V. de Arruda, S. Watson, C. S. Lin, J. Leavitt, and P. Matsudaira, "Fimbrin is a homologue of the cytoplasmic phosphoprotein plastin and has domains homologous with calmodulin and actin gelation proteins," *Journal of Cell Biology*, vol. 111, no. 3, pp. 1069–1079, 1990.

[4] Y. Namba, M. Ito, Y. Zu, K. Shigesada, and K. Maruyama, "Human T cell L-plastin bundles actin filaments in a calcium dependent manner," *Journal of Biochemistry*, vol. 112, no. 4, pp. 503–507, 1992.

[5] S. C. Goldsmith, N. Pokala, W. Shen, A. A. Fedorov, P. Matsudaira, and S. C. Almo, "The structure of an actin-crosslinking domain from human fimbrin," *Nature Structural Biology*, vol. 4, no. 9, pp. 708–712, 1997.

[6] H. Shinomiya, M. Shinjo, L. Fengzhi, Y. Asano, and H. Kihara, "Conformational analysis of the leukocyte-specific EF-hand protein p65/L-plastin by X-ray scattering in solution," *Biophysical Chemistry*, vol. 131, no. 1–3, pp. 36–42, 2007.

[7] M. G. Klein, W. Shi, U. Ramagopal et al., "Structure of the actin crosslinking core of fimbrin," *Structure*, vol. 12, no. 6, pp. 999–1013, 2004.

[8] H. Shinomiya, H. Hirata, and M. Nakano, "Purification and characterization of the 65-kDa protein phosphorylated in murine macrophages by stimulation with bacterial lipopolysaccharide," *Journal of Immunology*, vol. 146, no. 10, pp. 3617–3625, 1991.

[9] H. Shinomiya, H. Hirata, S. Saito, H. Yagisawa, and M. Nakano, "Identification of the 65-kDa phosphoprotein in murine macrophages as a novel protein: homology with human L-plastin," *Biochemical and Biophysical Research Communications*, vol. 202, no. 3, pp. 1631–1638, 1994.

[10] H. Shinomiya, A. Hagi, M. Fukuzumi, M. Mizobuchi, H. Hirata, and S. Utsumi, "Complete primary structure and phosphorylation site of the 65-kDa macrophage protein phosphorylated by stimulation with bacterial lipopolysaccharide," *Journal of Immunology*, vol. 154, no. 7, pp. 3471–3478, 1995.

[11] C. S. Lin, R. H. Aebersold, S. B. Kent, M. Varma, and J. Leavitt, "Molecular cloning and characterization of plastin, a human leukocyte protein expressed in transformed human fibroblasts," *Molecular and Cellular Biology*, vol. 8, no. 11, pp. 4659–4668, 1988.

[12] A. Hagi, H. Hirata, and H. Shinomiya, "Analysis of a bacterial lipopolysaccharide-activated serine kinase that phosphorylates p65/L-plastin in macrophages," *Microbiology and Immunology*, vol. 50, no. 4, pp. 331–335, 2006.

[13] H. Shinomiya, K. Nagai, H. Hirata et al., "Preparation and characterization of recombinant murine p65/L-plastin expressed in *Escherichia coli* and high-titer antibodies against the protein," *Bioscience, Biotechnology and Biochemistry*, vol. 67, no. 6, pp. 1368–1375, 2003.

[14] K. Toyooka, F. Liu, M. Ishii et al., "Generation and characterization of monoclonal antibodies that specifically recognize p65/L-plastin isoform but not T-plastin isoform," *Bioscience, Biotechnology and Biochemistry*, vol. 70, no. 6, pp. 1402–1407, 2006.

[15] F. Liu, H. Shinomiya, T. Kirikae, H. Hirata, and Y. Asano, "Characterization of murine grancalcin specifically expressed in leukocytes and its possible role in host defense against bacterial infection," *Bioscience, Biotechnology and Biochemistry*, vol. 68, no. 4, pp. 894–902, 2004.

[16] M. Shiroo and K. Matsushima, "Enhanced phosphorylation of 65 and 74 kDa proteins by tumor necrosis factor and interleukin-1 in human peripheral blood mononuclear cells," *Cytokine*, vol. 2, no. 1, pp. 13–20, 1990.

[17] J. G. Evans, I. Correia, O. Krasavina, N. Watson, and P. Matsudaira, "Macrophage podosomes assemble at the leading lamella by growth and fragmentation," *Journal of Cell Biology*, vol. 161, no. 4, pp. 697–705, 2003.

[18] P. Herbomel, B. Thisse, and C. Thisse, "Ontogeny and behaviour of early macrophages in the zebrafish embryo," *Development*, vol. 126, no. 17, pp. 3735–3745, 1999.

[19] P. Herbomel, B. Thisse, and C. Thisse, "Zebrafish early macrophages colonize cephalic mesenchyme and developing brain, retina, and epidermis through a M-CSF receptor-dependent invasive process," *Developmental Biology*, vol. 238, no. 2, pp. 274–288, 2001.

[20] J. R. Mathias, M. E. Dodd, K. B. Walters, S. K. Yoo, E. A. Ranheim, and A. Huttenlocher, "Characterization of zebrafish larval inflammatory macrophages," *Developmental and Comparative Immunology*, vol. 33, no. 11, pp. 1212–1217, 2009.

[21] D. H. Zhou, Z. G. Yuan, F. R. Zhao et al., "Modulation of mouse macrophage proteome induced by *Toxoplasma gondii* tachyzoites in vivo," *Parasitology Research*, vol. 109, no. 6, pp. 1637–1646, 2011.

[22] M. Shibata, Y. Yamakawa, T. Ohoka, S. Mizuno, and K. Suzuki, "Characterization of a 64-kd protein phosphorylated during chemotactic activation with IL-8 and fMLP of human polymorphonuclear leukocytes. II. Purification and amino

acid analysis of phosphorylated 64-kd protein," *Journal of Leukocyte Biology*, vol. 54, no. 1, pp. 10–16, 1993.

[23] S. L. Jones and E. J. Brown, "FcγRII-mediated adhesion and phagocytosis induce L-plastin phosphorylation in human neutrophils," *The Journal of Biological Chemistry*, vol. 271, no. 24, pp. 14623–14630, 1996.

[24] J. Wang and E. J. Brown, "Immune complex-induced integrin activation and L-plastin phosphorylation require protein kinase A," *The Journal of Biological Chemistry*, vol. 274, no. 34, pp. 24349–24356, 1999.

[25] S. L. Jones, J. Wang, C. W. Turck, and E. J. Brown, "A role for the actin-bundling protein L-plastin in the regulation of leukocyte integrin function," *Proceedings of the National Academy of Sciences of the United States of America*, vol. 95, no. 16, pp. 9331–9336, 1998.

[26] H. Chen, A. Mocsai, H. Zhang et al., "Role for plastin in host defense distinguishes integrin signaling from cell adhesion and spreading," *Immunity*, vol. 19, no. 1, pp. 95–104, 2003.

[27] K. Pazdrak, T. W. Young, C. Straub, S. Stafford, and A. Kurosky, "Priming of eosinophils by GM-CSF is mediated by protein kinase CbII-phosphorylated L-plastin," *Journal of Immunology*, vol. 186, no. 11, pp. 6485–6496, 2011.

[28] S. G. Babb, P. Matsudaira, M. Sato, I. Correia, and S. S. Lim, "Fimbrin in podosomes of monocyte-derived osteoclasts," *Cell Motility and the Cytoskeleton*, vol. 37, no. 4, pp. 308–325, 1997.

[29] T. Ma, K. Sadashivaiah, N. Madayiputhiya, and M. A. Chellaiah, "Regulation of sealing ring formation by L-plastin and cortactin in osteoclasts," *The Journal of Biological Chemistry*, vol. 285, no. 39, pp. 29911–29924, 2010.

[30] Y. Zu, K. Shigesada, E. Nishida et al., "65-kilodalton protein phosphorylated by interleukin 2 stimulation bears two putative actin-binding sites and two calcium-binding sites," *Biochemistry*, vol. 29, no. 36, pp. 8319–8324, 1990.

[31] S. W. Henning, S. C. Meuer, and Y. Samstag, "Serine phosphorylation of a 67-kDa protein in human T lymphocytes represents an accessory receptor-mediated signaling event," *Journal of Immunology*, vol. 152, no. 10, pp. 4808–4815, 1994.

[32] M. J. Frederick, L. V. Rodriguez, D. A. Johnston, B. G. Darnay, and E. A. Grimm, "Characterization of the Mr 65,000 lymphokine-activated killer proteins phosphorylated after tumor target binding: evidence that pp65a and pp65b are phosphorylated forms of L-plastin," *Cancer Research*, vol. 56, no. 1, pp. 138–144, 1996.

[33] G. H. Wabnitz, T. Köcher, P. Lohneis et al., "Costimulation induced phosphorylation of L-plastin facilitates surface transport of the T cell activation molecules CD69 and CD25," *European Journal of Immunology*, vol. 37, no. 3, pp. 649–662, 2007.

[34] S. C. Morley, C. Wang, W. L. Lo et al., "The actin-bundling protein L-plastin dissociates CCR7 proximal signaling from CCR7-induced motility," *Journal of Immunology*, vol. 184, no. 7, pp. 3628–3638, 2010.

[35] G. H. Wabnitz, P. Lohneis, H. Kirchgessner et al., "Sustained LFA-1 cluster formation in the immune synapse requires the combined activities of L-plastin and calmodulin," *European Journal of Immunology*, vol. 40, no. 9, pp. 2437–2449, 2010.

[36] W. Chen, S. C. Morley, D. Donermeyer et al., "Actin-bundling protein L-plastin regulates T cell activation," *Journal of Immunology*, vol. 185, no. 12, pp. 7487–7497, 2010.

[37] E. M. Todd, L. E. Deady, and S. C. Morley, "The actin-bundling protein L-plastin is essential for marginal zone B cell development," *Journal of Immunology*, vol. 187, no. 6, pp. 3015–3025, 2011.

[38] C. A. Lunn, E. P. Reich, J. S. Fine et al., "Biology and therapeutic potential of cannabinoid CB2 receptor inverse agonists," *British Journal of Pharmacology*, vol. 153, no. 2, pp. 226–239, 2008.

[39] G. H. Wabnitz, F. Michalke, C. Stober et al., "L-plastin phosphorylation: a novel target for the immunosuppressive drug dexamethasone in primary human T-cells," *European Journal of Immunology*, vol. 41, no. 11, pp. 3157–3169, 2011.

[40] B. Janji, A. Giganti, V. De Corte et al., "Phosphorylation on Ser5 increases the F-actin-binding activity of L-plastin and promotes its targeting to sites of actin assembly in cells," *Journal of Cell Science*, vol. 119, no. 9, pp. 1947–1960, 2006.

[41] G. E. Oprea, S. Kröber, M. L. McWhorter et al., "Plastin 3 is a protective modifier of autosomal recessive spinal muscular atrophy," *Science*, vol. 320, no. 5875, pp. 524–527, 2008.

[42] M. Ralser, U. Nonhoff, M. Albrecht et al., "Ataxin-2 and huntingtin interact with endophilin-A complexes to function in plastin-associated pathways," *Human Molecular Genetics*, vol. 14, no. 19, pp. 2893–2909, 2005.

[43] M. Bowerman, C. L. Anderson, A. Beauvais, P. P. Boyl, W. Witke, and R. Kothary, "SMN, profilin IIa and plastin 3: a link between the deregulation of actin dynamics and SMA pathogenesis," *Molecular and Cellular Neuroscience*, vol. 42, no. 1, pp. 66–74, 2009.

[44] D. Drenckhahn, K. Engel, D. Hofer, C. Merte, L. Tilney, and M. Tilney, "Three different actin filament assemblies occur in every hair cell: each contains a specific actin crosslinking protein," *Journal of Cell Biology*, vol. 112, no. 4, pp. 641–651, 1991.

[45] K. H. Lee and D. A. Cotanche, "Localization of the hair-cell-specific protein fimbrin during regeneration in the chicken cochlea," *Audiology and Neuro-Otology*, vol. 1, no. 1, pp. 41–53, 1996.

[46] P. Lawlor, W. Marcotti, M. N. Rivolta, C. J. Kros, and M. C. Holley, "Differentiation of mammalian vestibular hair cells from conditionally immortal, postnatal supporting cells," *Journal of Neuroscience*, vol. 19, no. 21, pp. 9445–9458, 1999.

[47] N. Daudet and M. C. Lebart, "Transient expression of the T-isoform of plastins/fimbrin in the stereocilia of developing auditory hair cells," *Cell Motility and the Cytoskeleton*, vol. 53, no. 4, pp. 326–336, 2002.

[48] A. Bretscher and K. Weber, "Fimbrin, a new microfilament-associated protein present in microvilli and other cell surface structures," *Journal of Cell Biology*, vol. 86, no. 1, pp. 335–340, 1980.

[49] A. Bretscher, "Purification of an 80,000-dalton protein that is a component of the isolated microvillus cytoskeleton, and its localization in nonmuscle cells," *Journal of Cell Biology*, vol. 97, no. 2, pp. 425–432, 1983.

[50] M. B. Heintzelman and M. S. Mooseker, "Assembly of the brush border cytoskeleton: changes in the distribution of microvillar core proteins during enterocyte differentiation in adult chicken intestine," *Cell Motility and the Cytoskeleton*, vol. 15, no. 1, pp. 12–22, 1990.

[51] C. S. Lin, W. Shen, Z. P. Chen, Y. H. Tu, and P. Matsudaira, "Identification of I-plastin, a human fimbrin isoform expressed in intestine and kidney," *Molecular and Cellular Biology*, vol. 14, no. 4, pp. 2457–2467, 1994.

[52] E. M. S. Grimm-Günter, C. Revenu, S. Ramos et al., "Plastin 1 binds to keratin and is required for terminal web assembly in the intestinal epithelium," *Molecular Biology of the Cell*, vol. 20, no. 10, pp. 2549–2562, 2009.

[53] J. W. Brown and C. J. McKnight, "Molecular model of the microvillar cytoskeleton and organization of the brush border," *PLoS ONE*, vol. 5, no. 2, Article ID e9406, 2010.

[54] C. S. Lin, T. Park, Z. P. Chen, and J. Leavitt, "Human plastin genes. Comparative gene structure, chromosome location, and differential expression in normal and neoplastic cells," *The Journal of Biological Chemistry*, vol. 268, no. 4, pp. 2781–2792, 1993.

[55] C. S. Lin, Z. P. Chen, T. Park, K. Ghosh, and J. Leavitt, "Characterization of the human L-plastin gene promoter in normal and neoplastic cells," *The Journal of Biological Chemistry*, vol. 268, no. 4, pp. 2793–2801, 1993.

[56] T. Park, Z. P. Chen, and J. Leavitt, "Activation of the leukocyte plastin gene occurs in most human cancer cells," *Cancer Research*, vol. 54, no. 7, pp. 1775–1781, 1994.

[57] C. S. Lin, A. Lau, C. C. Yeh, C. H. Chang, and T. F. Lue, "Upregulation of L-plastin gene by testosterone in breast and prostate cancer cells: identification of three cooperative androgen receptor-binding sequences," *DNA and Cell Biology*, vol. 19, no. 1, pp. 1–7, 2000.

[58] J. Leavitt, Z. P. Chen, C. J. Lockwood, and F. Schatz, "Regulation of synthesis of the transformation-induced protein, leukocyte plastin, by ovarian steroid hormones," *Cancer Research*, vol. 54, no. 13, pp. 3447–3454, 1994.

[59] J. Zheng, N. Rudra-Ganguly, G. J. Miller, K. A. Moffatt, R. J. Cote, and P. Roy-Burman, "Steroid hormone induction and expression patterns of L-plastin in normal and carcinomatous prostate tissues," *American Journal of Pathology*, vol. 150, no. 6, pp. 2009–2018, 1997.

[60] S. Galiègue-Zouitina, S. Quief, M. P. Hildebrand et al., "Nonrandom fusion of L-plastin(LCP1) and LAZ3(BCL6) genes by t(3;13)(q27;q14) chromosome translocation in two cases of B-cell non-Hodgkin lymphoma," *Genes Chromosomes and Cancer*, vol. 26, no. 2, pp. 97–105, 1999.

[61] A. Lapillonne, O. Coué, E. Friederich et al., "Expression patterns of L-plastin isoform in normal and carcinomatous breast tissues," *Anticancer Research*, vol. 20, no. 5, pp. 3177–3182, 2000.

[62] M. Otsuka, M. Kato, T. Yoshikawa et al., "Differential expression of the L-plastin gene in human colorectal cancer progression and metastasis," *Biochemical and Biophysical Research Communications*, vol. 289, no. 4, pp. 876–881, 2001.

[63] E. Foran, P. McWilliam, D. Kelleher, D. T. Croke, and A. Long, "The leukocyte protein L-plastin induces proliferation, invasion and loss of E-cadherin expression in colon cancer cells," *International Journal of Cancer*, vol. 118, no. 8, pp. 2098–2104, 2006.

[64] M. Klemke, M. T. Rafael, G. H. Wabnitz et al., "Phosphorylation of ectopically expressed L-plastin enhances invasiveness of human melanoma cells," *International Journal of Cancer*, vol. 120, no. 12, pp. 2590–2599, 2007.

[65] Y. Sasaki, F. Itoh, T. Kobayashi et al., "Increased expression of T-fimbrin gene after DNA damage in CHO cells and inactivation of T-fimbrin by CPG methylation in human colorectal cancer cells," *International Journal of Cancer*, vol. 97, no. 2, pp. 211–216, 2002.

[66] L. Kari, A. Loboda, M. Nebozhyn et al., "Classification and prediction of survival in patients with the leukemic phase of cutaneous T cell lymphoma," *Journal of Experimental Medicine*, vol. 197, no. 11, pp. 1477–1488, 2003.

[67] E. Capriotti, E. C. Vonderheid, C. J. Thoburn, M. Wasik, D. W. Bahler, and A. D. Hess, "Expression of T-plastin, FoxP3

and other tumor-associated markers by leukemic T-cells of cutaneous T-cell lymphoma," *Leukemia and Lymphoma*, vol. 49, no. 6, pp. 1190–1201, 2008.

[68] M. W. Su, I. Dorocicz, W. H. Dragowska et al., "Aberrant expression of T-plastin in sezary cells," *Cancer Research*, vol. 63, no. 21, pp. 7122–7127, 2003.

[69] N. Tang, H. Gibson, T. Germeroth, P. Porcu, H. W. Lim, and H. K. Wong, "T-plastin (PLS3) gene expression differentiates Sézary syndrome from mycosis fungoides and inflammatory skin diseases and can serve as a biomarker to monitor disease progression," *British Journal of Dermatology*, vol. 162, no. 2, pp. 463–466, 2010.

[70] R. Hussain and A. Bajoghli, "Primary cutaneous CD30-positive large T cell lymphoma in an 80-year-old man: a case report," *ISRN Dermatology*, vol. 2011, Article ID 634042, pp. 1–3, 2011.

[71] M. Hasselblatt, C. Böhm, L. Tatenhorst et al., "Identification of novel diagnostic markers for choroid plexus tumors: a microarray-based approach," *American Journal of Surgical Pathology*, vol. 30, no. 1, pp. 66–74, 2006.

[72] L. D. Harris, J. De La Cerda, T. Tuziak et al., "Analysis of the expression of biomarkers in urinary bladder cancer using a tissue microarray," *Molecular Carcinogenesis*, vol. 47, no. 9, pp. 678–685, 2008.

[73] S. Kang, H. S. Shim, J. S. Lee et al., "Molecular proteomics imaging of tumor interfaces by mass spectrometry," *Journal of Proteome Research*, vol. 9, no. 2, pp. 1157–1164, 2010.

[74] C. S. Ang and E. C. Nice, "Targeted in-gel MRM: a hypothesis driven approach for colorectal cancer biomarker discovery in human feces," *Journal of Proteome Research*, vol. 9, no. 9, pp. 4346–4355, 2010.

[75] I. Chung, P. E. Schwartz, R. G. Crystal, G. Pizzorno, J. Leavitt, and A. B. Deisseroth, "Use of L-plastin promoter to develop an adenoviral system that confers transgene expression in ovarian cancer cells but not in normal mesothelial cells," *Cancer Gene Therapy*, vol. 6, no. 2, pp. 99–106, 1999.

[76] X. Y. Peng, J. H. Won, T. Rutherford et al., "The use of the L-plastin promoter for adenoviral-mediated, tumor-specific gene expression in ovarian and bladder cancer cell lines," *Cancer Research*, vol. 61, no. 11, pp. 4405–4413, 2001.

[77] H. Akbulut, L. Zhang, Y. Tang, and A. Deisseroth, "Cytotoxic effect of replication-competent adenoviral vectors carrying L-plastin promoter regulated E1A and cytosine deaminase genes in cancers of the breast, ovary and colon," *Cancer Gene Therapy*, vol. 10, no. 5, pp. 388–395, 2003.

[78] J. Zheng, N. Rudra-Ganguly, W. C. Powell, and P. Roy-Burman, "Suppression of prostate carcinoma cell invasion by expression of antisense L-plastin gene," *American Journal of Pathology*, vol. 155, no. 1, pp. 115–122, 1999.

16

Alterations in Cell-Extracellular Matrix Interactions during Progression of Cancers

Rajeswari Jinka,[1] Renu Kapoor,[2] Pavana Goury Sistla,[2] T. Avinash Raj,[2] and Gopal Pande[2]

[1] *Department of Biochemistry, Acharya Nagarjuna University, Guntur 522510, India*
[2] *Centre for Cellular and Molecular Biology, Council of Scientific and Industrial Research (CSIR), Uppal Road, Hyderabad 500 007, India*

Correspondence should be addressed to Rajeswari Jinka, rjinka@yahoo.com and Gopal Pande, gpande@ccmb.res.in

Academic Editor: Jun Chung

Cancer progression is a multistep process during which normal cells exhibit molecular changes that culminate into the highly malignant and metastatic phenotype, observed in cancerous tissues. The initiation of cell transformation is generally associated with genetic alterations in normal cells that lead to the loss of intercellular- and/or extracellular-matrix- (ECM-) mediated cell adhesion. Transformed cells undergo rapid multiplication and generate more modifications in adhesion and motility-related molecules which allow them to escape from the original site and acquire invasive characteristics. Integrins, which are multifunctional adhesion receptors, and are present, on normal as well as transformed cells, assist the cells undergoing tumor progression in creating the appropriate environment for their survival, growth, and invasion. In this paper, we have briefly discussed the role of ECM proteins and integrins during cancer progression and described some unique conditions where adhesion-related changes could induce genetic mutations in anchorage-independent tumor model systems.

1. Introduction

Cancer afflicts an organ or a tissue by inducing abnormal and uncontrolled division of cells that either constitute it or migrate to it. At the cellular level, this is caused by genetic alterations in networks that regulate cell division and cell death. The increased rate of proliferation of transformed cells causes further mutations in genes that regulate other cellular processes. For example, transformed cells eventually gain the capacity to invade into other tissues by modulating their own kinetic properties without losing the capacity to divide rapidly and avoid cell death, despite internal and external perturbations.

Cancer cells adopt diverse mechanisms to cope with the various physiological insults, such as low oxygen and metabolic stress, that they encounter [1]. These mechanisms have been discussed in a recent review [2], and based upon that discussion, six important hall marks of cancer cells can be identified. These are (a) sustained proliferative signalling, (b) evasion of growth suppressors, (c) resistance to cell death, (d) replicative immortality, (e) copious angiogenesis, and (f) active invasion and metastasis. In addition to these, cancer cells can exhibit two other properties, that is, tumor promoting inflammation and gene instability that assist the cells in the transition from normal to oncogenic phenotype [2]. Eventually transformed cells undergo somatic evolution and generate diverse populations that tend to harbor genetic and epigenetic instabilities and alterations [3]. These changes also assist the cells in adapting to the variations in the surrounding microenvironment and even to alter it. As a consequence of these alterations, the tumor milieu or microenvironment becomes an "enabling element" for defining some characteristics of cancer cells. For example, the tumor microenvironment can induce cancer cells in acquiring anoikis resistance and in selecting new sites to colonize and grow. Sometimes these cells remain unresponsive until signals generated from the ECM reach the cell's nucleus and they determine whether the cell would proceed to the next stage in cancer progression or not. This response of cancer cells to ECM-generated signals similar to the "dynamic reciprocity" proposed by Bisell for normal cells. An example of such an adaptation of cancer cells to their microenvironment and the resultant clonal selection of invasive cells has been recently reported [4–6].

Metastatic invasion is generally the final phase of cancer progression, and it involves formation of new blood vessels either by neovasculogenesis, in which endothelial cell precursors (angioblasts) migrate to the tumor site and differentiate and assemble into primitive blood vessels, or by angiogenesis in which we observe sprouting of new blood vessels from preexisting ones, or their longitudinal bifurcation, in the tumor [7]. The invasive tumor cells migrate through these newly formed blood vessels to other sites such as lung and liver brain and this leads to the death of tumor-bearing patients or animals as the case may be.

Based on available evidence, the entire process of cancer formation can be divided into four different stages: initiation, progression, epithelial mesenchymal transition (EMT), and metastasis (see Figure 1). At the initiation stage, a normal cell acquires oncogenic properties mainly through genetic alterations, which lead to changes in cell structure, adhesion properties, and response to signals from ECM proteins. In the second stage, transformed cells respond to cues from the altered environmental conditions and acquire properties of adhesion-independent growth and colonization. The third stage is also referred to as a transitional or the EMT stage, and, in this stage, the fully transformed cells begin to exhibit mesenchymal gene expression patterns which induce them to invade into the neighbouring tissue and enter into blood circulation [8]. The fourth and prominent stage is metastasis in which the invasive mesenchymes like cells move from the primary site and colonize in a new location. This stage spreads the disease into different parts of the body and involves several alterations in the adhesion properties of cells.

From all earlier observations, it is clear that the cell adhesion in transformed cells plays an important role in all four stages of cancer formation. This paper highlights recent studies done on the integrin-mediated interaction of transformed cells with the ECM and discusses its role in cancer progression.

2. ECM Components and Properties

Over the past two decades, research in the field of cancer biology has focussed extensively on the role of ECM constituents during cancer progression. These molecules comprise the cell's microenvironment, and they can affect the mechanical and biophysical properties of cells as well as that of the ECM such as its mechanics, geometry, and topology [9].

In some tissues, mainly of epithelial origin, ECM constituents are present in the basement membrane that defines the boundaries of that tissue. In this location, the organization of these components is different than in the matrix. In the basement membrane, we notice molecules such as collagens, proteoglycans, laminins, and fibronectins associate strongly with certain carbohydrate polymers and generate a membrane-like structure which facilitates the formation of a framework of cells and ECM constituents [10]. Specific domains in ECM proteins that are created by partial gene duplication and exon shuffling during the process of evolution [11] play a critical role in keeping the cells attached to the ECM and the basement membrane and initiating signalling cascades in the cells.

Inside the cells, the ECM-induced signaling pathways are transmitted mainly through integrin molecules that are transmembrane multifunctional ECM receptors. Integrin-mediated signaling in association with many cofactors, for example, cytokines, growth factors, and intracellular adapter molecules, can significantly affect diverse cell processes such as cell cycle progression, migration, and differentiation. The interplay between the biophysical properties of the cell and ECM establishes a dynamic, mechanical reciprocity between the cell and the ECM in which the cell's ability to exert contractile stresses against the extracellular environment balances the elastic resistance of the ECM to that deformation [4, 9]. The ECM in association with the available growth factors activates a sequence of reactions with a complex network of proteases, sulfatases, and possibly other enzymes to liberate and activate various signalling pathways in a highly specific and localized fashion. The maintenance of ECM homeostasis therefore involves a tight balance between biosynthesis of ECM proteins, their 3D organization, cross-linking, and degradation.

3. Modulation of ECM-Generated Signaling in Cancer

During cancer progression, we observe significant changes in the structural and mechanical properties of ECM constituents. It has been reported that changes in matrix stiffness, which offers resistance to cell traction forces [12] and also influences "shape dependence" in cells, can contribute actively to the tumor formation [13]. Deregulation of cell shape and alterations in the interactions with the ECM are considered as important hallmarks of cancer cells. These changes in ECM homeostasis can be brought about by the properties of tumor cells themselves or by the secretions of other surrounding cells such as fibroblasts, macrophages, and leukocytes [14]. Integrin ECM interactions are significantly modulated by crosstalk with several other signal-generating molecules, some of them are receptor molecules on the cell surface whereas others are present in the cytoplasm as adaptor proteins and actin-binding proteins [15]. These signaling crosstalks in which integrin molecules lie at the center are very useful for the transition of transformed cells to metastatic cells [16]. Integrin-generated ECM remodelling is further controlled by the localization and activity of proteases [17]. One example of such an integrin-directed cancer progression is seen in breast cancers, where adhesion-independent mammary epithelial cells secrete laminin-5 and luminal cells secrete laminin-1. This leads to aberrant polarity in cells, causing upregulation of metalloproteinases (MMPs, such as MMP9) and induction of tumor invasion and metastasis [18–20].

4. Integrins: Its Ligands and Signalling

Integrins are heterodimeric cell-surface receptors that mediate adhesion to ECM and immunoglobulin superfamily molecules. At least 24 distinct integrin heterodimers are formed by the combination of 18 α-subunits and

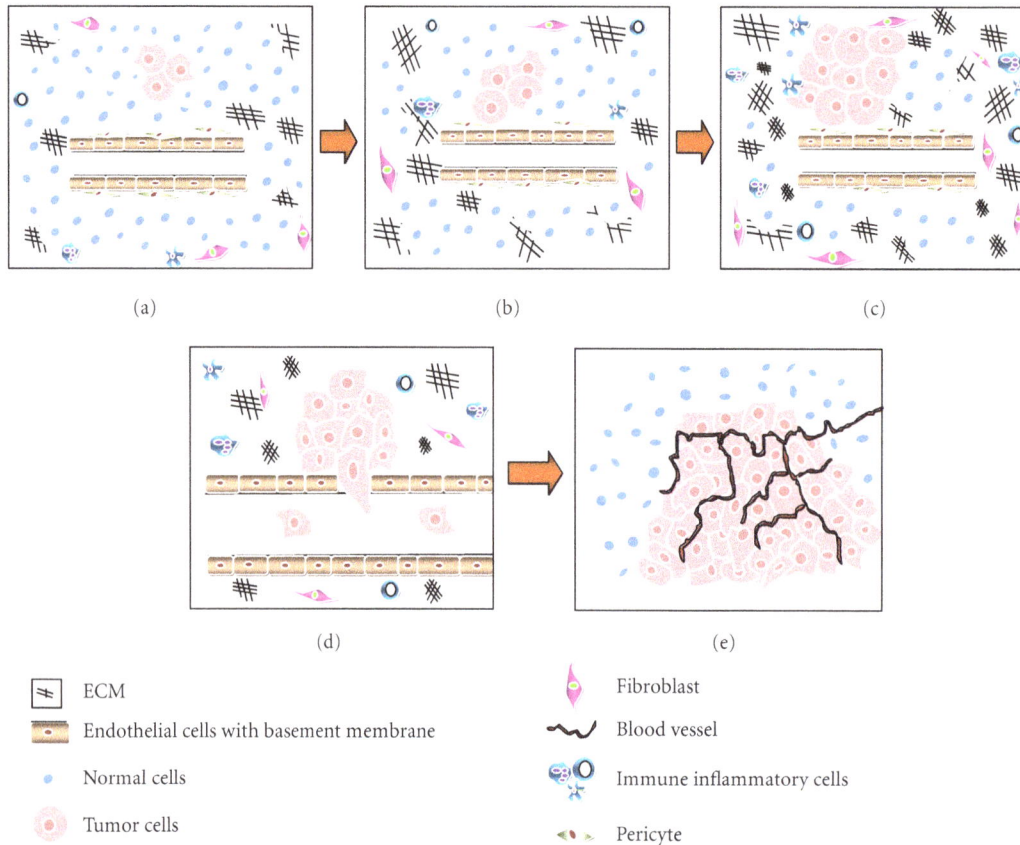

(a)   (b)   (c)

(d)   (e)

	ECM		Fibroblast
	Endothelial cells with basement membrane		Blood vessel
	Normal cells		Immune inflammatory cells
	Tumor cells		Pericyte

FIGURE 1: Various steps in tumor initiation and progression where panel (a) represents initiation of tumor by transforming normal cells, panel (b) shows the modulation of ECM proteins allowing transformed cells to multiply, panel (c) shows progression of cancer by replacing normal cells, panel (d) represents the invasion, where cancer cells migrate into the blood stream by modulating ECM and cell adhesion molecules, and panels (e) shows metastasis where the cancer cells are localized at different sites enabling angiogenesis.

8 β-subunits. Specific integrin heterodimers preferentially bind to distinct ECM proteins like laminin, collagen IV, fibronectin, and so forth. The level of integrin expression on the cell surface dictates the efficiency of cell adhesion and migration on different matrices. While some integrins selectively recognise primarily a single ECM protein ligand (e.g., $\alpha5\beta1$ recognises primarily fibronectin), others can bind several ligands (e.g., integrin $\alpha v\beta3$ binds vitronectin, fibronectin, fibrinogen, denatured or proteolysed collagen, and other matrix proteins). Several integrins recognise the tripeptide Arg-Gly-Asp (e.g., $\alpha v\beta3$, $\alpha5\beta1$, $\alpha IIb\beta3$), whereas others recognise alternative short peptide sequences (e.g., integrin $\alpha4\beta1$ recognises EILDV and REDV in alternatively spliced CS-1 fibronectin). Inhibitors of integrin function include function-blocking monoclonal antibodies, peptide antagonists, and small molecule peptide mimetics matrix [21–23].

The positioning of integrin receptors acts as a direct bridge between the extracellular matrix and the internal cell cytoskeleton by transducing key intracellular signals by associating with the clusters of kinases and adaptor proteins in focal adhesion complexes. Integrins thus act as mediators in transmitting different signals from "inside out" (intracellular to extracellular) and "outside in" (extracellular to intracellular) between ECM to cells and vice versa.

Through these pathways, ECM proteins are able to control proapoptotic and antiapoptotic cascades by regulating the activity of caspase 8 and caspase 3 [24–26]. ECM-integrin interactions thus determine the balance of apoptotic and cell survival signals and maintain the homeostasis of organs and tissues. Although integrins lack kinase activity, by inter- and intramolecular clustering, they recruit and activate kinases, such as focal adhesion kinases (FAKs) and src family kinases (sFKs) to a focal adhesion complex. In addition to scaffolding molecules, such as p130 CRK-associated substrate (p130CAs; also known as BCAR1), integrins also couple the ECM to actin cytoskeleton by recruiting cytoskeletal proteins, including talin, paxillin, α-actinin, tensin, and vinculin. Additionally, they form a ternary complex consisting of an integrin-linked kinase, PINCH, and parvin to regulate many scaffolding and signalling functions required for integrin-mediated effects on cell migration and survival [27].

5. Integrin Expression and Signalling in Cancer Progression

Although anchorage-independent growth is a hallmark of malignant transformation, integrins expression levels and activity are an important role in different steps of tumour progression including initiation [28]. Higher expression of $\alpha3$, $\alpha5$, $\alpha6$, αv, $\beta1$, $\beta4$, $\alpha6\beta4$, $\alpha9\beta1$, $\alpha v\beta5$, and $\alpha v\beta3$ integrins is directly correlated with the progression of the disease [10]. Several epithelial cell tumors showed the altered $\alpha6\beta4$, $\alpha6\beta1$, $\alpha v\beta5$, $\alpha2\beta1$, and $\alpha3\beta1$ integrin expression [29]. Integrin recruitment to membrane microdomains has been shown to be regulated by tetraspanins and crucially regulate integrin function in tumour cells [30]. Recent studies have demonstrated that cell signalling generated by growth factors and oncogenes in transformed cells requires collaboration with specific integrins, especially during tumour initiation. In tumor cells, several survival signals are upregulated upon integrin ligation, which includes increased expression of BCl-2 or FlIP (also known as CFlAR), activation of the PI3K-AKT pathway or nuclear factor-κB (nF-κB) signaling, and/or p53 inactivation [24].

Invasive cancer cells evacuate from the primary site and migrate to the secondary site by the process of tissue invasion and cell migration. Integrin-mediated pathways involving focal adhesion kinase (FAK) and src family kinase (SFK) signaling play a major role in this. In order to survive in the new location and to withstand the stressful conditions of hypoxia, nutrient deprivation, and inflammatory mediators the migratory cells increase the blood supply to themselves by neoangiogenesis. This is achieved by increased expression of $\alpha v\beta3$ and $\alpha v\beta5$ integrins and by deposition of provisional matrix proteins such as vitronectin, fibrinogen, von Willebrand factor, osteopontin, and fibronectin in the tumour microenvironment. Interaction between these molecules plays a critical role in the process of generating new blood vessels in the newly formed tumor site [31, 32].

Integrins found on tumour-associated normal cells, such as the vascular endothelium, perivascular cells, fibroblasts, bone-marrow-derived cells, and platelets, also have a profound effect in tumor progression via integrin-mediated pathways. A summary of these has been given in Table 1 [33–47].

6. Genetic and Chromosomal Aberrations at the Onset of Cancer

Neoplasia occurs when cells are exposed to cancer-promoting substances that cause single or multiple premalignant genetic/epigenetic changes which may coalesce to form a large lesion. These genetic changes may have neutral, deleterious, or advantageous effects on the proliferation of a clone or clones of cells. Neutral or deleterious genetic changes may result in stagnation or cell death, whereas the cell receiving advantageous events may result in higher proliferation, recruitment of blood vessels to the developing tumors, and gain the ability to metastasize [48]. The model of Braakhuis et al., 2004 [49], advanced this idea by suggesting that initial genetic alterations occur in stem cells, forming a patch and expanding field of cells with the original and subsequent genomic and or chromosomal alterations. Then, clonal selection of one or more cells within this field of preneoplastic cells leads to the development of a carcinoma. There is a considerable cytogenetic variability

among cells reflecting heterogeneity due to clonal evolution within the original tumor [50]. Initial heterogeneity or cell-to-cell differences in cancer are due to cytoskeletal alterations which result in defecting chromosomal segregation and lead to karyotypic variations during mitosis, causing chromosomal aberrations, for example, NUMA1 gene at 11q13, which results in multipolar spindles, leading to daughter cells that differ from each other and their mother cells [51]. Structural chromosome alterations also occur due to deletions, translocations, isochromosomes, dicentric chromosomes, and endoduplicated chromosomes. The gain or amplification of chromosomal segments is driven by more than one gene [52]. Structural rearrangements involving the cleavage and fusion of centromeres from participating chromosomes, also referred to as Robertsonian translocations, are the most frequently observed alterations. Chromosomal aberrations identified with the help of cytogenetic methods including FISH, cCGH, or aCGH showed the gain of the entire long arm of chromosome 3 which amplifies the EGFR gene in SCCHN [53], 8q24 gain to amplify MYC and PTK2 in primary tumors, 11q13 amplifications to amplify cyclin D1 gene, loss of 3q14 causes deletion of fragile site FRA3B/FHIT, necessary to protect cells from accumulation of DNA damage [48]. Aberrations mainly in chromosome 13 and also involving chromosomes 6, 11, 12, and 17 are associated with B-CLL [54].

7. Anchorage-Independent Tumor Model System

The wide range of *in vivo* tumor models like syngeneic, human tumor xenograft, orthotopic, metastatic, and genetically engineered mouse models is available from the basis of the compounds selected and treatments that go into clinical testing of patients [55]. The ability to exhibit anchorage-independent cell growth (colony-forming capacity in semisolid media) has been considered to be fundamental in cancer biology because it has been connected with tumor cell aggressiveness *in vivo* such as tumorigenic and metastatic potentials and also utilized as a marker for *in vitro* transformation. Although multiple genetic factors for anchorage-independence have been identified, the molecular basis for this capacity is still largely unknown [56, 57]. During the process of *in vitro* tumorigenesis, various oncogenes with distinct pathways have been shown to transform anchorage-dependent cells to anchorage-independent cells [5, 57]. For example, transfer of c-Myc (a transcription factor), v-Src (a tyrosine kinase), or H-Ras (a small GTPase) into spontaneously immortalized mouse embryonic fibroblasts (MEFs) provides the cells an ability to grow in an anchorage-independent manner [56, 57]. Anchorage-independent multicellular spheroids made by Ewing tumor cell lines were more closely related to primary tumors with respect to cell morphology, cell-cell junctions, proliferative index, and kinase activation [58].

However, changes in ECM and cell adhesion molecular interaction and genetic variations were observed till the date only with the primary or secondary tumors. We have

Table 1: Various integrins in association with different ligands to induce different signaling pathways in generation of tumor and metastasis.

Integrin type	Interacting ECM protein	Activated signaling cascade	Tumor/metastasis	Reference
$\alpha3\beta1$	Laminin	MMP9 and oncogenic Ras, VEGF, FAK-paxillin signaling cascade	Invasion in keratinocytes, Induces angiogenesis, Human hepatoma cells	[33, 34]
$\alpha6\beta1$	Laminin	Urokinase plasminogen activator and MMP-2, PI3Kinase, Src	Tumor invasion in pancreatic cells	[35, 36]
$\alpha7\beta1$	Laminin	Rho-A signaling cascade	Invasion in breast cancer	[37]
$\alpha2\beta1, \alpha1\beta1, \alpha10\beta1, \alpha II\beta1$	Collagen	FAK and src signaling	Invasion of melanoma cells, cancer progression, and invasion of lung adenocarcinoma	[38, 39]
$\alpha v\beta1, \alpha v\beta6, \alpha v\beta3$	Vitronectin, syndican, thromospondin-1	MMP9, urokinase signaling, MEK/Erk/NF-κB, PKCa, FAK	Metastatic breast Cancer, pancreatic, cervical, colon, lung/liver metastasis	[40–42]
$\alpha9\beta1$	CCN3, osteopontin	Src, P130 Cas, Rac, NOS signaling	Metastatic potential	[43]
$\alpha IIb3, \alpha vb3$	Von Willebrand factor	Interacts with thrombospondin-1 and induces VEGF/FGF signaling	Breast cancer	[44, 45]
$\alpha5$	Fibronectin	FAK, ERK, PI-3 K, ILK, and nuclear factor-kappa B -	Metastatic lung and cervical cancer	[46, 47]
$\alpha L\beta2$	Intercellular cell adhesion molecules		Breast cancer	[1]

developed a cellular model system by using normal, adherent rat fibroblast cell lines. These cells lose their cytoskeletal organization and specificity to fibronectin as $\alpha5\beta1$ integrins are constantly recycled between cytoplasm and plasma. In the drastic unfavorable stressful conditions, the mechanical, phenotypic, and genetic characteristics are altered/modified to sustain their identity [5, 26]. We observed that this cellular model system represents a tumor with all characteristics of cancer described by Hanahan and Weinberg 2011 as hallmarks of cancer.

The cells during the nonadhesion process can evade from cell death by caspase 3 interaction with unligated $\alpha5\beta1$ integrins inducing resistance to integrin-mediated death (IMD) and also gain the ability to metastatise. Mutational changes mainly with 2;6 Robertsonian's translocation and activated Ras, FAK, and PKC provide self-sufficient growth signals potential for uncontrollable growth of the cells (Figure 2). Upregulated Spp1, MMP3, Egfr, Rb1, Ddit3, Egln3, Vegfa, Stc1, Hif1a, MMP3, and altered pathways like glycolysis/gluconeogenesis and hypoxia (Pfkm, HK2, Pdk1, Adh1, aldh3a1, and Slc2a) lead the cells to invade, metastasize, and sustain angiogenesis. We observed another phenomenon of dedifferentiation by gaining the stem-cell-like and multidrug resistance properties by expressing Cd133 and ABCG-2 when the cells are exposed to unfavorable condition [5].

The anchorage-independent cellular model system represents a multicentric tumor model system apprehended with genes related to tumor progression, angiogenesis, and metastasis (Figure 2). It is very advantageous, convenient, and possible model system to study the effect of various cancer-mediated drugs at the initial stage itself for the proper diagnosis.

8. Integrin Signalling as a Target in Cancer Treatment

Several studies showed the correlation of integrin inhibition at any point of its action will lead to the inhibition of tumor progression [24]. Therefore, integrins are focused pharmacologically in the treatment and prevention of cancer. Antagonists of these integrins suppress cell migration and invasion of primary and transformed cells by inducing apoptosis in primary cells could block tumor angiogenesis and metastasis. Recycling integrins present on the surface of endothelial cells are targeted in the blood stream by exposing to the circulating drugs and agents [59]. Various antibodies, cyclic peptides, disintegrins, and peptidomimetics are meant to bind the targeted integrins to prevent integrin ligation. cRGD, cyclic arginine-glycine-aspartic acid; RGDK, arginine-glycine-aspartic acid-lysine; TRAIL, tumour necrosis factor-related apoptosis-inducing ligand are being used as antagonists integrins to hit the integrin ligand function. The function of upregulated $\alpha v\beta3$ integrin can be blocked by function-blocking monoclonal antibodies, such as LM 609 [60]. The human αv integrin specific monoclonal antibody CnTo 95, which targets both $\alpha v\beta3$ and $\alpha v\beta5$ integrins to induce endothelial apoptosis, also had antitumour and antiangiogenic effects in xenograft tumour models [15].

Cilengitide, inhibitor of both $\alpha v\beta3$ and $\alpha v\beta5$ integrins and volociximab, a function-blocking monoclonal antibody against integrin $\alpha5\beta1$, inhibits angiogenesis and impedes tumour growth [61, 62]. $\alpha v\beta3$ is targeted by various therapeutic antibodies like LM609, vitaxin, humanized mouse monoclonal derived from LM609, CNTO 95, humanized IgG1, c7E3, chimeric mouse human, 17E6, mouse monoclonal antibodies to inhibit tumour growth, and angiogenesis

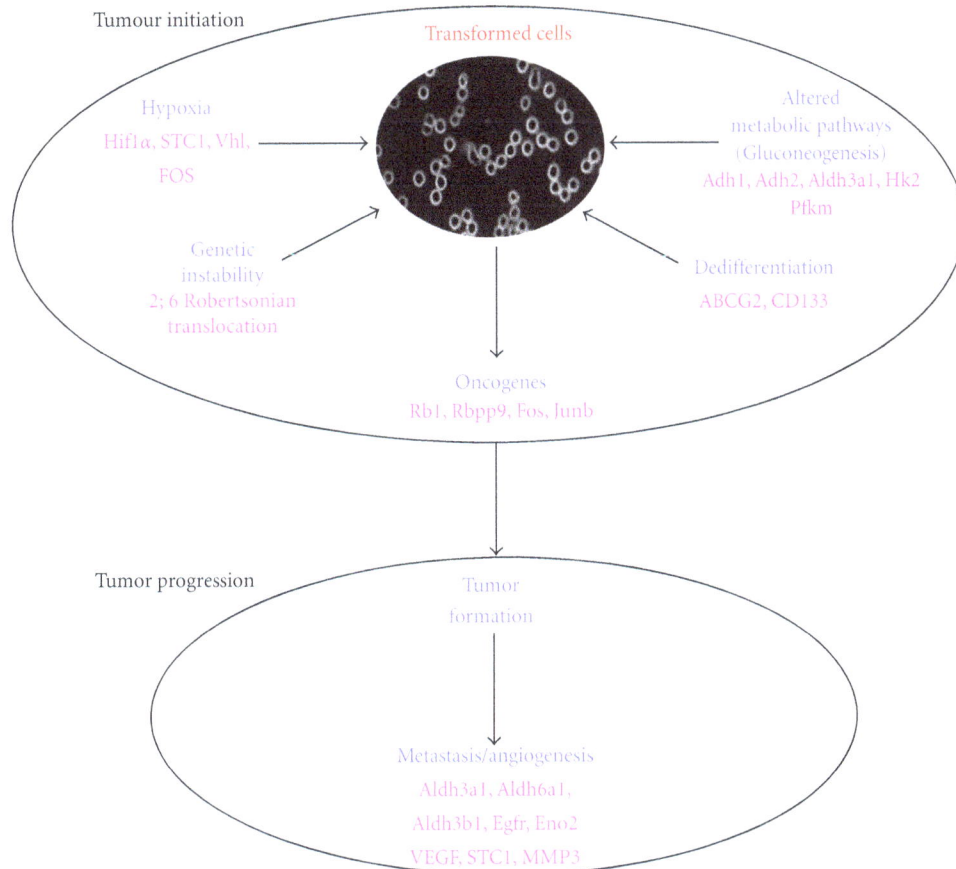

FIGURE 2: This figure shows the hallmark characteristics of transformed cells at initial stages.

in tumour xenografts and preclinical studies [63]. The prognostic $\alpha5\beta1$ integrins in ovarian cancer can be targeted effectively both *in vitro* and *in vivo* by using specific antibodies. Cell-mediated $\alpha5\beta1$ adhesion can be blocked with a small molecule antagonist (SJ479), a fibronectin-derived peptides in prostate, and colon cancer models [64, 65]. Recent activity is extended to detect tumors and angiogenesis and deliver the drugs to the site of cancer by coupling integrin antagonists to a paramagnetic contrast agent or radionuclide in rabbit and mouse tumour models [66, 67].

9. Conclusion

Normal cells lead to the transformation when exposed to adverse conditions such as anchorage independence and effects are found to be similar to the effect of carcinogens and mutagens. These cells could alter the ECM and other cell adhesion molecules by showing altered integrins expression on their surface and are associated with different kinds of growth factors and oncogene. ECM-integrin interactions along with other growth factors provide the diversified anchorage-independent signals to the transformed cells to progress as cancers, metastasis, and angiogenesis. Several tumours are sustained with diversified integrins found to be specific to the tumour-host microenvironment. In future, these integrins can be targeted at the initial stages of cancer

by using integrin antagonists to minimize the growth of tumour and metastasis.

Acknowledgment

This work was supported by Grant no. GAP0220 from Department of Science and Technology, Government of India, to G. Pande.

References

[1] T. Tian, S. Olson, J. M. Whitacre, and A. Harding, "The origins of cancer robustness and evolvability," *Integrative Biology*, vol. 3, no. 1, pp. 17–30, 2011.

[2] D. Hanahan and R. A. Weinberg, "Hallmarks of cancer: the next generation," *Cell*, vol. 144, no. 5, pp. 646–674, 2011.

[3] C. Coghlin and G. I. Murray, "Current and emerging concepts in tumour metastasis," *Journal of Pathology*, vol. 222, no. 1, pp. 1–15, 2010.

[4] C. M. Nelson and M. J. Bissell, "Modeling dynamic reciprocity: engineering three-dimensional culture models of breast architecture, function, and neoplastic transformation," *Seminars in Cancer Biology*, vol. 15, no. 5, pp. 342–352, 2005.

[5] J. Rajeswari, R. Kapoor, S. Pavuluri et al., "Differential gene expression and clonal selection during cellular transformation induced by adhesion deprivation," *BMC Cell Biology*, vol. 11, article 93, pp. 1–13, 2010.

[6] S. Shen, J. Fan, B. Cai et al., "Vascular endothelial growth factor enhances cancer cell adhesion to microvascular endothelium in vivo," *Experimental Physiology*, vol. 95, no. 2, pp. 369–379, 2010.

[7] E. E. Conway, "Central nervous system findings and intussusception: how are they related?" *Pediatric Emergency Care*, vol. 9, no. 1, pp. 15–18, 1993.

[8] J. P. Thiery and J. P. Sleeman, "Complex networks orchestrate epithelial-mesenchymal transitions," *Nature Reviews Molecular Cell Biology*, vol. 7, no. 2, pp. 131–142, 2006.

[9] S. Kumar and V. M. Weaver, "Mechanics, malignancy, and metastasis: the force journey of a tumor cell," *Cancer and Metastasis Reviews*, vol. 28, no. 1-2, pp. 113–127, 2009.

[10] R. Rathinam and S. K. Alahari, "Important role of integrins in the cancer biology," *Cancer and Metastasis Reviews*, vol. 29, no. 1, pp. 223–237, 2010.

[11] H. Hutter, B. E. Vogel, J. D. Plenefisch et al., "Conservation and novelty in the evolution of cell adhesion and extracellular matrix genes," *Science*, vol. 287, no. 5455, pp. 989–1010, 2000.

[12] D. E. Ingber, J. A. Madri, and J. D. Jamieson, "Role of basal lamina in neoplastic disorganization of tissue architecture," *Proceedings of the National Academy of Sciences of the United States of America*, vol. 78, no. 6 I, pp. 3901–3905, 1981.

[13] S. C. Wittelsberger, K. Kleene, and S. Penman, "Progressive loss of shape-responsive metabolic controls in cells with increasingly transformed phenotype," *Cell*, vol. 24, no. 3, pp. 859–866, 1981.

[14] H.-G. Zhang and W. E. Grizzle, "Exosomes and cancer: a newly described pathway of immune suppression," *Clinical Cancer Research*, vol. 17, no. 5, pp. 959–964, 2011.

[15] Q. Chen, C. D. Manning, H. Millar et al., "CNTO 95, a fully human anti αv integrin antibody, inhibits cell signaling, migration, invasion, and spontaneous metastasis of human breast cancer cells," *Clinical and Experimental Metastasis*, vol. 25, no. 2, pp. 139–148, 2008.

[16] D. Barkan, J. E. Green, and A. F. Chambers, "Extracellular matrix: a gatekeeper in the transition from dormancy to metastatic growth," *European Journal of Cancer*, vol. 46, no. 7, pp. 1181–1188, 2010.

[17] R. K. Assoian and E. A. Klein, "Growth control by intracellular tension and extracellular stiffness," *Trends in Cell Biology*, vol. 18, no. 7, pp. 347–352, 2008.

[18] N. Zahir, J. N. Lakins, A. Russell et al., "Autocrine laminin-5 ligates $\alpha 6\beta 4$ integrin and activates RAC and NFκB to mediate anchorage-independent survival of mammary tumors," *Journal of Cell Biology*, vol. 163, no. 6, pp. 1397–1407, 2003.

[19] T. Gudjonsson, L. Rønnov-Jessen, R. Villadsen, F. Rank, M. J. Bissell, and O. W. Petersen, "Normal and tumor-derived myoepithelial cells differ in their ability to interact with luminal breast epithelial cells for polarity and basement membrane deposition," *Journal of Cell Science*, vol. 115, no. 1, pp. 39–50, 2002.

[20] A. Beliveau, J. D. Mott, A. Lo et al., "Raf-induced MMP9 disrupts tissue architecture of human breast cells in three-dimensional culture and is necessary for tumor growth in vivo," *Genes and Development*, vol. 24, no. 24, pp. 2800–2811, 2010.

[21] R. Pytela, M. D. Pierschbacher, and E. Ruoslahti, "Identification and isolation of a 140 kd cell surface glycoprotein with properties expected of a fibronectin receptor," *Cell*, vol. 40, no. 1, pp. 191–198, 1985.

[22] R. O. Hynes, "Integrins: versatility, modulation, and signaling in cell adhesion," *Cell*, vol. 69, no. 1, pp. 11–25, 1992.

[23] J. S. Desgrosellier and D. A. Cheresh, "Integrins in cancer: biological implications and therapeutic opportunities," *Nature Reviews Cancer*, vol. 10, no. 1, pp. 9–22, 2010.

[24] A. E. Aplin, A. K. Howe, and R. L. Juliano, "Cell adhesion molecules, signal transduction and cell growth," *Current Opinion in Cell Biology*, vol. 11, no. 6, pp. 737–744, 1999.

[25] D. G. Stupack, X. S. Puente, S. Boutsaboualoy, C. M. Storgard, and D. A. Cheresh, "Apoptosis of adherent cells by recruitment of caspase-8 to unligated integrins," *Journal of Cell Biology*, vol. 155, no. 4, pp. 459–470, 2001.

[26] J. Rajeswari and G. Pande, "The significance of $\alpha 5\beta 1$ integrin dependent and independent actin cytoskelton organization in cell transformation and survival," *Cell Biology International*, vol. 26, no. 12, pp. 1043–1055, 2002.

[27] R. Zaidel-Bar and B. Geiger, "The switchable integrin adhesome," *Journal of Cell Science*, vol. 123, no. 9, pp. 1385–1388, 2010.

[28] S. Han, F. R. Khuri, and J. Roman, "Fibronectin stimulates non-small cell lung carcinoma cell growth through activation of Akt/mammalian target of rapamycin/S6 kinase and inactivation of LKB1/AMP-activated protein kinase signal pathways," *Cancer Research*, vol. 66, no. 1, pp. 315–323, 2006.

[29] A. Kren, V. Baeriswyl, F. Lehembre et al., "Increased tumor cell dissemination and cellular senescence in the absence of $\beta 1$-integrin function," *EMBO Journal*, vol. 26, no. 12, pp. 2832–2842, 2007.

[30] M. Zöller, "Tetraspanins: push and pull in suppressing and promoting metastasis," *Nature Reviews Cancer*, vol. 9, no. 1, pp. 40–55, 2009.

[31] D. Ribatti, "The contribution of Harold F. Dvorak to the study of tumor angiogenesis and stroma generation mechanisms," *Endothelium*, vol. 14, no. 3, pp. 131–135, 2007.

[32] P. C. Brooks, R. A. F. Clark, and D. A. Cheresh, "Requirement of vascular integrin $\alpha v\beta 3$ for angiogenesis," *Science*, vol. 264, no. 5158, pp. 569–571, 1994.

[33] P. C. Brooks, S. Strömblad, L. C. Sanders et al., "Localization of matrix metalloproteinase MMP-2 to the surface of invasive cells by interaction with integrin $\alpha v\beta 3$," *Cell*, vol. 85, no. 5, pp. 683–693, 1996.

[34] E. I. Deryugina, M. A. Bourdon, G. X. Luo, R. A. Reisfeld, and A. Strongin, "Matrix metalloproteinase-2 activation modulates glioma cell migration," *Journal of Cell Science*, vol. 110, no. 19, pp. 2473–2482, 1997.

[35] M. Kielosto, P. Nummela, K. Järvinen, M. Yin, and E. Hölttä, "Identification of integrins $\alpha 6$ and $\beta 7$ as c-Jun- and transformation-relevant genes in highly invasive fibrosarcoma cells," *International Journal of Cancer*, vol. 125, no. 5, pp. 1065–1073, 2009.

[36] Y. He, X. D. Liu, Z. Y. Chen et al., "Interaction between cancer cells and stromal fibroblasts is required for activation of the uPAR-uPA-MMP-2 cascade in pancreatic cancer metastasis," *Clinical Cancer Research*, vol. 13, no. 11, pp. 3115–3124, 2007.

[37] I. S. Vizirianakis, C. C. Yao, Y. Chen, B. L. Ziober, A. S. Tsiftsoglou, and R. H. Kramer, "Transfection of MCF-7 carcinoma cells with human integrin $\alpha 7$ cDNA promotes adhesion to laminin," *Archives of Biochemistry and Biophysics*, vol. 385, no. 1, pp. 108–116, 2001.

[38] I. Staniszewska, E. M. Walsh, V. L. Rothman et al., "Effect of VP12 and viperistatin on inhibition of collagen receptors-dependent melanoma metastasis," *Cancer Biology and Therapy*, vol. 8, no. 15, pp. 1507–1516, 2009.

[39] S. Van Slambrouck, C. Grijelmo, O. De Wever et al., "Activation of the FAK-src molecular scaffolds and p130Cas-JNK

signaling cascades by α1-integrins during colon cancer cell invasion," *International Journal of Oncology*, vol. 31, no. 6, pp. 1501–1508, 2007.

[40] M. Rolli, E. Fransvea, J. Pilch, A. Saven, and B. Felding-Habermann, "Activated integrin αvβ3 cooperates with metalloproteinase MMP-9 in regulating migration of metastatic breast cancer cells," *Proceedings of the National Academy of Sciences of the United States of America*, vol. 100, no. 16, pp. 9482–9487, 2003.

[41] J. Huang, R. Roth, J. E. Heuser, and J. E. Sadler, "Integrin alpha(v)beta(3) on human endothelial cells binds von Willebrand factor strings under fluid shear stress," *Blood*, vol. 113, no. 7, pp. 1589–1597, 2009.

[42] G. Y. Yang, K. S. Xu, Z. Q. Pan et al., "Integrin alphavbeta6 mediates the potential for colon cancer cells to colonize in and metastasize to the liver," *Cancer Science*, vol. 99, no. 5, pp. 879–887, 2008.

[43] S. K. Gupta and N. E. Vlahakis, "Integrin α9β1 mediates enhanced cell migration through nitric oxide synthase activity regulated by Src tyrosine kinase," *Journal of Cell Science*, vol. 122, no. 12, pp. 2043–2054, 2009.

[44] N. Gomes, C. Legrand, and F. Fauvel-Lafève, "Shear stress induced release of von Willebrand factor and thrombospondin-1 in HUVEC extracellular matrix enhances breast tumour cell adhesion," *Clinical and Experimental Metastasis*, vol. 22, no. 3, pp. 215–223, 2005.

[45] I. Gil-Bazo, V. Catalán, J. Páramo et al., "Von Willebrand factor as an intermediate between hemostasis and angiogenesis of tumor origin," *Revista de Medicina de la Universidad de Navarra*, vol. 47, no. 3, pp. 22–28, 2003.

[46] J. Roman, J. D. Ritzenthaler, S. Roser-Page, X. Sun, and S. Han, "α5β1-integrin expression is essential for tumor progression in experimental lung cancer," *American Journal of Respiratory Cell and Molecular Biology*, vol. 43, no. 6, pp. 684–691, 2010.

[47] G. Maity, S. Fahreen, A. Banerji et al., "Fibronectin-integrin mediated signaling in human cervical cancer cells (SiHa)," *Molecular and Cellular Biochemistry*, vol. 336, no. 1-2, pp. 65–74, 2010.

[48] F. Mitelman and S. Heim, *Cancer Cytogenetics*, John Wiley & Sons, Chichester, UK, 2009.

[49] B. J. M. Braakhuis, C. R. Leemans, and R. H. Brakenhoff, "A genetic progression model of oral cancer: current evidence and clinical implications," *Journal of Oral Pathology and Medicine*, vol. 33, no. 6, pp. 317–322, 2004.

[50] P. L. Martin, Q. Jiao, J. Cornacoff et al., "Absence of adverse effects in cynomolgus macaques treated with CNTO 95, a fully human anti-αv integrin monoclonal antibody, despite widespread tissue binding," *Clinical Cancer Research*, vol. 11, no. 19 I, pp. 6959–6965, 2005.

[51] N. J. Quintyne, J. E. Reing, D. R. Hoffelder, S. M. Gollin, and W. S. Saunders, "Spindle multipolarity is prevented by centrosomal clustering," *Science*, vol. 307, no. 5706, pp. 127–129, 2005.

[52] X. Huang, T. E. Godfrey, W. E. Gooding, K. S. McCarty Jr., and S. M. Gollin, "Comprehensive genome and transcriptome analysis of the 11q13 amplicon in human oral cancer and synteny to the 7F5 amplicon in murine oral carcinoma," *Genes Chromosomes and Cancer*, vol. 45, no. 11, pp. 1058–1069, 2006.

[53] S. Kalyankrishna and J. R. Grandis, "Epidermal growth factor receptor biology in head and neck cancer," *Journal of Clinical Oncology*, vol. 24, no. 17, pp. 2666–2672, 2006.

[54] B. Jahrsdörfer, J. E. Wooldridge, S. E. Blackwell, C. M. Taylor, B. K. Link, and G. J. Weiner, "Good prognosis cytogenetics in B-cell chronic lymphocytic leukemia is associated in vitro with low susceptibility to apoptosis and enhanced immunogenicity," *Leukemia*, vol. 19, no. 5, pp. 759–766, 2005.

[55] Beverly A. Teicher, *Cancer Drug Discovery and Development*, vol. 14, 2nd edition, 2011.

[56] M. A. Cifone and I. J. Fidler, "Correlation of patterns of anchorage-independent growth with in vivo behavior of cells from a murine fibrosarcoma," *Proceedings of the National Academy of Sciences of the United States of America*, vol. 77, no. 2, pp. 1039–1043, 1980.

[57] D. Hanahan and R. A. Weinberg, "The hallmarks of cancer," *Cell*, vol. 100, no. 1, pp. 57–70, 2000.

[58] E. R. Lawlor, C. Scheel, J. Irving, and P. H. B. Sorensen, "Anchorage-independent multi-cellular spheroids as an in vitro model of growth signaling in Ewing tumors," *Oncogene*, vol. 21, no. 2, pp. 307–318, 2002.

[59] A. Aparna and A. V. Judith, "Integrins in cancer,progression and therapy," *Science & Medicine*, vol. 10, no. 2, pp. 84–96, 2005.

[60] P. C. Brooks, S. Stromblad, R. Klemke, D. Visscher, F. H. Sarkar, and D. A. Cheresh, "Antiintegrin αvβ3 blocks human breast cancer growth and angiogenesis in human skin," *Journal of Clinical Investigation*, vol. 96, no. 4, pp. 1815–1822, 1995.

[61] J. W. Smith, Z. M. Ruggeri, T. J. Kunicki, and D. A. Cheresh, "Interaction of integrins αvβ3 and glycoprotein IIb-IIIa with fibrinogen. Differential peptide recognition accounts for distinct binding sites," *Journal of Biological Chemistry*, vol. 265, no. 21, pp. 12267–12271, 1990.

[62] V. Bhaskar, D. Zhang, M. Fox et al., "A function blocking anti-mouse integrin α5β1 antibody inhibits angiogenesis and impedes tumor growth in vivo," *Journal of Translational Medicine*, vol. 5, article 61, 2007.

[63] M. Millard, S. Odde, and N. Neamati, "Integrin targeted therapeutics," *Theranostics*, vol. 1, pp. 154–188, 2011.

[64] O. Stoeltzing, W. Liu, N. Reinmuth et al., "Inhibition of integrin α5β1 function with a small peptide (ATN-161) plus continuous 5-fu infusion reduces colorectal liver metastases and improves survival in mice," *International Journal of Cancer*, vol. 104, no. 4, pp. 496–503, 2003.

[65] D. L. Livant, R. K. Brabec, K. J. Pienta et al., "Anti-invasive, antitumorigenic, and antimetastatic activities of the PHSCN sequence in prostate carcinoma," *Cancer Research*, vol. 60, no. 2, pp. 309–320, 2000.

[66] J. H. Kim, Y. S. Kim, K. Park et al., "Self-assembled glycol chitosan nanoparticles for the sustained and prolonged delivery of antiangiogenic small peptide drugs in cancer therapy," *Biomaterials*, vol. 29, no. 12, pp. 1920–1930, 2008.

[67] M. K. Yu, J. Park, Y. Y. Jeong, W. K. Moon, and S. Jon, "Integrin-targeting thermally cross-linked superparamagnetic iron oxide nanoparticles for combined cancer imaging and drug delivery," *Nanotechnology*, vol. 21, no. 41, Article ID 415102, 2010.

17

Selective Types of Autophagy

Fulvio Reggiori,[1] Masaaki Komatsu,[2] Kim Finley,[3] and Anne Simonsen[4]

[1] Department of Cell Biology and Institute of Biomembranes, University Medical Center Utrecht, 3584 CX Utrecht, The Netherlands
[2] Protein Metabolism Project, Tokyo Metropolitan Institute of Medical Science, Tokyo 156-8506, Japan
[3] BioScience Center, San Diego State University, San Diego, CA 921825, USA
[4] Department of Biochemistry, Institute of Basic Medical Sciences, Faculty of Medicine, University of Oslo, 0317 Oslo, Norway

Correspondence should be addressed to Anne Simonsen, anne.simonsen@medisin.uio.no

Academic Editor: G. S. Stein

Autophagy is a catabolic pathway conserved among eukaryotes that allows cells to rapidly eliminate large unwanted structures such as aberrant protein aggregates, superfluous or damaged organelles, and invading pathogens. The hallmark of this transport pathway is the sequestration of the cargoes that have to be degraded in the lysosomes by double-membrane vesicles called autophagosomes. The key actors mediating the biogenesis of these carriers are the autophagy-related genes (*ATGs*). For a long time, it was assumed that autophagy is a bulk process. Recent studies, however, have highlighted the capacity of this pathway to exclusively eliminate specific structures and thus better fulfil the catabolic necessities of the cell. We are just starting to unveil the regulation and mechanism of these selective types of autophagy, but what it is already clearly emerging is that structures targeted to destruction are accurately enwrapped by autophagosomes through the action of specific receptors and adaptors. In this paper, we will briefly discuss the impact that the selective types of autophagy have had on our understanding of autophagy.

1. Introduction

Three different pathways can deliver cytoplasmic components into the lumen of the lysosome for degradation. They are commonly referred to as autophagy (cell "self-eating") and include chaperone-mediated autophagy (CMA), microautophagy, and macroautophagy. CMA involves the direct translocation of specific proteins containing the KFERQ pentapeptide sequence across the lysosome membrane [1, 2]. Microautophagy, on the other hand, entails the invagination and pinching off of the lysosomal limiting membrane, which allows the sequestration and elimination of cytoplasmic components. The molecular mechanism underlying this pathway remains largely unknown. The only cellular function that so far has been indisputably assigned to microautophagy is the turnover of peroxisomes under specific conditions in fungi [3]. Recently, it has been reported the existence of a microautophagy-like process at the late endosomes, where proteins are selectively incorporated into the vesicles that bud inward at the limiting membrane of these organelles during the multivesicular bodies biogenesis [4]. In contrast to CMA and microautophagy, macroautophagy (hereafter referred to as autophagy) entails the formation of a new organelle, the autophagosome, which allows the delivery of a large number of different cargo molecules into the lysosome.

Autophagy is a primordial and highly conserved intracellular process that occurs in most eukaryotic cells and participates in stress management. This pathway involves the *de novo* formation of vesicles called autophagosomes, which can engulf entire regions of the cytoplasm, individual organelles, protein aggregates, and invading pathogens (Figure 1). The autophagosomes fuse with endosomal compartments to form amphisomes prior to fusion with the lysosome, where their contents are degraded and the resulting metabolites are recycled back to the cytoplasm (Figure 1). Unique features of the pathway include the double-membrane structure of the autophagosomes, which were originally characterized over 50 years ago from detailed electron microscopy studies [5]. Starting in the 1990s yeast mutational studies began the genetic and molecular characterization of the key components required to initiate and build an autophagosome

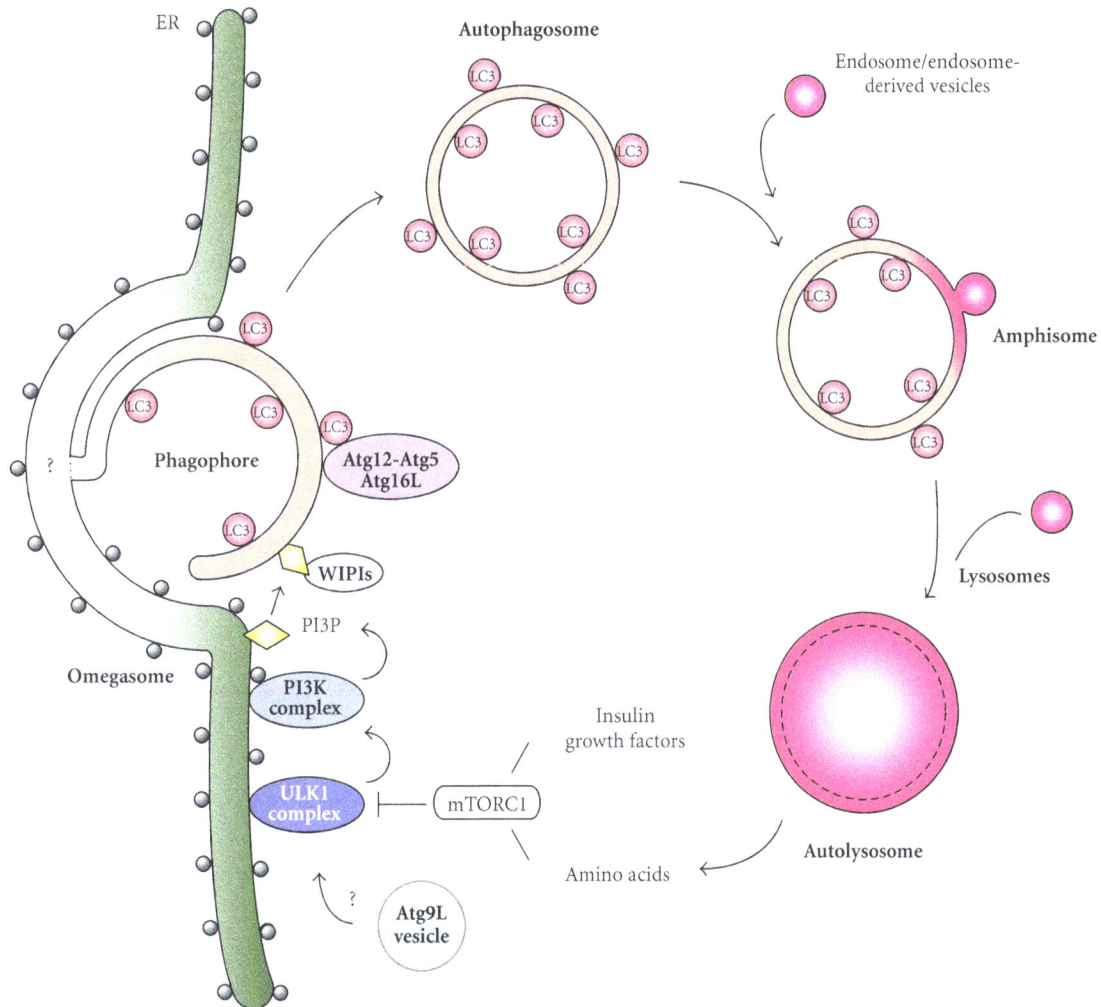

FIGURE 1: Multiple Atg proteins govern autophagosome formation. In response to inactivation of mTORC1 (but also other cellular and environmental cues), the ULK1 complex is activated and translocates in proximity of the endoplasmic reticulum (ER). Thereafter, the ULK1 complex regulates the class III PI3K complex. Atg9L, a multimembrane spanning protein, is also involved in an early stage of autophagosome formation by probably supplying part of the membranes necessary for the formation and/or expansion. Local formation of PI3P at sites called omegasomes promotes the formation of the phagophore, from which autophagosomes appear to be generated. The PI3P-binding WIPI proteins (yeast Atg18 homolog), as well as the Atg12-Atg5-Atg16L1 complex and the LC3-phosphatidylethanolamine (PE) conjugate play important roles in the elongation and closure of the isolation membrane. Finally, the complete autophagosome fuses with endosomes or endosome-derived vesicles forming the amphisome, which subsequently fuses with lysosomes to form autolysosomes. In the lysosomes, the cytoplasmic materials engulfed by the autophagosomes are degraded by resident hydrolases. The resulting amino acids and other basic cellular constituents are reused by the cell; when in high levels they also reactivate mTORC1 and then suppress autophagy.

[6]. Subsequently, genetic and transgenic studies in plants, worms, fruit flies, mice, and humans have underscored the pathway's conservation and have begun to unveil the intricate vital role that autophagy plays in the physiology of cells and multicellular organisms.

For a long time, autophagy was considered a non-selective pathway induced as a survival mechanism in response to cellular stresses. Over the past several years, however, it has become increasingly evident that autophagy also is a highly selective process involved in clearance of excess or dysfunctional organelles, protein aggregates and intracellular pathogens. In this introductory piece, we will briefly discuss the molecular mechanisms of selective types

of autophagy and their emerging importance as a quality control to maintain cellular and organismal health, aspects that will be presented in deep in the reviews of this special issue of the *International Journal of Cell Biology* and highlighted by the research papers.

2. The Mechanism of Autophagy

2.1. The Function of the Atg Proteins. Autophagosomes are formed by expansion and sealing of a small cistern known as the phagophore or isolation membrane (Figure 1). Once complete, they deliver their cargo into the hydrolytic lumen of lysosomes for degradation. A diverse set of components are involved in the biogenesis of autophagosomes, which primarily includes the proteins encoded by the autophagy-related genes (ATG). Most *ATG* genes have initially been identified and characterized in yeast. Subsequent studies in higher eukaryotes have revealed that these key factors are highly conserved. To date, 36 Atg proteins have been identified and 16 are part of the core Atg machinery essential for all autophagy-related pathways [7]. Upon autophagy induction, these proteins associate following a hierarchical order [8, 9] to first mediate the formation of the phagophore and then to expand it into an autophagosome [10, 11]. While their molecular functions and their precise contribution during the biogenesis of double-membrane vesicles remain largely unknown, they have been classified in 4 functional groups of genes: (1) the Atg1/ULK complex, (2) the phosphatidylinositol 3-kinase (PI3K) complex, (3) the Atg9 trafficking system, and (4) the two parallel ubiquitin-like conjugation systems (Figure 1).

The Atg1/ULK complex consists of Atg1, Atg13, and Atg17 in yeast, and ULK1/2, Atg13, FIP200 and Atg101 in mammals [12–15]. This complex is central in mediating the induction of autophagosome biogenesis and as a result it is the terminal target of various signaling cascades regulating autophagy, such as the TOR, insulin, PKA, and AMPK pathways [16] (Figure 1). Increased activity of the Atg1/ULK kinase is the primary event that determines the acute induction and upregulation of autophagy. It is important to note that ULK1 is part of a protein family and two other members, ULK2 and ULK3, have been shown play a role in autophagy induction as well [14, 17]. The expansion of this gene family may reflect the complex regulation and requirements of the pathway in multicellular long-lived organisms. Stimulation of the ULK kinases is achieved through an intricate network of phosphorylation and dephosphorylation modifications of the various subunits of the Atg1/ULK complex. For example, Atg13 is directly phosphorylated by TOR and the phosphorylation state of Atg13 modulates its binding to Atg1 and Atg17. Inactivation of TOR leads to a rapid dephosphorylation of Atg13, which increases Atg1–Atg13–Atg17 complex formation, stimulates the Atg1 kinase activity and induces autophagy [18, 19]. The mAtg13 is also essential for autophagy, but seems to directly interact with ULK1, ULK2 and FIP200 independently of its phosphorylation state [13, 14]. In addition, there are several phosphorylation events within this complex as well, including phosphorylation of mAtg13 by ULK1, ULK2, and TOR; phosphorylation of FIP200 by ULK1 and ULK2; phosphorylation of ULK1 and ULK2 by TOR [13, 14]. Additional studies are required to fully characterize the functional significance of these posttranslational modifications.

Autophagy is also regulated by the activity of PI3K complexes. Yeast contains a single PI3K, Vps34, which is present in two different tetrameric complexes that share 3 common subunits, Vps34, Vps15, and Atg6 [20]. Complex I is required for the induction of autophagy and through its fourth component, Atg14, associates to the autophagosomal membranes where the lipid kinase activity of Vps34 is essential for generating the phosphatidylinositol-3-phosphate (PI3P) that permits the recruitment of other Atg

proteins [9, 21] (Figure 1). Complex II contains Vps38 as the fourth subunit and it is involved in endosomal trafficking and vacuole biogenesis [20]. There are three types of PI3K in mammals: class I, II, and III. The functions of class II PI3K remains largely unknown, but both classes I and III PI3Ks are involved in autophagy. While class I PI3K is principally implicated in the modulation of signalling cascades, class III PI3K complexes regulate organelle biogenesis and, like yeast, contain three common components: hVps34, p150 (Vps15 ortholog), and Beclin 1 (Atg6 ortholog). The counterparts of Atg14 and Vps38 are called Atg14L/Barkor and UVRAG, respectively [22–24]. The Atg14L-containing complex plays a central role in autophagy and functions very similarly as the yeast complex I by directing the class III PI3K complex I to the phagophore to produce PI3P and initiate the recruitment of the Atg machinery (Figure 1). Atg14L is thought to be present on the ER irrespective of autophagy induction [25]. Upon starvation, Atg14L localizes to autophagosomal membranes [8]. Importantly, depletion of Atg14L reduces PI3P production, impairs the formation of autophagosomal precursor structures, and inhibits autophagy [8, 24, 26, 27]. The UVRAG-containing class III PI3K complex also regulates autophagy but it appears to act at the intersection between autophagy and the endosomal transport pathways. UVRAG initially associates with the BAR-domain protein Bif-1, which may regulate mAtg9 trafficking from the trans-Golgi network (TGN) [28, 29]. UVRAG then interacts with the class C Vps/HOPS protein complex, promoting the fusion of autophagosomes with late endosomes and/or lysosomes [30]. Finally, the UVRAG-containing class III protein complex binds to Rubicon, a late endosomal and lysosomal protein that suppresses autophagosome maturation by reducing hVps34 activity [26, 31]. Importantly, both the Atg14L- and UVRAG-containing complexes interact through Beclin 1 with Ambra1, which in turn tethers these protein complexes to the cytoskeleton via an interaction with dynein [32, 33]. Following the induction of autophagy, ULK1 phosphorylates Ambra1 thus releasing the class III PI3K complexes from dynein and their subsequent relocalization triggers autophagosome formation. Therefore, Ambra1 constitutes a direct regulatory link between the Atg1/ULK1 and the PI3K complexes [32].

Together with the Atg1/ULK and the PI3K complexes, Atg9 is one of the first factors localizing to the preautophagosomal structure or phagophore assembly site (PAS), the structure believed to be the precursor of the phagophore [9, 34] (Figure 1). Atg9 is the only conserved transmembrane protein that is essential for autophagy. It is distributed to the PAS and multiple additional cytoplasmic tubulovesicular compartments derived from the Golgi [35–37]. Atg9 cycles between these two locations and consequently it is thought to serve as a membrane carrier providing the lipid building blocks for the expanding phagophore [37]. One of the established functions of Atg9 is that it leads to the formation of the yeast PAS when at least one of the cytoplasmic tubulovesicular compartments translocates near the vacuole [34]. Atg9 is also essential to recruit the PI3K Complex I to the PAS [9]. Retrieval transport of yeast Atg9 from the PAS and/or complete autophagosome is mediated by the

Atg2-Atg18 complex [38] and appears to be regulated by the Atg1/ULK and PI3K complexes [37]. Mammalian Atg9 (mAtg9) has similar characteristics to its yeast counterpart. mAtg9 localizes to the TGN and late endosomes and redistributes to autophagosomal structures upon the induction of autophagy (Figure 1) [39], further promoting pathway activity [29, 40–42]. As in yeast, cycling of mAtg9 between locations also requires the Atg1/ULK complex and kinase activity hVps34 [39, 43].

The core Atg machinery also entails two ubiquitin-like proteins, Atg12 and Atg8/microtubule-associated protein 1 (MAP1)-light chain 3 (LC3), and their respective, partially overlapping, conjugation systems [44–46] (Figure 1). Atg12 is conjugated to Atg5 through the activity of the Atg7 (E1-like) and the Atg10 (E2-like) enzymes. The Atg12–Atg5 conjugate then interacts with Atg16, which oligomerizes to form a large multimeric complex. Atg8/LC3 is cleaved at its C terminus by the Atg4 protease to generate the cytosolic LC3-I with a C-terminal glycine residue, which is then conjugated to phosphatidylethanolamine (PE) in a reaction that requires Atg7 and the E2-like enzyme Atg3. This lipidated form of LC3 (LC3-II) is attached to both faces of the phagophore membrane. Once the autophagosome is completed, Atg4 removes LC3-II from the outer autophagosome surface. These two ubiquitination-like systems appear to be closely interconnected. On one hand, the multimeric Atg12-Atg5-Atg16 complex localizes to the phagophore and acts as an E3-like enzyme, determining the site of Atg8/LC3 lipidation [47, 48]. On the other hand, the Atg8/LC3 conjugation machinery seems to be essential for the optimal functioning of the Atg12 conjugation system. In Atg3-deficient mice, Atg12-Atg5 conjugation is markedly reduced, and normal dissociation of the Atg12-Atg5-Atg16 complex from the phagophore is delayed [49]. Some evidences suggest that these two conjugation systems also function together during the expansion and closure of the phagophore. For example, overexpression of an inactive mutant of Atg4 inhibits the lipidation of LC3 and leads to the accumulation of a number of nearly complete autophagosomes [47]. While controversial [50], it has been postulated that Atg8/LC3 also possesses fusogenic properties, thus mediating the assembly of the autophagic membrane [51, 52].

It has to be noted that mammals possess at least 7 genes coding for LC3/Atg8 proteins that can be grouped into three subfamilies: (1) the LC3 subfamily containing LC3A, LC3B, LC3B2 and LC3C; (2) the gammaaminobutyrate receptor-associated protein (GABARAP) subfamily comprising GABARAP and GABARAPL1 (also called GEC-1); (3) the Golgi-associated ATPase enhancer of 16 kDa (GATE-16) protein (also called GABARAP-L2/GEF2) [53]. Although *in vivo* studies show that they are all conjugated to PE, they appear to have evolved complex nonredundant functions [54].

2.2. The Autophagosomal Membranes. The origin of the membranes composing autophagosomes is a long-standing mystery in the field of autophagy. A major difficulty in addressing this question has been that phagophores as well as autophagosomes do not contain marker proteins of other subcellular compartments [55, 56]. A series of new studies has implicated several cellular organelles as the possible source for the autophagosomal lipid bilayers. The plasma membrane and elements of the trafficking machinery to the cell surface have been linked to the formation of an early autophagosomal intermediate, perhaps the phagophore [57–61]. It is possible that early endosomal- and/or Golgi-derived membranes are also key factors in the initial steps of autophagy [34, 36, 39]. The Golgi, moreover, appears also important for autophagy by supplying at least in part the extra lipids required for the phagophore expansion [29, 62–65]. The endoplasmic reticulum (ER) is also central in this latter event. While the relevance of the ER in autophagosome biogenesis was already pointed out a long time ago [5, 55, 66, 67], recently two electron tomography studies have demonstrated the existence of a physical connection between the ER and the forming autophagosomes [68, 69]. These analyses have revealed that the ER is connected to the outer as well as the inner membrane of the phagophore through points of contact, supporting the notion that lipids could be supplied via direct transfer at the sites of membrane contact. In line with this view, it has been found that Atg14L is associated to the ER and PI3P is generated on specific subdomains of this organelle from where autophagosomes emerge under autophagy-inducing conditions [25, 70] (Figure 1). It has also been proposed that the outer membrane of the mitochondria is the main source of the autophagosomal lipid bilayers, but while the experimental evidences appear to show that mitochondria are essential for the phagophore expansion, it remains unclear whether these organelles play a key role in the phagophore biogenesis [71]. The discrepancy between the conclusions of the various studies has not allowed yet drawing a model about the membrane dynamics during autophagosome biogenesis. The different results could be due to the different experimental conditions and model systems used by the various laboratories. Alternatively, the lipids forming the autophagosomes could have different sources depending on the cell and the conditions inducing autophagy [72, 73]. A third possibility is that the source of phagophore membrane could depend on the nature of the double-membrane vesicle cargo. Additional investigations are required to shed light on these issues.

2.3. Pharmacological Manipulation of Autophagy. Despite the potential of curing, quite a substantial range of specific pathological conditions by inducting autophagy, there are currently no small molecules that allow to exclusively stimulate this pathway [74]. Nevertheless, there is a variety of chemicals that by acting on signaling cascades that also regulate autophagy permit to trigger this degradative process. These agents fall into two distinct categories based on the mechanism of action; whether they work through an mTOR-dependent (Rapamycin or Torin) or mTOR-independent pathway (e.g., lithium or resveratrol) [74]. In addition to these compounds, there are biological molecules such as interferon γ (IFNγ) and vitamin D that can be used to stimulate autophagy especially in experimental setups [75, 76].

Inhibition of autophagy can also be beneficial in specific diseases but as for the inducers there are no compounds that exclusively block this pathway without affecting other cellular processes. The small molecules inhibiting autophagy include wortmannin and 3-methyladenine, which hamper the activity of the PI3K; Bafilomycin A and chloroquine, which impair the degradative activity of lysosomes [77]. They are currently solely used in the basic research on autophagy.

3. Selective Types of Autophagy

3.1. The Molecular Machinery of Selective Autophagy. It is becoming increasingly evident that autophagy is a highly selective quality control mechanism whose basal levels are important to maintain cellular homeostasis (see below). A number of organelles have been found to be selectively turned over by autophagy and cargo-specific names have been given to distinguish the various selective pathways, including the ER (reticulophagy or ERphagy), peroxisomes (pexophagy), mitochondria (mitophagy), lipid droplets (lipophagy), secretory granules (zymophagy), and even parts of the nucleus (nucleophagy). Moreover, pathogens (xenophagy), ribosomes (ribophagy), and aggregate-prone proteins (aggrephagy) are specifically targeted for degradation by autophagy [78].

Selective types of autophagy perform a cellular quality control function and therefore they must be able to distinguish their substrates, such as protein aggregates or dysfunctional mitochondria, from their functional counterparts. The molecular mechanisms underlying cargo selection and regulation of selective types of autophagy are still largely unknown. This has been an area of intense research during the last years and our understanding of the various selective types of autophagy is starting to unravel. A recent genome-wide small interfering RNA screen aimed at identifying mammalian genes required for selective autophagy found 141 candidate genes to be required for viral autophagy and 96 of those genes were also required for Parkin-mediated mitophagy [79].

In general, these pathways appear to rely upon specific cargo-recognizing autophagy receptors, which connect the cargo to the autophagic membranes. The autophagy receptors might also interact with specificity adaptors, which function as scaffolding proteins that bring the cargo-receptor complex in contact with the core Atg machinery to allow the specific sequestration of the substrate. The selective types of autophagy appear to rely on the same molecular core machinery as non-selective (starvation-induced) bulk autophagy. In contrast, the autophagy receptors and specificity adapters do not seem to be required for nonselective autophagy.

Autophagy receptors are defined as proteins being able to interact directly with both the autophagosome cargo and the Atg8/LC3 family members through a specific (WxxL) sequence [80], commonly referred to as the LC3-interacting region (LIR) motif [81] or the LC3 recognition sequences (LRS) [82]. Based on comparison of LIR domains from more than 20 autophagy receptors it was found that the LIR

consensus motif is an eight amino acids long sequence that can be written D/E-D/E-D/E-W/F/Y-X-X-L/I/V. Although not an absolute requirement, usually there is at least one acidic residue upstream of the W-site. The terminal L-site is occupied by a hydrophobic residue, either L, I, or V [83]. The LIR motifs of several autophagy receptors have been found to interact both with LC3 and GABARAP family members *in vitro*, but whether this reflects a physiological interaction remains to be clarified in most cases. It should be pointed out that not all LIR-containing proteins are autophagy cargo receptors. Some LIR-containing proteins, like Atg3 and Atg4B, are recruited to autophagic membranes to perform their function in autophagosome formation [84, 85], whereas others like FYVE and coiled-coil domain-containing protein 1 (FYCO1) interact with LC3 to facilitate autophagosome transport and maturation [86]. Others might use an LIR motif to become degraded, like Dishevelled, an adaptor protein in the Wnt signalling pathway [87]. The adaptor proteins are less well-described, but seem to interact with autophagy receptors and work as scaffold proteins recruiting and assembling the Atg machinery required to generate autophagosomes around the cargo targeted to degradation. Examples of autophagy adaptors are Atg11 and ALFY [88, 89].

The list of specific autophagy receptors is rapidly growing and the role of several of them in different types of selective autophagy will be described in detail in the reviews of this special issue. Here we will briefly discuss the best studied form of selective autophagy, the yeast cytosol to vacuole targeting (Cvt) pathway, as well as the best studied mammalian autophagy receptor, p62/sequestosome 1 (SQSTM1) (Figure 2).

The Cvt pathway is a biosynthetic process mediating the transport of the three vacuolar hydrolases, aminopeptidase 1 (Ape1), aminopeptidase 4 (Ape4) and α-mannosidase (Ams1), and the Ty1 transposome into the vacuole [90, 91]. Ape1 is synthesized as a cytosolic precursor (prApe1), which multimerizes into the higher order Ape1 oligomer, to which Ape4, Ams1, and Ty1 associate to form the so-called Cvt complex, prior to being sequestered into a small autophagosome-like Cvt vesicle. Sequestration of the Cvt complex into Cvt vesicles is a multistep process, which requires the autophagy receptor Atg19, which facilitates binding to Atg8 at the PAS, as well as the adaptor protein Atg11 (Figure 2(a)) [92]. Atg11 acts as a scaffold protein by directing the Cvt complex and Atg9 reservoirs translocation to the PAS in an actin-dependent way and then recruiting the Atg1/ULK complex [40, 93]. The PI3P-binding proteins Atg20, Atg21, and Atg24 are also required for the Cvt pathway [94, 95], but their precise function remains to be elucidated. Interestingly, Atg11 overexpression was found to recruit more Atg8 and Atg9 to the PAS resulting in more Cvt vesicles. This observation indicates that Atg11 levels could regulate the rate of selective autophagy, and maybe also the size of the cargo-containing autophagosomes in yeast [90, 96]. Indeed, a series of studies has revealed that Atg11 is also involved in other types of selective autophagy such as mitophagy and pexophagy. However, the autophagy receptors involved in the different Atg11-dependent types

(a)

(b)

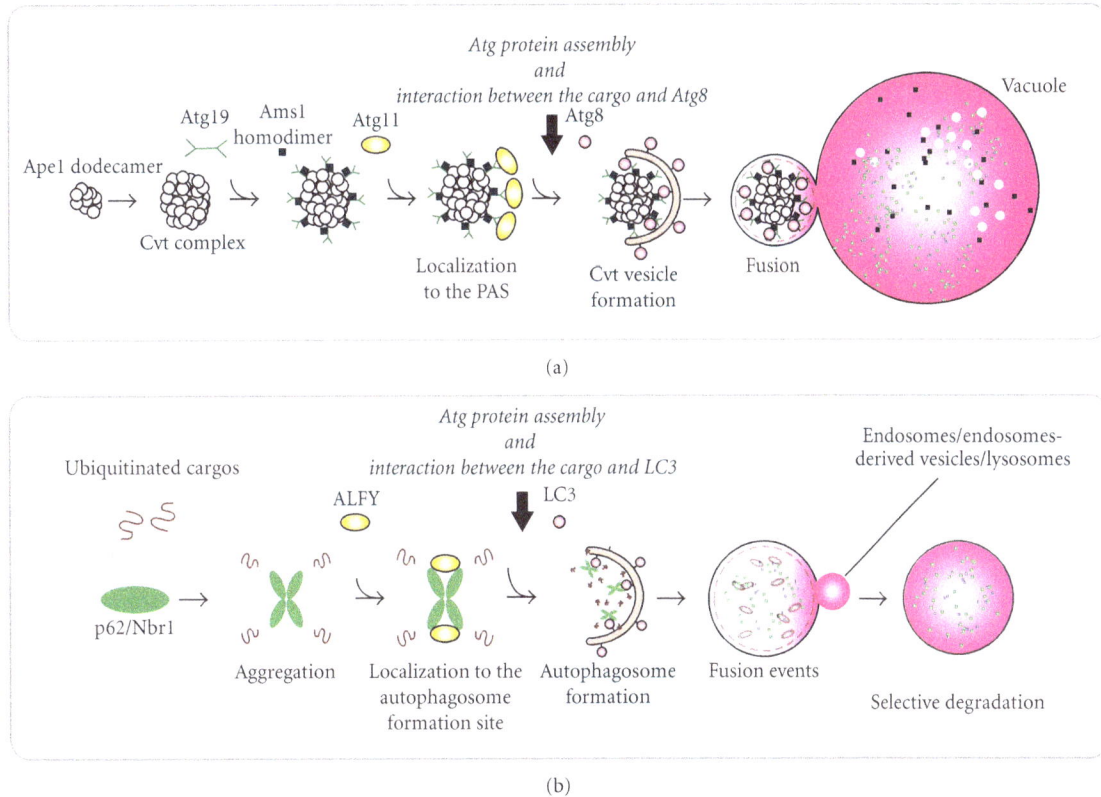

FIGURE 2: Representative selective autophagy. (a) The cytoplasm-to-vacuole targeting (Cvt) pathway. Ape1 is synthesized as a cytoplasmic precursor protein with a propeptide and rapidly oligomerizes into dodecamers that subsequently associate with each other to form a higher order complex. The autophagy receptor Atg19 directly binds to the complex and mediates the recruitment of another Cvt pathway cargo, Ams1, leading to the formation of the so-called Cvt complex. Atg19 also interacts with the autophagy adaptor Atg11 and this protein allows the transport of the Cvt complex to the site where the double-membrane vesicle will be generated. At this location, Atg11 tethers the Atg proteins essential for the Cvt vesicle formation and the direct binding of Atg19 to Atg8 permits the exclusive sequestration of the Cvt complex into the vesicle. (b) A model for p62 and NBR1 as autophagy receptors for ubiquitinated cargos. p62 and NBR1 bind with ubiquitinated cargos via their ubiquitin-associated (UBA) domain and this interaction triggers the aggregate formation through the oligomerization of p62 via its Phox and Bem1p (PB1) domain. Furthermore, p62 interacts with both autophagy-linked FYVE protein (ALFY), which serves to recruit Atg5 and to bind PI3P, and directly with LC3. This latter event appears to organize and activate the Atg machinery in close proximity of the ubiquitinated cargos, which allows to selectively sequester them in the autophagosomes in analogous to the Cvt pathway.

of selective autophagy are different as Atg32 is required for mitophagy [97, 98], whereas Atg30 is essential for pexophagy [99]. Like Atg19, these two proteins have an Atg8-binding LIR motif and directly interact with Atg11. Mammalian cells appear to not possess an Atg11 homologue, and further studies are necessary to delineate the molecular machinery involved in sequestration and targeting of different cargoes for degradation by autophagy in higher eukaryotes.

The mechanism of the Cvt pathway is reminiscent of the selective form of mammalian autophagy called aggrephagy, which involves degradation of misfolded and unwanted proteins by packing them into ubiquitinated aggregates. In both cases aggregation of the substrate (prApe1 or misfolded proteins) is required prior to sequestration into Cvt vesicles or autophagosomes, respectively [100–102]. Similar to Cvt vesicles, aggregate-containing autophagosomes appear to be largely devoid of cytosolic components suggesting that the vesicle membrane expands tightly around its cargo [88]. Aggrephagy also depends on proteins with exclusive functions in substrate selection and targeting [81, 88, 100, 103].

The autophagy receptors p62 and neighbour of BRCA1 gene (NBR1) bind both ubiquitinated protein aggregates through an ubiquitin-associated (UBA) domain and to LC3 via their LIR motifs and, thereby, promote the specific autophagic degradation of ubiquitinated proteins (Figure 2(b)) [81, 82, 100, 103, 104]. NBR1 and p62 also contain an N-terminal Phox and Bem1p (PB1) domain through which they can oligomerize, or interact with other PB1-containing binding partners [83]. In addition to being a cargo receptor for protein aggregates, p62 has been implicated in autophagic degradation of other ubiquitinated substrates such as intracellular bacteria [105], viral capsid proteins [106], the midbody remnant formed after cytokinesis [107], peroxisomes [108, 109], damaged mitochondria [110, 111], and bacteriocidal precursor proteins [112]. The PB1 domain was recently found to be required for p62 to localize to the autophagosome formation site adjacent to the ER [113], suggesting that it could target ubiquitinated cargo to the site of autophagosome formation or alternatively promote the assembly of the Atg machinery at this location.

The large scaffolding protein autophagy-linked FYVE (ALFY) appears to have a similar function as the specificity adaptor Atg11. ALFY is recruited to aggregate-prone proteins through its interaction with p62 [101] and through a direct interaction with Atg5 and PI3P it serves to recruit the core Atg machinery and allow formation of autophagic membranes around the protein aggregate [88] (Figure 2(b)). Interestingly, ALFY is recruited from the nucleus to cytoplasmic ubiquitin-positive structures upon cell stress suggesting that it might regulate the level of aggrephagy [114]. In line with this, it was found that overexpression of ALFY in mouse and fly models of Huntington's disease reduced the number of protein inclusions [88]. It will be interesting to determine whether ALFY, as p62, is involved in other selective types of autophagy such as the one eliminating midbody ring structures or mitochondria.

3.2. Regulation of Selective Autophagy. It is well known that posttranslational modifications like phosphorylation and ubiquitination are involved in the regulation of the activity of proteins involved in autophagy and degradation of autophagic cargo proteins, respectively. However, little is known about how these modifications may regulate selective autophagy. The fact that the core Atg machinery is required for both nonselective and selective types of autophagy gives raise to the question of whether these two types of autophagy may compete for the same molecular machinery. Such a competition could be detrimental for the cells undergoing starvation and to avoid this, there might be a tight regulation of the expression level and/or activity of the proteins specifically involved the selective autophagy. It has recently been proposed that phosphorylation of autophagy receptors might be a general mechanism for the regulation of selective autophagy. Dikic and coworkers noted that several autophagy receptors contain conserved serine residues adjacent to their LIR motifs and indeed, the TANK binding kinase 1 (TBK1) was found to phosphorylate a serine residue close to the LIR motif of the autophagy receptor optineurin. This modification enhances the LC3 binding affinity of optineurin and promotes selective autophagy of ubiquitinated cytosolic *Salmonella enterica* [115]. In yeast, phosphorylation of Atg32, the autophagy receptor for mitophagy, by mitogen-activated protein kinases was found to be required for mitophagy [116, 117].

The Atg8/LC3 proteins themselves have also been found to become phosphorylated and recent works have identified specific phosphorylation sites for protein kinase A (PKA) [118] and protein kinase C (PKC) [119] in the N-terminal region of LC3. Interestingly, the N-terminal of LC3 is involved in the binding of LC3 to LIR-containing proteins [120]. It is therefore tempting to speculate that phosphorylation of the PKA and PKC sites might facilitate or prevent the interaction of LC3 with LIR-containing proteins such as p62. It has been found that phosphorylation of the PKA site, which is conserved in all mammalian LC3 isoforms, but not in GABARAP, inhibits recruitment of LC3 into autophagosomes [118].

The role of ubiquitin in autophagy has so far been ascribed as a signal for cargo degradation. Ubiquitination of aggregate prone proteins, as well as bacteria and mitochondria, has been found to serve as a signal for recognition by autophagy receptors like p62 and NBR1, which are themselves also degraded together with the cargo that they associate with [83]. The *in vivo* specificity of p62 and NBR1 toward ubiquitin signals remains to be established under the different physiological conditions. Interestingly, it was recently found that casein kinase 2- (CK2-) mediated phosphorylation of the p62 UBA domain increases the binding affinity of this motif for polyubiquitin chains leading to more efficient targeting of polyubiquitinated proteins to autophagy [121]. CK2 overexpression or phosphatase inhibition reduced the formation of aggregates containing the polyglutamine-expanded huntingtin exon1 fragments in a p62-dependent manner. The E3 ligases involved in ubiquitination of different autophagic cargo largely remains to be identified. However, it is known that the E3 ligases Parkin and RNF185 both regulate mitophagy [122, 123]. SMURF1 (SMAD-specific E3 ubiquitin protein ligase 1) was recently also implicated in mitophagy, as well as in autophagic targeting of viral particles [79]. Interestingly, the role of SMURF1 in selective autophagy seems to be independent of its E3 ligase activity, but it rather depends on its membrane-targeting C2 domain, although the exact mechanism involved remains to be elucidated. It is also not clear whether ubiquitination could serve as a signal to regulate the activity or binding selectivity of proteins directly involved in autophagy, and whether this in some way could regulate selective autophagy. The role of ubiquitin-like proteins as SUMO and Nedd in autophagy is also unexplored.

Acetylation is another posttranslational modification that only recently has been implicated in selective autophagy. The histone de-acetylase 6 (HDAC6), initially found to mediate transport of misfolded proteins to the aggresome [124], was lately implicated in maturation of ubiquitin-positive autophagosomes [125]. The fact that HDAC6 overproduction in fly eyes expressing expanded polyQ proteins is neuroprotective further indicates that HDAC6 activity stimulates aggrephagy [126]. Furthermore, the acetylation of an aggrephagy cargo protein, muntant huntingtin, the protein causing Huntington's disease, is important for its degradation by autophagy [127]. HDAC6 has been also implicated in Parkin-mediated clearance of damaged mitochondria [128]. The acetyl transferase(s) involved in these forms of selective autophagy is currently unknown, but understanding the role of acetylation in relation to various aspects of autophagy is an emerging field and it will very likely provide more mechanistic insights into these pathways.

4. Pathophysiological Relevance of Selective Types of Autophagy

Basal autophagy acts as the quality control pathway for cytoplasmic components and it is crucial to maintain the homeostasis of various postmitotic cells [129]. While this quality control could be partially achieved by nonselective autophagy, growing lines of evidence have demonstrated that specific proteins, organelles, and invading bacteria are specifically degraded by autophagy (Figure 3).

4.1. Tissue Homeostasis. Mice deficient in autophagy die either *in utero* (e.g., *Beclin 1* and *Fip200* knockout mice) [130–132] or within 24 hours after birth due, at least in part, to a deficiency in the mobilization of amino acids from various tissues (e.g., *Atg3, Atg5, Atg7, Atg9,* and *Atg16L* knockout mice) [49, 133–136]. As a result, to investigate the physiological roles of autophagy, conditional knockout mice for *Atg5, Atg7,* or *FIP200* and various tissue-specific *Atg* knockout mice have been established and analyzed [133, 137, 138]. For example, the liver-specific *Atg7*-deficient mouse displayed severe hepatomegaly accompanied by hepatocyte hypertrophy, resulting in severe liver injuries [133]. Mice lacking *Atg5, Atg7,* or *FIP200* in the central nervous system exhibited behavioral deficits, such as abnormal limb-clasping reflexes and reduction of coordinated movement as well as massive neuronal loss in the cerebral and cerebellar cortices [137–139]. Loss of *Atg5* in cardiac muscle caused cardiac hypertrophy, left ventricular dilatation, and systolic dysfunction [140]. Skeletal muscle-specific *Atg5* or *Atg7* knockout mice showed age-dependent muscle atrophy [141, 142]. Pancreatic β cell-specific *Atg7* knockout animals exhibited degeneration of islets and impaired glucose tolerance with reduced insulin secretion [143, 144]. Podocyte-specific deletion of *Atg5* caused glomerulosclerosis in aging mice and these animals displayed increased susceptibility to proteinuric diseases caused by puromycin aminonucleoside and adriamycin [145]. Proximal tubule-specific *Atg5* knockout mice were susceptible to ischemia-reperfusion injury [146]. Finally, deletion of *Atg7* in bronchial epithelial cells resulted in hyperresponsiveness to cholinergic stimuli [147]. All together, these results undoubtedly indicate that basal autophagy prevents numerous life-threatening diseases.

How does impairment of autophagy lead to diseases? Ultrastructural analyses of the mutant mice revealed a marked accumulation of swollen and deformed mitochondria in the mutant hepatocytes [133], pancreatic β cells [143, 144], cardiac and skeletal myocytes [140, 141] and neurons [138], but also the appearance of concentric membranous structures consisting of ER or sarcoplasmic reticulum in hepatocytes [133], neuronal axons [137, 139] and skeletal myocytes [141], as well as an increased number of peroxisomes and lipid droplets in hepatocytes [133, 148]. In addition to the accumulation of aberrant organelles, histological analyses of tissues with defective autophagy showed the amassment of polyubiquitylated proteins in almost all tissues (although the level varied from one region to another) forming inclusion bodies whose size and number increased with aging [149]. Consequently, basal autophagy also acts as the quality control machinery for cytoplasmic organelles (Figure 3(a)). Although this could be partially achieved by bulk autophagy, these observations point to the existence of selective types of autophagy, a notion that is now supported by experimental data.

4.2. Implications of Selective Degradation of p62 by Autophagy. p62/SQSTM1 is the best-characterized disease-related autophagy receptor and a ubiquitously expressed cellular protein conserved among metazoan but not in plants and fungi [83]. Besides a role of p62 as the receptor, this protein itself is specific substrate for autophagy. Suppression of autophagy is usually accompanied by an accumulation of p62 mostly in large aggregates also positive for ubiquitin (Figure 3(a)) [104, 150]. Ubiquitin and p62-positive inclusion bodies have been detected in numerous neurodegenerative diseases (i.e., Alzheimer's disease, Parkinson's disease, and amyotrophic lateral sclerosis), liver disorders (i.e., alcoholic hepatitis and steatohepatitis), and cancers (i.e., malignant glioma and hepatocellular carcinoma) [151]. Very interestingly, the p62-positive aggregates observed in hepatocytes and neurons of liver- and brain-specific *Atg7* deficient mice, respectively, as well as in human hepatocellular carcinoma cells, are completely dispersed by the additional loss of p62 strongly implicating involvement of p62 in the formation of disease-related inclusion bodies [104, 152].

Through its self-oligomerization, p62 is involved in several signal transduction pathways. For example, this protein functions as a signaling hub that may determine whether cells survive by activating the TRAF6–NF-κB pathway, or die by facilitating the aggregation of caspase 8 and the downstream effector caspases [153, 154]. On the other hand, p62 interacts with the Nrf2-binding site on Keap1, a component of the Cullin 3-type ubiquitin ligase for Nrf2, resulting in stabilization of Nrf2 and transcriptional activation of Nrf2 target genes including a battery of antioxidant proteins [155–159]. It is thus plausible that excess accumulation or mutation of p62 leads to hyperactivation of these signaling pathways, resulting in a disease onset (Figure 3(b)).

Paget's disease of bone is a chronic and metabolic bone disorder that is characterized by an increased bone turnover within discrete lesions throughout the skeleton. Mutations in the *p62* gene, in particular in its UBA domain, can cause this illness [160]. A proposed model explaining how p62 mutations lead to the Paget's disease of bone is the following: mutations of the UBA domain cause an impairment in the interaction between p62 and ubiquitinated TRAF6 and/or CYLD, an enzyme deubiquitinating TRAF6, which in turn enhances the activation of the NF-κB signaling pathway and the resulting increased osteoclastogenesis (Figure 3(b)) [160]. If proven, this molecular scenario could open the possibility of using autophagy enhancers as a therapy to cure Paget's disease of bone.

It is established that autophagy has a tumor-suppressor role and several autophagy gene products including Beclin1 and UVRAG are known to function as tumor suppressor proteins [161]. The tumor-suppressor role of autophagy appears to be important particularly in the liver. Spontaneous tumorigenesis is observed in the livers of mice with either a systemic mosaic deletion of *Atg5* or a hepatocyte-specific *Atg7* deletion [152, 162]. Importantly, no tumors are formed in other organs in *Atg5* mosaically deleted mice. Enlarged mitochondria, whose functions are at least partially impaired, accumulate in *Atg5*- or *Atg7*-deficient hepatocytes [152, 162]. This observation is in line with the previous

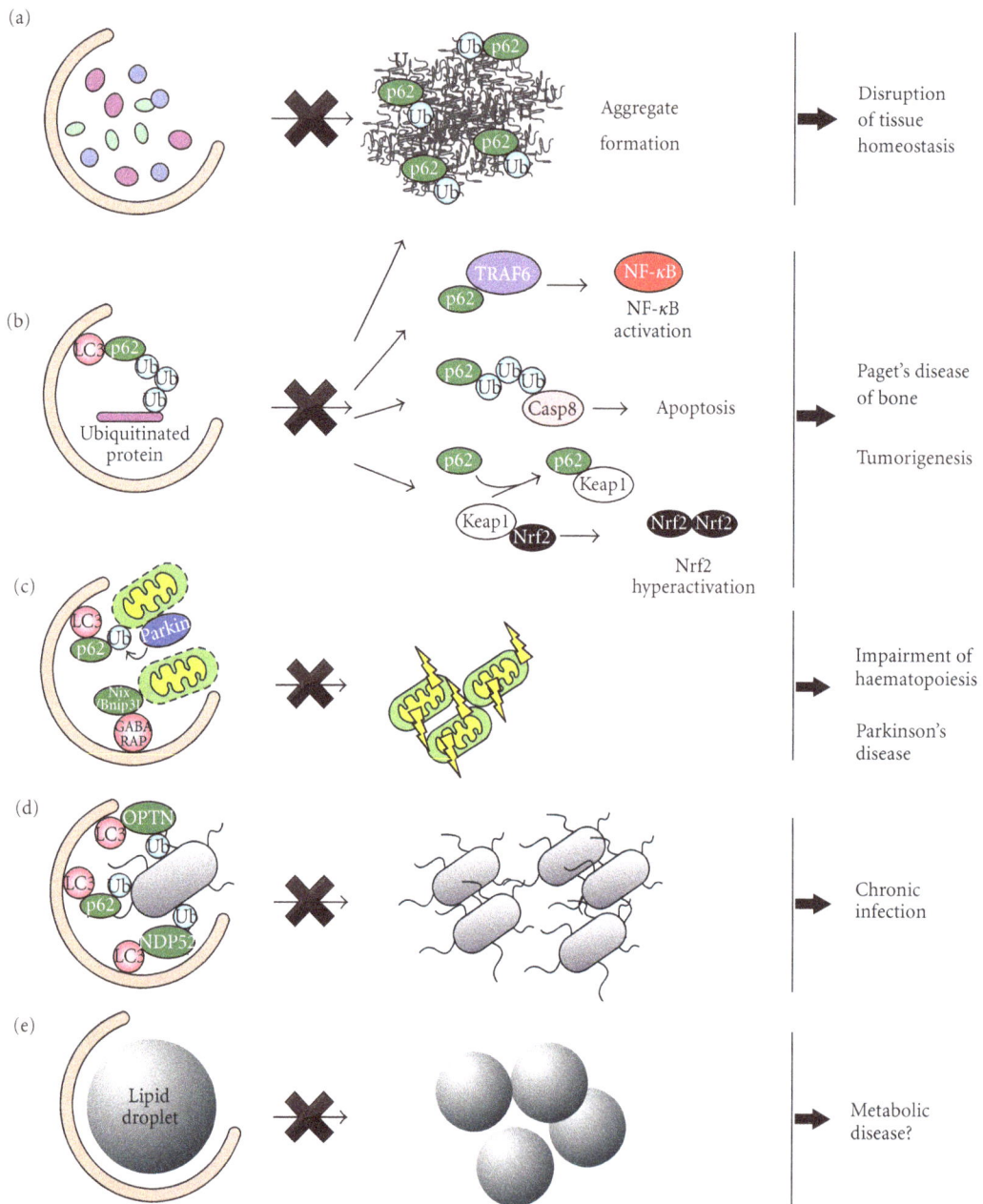

FIGURE 3: Pathophysiological relevance of selective autophagy. (a, b) Selective types of autophagy operates constitutively at low levels even under nutrient-rich conditions and mediates turnover of selected cytoplasmic materials through the action of autophagy receptors such as p62 and NBR1. These proteins mediate the elimination of ubiquitinated structures, including protein aggregates (a) and defects in these pathways lead to the disruption of tissue homeostasis, resulting in life-threatening diseases. Defective autophagy is usually accompanied by extensive accumulation of p62-containing aggregates, which enhances its function as a scaffold protein in several signaling cascades such as NF-κB signaling, apoptosis, and Nrf2 activation (b). Such abnormalities might be involved in tumorigenesis and Paget's disease of bone. (c) During erythroid differentiation, Nix/Bnip3L relocalization to mitochondria leads to their depolarization, which triggers mitophagy. Loss of *Nix/Bnip3L* causes an arrest in the erythroid maturation arrest, leading to severe anaemia. In response to loss of the mitochondrial membrane potential, Parkin translocates onto the damaged mitochondria in a PINK1-dependent manner, and ubiquitinated proteins present on the outer mitochondrial membrane, which induces mitophagy. Parkinson's disease-related mutations in the *Parkin* and *PINK1* genes provoke a defect in mitophagy, suggesting this selective type of autophagy has a role in preventing the pathogenesis of the Parkinson's disease. (d) Specific bacteria invading the cytosol get ubiquitinated and are recognized by autophagy receptors such as p62, NDP52, and optineurin (OPTN). This allows the specific sequestration of the microbes into autophagosomes and their delivery into the lysosomes for degradation. (e) The lipid droplets are probably degraded by autophagy selectively. This selective type of autophagy, lipophagy, supplies free-fatty acids utilized to generate energy through the β-oxidation. Impairments in lipophagy are known to cause accumulation of lipid droplets in hepatocytes and reduced production of AgRP in neurons.

data obtained in iBMK cell lines showing that both the oxidative stress and genomic damage responses are activated by loss of autophagy [163, 164]. Again, it is clear that accumulation of p62, at least partially, contributes to tumor growth because the size of the $Atg7^{-/-}$ liver tumors is reduced by the additional deletion of *p62* [162], which may cause a dysregulation of NF-κB signaling [165] and/or a persistent activation of Nrf2 [166].

4.3. Selective Degradation of Ubiquitinated Proteins. Almost all tissues with defective autophagy are usually displaying an accumulation of polyubiquitinated proteins [149]. Loss of autophagy is considered to lead to a delay in the global turnover of cytoplasmic components [137] and/or to an impaired degradation of substrates destined for the proteasome [167]. Both observations could partially explain the accumulation of misfolded and/or unfolded proteins that is followed by the formation of inclusion bodies.

As discussed above, p62 and NBR1 act as autophagy receptors for ubiquitinated cargos such as protein aggregates, mitochondria, midbody rings, bacteria, ribosomal proteins and virus capsids [83, 168] (Figure 3). Although these studies suggest the role of p62 as an ubiquitin receptor, it remains to be established whether soluble ubiquitinated proteins are also degraded one-by-one by p62 and possibly NBR1. A mass spectrometric analysis has clearly demonstrated the accumulation of all detectable topologies of ubiquitin chain in *Atg* deficient livers and brains, indicating that specific polyubiquitin chain linkage is not the decisive signal for autophagic degradation [169]. Because the increase in ubiquitin conjugates in the *Atg7* deficient liver and brain is completely suppressed by additional knockout of either *p62* or *Nrf2* [169], accumulation of ubiquitinated proteins in tissues defective in autophagy might be attributed to p62-mediated activation of Nrf2, resulting in global transcriptional changes to ubiquitin-associated genes. Further studies are needed to precisely elucidate the degradation mechanism of soluble ubiquitinated proteins by autophagy.

4.4. Mitophagy. Concomitant with the energy production through oxidative phosphorylation, mitochondria also generate reactive oxygen species (ROS), which cause damage through the oxidation of proteins, lipids and DNA often inducing cell death. Therefore, the quality control of mitochondria is essential to maintain cellular homeostasis and this process appears to be achieved via autophagy.

It has been postulated that mitophagy contributes to differentiation and development by participating to the intracellular remodelling that occurs for example during haematopoiesis and adipogenesis. In mammalian red blood cells, the expulsion of the nucleus followed by the removal of other organelles, such as mitochondria, are necessary differentiation steps. Nix/Bnip3L, an autophagy receptor whose structure resembles that of Atg32, is also an outer mitochondrial membrane protein that interacts with GABARAP [170, 171] and plays an important role in mitophagy during erythroid differentiation [172, 173] (Figure 3(c)). Although autophagosome formation probably

still occurs in *Nix/Bnip3L* deficient reticulocytes, mitochondrial elimination is severely impaired. Consequently, mutant reticulocytes are exposed to increased levels of ROS and die, and *Nix/Bnip3L* knockout mice suffer severe anemia. Depolarization of the mitochondrial membrane potential of mutant reticulocytes by treatment with an uncoupling agent results in restoration of mitophagy [172], emphasizing the importance of Nix/Binp3L for the mitochondrial depolarization and implying that mitophagy targets uncoupled mitochondria. Haematopoietic-specific *Atg7* knockout mice also exhibited severe anaemia as well as lymphopenia, and the mutant erythrocytes markedly accumulated degenerated mitochondria but not other organelles [174]. The mitochondrial content is regulated during the development of the T cells as well; that is, the high mitochondrial content in thymocytes is shifted to a low mitochondrial content in mature T cells. *Atg5* or *Atg7* deleted T cells fail to reduce their mitochondrial content resulting in increased ROS production as well as an imbalance in pro- and antiapoptotic protein expression [175–177]. All together, these evidences demonstrate the essential role of mitophagy in haematopoiesis.

Recent studies have described the molecular mechanism by which damaged mitochondria are selectively targeted for autophagy, and have suggested that the defect is implicated in the familial Parkinson's disease (PD) [178] (Figure 3(c)). PINK1, a mitochondrial kinase, and Parkin, an E3 ubiquitin ligase, have been genetically linked to both PD and a pathway that prevents progressive mitochondrial damage and dysfunction. When mitochondria are damaged and depolarized, PINK1 becomes stabilized and recruits Parkin to the damaged mitochondria [122, 179–181]. Various mitochondrial outer membrane proteins are ubiquitinated by Parkin and mitophagy is then induced. Of note, PD-related mutations in *PINK1* and *Parkin* impair mitophagy [122, 179–181], suggesting that there is a link between defective mitophagy and PD. How these ubiquitinated mitochondria are recognized by the autophagosome remains unknown. Although p62 has been implicated in the recognition of ubiquitinated mitochondria, elimination of the mitochondria occurs normally in *p62*-deficient cells [182, 183].

4.5. Elimination of Invading Microbes. When specific bacteria invade host cells through endocytosis/phagocytosis, a selective type of autophagy termed xenophagy, engulfs them to restrict their growth [184] (Figure 3(d)). Although neither the target proteins nor the E3 ligases have yet been identified, invading bacteria such as *Salmonella enterica*, *Listeria monocytogenes*, or *Shigella flexneri* become positive for ubiquitin when they access the cytosol by rupturing the endosome/phagosome limiting membrane [185, 186]. These findings raise the possibility that ubiquitin also serves as a tag during xenophagy. In fact, to date, three proteins, p62 [105, 185, 187], NDP52 [188], and optineurin [115] have been proposed to be autophagy receptors linking ubiquitinated bacteria and LC3. An ubiquitin-independent mechanism has recently been revealed; recognition of a *Shigella* mutant that lacks the icsB gene requires the tectonin domain-containing protein 1 (Tecpr1), which appears to be a new type of autophagy adaptor targeting *Shigella* to Atg5- and WIPI-2-positive membranes [189]. Interestingly, the *Shigella* icsB normally prevents autophagic sequestration of this bacterium by inhibiting the interaction of *Shigella* VirG with Atg5 indicating that some bacteria have developed mechanism to inhibit or subvert autophagy to their advantage [190]. This latter category of pathogens also includes viruses such as Herpes simplex virus-1 (HSV-1), which express an inhibitor (ICP34.5) of Atg6/Beclin1 [106]. However, it was recently shown that a mutant HSV-1 strain lacking ICP34.5 becomes degraded by selective autophagy in a SMURF1-dependent manner [79], suggesting that selective autophagy plays an important role in our immune system.

Recently, a different antimicrobial function has been assigned to autophagy and this function appears to be selective. During infection, ribosomal protein precursors are transported by autophagy in a p62-dependent manner into lysosomes [112]. These ribosomal protein precursors are subsequently processed by lysosomal protease into small antimicrobial peptides. Importantly, it has been shown that induction of autophagy during a *Mycobacterium tuberculosis* infection leads to the fusion between phagosomes containing this bacterium and autophagosomes, and the production of the antimicrobial peptides in this compartment kills *M. tuberculosis* [112].

4.6. Lipophagy. While the molecular mechanism is largely unknown, autophagy contributes at least partially to the supply of free fatty acids in response to fasting (Figure 3(e)). Fasting provokes the increase of the levels of free fatty acids circulating in the blood, which are mobilized from adipose tissues. These free fatty acids are rapidly captured by various organs including hepatocytes and then transformed into triglycerides by esterification within lipid droplets. These lipid droplets appear to be turned over by a selective type of autophagy that has been named lipophagy in order to provide endogenous free fatty acids for energy production through β-oxidation [148]. Indeed, liver-specific *Atg7* deficient mice display massive accumulation of triglycerides and cholesterol in the form of lipid droplets [191]. Agouti-related peptide- (AgRP-) expressing neurons also respond to increased circulating levels of free fatty acids after fasting and then induce autophagy to degrade the lipid droplets [192]. Similar to the case in hepatocytes, autophagy in the neurons supplies endogenous free fatty acids for energy production and seems to be necessary for gene expression of AgPR, which is a neuropeptide that increases appetite and decreases metabolism and energy expenditure [192].

5. Conclusions and Perspectives

Originally, it was assumed that autophagy was exclusively a bulk process. Recent experimental evidences have demonstrated that through the use of autophagy receptors and adaptors, this pathway can be selective by exclusively degrading specific cellular constituents. The list of physiological and pathological situations where autophagy is selective is constantly growing and this fact challenges the earliest concept whether autophagy can be nonselective. It is believe that under starvation, cytoplasmic structures are randomly engulfed by autophagosomes and delivered into the lysosome to be degraded and thus generate an internal pool of nutrients. In yeast *Saccharomyces cerevisiae*, however, the degradation of ribosomes, for example, ribophagy, as well as mitophagy and pexophagy, and the transport of the prApe1 oligomer into the vacuole under the same conditions requires the presence of autophagy receptors [97, 193–195]. As a result, these observations suggest that autophagy could potentially always operate selectively. This is a conceivable hypothesis because this process allows the cell to survive stress conditions and the casual elimination of cytoplasmic structure in the same scenario could lead to the lethal depletion of an organelle crucial for cell survival. Future studies will certainly provide more molecular insights into the regulation and mechanism of the selective types of autophagy, and this information will also be important to determine if indeed bulk autophagy exists.

Abbreviations

AgRP:	Agouti-related peptide
AMPK:	AMP-activated protein kinase
ALFY:	Autophagy-linked FYVE protein
Ams1:	α-mannosidase 1
Ape1:	Aminopeptidase 1
Ape4:	Aminopeptidase 4
Atg:	Autophagy-related gene
Bnip3L:	B-cell leukemia/lymphoma 2 (BCL-2)/adenovirus E1B interacting protein 3
CK2:	Casein kinase 2
CMA:	Chaperone-mediated autophagy
Cvt:	Cytoplasm to vacuole targeting
ER:	Endoplasmic reticulum
FIP200:	Focal adhesion kinase family interacting protein of 200 kD
FYCO1:	FYVE and coiled-coil domain-containing protein 1
GABARAP:	Gamma-aminobutyrate receptor-associated protein
GATE-16:	Golgi-associated ATPase enhancer of 16 kDa
HDAC6:	Histone de-acetylase 6
HOPS:	Homotypic fusion and protein sorting
HSV-1:	Herpes simplex virus-1
Keap1:	Kelch-like ECH-associated protein 1
LC3:	Microtubule-associated protein 1 (MAP1)-light chain 3
LIR:	LC3-interacting region
LRS:	LC3 recognition sequences
NBR1:	Neighbour of BRCA1 gene
NDP52:	Nuclear dot protein (NDP) 52
NF-κB:	Nuclear factor κB
NIX:	Nip-like protein X
Nrf2:	NF-E2 related factor 2
PAS:	Phagophore assembly site
PB1:	Phox and Bem1p

PE: Phosphatidylethanolamine
PD: Parkinson's disease
PI3K: Phosphatidylinositol 3-kinase
PI3P: Phosphatidylinositol 3-phosphate
PKA: Protein kinase A
PKC: Protein kinase C
ROS: Reactive oxygen species
Rubicon: RUN domain and cysteine-rich domain containing Beclin 1-interacting protein
SMURF1:   SMAD-specific E3 ubiquitin protein ligase 1
SUMO: Small ubiquitin-like modifier
SQSTM1:   p62/sequestosome 1
TBK1:   TANK binding kinase 1
Tecpr1:   Tectonin domain-containing protein 1
TRAF6:   Tumour necrosis factor receptor-associated factor 6
TOR: Target of Rapamycin
TGN: *Trans*-Golgi network
UBA: Ubiquitin associated
ULK1:   Unc-51-like kinase 1
UVRAG: UV-resistance associated gen
Vps: Vacuolar protein sorting.

Acknowledgments

The authors thank Shun Kageyama (Tokyo Metropolitan Institute of Medical Science) for helping in the creation of the figures used in the paper. F. Reggiori is supported by The Netherlands Organization for Health Research and Development (ZonMW-VIDI-917.76.329), by The Netherlands Organization for Scientific Research (Chemical Sciences, ECHO grant-700.59.003). M. Komatsu is supported by Funding Program for Next Generation World-Leading Researchers. K. Finley is supported by the NIH. A. Simonsen is supported by the Research Council of Norway, by the Norwegian Cancer Society, and by the South-Eastern Norway Regional Health Authority.

References

[1] E. Arias and A. M. Cuervo, "Chaperone-mediated autophagy in protein quality control," *Current Opinion in Cell Biology*, vol. 23, no. 2, pp. 184–189, 2011.

[2] S. J. Orenstein and A. M. Cuervo, "Chaperone-mediated autophagy: molecular mechanisms and physiological relevance," *Seminars in Cell and Developmental Biology*, vol. 21, no. 7, pp. 719–726, 2010.

[3] D. Mijaljica, M. Prescott, and R. J. Devenish, "Microautophagy in mammalian cells: revisiting a 40-year-old conundrum," *Autophagy*, vol. 7, no. 7, pp. 673–682, 2011.

[4] R. Sahu, S. Kaushik, C. C. Clement et al., "Microautophagy of cytosolic proteins by late endosomes," *Developmental Cell*, vol. 20, no. 1, pp. 131–139, 2011.

[5] E.-L. Eskelinen, F. Reggiori, M. Baba, A. L. Kovács, and P. O. Seglen, "Seeing is believing: the impact of electron microscopy on autophagy research," *Autophagy*, vol. 7, no. 9, pp. 935–956, 2011.

[6] Z. Yang and D. J. Klionsky, "Eaten alive: a history of macroautophagy," *Nature Cell Biology*, vol. 12, no. 9, pp. 814–822, 2010.

[7] H. Nakatogawa, K. Suzuki, Y. Kamada, and Y. Ohsumi, "Dynamics and diversity in autophagy mechanisms: lessons from yeast," *Nature Reviews Molecular Cell Biology*, vol. 10, no. 7, pp. 458–467, 2009.

[8] E. Itakura and N. Mizushima, "Characterization of autophagosome formation site by a hierarchical analysis of mammalian Atg proteins," *Autophagy*, vol. 6, no. 6, pp. 764–776, 2010.

[9] K. Suzuki, Y. Kubota, T. Sekito, and Y. Ohsumi, "Hierarchy of Atg proteins in pre-autophagosomal structure organization," *Genes to Cells*, vol. 12, no. 2, pp. 209–218, 2007.

[10] Z. Xie and D. J. Klionsky, "Autophagosome formation: core machinery and adaptations," *Nature Cell Biology*, vol. 9, no. 10, pp. 1102–1109, 2007.

[11] T. Yoshimori and T. Noda, "Toward unraveling membrane biogenesis in mammalian autophagy," *Current Opinion in Cell Biology*, vol. 20, no. 4, pp. 401–407, 2008.

[12] I. G. Ganley, D. H. Lam, J. Wang, X. Ding, S. Chen, and X. Jiang, "ULK1·ATG13·FIP200 complex mediates mTOR signaling and is essential for autophagy," *Journal of Biological Chemistry*, vol. 284, no. 18, pp. 12297–12305, 2009.

[13] N. Hosokawa, T. Hara, T. Kaizuka et al., "Nutrient-dependent mTORCl association with the ULK1-Atg13-FIP200 complex required for autophagy," *Molecular Biology of the Cell*, vol. 20, no. 7, pp. 1981–1991, 2009.

[14] C. H. Jung, C. B. Jun, S. H. Ro et al., "ULK-Atg13-FIP200 complexes mediate mTOR signaling to the autophagy machinery," *Molecular Biology of the Cell*, vol. 20, no. 7, pp. 1992–2003, 2009.

[15] C. A. Mercer, A. Kaliappan, and P. B. Dennis, "A novel, human Atg13 binding protein, Atg101, interacts with ULK1 and is essential for macroautophagy," *Autophagy*, vol. 5, no. 5, pp. 649–662, 2009.

[16] C. He and D. J. Klionsky, "Regulation mechanisms and signaling pathways of autophagy," *Annual Review of Genetics*, vol. 43, pp. 67–93, 2009.

[17] A. R. J. Young, M. Narita, M. Ferreira et al., "Autophagy mediates the mitotic senescence transition," *Genes and Development*, vol. 23, no. 7, pp. 798–803, 2009.

[18] Y. Kamada, T. Funakoshi, T. Shintani, K. Nagano, M. Ohsumi, and Y. Ohsumi, "Tor-mediated induction of autophagy via an Apg1 protein kinase complex," *Journal of Cell Biology*, vol. 150, no. 6, pp. 1507–1513, 2000.

[19] Z. Yang and D. J. Klionsky, "An overview of the molecular mechanism of autophagy," *Current Topics in Microbiology and Immunology*, vol. 335, no. 1, pp. 1–32, 2009.

[20] A. Kihara, T. Noda, N. Ishihara, and Y. Ohsumi, "Two distinct Vps34 phosphatidylinositol 3-kinase complexes function in autophagy and carboxypeptidase y sorting in Saccharomyces cerevisiae," *Journal of Cell Biology*, vol. 153, no. 3, pp. 519–530, 2001.

[21] K. Obara, T. Sekito, and Y. Ohsumi, "Assortment of phosphatidylinositol 3-kinase complexes-Atg14p directs association of complex I to the pre-autophagosomal structure in Saccharomyces cerevisiae," *Molecular Biology of the Cell*, vol. 17, no. 4, pp. 1527–1539, 2006.

[22] E. Itakura, C. Kishi, K. Inoue, and N. Mizushima, "Beclin 1 forms two distinct phosphatidylinositol 3-kinase complexes with mammalian Atg14 and UVRAG," *Molecular Biology of the Cell*, vol. 19, no. 12, pp. 5360–5372, 2008.

[23] C. Liang, P. Feng, B. Ku et al., "Autophagic and tumour suppressor activity of a novel Beclin1-binding protein UVRAG," *Nature Cell Biology*, vol. 8, no. 7, pp. 688–698, 2006.

[24] Q. Sun, W. Fan, K. Chen, X. Ding, S. Chen, and Q. Zhong, "Identification of Barkor as a mammalian autophagy-specific factor for Beclin 1 and class III phosphatidylinositol 3-kinase," *Proceedings of the National Academy of Sciences of the United States of America*, vol. 105, no. 49, pp. 19211–19216, 2008.

[25] K. Matsunaga, E. Morita, T. Saitoh et al., "Autophagy requires endoplasmic reticulum targeting of the PI3-kinase complex via Atg14L," *Journal of Cell Biology*, vol. 190, no. 4, pp. 511–521, 2010.

[26] K. Matsunaga, T. Saitoh, K. Tabata et al., "Two Beclin 1-binding proteins, Atg14L and Rubicon, reciprocally regulate autophagy at different stages," *Nature Cell Biology*, vol. 11, no. 4, pp. 385–396, 2009.

[27] X. Zhong, Z. Guo, H. Yang et al., "Amino terminus of the SARS coronavirus protein 3a elicits strong, potentially protective humoral responses in infected patients," *Journal of General Virology*, vol. 87, no. 2, pp. 369–374, 2006.

[28] Y. Takahashi, D. Coppola, N. Matsushita et al., "Bif-1 interacts with Beclin 1 through UVRAG and regulates autophagy and tumorigenesis," *Nature Cell Biology*, vol. 9, no. 10, pp. 1142–1151, 2007.

[29] Y. Takahashi, C. L. Meyerkord, T. Hori et al., "Bif-1 regulates Atg9 trafficking by mediating the fission of Golgi membranes during autophagy," *Autophagy*, vol. 7, no. 1, pp. 61–73, 2011.

[30] C. Liang, J. S. Lee, K. S. Inn et al., "Beclin1-binding UVRAG targets the class C Vps complex to coordinate autophagosome maturation and endocytic trafficking," *Nature Cell Biology*, vol. 10, no. 7, pp. 776–787, 2008.

[31] Y. Zhong, Q. J. Wang, X. Li et al., "Distinct regulation of autophagic activity by Atg14L and Rubicon associated with Beclin 1-phosphatidylinositol-3-kinase complex," *Nature Cell Biology*, vol. 11, no. 4, pp. 468–476, 2009.

[32] S. Di Bartolomeo, M. Corazzari, F. Nazio et al., "The dynamic interaction of AMBRA1 with the dynein motor complex regulates mammalian autophagy," *Journal of Cell Biology*, vol. 191, no. 1, pp. 155–168, 2010.

[33] G. Maria Fimia, A. Stoykova, A. Romagnoli et al., "Ambra1 regulates autophagy and development of the nervous system," *Nature*, vol. 447, no. 7148, pp. 1121–1125, 2007.

[34] M. Mari, J. Griffith, E. Rieter, L. Krishnappa, D. J. Klionsky, and F. Reggiori, "An Atg9-containing compartment that functions in the early steps of autophagosome biogenesis," *Journal of Cell Biology*, vol. 190, no. 6, pp. 1005–1022, 2010.

[35] T. Noda, J. Kim, W. P. Huang et al., "Apg9p/Cvt7p is an integral membrane protein required for transport vesicle formation in the Cvt and autophagy pathways," *Journal of Cell Biology*, vol. 148, no. 3, pp. 465–479, 2000.

[36] Y. Ohashi and S. Munro, "Membrane delivery to the yeast autophagosome from the golgi-endosomal system," *Molecular Biology of the Cell*, vol. 21, no. 22, pp. 3998–4008, 2010.

[37] F. Reggiori, K. A. Tucker, P. E. Stromhaug, and D. J. Klionsky, "The Atg1-Atg13 complex regulates Atg9 and Atg23 retrieval transport from the pre-autophagosomal structure," *Developmental Cell*, vol. 6, no. 1, pp. 79–90, 2004.

[38] K. Obara, T. Sekito, K. Niimi, and Y. Ohsumi, "The Atg18-Atg2 complex is recruited to autophagic membranes via phosphatidylinositol 3-phosphate and exerts an essential function," *Journal of Biological Chemistry*, vol. 283, no. 35, pp. 23972–23980, 2008.

[39] A. R. J. Young, E. Y. W. Chan, X. W. Hu et al., "Starvation and ULK1-dependent cycling of mammalian Atg9 between the TGN and endosomes," *Journal of Cell Science*, vol. 119, no. 18, pp. 3888–3900, 2006.

[40] C. He, H. Song, T. Yorimitsu et al., "Recruitment of Atg9 to the preautophagosomal structure by Atg11 is essential for selective autophagy in budding yeast," *Journal of Cell Biology*, vol. 175, no. 6, pp. 925–935, 2006.

[41] T. Sekito, T. Kawamata, R. Ichikawa, K. Suzuki, and Y. Ohsumi, "Atg17 recruits Atg9 to organize the pre-autophagosomal structure," *Genes to Cells*, vol. 14, no. 5, pp. 525–538, 2009.

[42] J. L. Webber and S. A. Tooze, "Coordinated regulation of autophagy by p38alpha MAPK through mAtg9 and p38IP," *The EMBO journal*, vol. 29, no. 1, pp. 27–40, 2010.

[43] E. Y. Chan and S. A. Tooze, "Evolution of Atg1 function and regulation," *Autophagy*, vol. 5, no. 6, pp. 758–765, 2009.

[44] Y. Ichimura, T. Kirisako, T. Takao et al., "A ubiquitin-like system mediates protein lipidation," *Nature*, vol. 408, no. 6811, pp. 488–492, 2000.

[45] N. Mizushima, T. Noda, T. Yoshimori et al., "A protein conjugation system essential for autophagy," *Nature*, vol. 395, no. 6700, pp. 395–398, 1998.

[46] Z. Yang and D. J. Klionsky, "Mammalian autophagy: core molecular machinery and signaling regulation," *Current Opinion in Cell Biology*, vol. 22, no. 2, pp. 124–131, 2010.

[47] N. Fujita, T. Itoh, H. Omori, M. Fukuda, T. Noda, and T. Yoshimori, "The Atg16L complex specifies the site of LC3 lipidation for membrane biogenesis in autophagy," *Molecular Biology of the Cell*, vol. 19, no. 5, pp. 2092–2100, 2008.

[48] T. Hanada, N. N. Noda, Y. Satomi et al., "The Atg12-Atg5 conjugate has a novel E3-like activity for protein lipidation in autophagy," *Journal of Biological Chemistry*, vol. 282, no. 52, pp. 37298–37302, 2007.

[49] Y. S. Sou, S. Waguri, J. I. Iwata et al., "The Atg8 conjugation system is indispensable for proper development of autophagic isolation membranes in mice," *Molecular Biology of the Cell*, vol. 19, no. 11, pp. 4762–4775, 2008.

[50] U. Nair, A. Jotwani, J. Geng et al., "SNARE proteins are required for macroautophagy," *Cell*, vol. 146, no. 2, pp. 290–302, 2011.

[51] H. Nakatogawa, Y. Ichimura, and Y. Ohsumi, "Atg8, a ubiquitin-like protein required for autophagosome formation, mediates membrane tethering and hemifusion," *Cell*, vol. 130, no. 1, pp. 165–178, 2007.

[52] H. Weidberg, T. Shpilka, E. Shvets, A. Abada, F. Shimron, and Z. Elazar, "LC3 and GATE-16 N termini mediate membrane fusion processes required for autophagosome biogenesis," *Developmental Cell*, vol. 20, no. 4, pp. 444–454, 2011.

[53] T. Shpilka, H. Weidberg, S. Pietrokovski, and Z. Elazar, "Atg8: an autophagy-related ubiquitin-like protein family," *Genome Biology*, vol. 12, no. 7, 2011.

[54] H. Weidberg, E. Shvets, T. Shpilka, F. Shimron, V. Shinder, and Z. Elazar, "LC3 and GATE-16/GABARAP subfamilies are both essential yet act differently in autophagosome biogenesis," *EMBO Journal*, vol. 29, no. 11, pp. 1792–1802, 2010.

[55] A. U. Arstila and B. F. Trump, "Studies on cellular autophagocytosis. The formation of autophagic vacuoles in the liver after glucagon administration," *American Journal of Pathology*, vol. 53, no. 5, pp. 687–733, 1968.

[56] P. E. Stromhaug, T. O. Berg, M. Fengsrud, and P. O. Seglen, "Purification and characterization of autophagosomes from rat hepatocytes," *Biochemical Journal*, vol. 335, no. 2, pp. 217–224, 1998.

[57] B. O. Bodemann, A. Orvedahl, T. Cheng et al., "RalB and the exocyst mediate the cellular starvation response by direct

activation of autophagosome assembly," *Cell*, vol. 144, no. 2, pp. 253–267, 2011.

[58] J. Geng, M. Baba, U. Nair, and D. J. Klionsky, "Quantitative analysis of autophagy-related protein stoichiometry by fluorescence microscopy," *Journal of Cell Biology*, vol. 182, no. 1, pp. 129–140, 2008.

[59] K. Moreau, B. Ravikumar, M. Renna, C. Puri, and D. C. Rubinsztein, "Autophagosome precursor maturation requires homotypic fusion," *Cell*, vol. 146, no. 2, pp. 303–317, 2011.

[60] U. Nair and D. J. Klionsky, "Molecular mechanisms and regulation of specific and nonspecific autophagy pathways in yeast," *Journal of Biological Chemistry*, vol. 280, no. 51, pp. 41785–41788, 2005.

[61] B. Ravikumar, K. Moreau, L. Jahreiss, C. Puri, and D. C. Rubinsztein, "Plasma membrane contributes to the formation of pre-autophagosomal structures," *Nature Cell Biology*, vol. 12, no. 8, pp. 747–757, 2010.

[62] A. Van Der Vaart, J. Griffith, and F. Reggiori, "Exit from the golgi is required for the expansion of the autophagosomal phagophore in yeast Saccharomyces cerevisiae," *Molecular Biology of the Cell*, vol. 21, no. 13, pp. 2270–2284, 2010.

[63] A. Van Der Vaart and F. Reggiori, "The Golgi complex as a source for yeast autophagosomal membranes," *Autophagy*, vol. 6, no. 6, pp. 800–802, 2010.

[64] A. Yamamoto, R. Masaki, and Y. Tashiro, "Characterization of the isolation membranes and the limiting membranes of autophagosomes in rat hepatocytes by lectin cytochemistry," *Journal of Histochemistry and Cytochemistry*, vol. 38, no. 4, pp. 573–580, 1990.

[65] W. L. Yen, T. Shintani, U. Nair et al., "The conserved oligomeric Golgi complex is involved in double-membrane vesicle formation during autophagy," *Journal of Cell Biology*, vol. 188, no. 1, pp. 101–114, 2010.

[66] W. A. Dunn, "Studies on the mechanisms of autophagy: maturation of the autophagic vacuole," *Journal of Cell Biology*, vol. 110, no. 6, pp. 1935–1945, 1990.

[67] J. L. E. Ericsson, "Studies on induced cellular autophagy. I. Electron microscopy of cells with in vivo labelled lysosomes," *Experimental Cell Research*, vol. 55, no. 1, pp. 95–106, 1969.

[68] M. Hayashi-Nishino, N. Fujita, T. Noda, A. Yamaguchi, T. Yoshimori, and A. Yamamoto, "A subdomain of the endoplasmic reticulum forms a cradle for autophagosome formation," *Nature Cell Biology*, vol. 11, no. 12, pp. 1433–1437, 2009.

[69] P. Ylä-Anttila, H. Vihinen, E. Jokitalo, and E. L. Eskelinen, "3D tomography reveals connections between the phagophore and endoplasmic reticulum," *Autophagy*, vol. 5, no. 8, pp. 1180–1185, 2009.

[70] E. L. Axe, S. A. Walker, M. Manifava et al., "Autophagosome formation from membrane compartments enriched in phosphatidylinositol 3-phosphate and dynamically connected to the endoplasmic reticulum," *Journal of Cell Biology*, vol. 182, no. 4, pp. 685–701, 2008.

[71] D. W. Hailey, A. S. Rambold, P. Satpute-Krishnan et al., "Mitochondria supply membranes for autophagosome biogenesis during starvation," *Cell*, vol. 141, no. 4, pp. 656–667, 2010.

[72] F. Reggiori and S. A. Tooze, "The EmERgence of autophagosomes," *Developmental Cell*, vol. 17, no. 6, pp. 747–748, 2009.

[73] S. A. Tooze and T. Yoshimori, "The origin of the autophagosomal membrane," *Nature Cell Biology*, vol. 12, no. 9, pp. 831–835, 2010.

[74] A. Fleming, T. Noda, T. Yoshimori, and D. C. Rubinsztein, "Chemical modulators of autophagy as biological probes and potential therapeutics," *Nature Chemical Biology*, vol. 7, no. 1, pp. 9–17, 2011.

[75] J. Harris, "Autophagy and cytokines," *Cytokine*, vol. 56, no. 2, pp. 140–144, 2011.

[76] S. Wu and J. Sun, "Vitamin D, vitamin D receptor, and macroautophagy in inflammation and infection," *Discovery medicine*, vol. 11, no. 59, pp. 325–335, 2011.

[77] D. J. Klionsky, H. Abeliovich, P. Agostinis et al., "Guidelines for the use and interpretation of assays for monitoring autophagy in higher eukaryotes," *Autophagy*, vol. 4, no. 2, pp. 151–175, 2008.

[78] D. J. Klionsky, A. M. Cuervo, W. A. Dunn, B. Levine, I. Van Der Klei, and P. O. Seglen, "How shall i eat thee?" *Autophagy*, vol. 3, no. 5, pp. 413–416, 2007.

[79] A. Orvedahl, R. Sumpter Jr., G. Xiao et al., "Image-based genome-wide siRNA screen identifies selective autophagy factors," *Nature*, vol. 480, no. 7375, pp. 113–117, 2011.

[80] N. N. Noda, H. Kumeta, H. Nakatogawa et al., "Structural basis of target recognition by Atg8/LC3 during selective autophagy," *Genes to Cells*, vol. 13, no. 12, pp. 1211–1218, 2008.

[81] S. Pankiv, T. H. Clausen, T. Lamark et al., "p62/SQSTM1 binds directly to Atg8/LC3 to facilitate degradation of ubiquitinated protein aggregates by autophagy," *Journal of Biological Chemistry*, vol. 282, no. 33, pp. 24131–24145, 2007.

[82] Y. Ichimura, T. Kumanomidou, Y. S. Sou et al., "Structural basis for sorting mechanism of p62 in selective autophagy," *Journal of Biological Chemistry*, vol. 283, no. 33, pp. 22847–22857, 2008.

[83] T. Johansen and T. Lamark, "Selective autophagy mediated by autophagic adapter proteins," *Autophagy*, vol. 7, no. 3, pp. 279–296, 2011.

[84] K. Satoo, N. N. Noda, H. Kumeta et al., "The structure of Atg4B-LC3 complex reveals the mechanism of LC3 processing and delipidation during autophagy," *EMBO Journal*, vol. 28, no. 9, pp. 1341–1350, 2009.

[85] M. Yamaguchi, N. N. Noda, H. Nakatogawa, H. Kumeta, Y. Ohsumi, and F. Inagaki, "Autophagy-related protein 8 (Atg8) family interacting motif in Atg3 mediates the Atg3-Atg8 interaction and is crucial for the cytoplasm-to-vacuole targeting pathway," *Journal of Biological Chemistry*, vol. 285, no. 38, pp. 29599–29607, 2010.

[86] S. Pankiv, E. A. Alemu, A. Brech et al., "FYCO1 is a Rab7 effector that binds to LC3 and PI3P to mediate microtubule plus end—directed vesicle transport," *Journal of Cell Biology*, vol. 188, no. 2, pp. 253–269, 2010.

[87] C. Gao, W. Cao, L. Bao et al., "Autophagy negatively regulates Wnt signalling by promoting Dishevelled degradation," *Nature Cell Biology*, vol. 12, no. 8, pp. 781–790, 2010.

[88] M. Filimonenko, P. Isakson, K. D. Finley et al., "The selective acroautophagic degradation of aggregated proteins requires the PI3P-binding protein alfy," *Molecular Cell*, vol. 38, no. 2, pp. 265–279, 2010.

[89] T. Shintani, W. P. Huang, P. E. Stromhaug, and D. J. Klionsky, "Mechanism of cargo selection in the cytoplasm to vacuole targeting pathway," *Developmental Cell*, vol. 3, no. 6, pp. 825–837, 2002.

[90] M. A. Lynch-Day and D. J. Klionsky, "The Cvt pathway as a model for selective autophagy," *FEBS Letters*, vol. 584, no. 7, pp. 1359–1366, 2010.

[91] K. Suzuki, M. Morimoto, C. Kondo, and Y. Ohsumi, "Selective Autophagy Regulates Insertional Mutagenesis by the Ty1

Retrotransposon in Saccharomyces cerevisiae," *Developmental Cell*, vol. 21, no. 2, pp. 358–365, 2011.

[92] I. Monastyrska and D. J. Klionsky, "Autophagy in organelle homeostasis: peroxisome turnover," *Molecular Aspects of Medicine*, vol. 27, no. 5-6, pp. 483–494, 2006.

[93] T. Yorimitsu and D. J. Klionsky, "Autophagy: molecular machinery for self-eating," *Cell Death and Differentiation*, vol. 12, supplement 2, pp. 1542–1552, 2005.

[94] D. C. Nice, T. K. Sato, P. E. Stromhaug, S. D. Emr, and D. J. Klionsky, "Cooperative binding of the cytoplasm to vacuole targeting pathway proteins, Cvt13 and Cvt20, to phosphatidylinositol 3-phosphate at the pre-autophagosomal structure is required for selective autophagy," *Journal of Biological Chemistry*, vol. 277, no. 33, pp. 30198–30207, 2002.

[95] P. E. Strømhaug, F. Reggiori, J. Guan, C. W. Wang, and D. J. Klionsky, "Atg21 is a phosphoinositide binding protein required for efficient lipidation and localization of Atg8 during uptake of aminopeptidase I by selective autophagy," *Molecular Biology of the Cell*, vol. 15, no. 8, pp. 3553–3566, 2004.

[96] T. Y. Nazarko, J. C. Farré, and S. Subramani, "Peroxisome size provides insights into the function of autophagy-related proteins," *Molecular Biology of the Cell*, vol. 20, no. 17, pp. 3828–3839, 2009.

[97] T. Kanki, K. Wang, Y. Cao, M. Baba, and D. J. Klionsky, "Atg32 is a mitochondrial protein that confers selectivity during mitophagy," *Developmental Cell*, vol. 17, no. 1, pp. 98–109, 2009.

[98] K. Okamoto, N. Kondo-Okamoto, and Y. Ohsumi, "Mitochondria-anchored receptor Atg32 mediates degrada-tion of mitochondria via selective autophagy," *Developmental Cell*, vol. 17, no. 1, pp. 87–97, 2009.

[99] J. C. Farré, R. Manjithaya, R. D. Mathewson, and S. Subramani, "PpAtg30 tags peroxisomes for turnover by selective autophagy," *Developmental Cell*, vol. 14, no. 3, pp. 365–376, 2008.

[100] G. Bjørkøy, T. Lamark, A. Brech et al., "p62/SQSTM1 forms protein aggregates degraded by autophagy and has a protective effect on huntingtin-induced cell death," *Journal of Cell Biology*, vol. 171, no. 4, pp. 603–614, 2005.

[101] T. H. Clausen, T. Lamark, P. Isakson et al., "p62/SQSTM1 and ALFY interact to facilitate the formation of p62 bodies/ALIS and their degradation by autophagy," *Autophagy*, vol. 6, no. 3, pp. 330–344, 2010.

[102] M. M. Quinones and P. E. Stromhaug, "The propeptide of Aminopeptidase 1 mediates aggregation and vesicle formation in the Cytoplasm-to-vacuole targeting pathway," *Journal of Biological Chemistry*, 2011.

[103] V. Kirkin, T. Lamark, Y. S. Sou et al., "A Role for NBR1 in autophagosomal degradation of ubiquitinated substrates," *Molecular Cell*, vol. 33, no. 4, pp. 505–516, 2009.

[104] M. Komatsu, S. Waguri, M. Koike et al., "Homeostatic levels of p62 control cytoplasmic inclusion body formation in autophagy-deficient mice," *Cell*, vol. 131, no. 6, pp. 1149–1163, 2007.

[105] Y. T. Zheng, S. Shahnazari, A. Brech, T. Lamark, T. Johansen, and J. H. Brumell, "The adaptor protein p62/SQSTM1 targets invading bacteria to the autophagy pathway," *Journal of Immunology*, vol. 183, no. 9, pp. 5909–5916, 2009.

[106] A. Orvedahl, S. MacPherson, R. Sumpter, Z. Tallóczy, Z. Zou, and B. Levine, "Autophagy protects against sindbis virus infection of the central nervous system," *Cell Host and Microbe*, vol. 7, no. 2, pp. 115–127, 2010.

[107] C. Pohl and S. Jentsch, "Midbody ring disposal by autophagy is a post-abscission event of cytokinesis," *Nature Cell Biology*, vol. 11, no. 1, pp. 65–70, 2009.

[108] P. K. Kim, D. W. Hailey, R. T. Mullen, and J. Lippincott-Schwartz, "Ubiquitin signals autophagic degradation of cytosolic proteins and peroxisomes," *Proceedings of the National Academy of Sciences of the United States of America*, vol. 105, no. 52, pp. 20567–20574, 2008.

[109] H. W. Platta and R. Erdmann, "Peroxisomal dynamics," *Trends in Cell Biology*, vol. 17, no. 10, pp. 474–484, 2007.

[110] W. X. Ding, H. M. Ni, M. Li et al., "Nix is critical to two distinct phases of mitophagy, reactive oxygen species-mediated autophagy induction and Parkin-ubiquitin-p62-mediated mitochondrial priming," *Journal of Biological Chemistry*, vol. 285, no. 36, pp. 27879–27890, 2010.

[111] S. Geisler, K. M. Holmström, D. Skujat et al., "PINK1/Parkin-mediated mitophagy is dependent on VDAC1 and p62/SQSTM1," *Nature Cell Biology*, vol. 12, no. 2, pp. 119–131, 2010.

[112] M. Ponpuak, A. S. Davis, E. A. Roberts et al., "Delivery of cytosolic components by autophagic adaptor protein p62 endows autophagosomes with unique antimicrobial properties," *Immunity*, vol. 32, no. 3, pp. 329–341, 2010.

[113] E. Itakura and N. Mizushima, "p62 targeting to the autophagosome formation site requires self-oligomerization but not LC3 binding," *Journal of Cell Biology*, vol. 192, no. 1, pp. 17–27, 2011.

[114] A. Simonsen, H. C. G. Birkeland, D. J. Gillooly et al., "Alfy, a novel FYVE-domain-containing protein associated with protein granules and autophagic membranes," *Journal of Cell Science*, vol. 117, no. 18, pp. 4239–4251, 2004.

[115] P. Wild, H. Farhan, D. G. McEwan et al., "Phosphorylation of the autophagy receptor optineurin restricts Salmonella growth," *Science*, vol. 333, no. 6039, pp. 228–233, 2011.

[116] Y. Aoki, T. Kanki, Y. Hirota et al., "Phosphorylation of serine 114 on Atg32 mediates mitophagy," *Molecular Biology of the Cell*, vol. 22, no. 17, pp. 3206–3217, 2011.

[117] K. Mao, K. Wang, M. Zhao, T. Xu, and D. J. Klionsky, "Two MAPK-signaling pathways are required for mitophagy in Saccharomyces cerevisiae," *Journal of Cell Biology*, vol. 193, no. 4, pp. 755–767, 2011.

[118] S. J. Cherra, S. M. Kulich, G. Uechi et al., "Regulation of the autophagy protein LC3 by phosphorylation," *Journal of Cell Biology*, vol. 190, no. 4, pp. 533–539, 2010.

[119] H. Jiang, D. Cheng, W. Liu, J. Peng, and J. Feng, "Protein kinase C inhibits autophagy and phosphorylates LC3," *Biochemical and Biophysical Research Communications*, vol. 395, no. 4, pp. 471–476, 2010.

[120] E. Shvets, E. Fass, R. Scherz-Shouval, and Z. Elazar, "The N-terminus and Phe52 residue of LC3 recruit p62/SQSTM1 into autophagosomes," *Journal of Cell Science*, vol. 121, no. 16, pp. 2685–2695, 2008.

[121] G. Matsumoto, K. Wada, M. Okuno, M. Kurosawa, and N. Nukina, "Serine 403 phosphorylation of p62/SQSTM1 regulates selective autophagic clearance of ubiquitinated proteins," *Molecular Cell*, vol. 44, no. 2, pp. 279–289, 2011.

[122] D. Narendra, A. Tanaka, D. F. Suen, and R. J. Youle, "Parkin is recruited selectively to impaired mitochondria and promotes their autophagy," *Journal of Cell Biology*, vol. 183, no. 5, pp. 795–803, 2008.

[123] F. Tang, B. Wang, N. Li et al., "RNF185, a novel mitochondrial ubiquitin E3 ligase, regulates autophagy through interaction with BNIP1," *PLoS ONE*, vol. 6, no. 9, 2011.

[124] Y. Kawaguchi, J. J. Kovacs, A. McLaurin, J. M. Vance, A. Ito, and T. P. Yao, "The deacetylase HDAC6 regulates aggresome formation and cell viability in response to misfolded protein stress," *Cell*, vol. 115, no. 6, pp. 727–738, 2003.

[125] J. Y. Lee, H. Koga, Y. Kawaguchi et al., "HDAC6 controls autophagosome maturation essential for ubiquitin-selective quality-control autophagy," *EMBO Journal*, vol. 29, no. 5, pp. 969–980, 2010.

[126] U. B. Pandey, Z. Nie, Y. Batlevi et al., "HDAC6 rescues neurodegeneration and provides an essential link between autophagy and the UPS," *Nature*, vol. 447, no. 7146, pp. 859–863, 2007.

[127] H. Jeong, F. Then, T. J. Melia et al., "Acetylation targets mutant huntingtin to autophagosomes for degradation," *Cell*, vol. 137, no. 1, pp. 60–72, 2009.

[128] J. Y. Lee, Y. Nagano, J. P. Taylor, K. L. Lim, and T. P. Yao, "Disease-causing mutations in Parkin impair mitochondrial ubiquitination, aggregation, and HDAC6-dependent mitophagy," *Journal of Cell Biology*, vol. 189, no. 4, pp. 671–679, 2010.

[129] N. Mizushima and M. Komatsu, "Autophagy: renovation of cells and tissues," *Cell*, vol. 147, no. 4, pp. 728–741, 2011.

[130] B. Gan, X. Peng, T. Nagy, A. Alcaraz, H. Gu, and J. L. Guan, "Role of FIP200 in cardiac and liver development and its regulation of TNFα and TSC-mTOR signaling pathways," *Journal of Cell Biology*, vol. 175, no. 1, pp. 121–133, 2006.

[131] X. Qu, J. Yu, G. Bhagat et al., "Promotion of tumorigenesis by heterozygous disruption of the beclin 1 autophagy gene," *Journal of Clinical Investigation*, vol. 112, no. 12, pp. 1809–1820, 2003.

[132] Z. Yue, S. Jin, C. Yang, A. J. Levine, and N. Heintz, "Beclin 1, an autophagy gene essential for early embryonic development, is a haploinsufficient tumor suppressor," *Proceedings of the National Academy of Sciences of the United States of America*, vol. 100, no. 25, pp. 15077–15082, 2003.

[133] M. Komatsu, S. Waguri, T. Ueno et al., "Impairment of starvation-induced and constitutive autophagy in Atg7-deficient mice," *Journal of Cell Biology*, vol. 169, no. 3, pp. 425–434, 2005.

[134] A. Kuma, M. Hatano, M. Matsui et al., "The role of autophagy during the early neonatal starvation period," *Nature*, vol. 432, no. 7020, pp. 1032–1036, 2004.

[135] T. Saitoh, N. Fujita, T. Hayashi et al., "Atg9a controls dsDNA-driven dynamic translocation of STING and the innate immune response," *Proceedings of the National Academy of Sciences of the United States of America*, vol. 106, no. 49, pp. 20842–20846, 2009.

[136] T. Saitoh, N. Fujita, M. H. Jang et al., "Loss of the autophagy protein Atg16L1 enhances endotoxin-induced IL-1β production," *Nature*, vol. 456, no. 7219, pp. 264–268, 2008.

[137] T. Hara, K. Nakamura, M. Matsui et al., "Suppression of basal autophagy in neural cells causes neurodegenerative disease in mice," *Nature*, vol. 441, no. 7095, pp. 885–889, 2006.

[138] C. C. Liang, C. Wang, X. Peng, B. Gan, and J. L. Guan, "Neural-specific deletion of FIP200 leads to cerebellar degeneration caused by increased neuronal death and axon degeneration," *Journal of Biological Chemistry*, vol. 285, no. 5, pp. 3499–3509, 2010.

[139] M. Komatsu, S. Waguri, T. Chiba et al., "Loss of autophagy in the central nervous system causes neurodegeneration in mice," *Nature*, vol. 441, no. 7095, pp. 880–884, 2006.

[140] A. Nakai, O. Yamaguchi, T. Takeda et al., "The role of autophagy in cardiomyocytes in the basal state and in response to hemodynamic stress," *Nature Medicine*, vol. 13, no. 5, pp. 619–624, 2007.

[141] E. Masiero, L. Agatea, C. Mammucari et al., "Autophagy is required to maintain muscle mass," *Cell Metabolism*, vol. 10, no. 6, pp. 507–515, 2009.

[142] N. Raben, V. Hill, L. Shea et al., "Suppression of autophagy in skeletal muscle uncovers the accumulation of ubiquitinated proteins and their potential role in muscle damage in Pompe disease," *Human Molecular Genetics*, vol. 17, no. 24, pp. 3897–3908, 2008.

[143] C. Ebato, T. Uchida, M. Arakawa et al., "Autophagy is important in islet homeostasis and compensatory increase of beta cell mass in response to high-fat diet," *Cell Metabolism*, vol. 8, no. 4, pp. 325–332, 2008.

[144] H. S. Jung, K. W. Chung, J. Won Kim et al., "Loss of autophagy diminishes pancreatic β cell mass and function with resultant hyperglycemia," *Cell Metabolism*, vol. 8, no. 4, pp. 318–324, 2008.

[145] B. Hartleben, M. Gödel, C. Meyer-Schwesinger et al., "Autophagy influences glomerular disease susceptibility and maintains podocyte homeostasis in aging mice," *Journal of Clinical Investigation*, vol. 120, no. 4, pp. 1084–1096, 2010.

[146] T. Kimura, Y. Takabatake, A. Takahashi et al., "Autophagy protects the proximal tubule from degeneration and acute ischemic injury," *Journal of the American Society of Nephrology*, vol. 22, no. 5, pp. 902–913, 2011.

[147] D. Inoue, H. Kubo, K. Taguchi et al., "Inducible disruption of autophagy in the lung causes airway hyper-responsiveness," *Biochemical and Biophysical Research Communications*, vol. 405, no. 1, pp. 13–18, 2011.

[148] R. Singh, S. Kaushik, Y. Wang et al., "Autophagy regulates lipid metabolism," *Nature*, vol. 458, no. 7242, pp. 1131–1135, 2009.

[149] N. Mizushima and B. Levine, "Autophagy in mammalian development and differentiation," *Nature Cell Biology*, vol. 12, no. 9, pp. 823–830, 2010.

[150] I. P. Nezis, A. Simonsen, A. P. Sagona et al., "Ref(2)P, the Drosophila melanogaster homologue of mammalian p62, is required for the formation of protein aggregates in adult brain," *Journal of Cell Biology*, vol. 180, no. 6, pp. 1065–1071, 2008.

[151] K. Zatloukal, C. Stumptner, A. Fuchsbichler et al., "p62 is a common component of cytoplasmic inclusions in protein aggregation diseases," *American Journal of Pathology*, vol. 160, no. 1, pp. 255–263, 2002.

[152] Y. Inami, S. Waguri, A. Sakamoto et al., "Persistent activation of Nrf2 through p62 in hepatocellular carcinoma cells," *Journal of Cell Biology*, vol. 193, no. 2, pp. 275–284, 2011.

[153] Z. Jin, Y. Li, R. Pitti et al., "Cullin3-based polyubiquitination and p62-dependent aggregation of caspase-8 mediate extrinsic apoptosis signaling," *Cell*, vol. 137, no. 4, pp. 721–735, 2009.

[154] J. Moscat and M. T. Diaz-Meco, "p62 at the crossroads of autophagy, apoptosis, and cancer," *Cell*, vol. 137, no. 6, pp. 1001–1004, 2009.

[155] I. M. Copple, A. Lister, A. D. Obeng et al., "Physical and functional interaction of sequestosome 1 with Keap1 regulates the Keap1-Nrf2 cell defense pathway," *Journal of Biological Chemistry*, vol. 285, no. 22, pp. 16782–16788, 2010.

[156] W. Fan, Z. Tang, D. Chen et al., "Keap1 facilitates p62-mediated ubiquitin aggregate clearance via autophagy," *Autophagy*, vol. 6, no. 5, pp. 614–621, 2010.

[157] A. Jain, T. Lamark, E. Sjøttem et al., "p62/SQSTM1 is a target gene for transcription factor NRF2 and creates a positive

feedback loop by inducing antioxidant response element-driven gene transcription," *Journal of Biological Chemistry*, vol. 285, no. 29, pp. 22576–22591, 2010.

[158] M. Komatsu, H. Kurokawa, S. Waguri et al., "The selective autophagy substrate p62 activates the stress responsive transcription factor Nrf2 through inactivation of Keap1," *Nature Cell Biology*, vol. 12, no. 3, pp. 213–223, 2010.

[159] A. Lau, X. J. Wang, F. Zhao et al., "A noncanonical mechanism of Nrf2 activation by autophagy deficiency: direct interaction between keap1 and p62," *Molecular and Cellular Biology*, vol. 30, no. 13, pp. 3275–3285, 2010.

[160] A. Goode and R. Layfield, "Recent advances in understanding the molecular basis of Paget disease of bone," *Journal of Clinical Pathology*, vol. 63, no. 3, pp. 199–203, 2010.

[161] E. J. White, V. Martin, J. L. Liu et al., "Autophagy regulation in cancer development and therapy," *American Journal of Cancer Research*, vol. 1, pp. 362–372, 2011.

[162] A. Takamura, M. Komatsu, T. Hara et al., "Autophagy-deficient mice develop multiple liver tumors," *Genes and Development*, vol. 25, no. 8, pp. 795–800, 2011.

[163] V. Karantza-Wadsworth, S. Patel, O. Kravchuk et al., "Autophagy mitigates metabolic stress and genome damage in mammary tumorigenesis," *Genes and Development*, vol. 21, no. 13, pp. 1621–1635, 2007.

[164] R. Mathew, S. Kongara, B. Beaudoin et al., "Autophagy suppresses tumor progression by limiting chromosomal instability," *Genes and Development*, vol. 21, no. 11, pp. 1367–1381, 2007.

[165] R. Mathew, C. M. Karp, B. Beaudoin et al., "Autophagy suppresses tumorigenesis through elimination of p62," *Cell*, vol. 137, no. 6, pp. 1062–1075, 2009.

[166] G. M. Denicola, F. A. Karreth, T. J. Humpton et al., "Oncogene-induced Nrf2 transcription promotes ROS detoxification and tumorigenesis," *Nature*, vol. 475, no. 7354, pp. 106–110, 2011.

[167] V. I. Korolchuk, A. Mansilla, F. M. Menzies, and D. C. Rubinsztein, "Autophagy inhibition compromises degradation of ubiquitin-proteasome pathway substrates," *Molecular Cell*, vol. 33, no. 4, pp. 517–527, 2009.

[168] H. Weidberg, E. Shvets, and Z. Elazar, "Biogenesis and cargo selectivity of autophagosomes," *Annual Review of Biochemistry*, vol. 80, pp. 125–156, 2011.

[169] B. E. Riley, S. E. Kaiser, T. A. Shaler et al., "Ubiquitin accumulation in autophagy-deficient mice is dependent on the Nrf2-mediated stress response pathway: a potential role for protein aggregation in autophagic substrate selection," *Journal of Cell Biology*, vol. 191, no. 3, pp. 537–552, 2010.

[170] I. Novak, V. Kirkin, D. G. McEwan et al., "Nix is a selective autophagy receptor for mitochondrial clearance," *EMBO Reports*, vol. 11, no. 1, pp. 45–51, 2010.

[171] M. Schwarten, J. Mohrlüder, P. Ma et al., "Nix directly binds to GABARAP: a possible crosstalk between apoptosis and autophagy," *Autophagy*, vol. 5, no. 5, pp. 690–698, 2009.

[172] H. Sandoval, P. Thiagarajan, S. K. Dasgupta et al., "Essential role for Nix in autophagic maturation of erythroid cells," *Nature*, vol. 454, no. 7201, pp. 232–235, 2008.

[173] R. L. Schweers, J. Zhang, M. S. Randall et al., "NIX is required for programmed mitochondrial clearance during reticulocyte maturation," *Proceedings of the National Academy of Sciences of the United States of America*, vol. 104, no. 49, pp. 19500–19505, 2007.

[174] M. Mortensen, E. J. Soilleux, G. Djordjevic et al., "The autophagy protein Atg7 is essential for hematopoietic stem cell maintenance," *Journal of Experimental Medicine*, vol. 208, no. 3, pp. 455–467, 2011.

[175] H. H. Pua, I. Dzhagalov, M. Chuck, N. Mizushima, and Y. W. He, "A critical role for the autophagy gene Atg5 in T cell survival and proliferation," *Journal of Experimental Medicine*, vol. 204, no. 1, pp. 25–31, 2007.

[176] H. H. Pua, J. Guo, M. Komatsu, and Y. W. He, "Autophagy is essential for mitochondrial clearance in mature T lymphocytes," *Journal of Immunology*, vol. 182, no. 7, pp. 4046–4055, 2009.

[177] L. M. Stephenson, B. C. Miller, A. Ng et al., "Identification of Atg5-dependent transcriptional changes and increases in mitochondrial mass in Atg5-deficient T lymphocytes," *Autophagy*, vol. 5, no. 5, pp. 625–635, 2009.

[178] R. J. Youle and D. P. Narendra, "Mechanisms of mitophagy," *Nature Reviews Molecular Cell Biology*, vol. 12, no. 1, pp. 9–14, 2011.

[179] N. Matsuda, S. Sato, K. Shiba et al., "PINK1 stabilized by mitochondrial depolarization recruits Parkin to damaged mitochondria and activates latent Parkin for mitophagy," *Journal of Cell Biology*, vol. 189, no. 2, pp. 211–221, 2010.

[180] D. P. Narendra, S. M. Jin, A. Tanaka et al., "PINK1 is selectively stabilized on impaired mitochondria to activate Parkin," *PLoS Biology*, vol. 8, no. 1, 2010.

[181] C. Vives-Bauza, C. Zhou, Y. Huang et al., "PINK1-dependent recruitment of Parkin to mitochondria in mitophagy," *Proceedings of the National Academy of Sciences of the United States of America*, vol. 107, no. 1, pp. 378–383, 2010.

[182] D. P. Narendra, L. A. Kane, D. N. Hauser, I. M. Fearnley, and R. J. Youle, "p62/SQSTM1 is required for Parkin-induced mitochondrial clustering but not mitophagy; VDAC1 is dispensable for both," *Autophagy*, vol. 6, no. 8, pp. 1090–1106, 2010.

[183] K. Okatsu, K. Saisho, M. Shimanuki et al., "P62/SQSTM1 cooperates with Parkin for perinuclear clustering of depolarized mitochondria," *Genes to Cells*, vol. 15, no. 8, pp. 887–900, 2010.

[184] N. Fujita and T. Yoshimori, "Ubiquitination-mediated autophagy against invading bacteria," *Current Opinion in Cell Biology*, vol. 23, no. 4, pp. 492–497, 2011.

[185] N. Dupont, S. Lacas-Gervais, J. Bertout et al., "Shigella phagocytic vacuolar membrane remnants participate in the cellular response to pathogen invasion and are regulated by autophagy," *Cell Host and Microbe*, vol. 6, no. 2, pp. 137–149, 2009.

[186] A. J. Perrin, X. Jiang, C. L. Birmingham, N. S. Y. So, and J. H. Brumell, "Recognition of bacteria in the cytosol of mammalian cells by the ubiquitin system," *Current Biology*, vol. 14, no. 9, pp. 806–811, 2004.

[187] Y. Yoshikawa, M. Ogawa, T. Hain et al., "Listeria monocytogenes ActA-mediated escape from autophagic recognition," *Nature Cell Biology*, vol. 11, no. 10, pp. 1233–1240, 2009.

[188] T. L. Thurston, G. Ryzhakov, S. Bloor, N. von Muhlinen, and F. Randow, "The TBK1 adaptor and autophagy receptor NDP52 restricts the proliferation of ubiquitin-coated bacteria," *Nature immunology*, vol. 10, no. 11, pp. 1215–1221, 2009.

[189] M. Ogawa, Y. Yoshikawa, T. Kobayashi et al., "A Tecpr1-dependent selective autophagy pathway targets bacterial pathogens," *Cell Host and Microbe*, vol. 9, no. 5, pp. 376–389, 2011.

[190] M. Ogawa, T. Yoshimori, T. Suzuki, H. Sagara, N. Mizushima, and C. Sasakawa, "Escape of intracellular Shigella from autophagy," *Science*, vol. 307, no. 5710, pp. 727–731, 2005.

[191] R. Singh, Y. Wang, J. M. Schattenberg, Y. Xiang, and M. J. Czaja, "Chronic oxidative stress sensitizes hepatocytes to death from 4-hydroxynonenal by JNK/c-Jun overactivation," *American Journal of Physiology*, vol. 297, no. 5, pp. G907–G917, 2009.

[192] S. Kaushik, J. A. Rodriguez-Navarro, E. Arias et al., "Autophagy in hypothalamic agrp neurons regulates food intake and energy balance," *Cell Metabolism*, vol. 14, no. 2, pp. 173–183, 2011.

[193] J. Kim, Y. Kamada, P. E. Stromhaug et al., "Cvt9/Gsa9 functions in sequestering selective cytosolic cargo destined for the vacuole," *Journal of Cell Biology*, vol. 153, no. 2, pp. 381–396, 2001.

[194] C. Kraft, A. Deplazes, M. Sohrmann, and M. Peter, "Mature ribosomes are selectively degraded upon starvation by an autophagy pathway requiring the Ubp3p/Bre5p ubiquitin protease," *Nature Cell Biology*, vol. 10, no. 5, pp. 602–610, 2008.

[195] S. V. Scott, J. Guan, M. U. Hutchins, J. Kim, and D. J. Klionsky, "Cvt19 is a receptor for the cytoplasm-to-vacuole targeting pathway," *Molecular Cell*, vol. 7, no. 6, pp. 1131–1141, 2001.

18

Fibrin and Collagen Differentially but Synergistically Regulate Sprout Angiogenesis of Human Dermal Microvascular Endothelial Cells in 3-Dimensional Matrix

Xiaodong Feng,[1] Marcia G. Tonnesen,[2,3] Shaker A. Mousa,[4] and Richard A. F. Clark[5]

[1] Department of Clinical and Administrative Sciences, California Northstate University College of Pharmacy, Rancho Cordova, CA 95670, USA
[2] Department of Dermatology, School of Medicine, State University of New York at Stony Brook, Stony Brook, NY 11794, USA
[3] Dermatology Section, Veterans Affairs Medical Center, Northport, NY 11768, USA
[4] Pharmaceutical Research Institute, Albany College of Pharmacy and Health Sciences, Albany, NY 12208, USA
[5] Center for Tissue Engineering, State University of New York at Stony Brook, Stony Brook, NY 11794, USA

Correspondence should be addressed to Xiaodong Feng; xfeng@cnsu.edu

Academic Editor: Michael Peter Sarras

Angiogenesis is a highly regulated event involving complex, dynamic interactions between microvascular endothelial cells and extracellular matrix (ECM) proteins. Alteration of ECM composition and architecture is a hallmark feature of wound clot and tumor stroma. We previously reported that during angiogenesis, endothelial cell responses to growth factors are modulated by the compositional and mechanical properties of a surrounding three-dimensional (3D) extracellular matrix (ECM) that is dominated by either cross-linked fibrin or type I collagen. However, the role of 3D ECM in the regulation of angiogenesis associated with wound healing and tumor growth is not well defined. This study investigates the correlation of sprout angiogenesis and ECM microenvironment using in vivo and in vitro 3D angiogenesis models. It demonstrates that fibrin and type I collagen 3D matrices differentially but synergistically regulate sprout angiogenesis. Thus blocking both integrin alpha v beta 3 and integrin alpha 2 beta 1 might be a novel strategy to synergistically block sprout angiogenesis in solid tumors.

1. Introduction

Angiogenesis, the development of new blood vessels from preexisting vessels, is critical for a wide array of complex normal and pathological processes including morphogenesis, wound healing, and tumor growth [1]. Under normal physiologic conditions, angiogenesis is well controlled by the local balance between endogenous angiogenesis stimulators and angiogenesis inhibitors, although the regulatory mechanism is still not clearly defined. Sustained tumor angiogenesis is one of the hallmark features of solid tumor development. It is essential for tumor development and tumor metastasis. Almost four decades ago Dr. Judah Folkman pioneered the strategy of stopping tumor growth and metastasis by blocking tumor angiogenesis. With the 2004 FDA approval of bevacizumab (Avastin), a humanized monoclonal antibody against vascular endothelial growth factor (VEGF),

to treat metastatic colorectal cancer in combination with 5-fluorouracil (5-FU), antiangiogenesis therapy has emerged as an essential new strategy for cancer treatment [2].

Angiogenesis is a highly regulated event that involves complex, dynamic interactions between microvascular endothelial cells and ECM proteins. In developing capillary sprouts, endothelial cells digest the surrounding extracellular matrix (ECM) and invade the matrix as a cylindrical aggregate of cells. These events clearly require an integrated response of endothelial cells to angiogenic factors and ECM proteins [3]. Alteration of ECM composition and architecture is a hallmark of wound clot and tumor stroma. ECM matrices induce multiple dynamic interactions with endothelial cells and stimulate the transduction of signals by cross-linking integrin receptors on endothelial cells. Initially viewed as merely a physical barrier, the ECM is now recognized as having a profound effect on the angiogenic phenotype.

However, the integrated regulatory mechanism of microvascular endothelial cell response to ECM and angiogenic factors is poorly defined [4, 5]. In addition, numerous evidences indicate that the in vitro cellular regulations of many cell types in 2D environment are significantly different than those of cells in 3D environment. Since 3D environment is more close to the in vivo microenvironment of cell functions, it suggests that reproducible and quantifiable in vitro 3D assays play an important role to study the regulation of cellular behaviors during physiological and pathological processes [6].

Fibrin and type I collagen are two major components of extracellular matrix microenvironment. Fibrin deposition is commonly observed in angiogenesis associated with wound healing and tumor growth. It has been reported that fibrin enhances angiogenesis of wound healing in vitro [7] and in vivo [8]. In contrast, type I collagen is a major component of normal dermis which has minimal angiogenesis activities, although some in vitro studies demonstrate that type I collagen gel supports angiogenesis as well as fibrin gel. The results of these in vitro studies are not consistent with the in vivo data reported by Dvorak et al. that fibrin but not type I collagen induces angiogenesis in vivo[8].

Integrin alpha v beta 3 is the receptor for fibrin matrix. Expression of integrin alpha v beta 3 is one of the hallmark features of sprout angiogenesis. Remarkably, integrin beta 3 expression was highly upregulated in vascular endothelial cells found in fibrin rich but not in collagen rich matrix environment in vivo and in vitro. We recently demonstrated that fibrin and collagen differentially regulated integrin expression in human dermal microvascular endothelial cells (HDMEC) [4] and in human dermal fibroblasts [9]. In particular, fibrin, but not collagen, increased the expression of integrin alpha v beta 3 in HDMEC [4]. Since integrin alpha v beta 3 expression is differentially regulated by ECM and it is required for an angiogenic response to certain angiogenic factors, such as VEGF and bFGF [10], we hypothesized that fibrin and collagen differentially regulate angiogenesis.

Angiogenesis is a tightly regulated event, which visually includes endothelial invasion, migration, capillary tube formation, and capillary network formation. It is essential to have a reproducible and quantifiable in vitro assay of human sprout angiogenesis to investigate the integrated response of human microvascular endothelial cells to angiogenic factors and 3D ECM. Using a modified microcarrier-based 3D angiogenesis assay [11–13], we demonstrated in vitro that fibrin and collagen differentially regulate sprout angiogenesis. Our in vitro data also indicated that fibrin was essential for sprout angiogenesis of human microvascular endothelial cells in response to angiogenic factors. This was consistent with our in vivo data that there was a close correlation between fibrin presence and sprout angiogenesis occurrence in porcine wound. Inversely, when fibrin was almost totally replaced by collagen, sprout angiogenesis regressed. In addition, for the first time we demonstrated that integrin alpha v beta 3 (receptor for fibrin) and integrin alpha 2 beta 1 (receptor for collagen) differentially but synergistically regulate sprout angiogenesis.

2. Materials and Methods

2.1. Materials. Gelatin-coated microcarrier beads (Cytodex-3) were purchased from Pharmacia (Uppsala, Sweden). Dimethyl dichlorosilane, aprotinin, dibutyryl cyclic AMP, hydrocortisone, trypsin, soybean trypsin inhibitor, and EDTA were obtained from Sigma Chemical Co. (St. Louis, MO, USA). Cyclic RGD, Gly-Pen-Gly-Arg-AspPro-Cys-Ala (GpenGRGDSPCA), Vitronectin specific GPenGRGDSPCA peptides (RGD (VN)), and inactive sham control peptide, Gly-Arg-Ala-Asp-Ser-Pro (GRADSP), from Life Technologies (Carlsbad, CA, USA). Endothelial cell basal medium (EBM), endothelial cell growth medium bulletkit-2 (EGM-2 Bulletkit), bovine brain extract, and epidermal growth factor were obtained from Clonetics Corp. (San Diego, CA, USA). Normal human serum was obtained from BioWhittaker, Inc. (Walkersville, MD, USA). Vascular endothelial cell growth factor (VEGF) was purchased from Peprotech (Rocky Hill, NJ, USA). Basic fibroblast growth factor (bFGF) was obtained from Scios Nova, Inc. (Mountainvale, CA, USA). Human thrombin was obtained from Calbiochem (San Diego, CA, USA). Propidium iodide (PI) was obtained from Molecular Probes (Eugene, OR, USA). Integrin alpha 2 beta 1 blocking monoclonal antibody MAB1998 was obtained from Chemicon. Disintegrin ELP12 was a kindly gift from Cezary Marcinkiewicz of Temple University, PA, USA. Echistatin was obtained from Sigma Chemical Co. (St. Louis, MO, USA).

2.2. Cell Culture. Human dermal microvascular endothelial cells (HDMEC) were isolated from human neonatal foreskins as previously reported [4]. Briefly, after initial harvest from minced trypsinized human foreskins, microvascular endothelial cells were further purified on a Percoll density gradient. HDMEC were cultured on collagen type 1 coated tissue culture flasks in EGM (endothelial cell growth medium) consisting of EBM supplemented with 10 ng/mL epidermal growth factor, 0.4% bovine brain extract, 17.5 microgram/mL dibutyryl cyclic AMP, and 1 microg/mL hydrocortisone in the presence of 30% normal human serum. Endothelial cell cultures were characterized and determined to be >99% pure on the basis of formation of typical cobblestone monolayers in culture, positive immunostaining for factor VIII-related antigen, and selective uptake of acetylated low density lipoprotein. All experiments were done with HDMEC below passage 8.

2.3. Preparation of Endothelial Cell-Loaded Microcarrier Beads (EC-Beads). Gelatin-coated cytodex-3 microcarrier beads were prepared as described by the manufacturer. Approximately 80,000 sterile microcarrier beads were washed, resuspended in EGM, and added to approximately 4.5 million endothelial cells (HDMEC). The beads and cells were mixed by gentle swirling, incubated at 37°C for 6 hr, and then rotated for 24–36 hr on an orbital mixer in a 37°C oven to generate endothelial cell-loaded microcarrier beads (EC-beads).

2.4. Cell Migration and Capillary Sprout Formation in Fibrin Gels and Type I Collagen Gels. A microcarrier in vitro

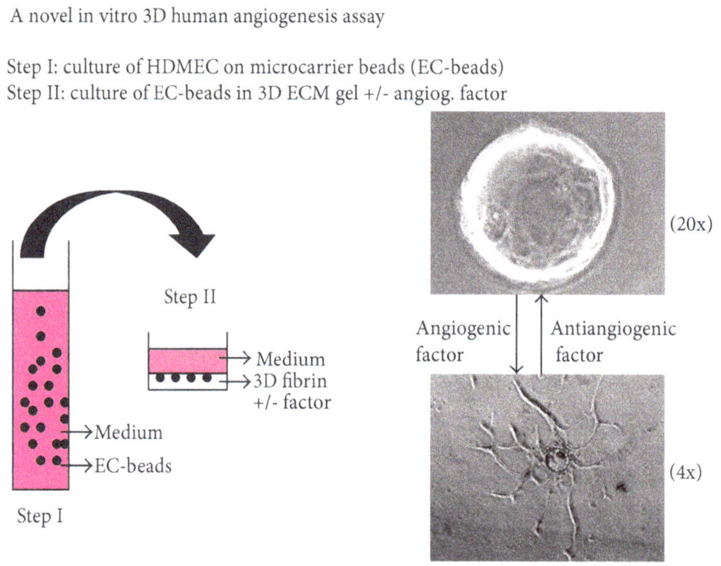

A novel in vitro 3D human angiogenesis assay

Step I: culture of HDMEC on microcarrier beads (EC-beads)
Step II: culture of EC-beads in 3D ECM gel +/- angiog. factor

FIGURE 1: In vitro three-dimensional human angiogenesis model for assaying cell invasion, migration, and sprout angiogenesis formation of human dermal microvascular endothelial cells (HDMEC) (a). Step I: HDMEC are cultured on the surface of microcarrier beads to generate EC-beads. Step II: EC-beads are embedded in fibrin gel, collagen gel, or fibrin/collagen gel, with or without the presence of angiogenesis factor.

angiogenesis assay previously designed to investigate bovine pulmonary artery endothelial cell angiogenic behavior in bovine fibrin gels [11–13] was modified for the study of human microvascular endothelial cell angiogenesis in different ECM environment (Figure 1(a)). Briefly, human fibrinogen, isolated as previously described [4], was dissolved in M199 medium at a concentration of 1 mg/mL (pH 7.4) and sterilized by filtering through a 0.22-micron filter. Pepsin-solubilized bovine dermal collagen dissolved in 0.012 M HCl was 99.9% pure containing 95–98% type I collagen and 2–5% type III collagen (Vitrogen 100, Collagen Biomaterials, Palo Alto, CA, USA). An isotonic 1.5 mg/mL collagen solution was prepared by mixing sterile Vitrogen 100 in 5X M199 medium and distilled water. pH was adjusted to 7.4 by 1 N NaOH. In certain experiments, angiogenic stimulators and/or inhibitors, such as VEGF, bFGF, RGD peptides, ELP12, and Echistatin, were added to the fibrinogen or collagen solutions (Figure 1). The angiogenic response was monitored visually and recorded by video image capture. Specifically, capillary sprout formation was observed and recorded with a Nikon Diaphot-TMD inverted microscope (Nikon Inc., Melville, NY, USA), equipped with an incubator housing with a Nikon NP-2 thermostat and Sheldon #2004 carbon dioxide flow mixer. The microscope was directly interfaced to a video system consisting of a Dage-MTI CCD-72S video camera and Sony 12″ PVM-122 video monitor linked to a Macintosh G3 computer. The images were captured at various magnifications using Adobe Photoshop. The effect of angiogenic factors on sprout angiogenesis was quantified visually by determining the number and percent of EC-beads with capillary sprouts. One hundred beads (five to six random low power fields) in each of triplicate wells were counted for each experimental condition. All experiments were repeated at least three times. To locate the nucleus of HDMEC, the

fibrin or collagen gel was fixed by methanol/acetone (1 : 1) and stained by 0.001% PI.

2.5. Porcine Cutaneous Wounds and Immunofluorescence Staining. Porcine cutaneous wounds were harvested at various times and then immunoprobed for expression of integrin receptors as previously described [14]. Briefly, full-thickness wounds were made with an 8 mm punch on the backs of White Yorkshire pigs and harvested at the times indicated. Specimens were bisected; one half was fixed in formalin and stained with Masson trichrome; the other half was frozen in liquid nitrogen for immunofluorescence studies. Antilaminin antibodies that were conjugated with biotin were used to identify wound vasculature. All antibodies were used at dilutions that gave maximal specific fluorescence and minimal background fluorescence on frozen tissue specimens. Bound antibody was detected by the avidin-biotin-complex (ABC) technique. Stained specimens were observed and photographed using a Nikon Microphot FXA epifluorescence microscope equipped with a Nikon FX-35DX 35 mm camera.

2.6. Confocal Microscopy. Confocal microscopy was done at the University Microscopy Imaging Center, Health Sciences Center, SUNY at Stony Brook, to confirm that sprouts emanating from the EC-beads formed tubes and to investigate the expression of integrin alpha v beta 3 on endothelial surface in 3D ECM. For these studies EC-beads (150) were suspended in a three-dimensional fibrin gel with VEGF (30 ng/mL) and bFGF (25 ng/mL) in a 4-well Lab-Tek Chambered Coverglass and incubated for 5 days. For experiments of tube formation, samples were washed in 2XPBS and then fixed in 2% paraformaldehyde. With the kind assistance of David Colflesh at the University Microscopy Imaging Center, images were sequentially obtained (1 micron

Fibrin and Collagen Differentially but Synergistically Regulate Sprout Angiogenesis of Human Dermal Microvascular Endothelial Cells in 3-Dimensional Matrix

187

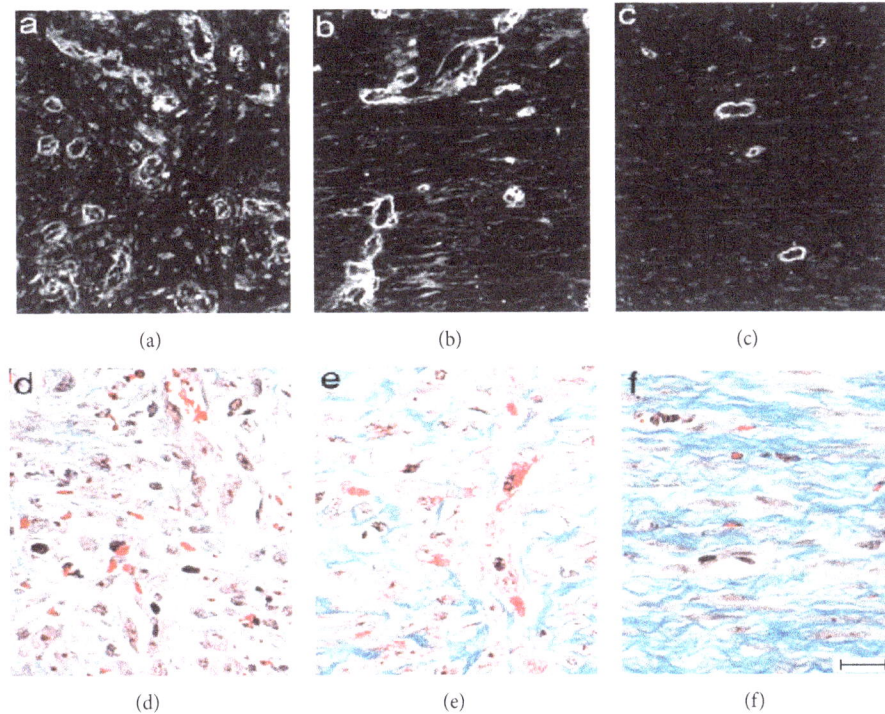

FIGURE 2: Angiogenic blood vessels in early granulation tissue mature and then regress as wound repair progresses. Porcine wounds at 5 days (a, d), 7 days (b, e), and 10 days (c, f) were stained with Masson trichrome (d, e, f) and antibody to laminin (a, b, c). At 5 days (a, d), the wound space is almost filled with granulation tissue, rich in newly forming microvessels. At 7 days (b, e), the neoepidermis has completely formed, the granulation tissue has organized, and the neovessels have matured and assumed a vertical orientation. At 10 days (c, f), wound contraction is underway, and blood vessel regression is apparent. Bar = 80 micron in (a, b, c) and 50 micron in (d, e, f).

contiguous tangential cross sections) using a Noran laser scanning confocal system (Odyssey; Noran Instruments Inc., Middleton, WI, USA) attached to a Nikon Diaphot-TMD microscope. A Silicon Graphics Iris workstation was used for processing digitized micrographs and assembling three-dimensional renditions from confocal images using Voxelview software (Voxelview; Vital Images). For experiments of immunostaining, 23C6, a monoclonal antibody to integrin beta 3, was used.

3. Results

3.1. Positive Correlation of Fibrin and Collagen with Angiogenesis Sprouting and Regression in Wound Healing.

To understand the relationship between angiogenesis and different ECM components during granulation tissue formation of wound repair, we analyzed tissue specimens from 5-, 7-, and 10-day porcine wounds (Figure 2). The 5-day wounds are mainly composed of a fibrin-rich provisional matrix, whereas 7-day wounds have a substantial organized collagen fiber network, and 10-day wounds have developed a compacted contracted collagen scar [15, 16]. Staining of 5-, 7-, and 10-day wound specimens with Masson trichrome (Figures 2(d), 2(e), and 2(f)) and with antilaminin (Figures 2(a), 2(b), and 2(c)) revealed that the fibrin-rich early granulation tissue in 5-day wounds is filled with newly formed vessels (Figures 2(a) and 2(d)). These neovessels consistently stained weakly for laminin, most likely as a result of blood vessel immaturity.

Such weak staining for laminin in immature blood vessels was previously observed by us in the microvasculature of human fetal skin [15]. By 7 days, the maturing blood vessels form an organized vertical array as collagen accumulates in the wound ECM (Figures 2(b) and 2(e)). At 10 days, as the collagen bundles thicken to produce scar, many blood vessels are regressing (Figures 2(c) and 2(f)). There is a close correlation between fibrin presence and sprout angiogenesis occurrence in porcine wound. Sprout angiogenesis mainly occurred in day 5 when wound clot was mainly filled with fibrin (Figures 2(a) and 2(d)). Inversely, when fibrin was almost totally replaced by collagen on day 10, sprout angiogenesis regressed (Figures 2(c) and 2(f)). Thus during wound repair in vivo, the angiogenic neovessels in early granulation tissue invade the fibrin clot as capillary sprouts, mature, and then regress as fibrin is replaced by collagen in the wound space.

3.2. 3D Fibrin Matrices with Angiogenic Factors Induce Sprout Angiogenesis of HDMEC In Vitro.

To delineate why angiogenesis of early wound granulation tissue started by invading the fibrin clot as capillary sprouts, we studied the regulatory effect of fibrin microenvironment on sprout angiogenesis. In our modified microcarrier-based in vitro 3D human sprout angiogenesis system, when VEGF (30 ng/mL) and bFGF (25 ng/mL) were added to serum free fibrin gel containing EC-beads, HDMEC formed capillary sprouts, which projected from the surface of the EC-beads and

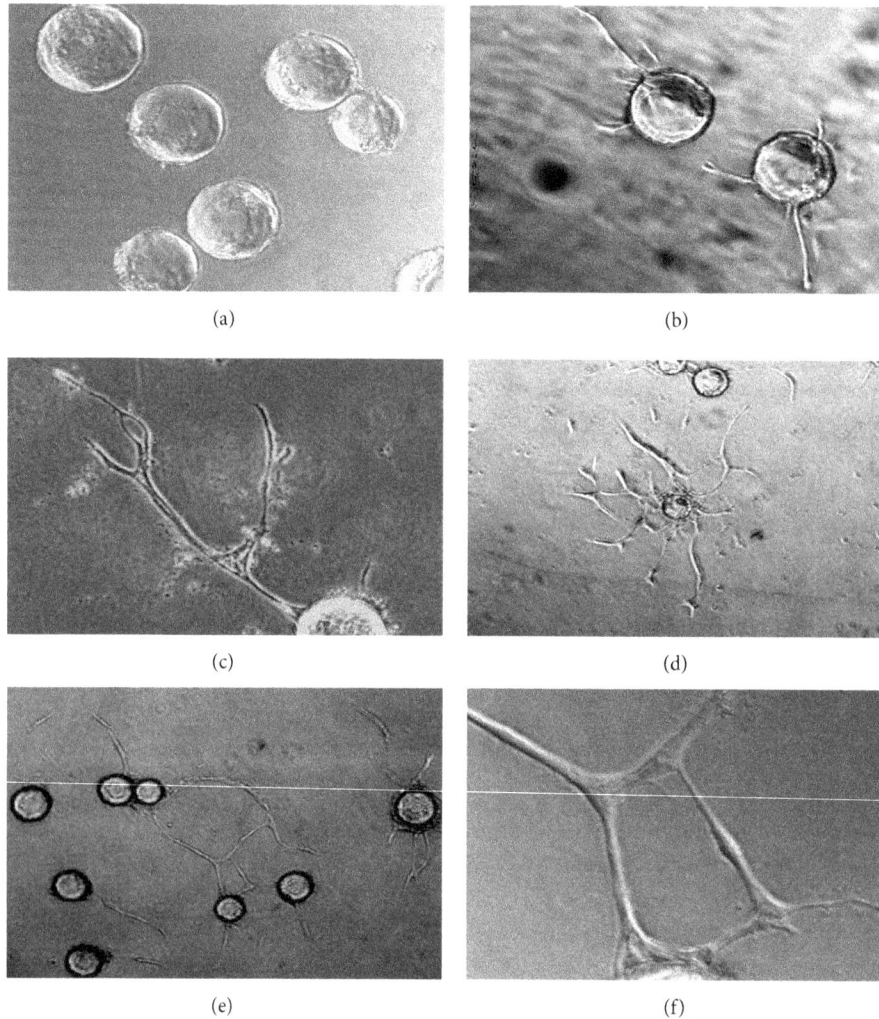

(a)

(b)

(c)

(d)

(e)

(f)

FIGURE 3: Formation of angiogenic sprouts and capillary tube-like structures by HDMEC in fibrin stimulated by VEGF (30 ng/mL) and bFGF (25 ng/mL). (a) In control fibrin gel without addition of angiogenic factors, no significant sprout formation occurred after 48 hr. (b) In the presence of VEGF (30 ng/mL) and bFGF (25 ng/mL), HDMEC formed angiogenic sprouts, and invaded and migrated into the fibrin gel within 48 hr. (c) By 5 days, VEGF and bFGF stimulated HDMEC forming branching capillary tube-like structures in the fibrin gel. (d) Formation of local capillary arcades and networks by HDMEC in fibrin with VEGF and bFGF for 5 days. (e) Formation of wild capillary networks by HDMEC in fibrin with VEGF and bFGF for 5 days. (f) Higher magnification of a typical capillary network illustrates that networks formed as a result of branching and fusion of capillary tubes from neighboring beads. Magnification: (a), (b), and (c) 100x, (d) 40x, (e) 40x, and (f) 200x.

invaded into the fibrin gel within 48 hr (Figure 3(b)). By 5 days the endothelial sprouts had elongated and in some cases formed branching capillary sprouts (Figure 3(c)). Local capillary networks formed by branching and fusion of capillary sprouts from the same bead (Figure 3(d)), and wild capillary networks formed by fusion of capillary sprouts from adjacent beads (Figures 3(e) and 3(f)). In contrast, without the addition of an angiogenesis factor, no significant HDMEC sprout formation occurred from the surface of EC-beads in fibrin gel, despite the presence of 20% normal human serum in the medium above the gels (Figure 3(a)). Fluorescence of cell nuclei staining clearly revealed that 5-day capillary sprouts (Figure 4(a)) were composed by multiple cells (Figure 4(b)). To demonstrate whether capillary tube-like structures (Figure 4(c)) had lumina, we used reflective

confocal microscopic analysis. Computer-assisted sectioning clearly revealed the presence of a lumen (Figure 4(d)) in the capillary-like structure shown in Figure 4(a). As the tube was cut through by a series of 1 micron contiguous tangential cross sections, first the top and then the central lumen with two walls become visible (Figure 4(d)).

3.3. Fibrin and Collagen 3D Microenvironment Differentially Fine-Regulate Angiogenesis of HDEMC. To determine whether ECM environment, fibrin and collagen in particular, regulates human sprout angiogenesis, a modified in vitro 3D angiogenesis system, which can differentiate sprouting angiogenesis from invasive migration, was used to compare the angiogenic response of HDMEC in 3D fibrin and collagen gels. EC-beads were embedded in fibrin gel or in collagen

Fibrin and Collagen Differentially but Synergistically Regulate Sprout Angiogenesis of Human Dermal Microvascular
Endothelial Cells in 3-Dimensional Matrix

189

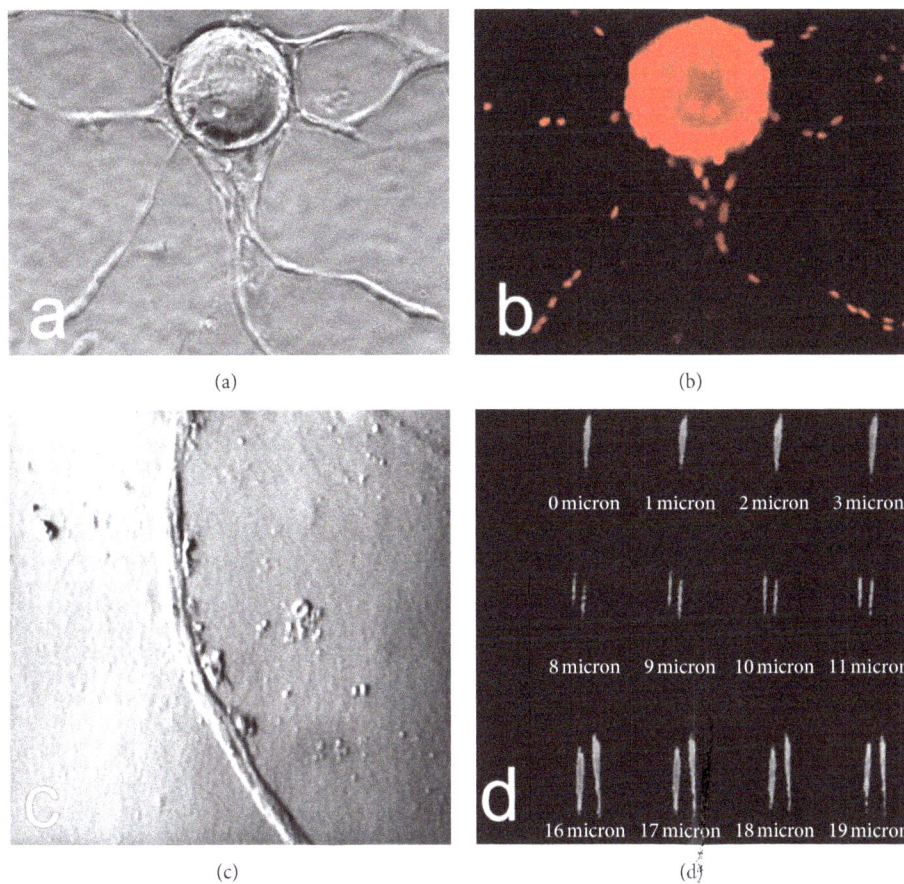

FIGURE 4: Presence of multiple cells in capillary tube-like structure by nuclei staining and presence of lumen demonstrated by confocal microscopic analysis. (a) HDMEC formed elongated and branched capillary sprouts in fibrin gel after 5 days in presence of VEGF (30 ng/mL) and bFGF (25 ng/mL). (b) The culture in (a) was fixed and stained with propidium iodide (PI) to reveal nuclei. (c) Phase-contrast photomicrograph of typical capillary tube-like structure formed by HDMEC in fibrin with VEGF (30 ng/mL) and bFGF (25 ng/mL). (d) Series of contiguous 1 micron thick tangential cross sections of the capillary tube-like structure, obtained by reflective confocal microscopy with computerized imaging, initially revealed the top of the tube (0–3 microns) and then the presence of a central lumen (8–19 microns) between two walls. (a) and (b) 100x and (c) and (d) 400x magnification.

gel, with or without presence of VEGF (30 ng/mL) and bFGF (25 ng/mL) for 48 hr. In the absence of angiogenic stimulators, HDMEC remained on the surface of EC-beads and did not invade either fibrin gel (Figure 5(a)) or collagen gel (Figure 5(b)). In the presence of VEGF in fibrin gel, HDMEC formed capillary-like sprouts from the surface of EC-beads and invaded and migrated into the surrounding fibrin (Figure 5(c)). In contrast, when angiogenic factors were added to collagen gel, HDMEC invaded and migrated into the surrounding collagen as individual cells but did not form sprouts (Figure 5(d)). After 5 days, the capillary sprouts in fibrin elongated and fused into networks (Figure 5(e)), but in collagen individual HDMEC randomly migrated without capillary sprouts and networks (Figure 5(f)). Nucleus staining by PI confirmed that in fibrin the capillary tubes were composed of multiple cells (Figure 5(g)), but in collagen HDMEC randomly distributed in the gel (Figure 5(h)).

3.4. Regulating Angiogenesis of HDMEC in Collagen by Modulating ECM Environment.
To determine if presence of fibrin in collagen gels will enable angiogenic factors to

induce capillary sprouts in collagen gels, we embedded EC-beads in collagen admixed with fibrin and compared the angiogenic response to that in either pure fibrin or pure collagen gels. Combination of VEGF (30 ng/mL) and bFGF (25 ng/mL) induced individual cell invasion and migration in pure collagen gels (Figure 6(a)) and induced capillary sprouts in pure fibrin gels (Figure 6(d)) as previously observed. With addition of fibrin to collagen gel (Figure 6(b), 20% fibrin, and Figure 6(c), 30% fibrin), tube-like capillary sprout formation occurred together with individual cell invasion and migration. Thus the presence of fibrin appeared to be essential for HDMEC sprout angiogenesis induced by angiogenic factors.

3.5. Antagonists for Integrin Alpha v Beta 3 and Integrin Alpha 2 Beta 1 Synergistically Inhibit Sprout Angiogenesis of HDMEC in 3D Fibrin.
We recently demonstrated that integrin alpha v beta 3 is essential for sprout angiogenesis and integrin beta 3 expression is highly upregulated in fibrin rich but not in collagen rich matrix environment in vitro and in vivo [4, 13]. Using the modified in vitro sprout angiogenesis model, we demonstrated that GPenGRGDSPCA peptide (RGD (VN)),

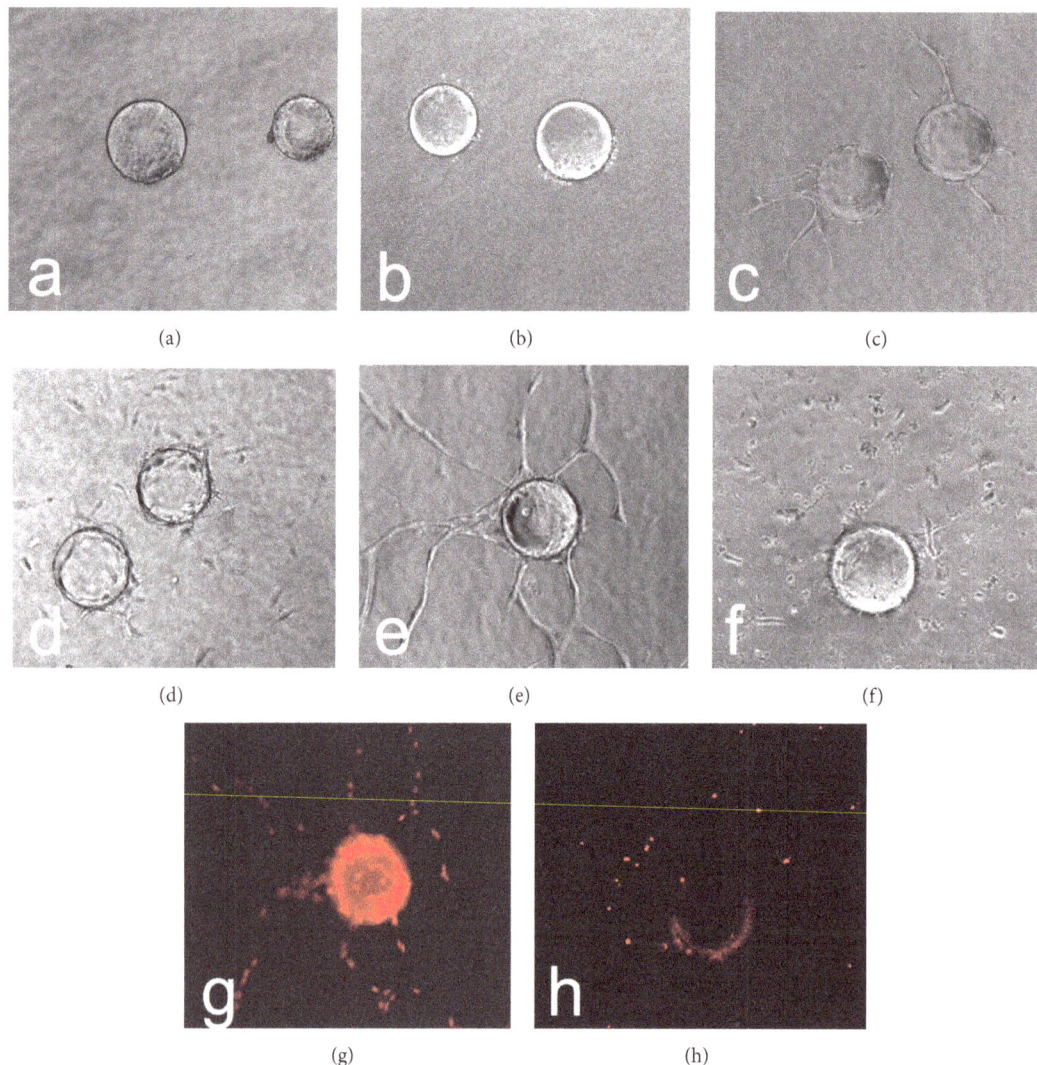

FIGURE 5: Fibrin gels (a, c, e, g) and collagen gels (b, d, f, h) differentially regulated HDMEC response to VEGF (30 ng/mL) and bFGF (25 ng/mL). Without the addition of angiogenesis factors, little or no sprout formation, invasion, or migration of HDMEC occurred either in fibrin gel (a) or in collagen gel (b). In the presence of VEGF (30 ng/mL) and bFGF (25 ng/mL), HDMEC formed capillary sprouts which invaded and migrated into fibrin gel within 48 hr (c). In contrast, in collagen gel in the presence of VEGF (30 ng/mL) and bFGF (25 ng/mL), HDMEC invaded and migrated into the gel individually without forming capillary sprouts (d). After 5 days, the capillary sprouts in fibrin elongated and branched (e). IF staining revealed that the capillary sprouts in fibrin were formed by multiple, lineal aligned cells (g). After 5 days, HDMEC randomly migrated into collagen gel without forming capillary sprouts (f). IF staining revealed that the nuclei were randomly distributed in collagen gel (h) (100x magnification).

which is a specific antagonist for integrin alpha v beta 3, significantly blocked sprout angiogenesis of HDMEC in fibrin compared to the control RGE peptide (Figure 7). Echistatin (10 microgram/mL, 3 microgram/mL, 1 microgram/mL), a disintegrin specific for alpha v beta 3, dose dependently inhibits sprout angiogenesis of HDMEC in fibrin (data not shown). In contrast, ECL12, inhibitor to collagen receptor integrin alpha 3 beta 1, had no inhibitory effect on sprout angiogenesis of HDMEC in fibrin, even at a concentration 10 microgram/mL (Figure 8). At a concentration of 0.1 microgram/mL, Echistatin has minimal inhibitory effect on sprout angiogenesis of HDMEC in fibrin. However, the combination of 0.1 microgram/mL Echistatin and 10 microgram/mL ECL12 completely inhibited sprout angiogenesis of HDEMC

in fibrin induced by VEGF (30 ng/mL) and bFGF (25 ng/mL). These data indicates that integrin alpha v beta 3 and integrin alpha 2 beta 1 differentially but synergistically inhibit sprout angiogenesis of HDMEC in 3D fibrin. It also suggests that collagen and fibrin differentially but synergistically regulate sprout angiogenesis.

4. Discussion

Alteration of fibrin and type I collagen matrices in the ECM microenvironment is a hallmark feature of wound clot and tumor stroma; however its role in the regulation of sprout angiogenesis is still poorly defined. Using in vitro 3D angiogenesis assay and in vivo wound healing model,

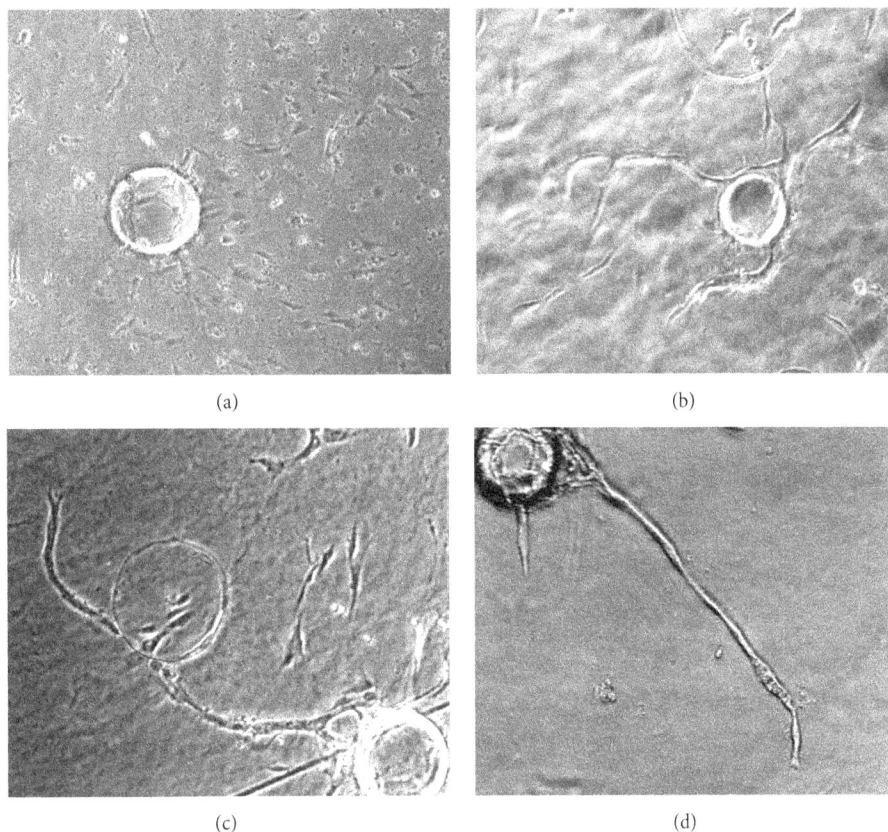

FIGURE 6: Fibrin facilitated sprout angiogenesis of HDMEC in collagen induced by angiogenic factors combo (30 ng/mL VEGF + 25 ng/mL bFGF) at 48 hr. Angiogenic factors stimulated HDMEC to invade and migrate into pure collagen gel individually without forming capillary sprouts (a). With the addition of 20% fibrin fibrils, by volume (b), or 30% fibrin fibrils, by volume (c), to collagen gel in the presence of angiogenic factors, sprout angiogenesis occurred together with individual cell invasion and migration. Angiogenic factors induced sprout angiogenesis of HDMEC in pure fibrin gel (d) (100x magnification).

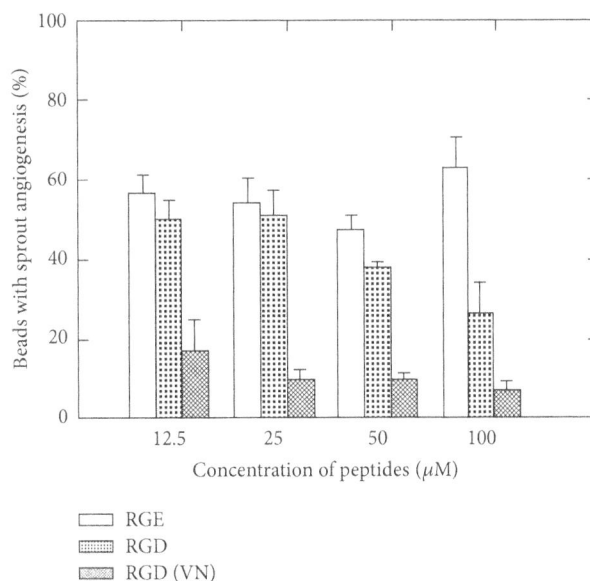

FIGURE 7: Effect of GPenGRGDSPCA peptides (RGD (VN)) on sprout angiogenesis of HDMEC in fibrin. HDMEC cultured on microcarrier beads were incubated with various doses of RGD (VN) (12.5 micro M–100 micro M) peptides at RT for 1 hr. The EC-beads were added to fibrinogen solution with presence of VEGF (30 ng/mL) and bFGF (25 ng/mL) and then polymerized with thrombin. RGD (VN) significantly inhibited sprout angiogenesis of HDMEC in fibrin compared to GRGESP(RGE) control peptides and nonspecific integrin antagonist cyclic RGD peptide (GRGDNP).

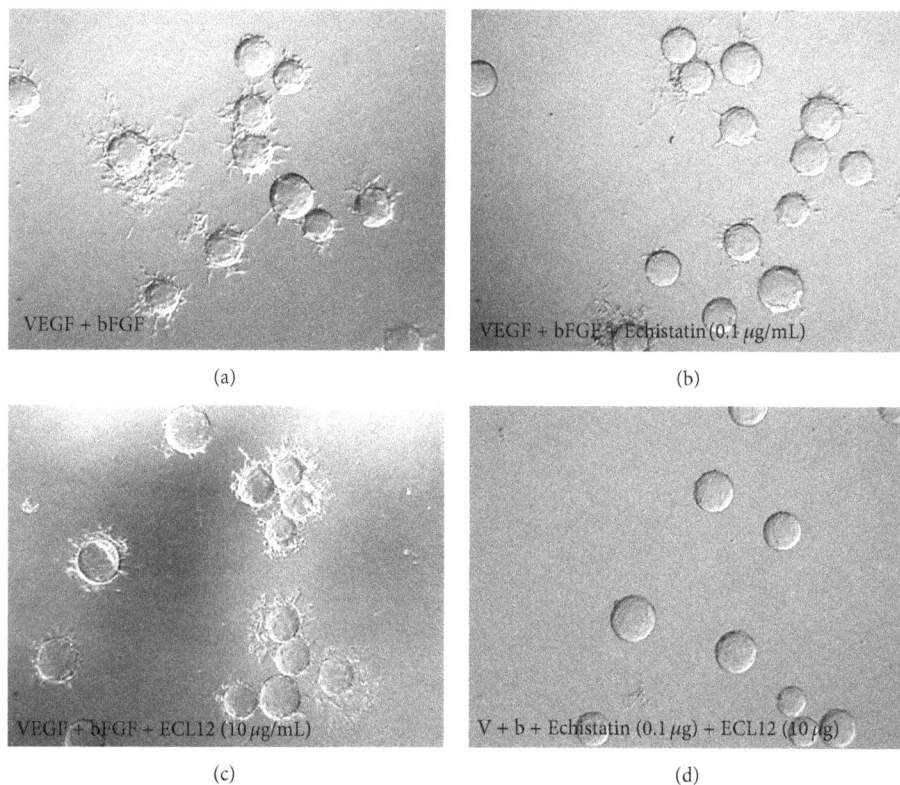

FIGURE 8: Integrin alpha v beta3 antagonist Echistatin and integrin alpha 2 beta 1 antagonist ECL12 differentially but synergistically inhibit sprout angiogenesis of HDMEC in fibrin. VEGF (30 ng/mL) and bFGF (25 ng/mL) induced sprout angiogenesis in fibrin (a). Echistatin at very low dose (0.1 microgram/mL) had mild inhibitory effect on sprout angiogenesis of HDMEC in fibrin (b). While disintegrin ECL12 had no inhibitory effect on sprout angiogenesis in vitro (c), it synergistically enhanced the inhibition of sprout angiogenesis by Echistatin (d).

we demonstrate in this study that robust angiogenic vessels invaded the fibrin clot of early wounds but matured and regressed as fibrin was replaced by type I collagen (Figure 2). These results are consistent with the finding that fibrin deposition is commonly observed in angiogenesis associated with wound healing and tumor growth. It has been reported that fibrin enhances angiogenesis of wound healing in vitro and in vivo [7, 8]. In contrast, type I collagen is a major component of normal dermis which has minimal angiogenesis activities. In our in vitro 3D angiogenesis assay, angiogenic stimulators induce endothelial cell migration but not sprout and capillary network formation in 3D type I collagen matrices (Figure 5). Interestingly once fibrin was added to type I collagen gel, the sprout angiogenesis is restored (Figure 6). Dvorak and colleagues also demonstrate that implanted fibrin gels themselves induce an angiogenic response in the subcutaneous space of guinea pigs in the absence of tumor cells or platelets, while type I collagen or agarose does not induce new blood vessel formation [8].

After tissue injury, type I collagen is removed with normal dermis and fibrinogen leaks from ligated blood vessels to form the fibrin clot, which fills the wound space [16–18]. Using in vivo wound healing model, we demonstrated that after a 3-day lag, endothelial cells migrate from the periphery of the wound and invade the fibrin clot as sprout angiogenesis to form nascent granulation tissue (Figure 2). Type I collagen

accumulation begins as the granulation tissue matures, and in small excisional wounds, type I collagen replaces fibrin in the wound space by 7 days and form contracted wound scar in 10 days [19, 20]. We demonstrate here a substantial correlation between sprout angiogenesis and presence of fibrin matrices and the regression of sprout angiogenesis when fibrin is replaced by type I collagen. However, type I collagen supports sprout angiogenesis in the presence of fibrin fibrils. The regulatory mechanism of type I collagen in sprout angiogenesis is still not clear. Recently two teams of researchers demonstrated that an amino terminal peptide of angiocidin binds type I collagen and inhibits tumor growth and angiogenesis [21, 22].

Although fibrin is actively involved in regulating angiogenesis, we demonstrated that pure fibrin matrices themselves do not induce sprout angiogenesis without presence of angiogenic factors (Figure 5(a)). Roy et al. demonstrated that platelet-rich fibrin matrix improves wound angiogenesis via inducing vascular endothelial cell proliferation [23]. Lafleur and colleagues demonstrated that membrane-type-matrix metalloproteinases (MT-MMPs) are essential for endothelial angiogenesis in fibrin matrix [24]. It has been argued that if fibrin itself was sufficient to induce angiogenesis, it must perform at least two functions: (1) providing a three-dimensional matrix that supports cell migration and (2) expressing selective chemotactic and/or chemokinetic

Fibrin and Collagen Differentially but Synergistically Regulate Sprout Angiogenesis of Human Dermal Microvascular
Endothelial Cells in 3-Dimensional Matrix

193

activity such that endothelial cells migrate into fibrin clot [8]. It is not surprising that fibrin might provide a three-dimensional matrix capable of supporting cell migration. However, it is unexpected that fibrin might directly induce cell migration without the presence of any soluble chemotactic factors. Greiling and Clark demonstrated that human dermal fibroblasts fail to migrate from a collagen matrix into a fibrin gel in the absence of platelet releasate or PDGF-BB [25]. Our in vitro data also demonstrated that there was no sprout angiogenesis of HDMEC in fibrin without presence of angiogenesis growth factors, such as VEGF and bFGF, in the matrix. Thus fibrin is essential to support sprout angiogenesis of microvascular endothelials induced by angiogenic stimulators in 3D environment.

Integrin alpha v beta 3 is the receptor for fibrin matrix. Expression of integrin alpha v beta 3 is one of the hallmark features of sprout angiogenesis. Accumulating evidence indicates that ECM microenvironment regulates cell functions through the integrin receptors. We recently demonstrated that fibrin and type I collagen differentially regulated integrin expression in human dermal microvascular endothelial cells (HDMEC) [4] and in human dermal fibroblasts [9]. In particular, fibrin, but not collagen, increased the expression of integrin alpha v beta 3 in HDMEC [4]. Since integrin alpha v beta 3 is the marker for sprout angiogenesis associated with wound healing and tumor growth [14, 25], this suggests that fibrin and collagen differentially regulate angiogenesis, in part, by altering endothelial cell integrin expression. Compared to highly regulated angiogenesis in wound healing, angiogenesis persists in solid tumor growth, as does the blood vessel leak of fibrinogen and resultant interstitial clotting [26]. Numerous studies indicated that integrin alpha v beta 3 has potential to be a novel target for cancer treatment [27–29]. Our results indicate that GPenGRGDSPCA peptide (RGD (VN)), which is a specific antagonist for integrin alpha v beta 3, significantly blocked sprout angiogenesis of HDMEC in fibrin induced by VEGF (Figure 7). Currently many new investigational cancer drugs based on RGD peptide are tested in clinical trials.

Integrin alpha 2 beta 1 is the integrin receptor for native collagens which mediate many important cellular functions, such as adhesion, migration, invasion, and contraction of collagen lattices. The role of integrin alpha 2 beta 1 in angiogenesis is controversial and is still not clearly defined [30]. Using in vitro 2D models, Senger and colleagues demonstrated that blocking antibodies for integrin alpha 2 beta 1 inhibit endothelial cell migration on type I collagen. In addition, they demonstrated that blocking antibodies to integrin alpha 2 beta 1 also potently inhibit VEGF induced angiogenesis using Matrigel transplant in mice skin [31]. In contrast, Zweers and colleagues found strong enhancement of angiogenesis in cutaneous wound and implanted sponges in alpha 2 null mice [32]. Furthermore, Zutter and colleagues demonstrated that poor expression of integrin alpha 2 beta 1 might play an essential role in cancer progression [33]. This is confirmed by Ramirez and colleagues in spontaneous mouse model of breast cancer showing alpha 2 beta 1 integrin suppresses metastasis [34]. Haidari and colleagues recently reported evidence that integrin alpha 2 beta 1 mediates

tyrosine phosphorylation of vascular endothelial cadherin induced by invasive breast cancer cells, which is essential for transendothelial migration (TEM) of invasive cancer cells [35]. In our in vitro 3D angiogenesis assay, antagonists for integrin alpha 2 beta 1 demonstrate minimal inhibitory effect on sprout angiogenesis of HDMEC in 3D fibrin matrices. Interestingly, when combined with Echistatin, a disintegrin antagonist specific for integrin alpha v beta 3 and antagonists for integrin alpha 2 beta 1 synergistically inhibit sprout angiogenesis of HDMEC in fibrin. Taken together, these results suggest that fibrin and collagen differentially but synergistically regulate sprout angiogenesis through controlling integrin alpha v beta 3 and integrin alpha 2 beta 1 functions.

This study indicated that growth factor-rich fibrin is essential to promote sprout angiogenesis. Fibrin and type I collagen 3D matrices differentially but synergistically regulate sprout angiogenesis in wound healing and solid tumors. Thus blocking both integrin alpha v beta 3 and integrin alpha 2 beta 1 might be a novel strategy to synergistically block sprout angiogenesis in solid tumors.

Conflict of Interests

The authors report no conflict of interests.

References

[1] M. G. Tonnesen, X. Feng, and R. A. F. Clark, "Angiogenesis in wound healing," *Journal of Investigative Dermatology*, vol. 5, no. 1, pp. 40–46, 2000.

[2] X. Feng, W. Ofstad, and D. Hawkins, "Antiangiogenesis therapy: a new strategy for cancer treatment," *US Pharmacist*, vol. 35, no. 7, pp. 4–9, 2010.

[3] G. E. Davis and D. R. Senger, "Endothelial extracellular matrix: biosynthesis, remodeling, and functions during vascular morphogenesis and neovessel stabilization," *Circulation Research*, vol. 97, no. 11, pp. 1093–1107, 2005.

[4] X. Feng, R. A. F. Clark, D. Galanakis, and M. G. Tonnesen, "Fibrin and collagen differentially regulate human dermal microvascular endothelial cell integrins: stabilization of $\alpha v/\beta 3$ mRNA by fibrin," *Journal of Investigative Dermatology*, vol. 113, no. 6, pp. 913–919, 1999.

[5] J. M. Rhodes and M. Simons, "The extracellular matrix and blood vessel formation: not just a scaffold: angiogenesis review series," *Journal of Cellular and Molecular Medicine*, vol. 11, no. 2, pp. 176–205, 2007.

[6] S. A. Carolyn, M. W. R. Reed, and N. J. Brown, "A critical analysis of current in vitro and in vivo angiogenesis assays," *International Journal of Experimental Pathology*, vol. 90, no. 3, pp. 195–221, 2009.

[7] A. Takei, Y. Tashiro, Y. Nakashima, and K. Sueishi, "Effects of fibrin on the angiogenesis in vitro of bovine endothelial cells in collagen gel," *In Vitro Cellular and Developmental Biology. Animal*, vol. 31, no. 6, pp. 467–472, 1995.

[8] H. F. Dvorak, V. S. Harvey, P. Estrella, L. F. Brown, J. McDonagh, and A. M. Dvorak, "Fibrin containing gels induce angiogenesis. Implications for tumor stroma generation and wound healing," *Laboratory Investigation*, vol. 57, no. 6, pp. 673–686, 1987.

[9] J. Xu, M. M. Zutter, S. A. Santoro, and R. A. F. Clark, "A three-dimensional collagen lattice activates NF-κB in human

fibroblasts: role in integrin $\alpha 2$ gene expression and tissue remodeling," *Journal of Cell Biology*, vol. 140, no. 3, pp. 709–719, 1998.

[10] M. Friedlander, P. C. Brooks, R. W. Shaffer, C. M. Kincaid, J. A. Varner, and D. A. Cheresh, "Definition of two angiogenic pathways by distinct αv integrins," *Science*, vol. 270, no. 5241, pp. 1500–1502, 1995.

[11] V. Nehls and D. Drenckhahn, "A novel, microcarrier-based in vitro assay for rapid and reliable quantification of three-dimensional cell migration and angiogenesis," *Microvascular Research*, vol. 50, no. 3, pp. 311–322, 1995.

[12] V. Nehls and R. Herrmann, "The configuration of fibrin clots determines capillary morphogenesis and endothelial cell migration," *Microvascular Research*, vol. 51, no. 3, pp. 347–364, 1996.

[13] S. A. Mousa, X. Feng, J. Xie et al., "Synthetic oligosaccharide stimulates and stabilizes angiogenesis: structure-function relationships and potential mechanisms," *Journal of Cardiovascular Pharmacology*, vol. 48, no. 2, pp. 6–13, 2006.

[14] R. A. F. Clark, M. G. Tonnesen, J. Gailit, and D. A. Cheresh, "Transient functional expression of $\alpha v \beta 3$ on vascular cells during wound repair," *American Journal of Pathology*, vol. 148, no. 5, pp. 1407–1421, 1996.

[15] M. G. Tonnesen, D. Jenkins Jr., S. L. Siegal, L. A. Lee, J. C. Huff, and R. A. Clark, "Expression of fibronectin, laminin, and factor VIII-related antigen during development of the human cutaneous microvasculature," *Journal of Investigative Dermatology*, vol. 85, no. 6, pp. 564–568, 1985.

[16] R. A. F. Clark, "Wound repair overview and general considerations," in *The Molecular and Cellular Biology of Wound Repair*, R. A. F. Clark, Ed., pp. 3–50, Plenum Press, New York, NY, USA, 2nd edition, 1996.

[17] R. A. F. Clark, L. D. Nielsen, M. P. Welch, and J. M. McPherson, "Collagen matrices attenuate the collagen-synthetic response of cultured fibroblasts to TGF-β," *Journal of Cell Science*, vol. 108, no. 3, pp. 1251–1261, 1995.

[18] S. A. McClain, M. Simon, E. Jones et al., "Mesenchymal cell activation is the rate-limiting step of granulation tissue induction," *American Journal of Pathology*, vol. 149, no. 4, pp. 1257–1270, 1996.

[19] R. A. F. Clark, J. M. Lanigan, P. DellaPelle, E. Manseau, H. F. Dvorak, and R. B. Colvin, "Fibronectin and fibrin provide a provisional matrix for epidermal cell migration during wound reepithelialization," *Journal of Investigative Dermatology*, vol. 79, no. 5, pp. 264–269, 1982.

[20] M. P. Welch, G. F. Odland, and R. A. F. Clark, "Temporal relationships of F-actin bundle formation, collagen and fibronectin matrix assembly, and fibronectin receptor expression to wound contraction," *Journal of Cell Biology*, vol. 110, no. 1, pp. 133–145, 1990.

[21] A. Gaurnier-Hausser, V. L. Rothman, S. Dimitrov, and G. P. Tuszynski, "The novel angiogenic inhibitor, angiocidin, induces differentiation of monocytes to macrophages," *Cancer Research*, vol. 68, no. 14, pp. 5905–5914, 2008.

[22] Y. Sabherwal, V. L. Rothman, S. Dimitrov et al., "Integrin $\alpha 2 \beta 1$ mediates the anti-angiogenic and anti-tumor activities of angiocidin, a novel tumor-associated protein," *Experimental Cell Research*, vol. 312, no. 13, pp. 2443–2453, 2006.

[23] S. Roy, J. Driggs, H. Elgharably et al., "Platelet-rich fibrin matrix improves wound angiogenesis via inducing endothelial cell proliferation," *Wound Repair and Regeneration*, vol. 19, no. 6, pp. 753–766, 2011.

[24] M. A. Lefleur, M. M. Handsley, V. Knäuper, G. Murphy, and D. R. Edwards, "Endothelial tubulogenesis within fibrin gels specifically requires the activity of membrane-type-matrix metalloproteinases (MT-MMPs)," *Journal of Cell Science*, vol. 115, no. 17, pp. 3427–3438, 2002.

[25] D. Greiling and R. A. F. Clark, "Fibronectin provides a conduit for fibroblast transmigration from collagenous stroma into fibrin clot provisional matrix," *Journal of Cell Science*, vol. 110, no. 7, pp. 861–870, 1997.

[26] H. F. Dvorak, "Tumors: wounds that do not heal: similarities between tumor stroma generation and wound healing," *The New England Journal of Medicine*, vol. 315, no. 26, pp. 1650–1659, 1986.

[27] P. C. Brooks, R. A. F. Clark, and D. A. Cheresh, "Requirement of vascular integrin $\alpha(v)\beta 3$ for angiogenesis," *Science*, vol. 264, no. 5158, pp. 569–571, 1994.

[28] Z. Liu, F. Wang, and X. Chen, "Integrin $\alpha v \beta 3$-targeted cancer therapy," *Drug Development Research*, vol. 69, no. 6, pp. 329–339, 2008.

[29] S. J. Desgrosellier and D. A. Cheresh, "Integrins in cancer: biological implications and therapeutic opportunities," *Nature Reviews Cancer*, vol. 10, no. 1, pp. 9–22, 2010.

[30] M. B. Srichai and R. Zent, "Integrin structure and function," in *Cell-Extracellular Matrix Interactions in Cancer*, R. Zent and A. Pozzi, Eds., pp. 19–41, Springer Science+Business Media, LLC., New York, NY, USA, 2010.

[31] D. R. Senger, K. P. Claffey, J. E. Benes, C. A. Perruzzi, A. P. Sergiou, and M. Detmar, "Angiogenesis promoted by vascular endothelial growth factor: regulation through $\alpha 1 \beta 1$ and $\alpha 2 \beta 1$ integrins," *Proceedings of the National Academy of Sciences of the United States of America*, vol. 94, no. 25, pp. 13612–13617, 1997.

[32] M. C. Zweers, J. M. Davidson, A. Pozzi et al., "Integrin $\alpha 2 \beta 1$ is required for regulation of murine wound angiogenesis but is dispensable for reepithelialization," *Journal of Investigative Dermatology*, vol. 127, no. 2, pp. 467–478, 2007.

[33] M. M. Zutter, G. Mazoujian, and S. A. Santoro, "Decreased expression of integrin adhesive protein receptors in adenocarcinoma of the breast," *American Journal of Pathology*, vol. 137, no. 4, pp. 863–870, 1990.

[34] N. E. Ramirez, Z. Zhang, A. Madamanchi et al., "The $\alpha 2 \beta 1$ integrin is a metastasis suppressor in mouse models and human cancer," *Journal of Clinical Investigation*, vol. 121, no. 1, pp. 226–237, 2011.

[35] M. Haidari, W. Zhang, A. Caivano et al., "Integrin alpha 2 beta 1 mediates tyrosine phosphorylation of vascular endothelial cadherin induced by invasive breast cancer cells," *The Journal of Biological Chemistry*, vol. 287, no. 39, pp. 32981–32992, 2012.

19

PKM2, a Central Point of Regulation in Cancer Metabolism

Nicholas Wong,[1,2,3,4] **Jason De Melo,**[1,2,3,4] **and Damu Tang**[1,2,3,4]

[1] *Division of Nephrology, Department of Medicine, McMaster University, Hamilton, ON, Canada L8S 4L8*
[2] *Division of Urology, Department of Surgery, McMaster University, Hamilton, ON, Canada L8S 4L8*
[3] *Father Sean O'Sullivan Research Centre, St. Joseph's Healthcare Hamilton, Hamilton, ON, Canada L8N 4A6*
[4] *The Hamilton Center for Kidney Research, St. Joseph's Healthcare Hamilton, Hamilton, ON, Canada L8N 4A6*

Correspondence should be addressed to Damu Tang; damut@mcmaster.ca

Academic Editor: Claudia Cerella

Aerobic glycolysis is the dominant metabolic pathway utilized by cancer cells, owing to its ability to divert glucose metabolites from ATP production towards the synthesis of cellular building blocks (nucleotides, amino acids, and lipids) to meet the demands of proliferation. The M2 isoform of pyruvate kinase (PKM2) catalyzes the final and also a rate-limiting reaction in the glycolytic pathway. In the PK family, PKM2 is subjected to a complex regulation by both oncogenes and tumour suppressors, which allows for a fine-tone regulation of PKM2 activity. The less active form of PKM2 drives glucose through the route of aerobic glycolysis, while active PKM2 directs glucose towards oxidative metabolism. Additionally, PKM2 possesses protein tyrosine kinase activity and plays a role in modulating gene expression and thereby contributing to tumorigenesis. We will discuss our current understanding of PKM2's regulation and its many contributions to tumorigenesis.

1. Introduction

Metabolism lies in the heart of cell biology. Understanding how cancer cells cope with metabolic needs for their unique biology has been a focus of cancer research for many years. It began with the landmark observation reported more than 80 years ago by Otto Warburg that cancer cells consumed more glucose and produced a large amount of lactate even in a well-oxygenized environment, a process known as aerobic glycolysis or the Warburg effect [1, 2]. While normal differentiated cells maximize ATP production by mitochondrial oxidative phosphorylation of glucose under normoxic conditions, cancer cells generate much less ATP from glucose by aerobic glycolysis. Despite being less efficient in ATP production, glycolysis is a much more rapid process [3, 4]. Cancers commonly deregulate pathways that enhance glycolysis, including activation of the PI3 K-ATK-mTOR pathway and upregulation of HIF-1 and c-Myc [5, 6]. The increase in aerobic glycolysis together with its dynamic process in cancer cells enables glycolytic intermediates to be redirected for the biosynthesis of cellular building blocks (nucleotides, amino acids, and lipids) while also producing ATP. Therefore,

the Warburg effect/aerobic glycolysis meets the demands of cancer and proliferating cells for macromolecular synthesis and energy production [7, 8]. As a result, cancer cells display enhanced glucose uptake and produce higher levels of lactate [1, 2]. The Warburg effect was explored for the common clinical detection of tumors by fluorodeoxyglucose (2-deoxy-2-(^{18}F)fluoro-D-glucose) positron emission tomography (FDG-PET) [7]. In the glycolytic process, pyruvate kinase (PK) catalyzes the last reaction, transfer of a high-energy phosphate group from phosphoenolpyruvate (PEP) to ADP, producing ATP and pyruvate [9]. Pyruvate is then either reduced to lactate by lactate dehydrogenase (LDH) in the cytosol or enters the mitochondria to produce ATP through the tricarboxylic acid (TCA) cycle (Figure 1). Along the glycolysis pathway, intermediate metabolites can be channeled to synthesize amino acids, nucleotides, and lipids (Figure 1) if the rate of flux through the pathway is controlled. PK is an ideal candidate for this control [10] because (1) PK catalyzes the last reaction of the pathway (Figure 1) and (2) the reaction is essentially irreversible (Figure 1) [9, 11]. Therefore, lowering PK activity is expected to produce less pyruvate (Figure 1) or prevent complete

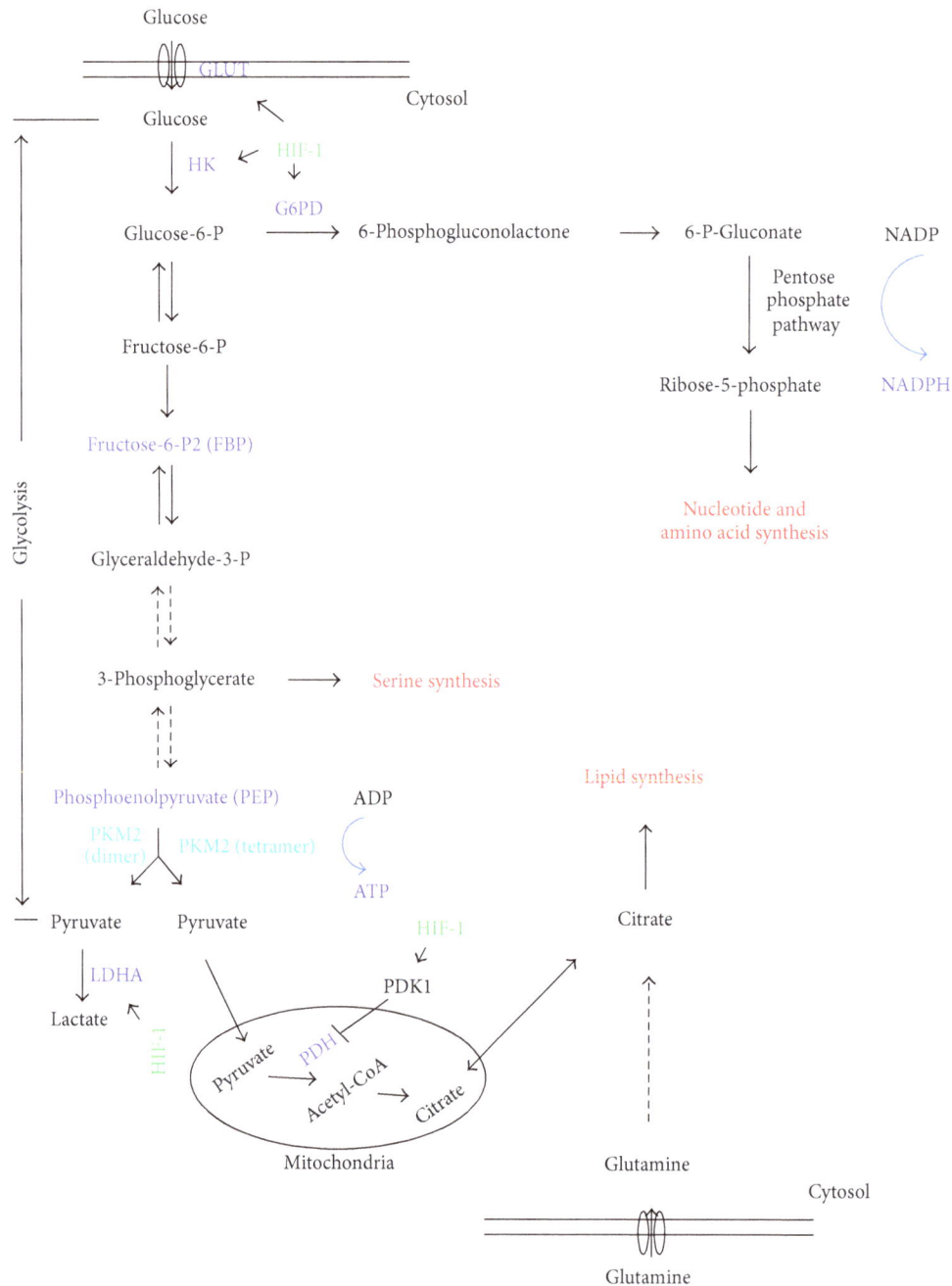

FIGURE 1: Schematic illustrating the cancer utilization of the metabolic pathways. Pyruvate kinase catalyzes the last step of glycolysis by converting PEP and ADP to pyruvate and ATP, respectively. PKM2 dimers and tetramers possess low and high levels of Pyruvate kinase activity, respectively. With reduced enzymatic activity, PKM2 dimer drives aerobic glycolysis, which allows the intermediate metabolites to be used for the synthesis of nucleotides, amino acids, and lipids and the production of reduced NADPH (see the pentose phosphate pathway). HIF-1 upregulates the indicated proteins. GLUT: glucose transporter, HK: hexokinase, G6PD: glucose-6-phosphate dehydrogenase, HIF-1: hypoxia-inducible factor 1, LDHA: lactate dehydrogenase A, PDK1: pyruvate dehydrogenase kinase isoenzyme 1, and PDH: pyruvate dehydrogenase.

conversion of glucose to pyruvate (1 molecule of glucose to 2 molecules of pyruvate). This enables the upstream glycolytic intermediates to accumulate and thus contribute to the shift of metabolism towards the anabolic phase for amino acids, nucleotides, and lipid production (Figure 1). Cancer cells explore this logic by predominantly using PKM2, an isoform

of PK, as its activity can be dynamically regulated between the less active PKM2 dimer and the highly active PKM2 tetramer [12].

PK consists of four isoforms: the L (PKL) and R (PKR) isoforms encoded by the *PKLR* (1q22) gene and the M1 (PKM1) and M2 (PKM2) isoforms encoded by the *PKM2*

(15q23) gene. The *PKLR* gene is regulated by tissue-specific promoters. The full-length PKR isoform is expressed in red blood cells while the PKL isoform missing exon 1 is detected in liver and kidney [5, 13, 14]. PKM1 and PKM2 are produced from the *PKM2* gene by alternative splicing [15]. The highly active PKM1 is expressed in tissues that consistently need high levels of energy, like skeletal muscle, heart, and brain [5, 10]. PKM2 is expressed in most cells except adult muscle, brain, and liver [12, 16, 17] and is the predominant PK in proliferating and cancer cells [18]. While PKL, PKR, and PKM1 form stable tetramers (the active form of PK), PKM2 exists as both dimers and tetramers [18, 19]. The PKM2 dimer has a higher K_m towards PEP than the tetramer and thus is less active in converting PEP to ATP and pyruvate [19, 20]. While tetrameric PKM2 favors ATP production through the TCA cycle, dimeric PKM2 plays a critical role in aerobic glycolysis (Figure 1) [19]. Therefore, the dynamic equilibrium between dimer and tetramer PKM2 allows proliferating cells to regulate their needs for anabolic and catabolic metabolism. This not only explains why cancer cells predominantly express PKM2 but also reveals the existence of mechanisms that regulate this dynamic equilibrium. To ensure PKM2 expression, cancer cells also develop mechanisms for alternative splicing to produce PKM2 rather than PKM1. These mechanisms are regulated by oncogenes and tumor suppressors [21–25]. Surprisingly, dimeric PKM2 has additional functions in regulating gene expression in the nucleus [26].

2. PKM2 Contributes to Tumorigenesis

A large body of evidence supports the notion that cancers predominantly express PKM2 [14]. Immunohistochemical analysis revealed that PKM2 is commonly expressed in colon cancer [12], renal cell carcinoma (RCC) [27], and lung cancer [28]. PKM2 has been suggested to be a marker for RCC [29, 30] and testicular cancer [31]. Elevation of serum PKM2 levels was reported in patients with colon cancer [32], breast cancer [33], urological tumors [34], lung carcinoma, cervical cancer, and gastrointestinal tumor [18]. PKM2 was detected in the feces of patients with gastric and colorectal cancers [35]. Recently, mass spectrometry has demonstrated increases in PKM2, and the predominant presence of PKM2 was confirmed in RCC, bladder carcinoma, hepatocellular carcinoma, colorectal cancer, lung carcinoma, and follicular thyroid adenoma [16].

PKM2 expression correlates with tumorigenesis. High levels of PKM2 associate with poor prognosis for patients with signet ring cell gastric cancer [36]. A unique pattern of four expressed genes, including PKM2, was reported to predict outcomes for mesothelioma patients undergoing surgery [37]. Events that negatively impact tumorigenesis can also reduce PKM2 function. Vitamins K3 and K5 inhibit tumorigenesis along with potently inhibiting PKM2 activity [38]. Butyrate displays anticolon cancer effects along with the inhibition of PKM2 expression in neoplastic but not nontumor colon tissues [39]. Shikonin, a derivative of a Chinese herb with antitumor activities, induces necrosis and

inhibits PKM2 expression in cancer cell lines [40]. A reverse correlation was observed between antitumor microRNA-326 and PKM2 in glioma [41]. Finally, the Spry2 tumor suppressor was reported to inhibit hepatocarcinogenesis via the MAPK and PKM2 pathways [42].

Furthermore, PKM2 possesses activities that directly promote tumorigenesis. Overexpression of PKM2 upregulates Bcl-xL in gastric cancer and promotes the proliferation and migration of colon cancer cells [43, 44]. Knockdown of PKM2 using specific siRNA inhibited cancer cell's proliferation and invasion in vitro and the formation of xenograft tumors in vivo [41, 45].

3. PKM2 Promotes Tumorigenesis via Regulating the Warburg Effect

The needs of energy production (ATP) and synthesis of cellular building blocks for proliferating cancer cells dictate the shift from oxidative to glycolytic metabolism even under normoxic conditions, the Warburg effect or aerobic glycolysis [2, 7, 8]. Under hypoxic conditions, cells metabolize glucose by anaerobic glycolysis, a process that is regulated by two master transcription factors, hypoxia-inducible factor (HIFs), and c-Myc [46]. Both transcriptional factors are also critical for aerobic glycolysis in cancer cells. Consistent with PKM2 being essential for aerobic glycolysis, a relationship exists among HIF-1, c-Myc, and PKM2. We will discuss the current understanding of these relationships.

3.1. Positive Feedback Regulation between PKM2 and HIF-1. It was first demonstrated by Christofk and colleagues in 2008 that knockdown of PKM2 in a panel of cancer cell lines decreased the rate of glycolysis and proliferation. Introducing PKM2 but not PKM1 to the knockdown cells not only enhanced glycolysis but also increased the ability to form xenograft tumors [12]. This research elegantly revealed that PKM2 is important and that the level of PK activity is essential, as the defects in PKM2 knockdown cells in supporting tumorigenesis could not be corrected by overexpression of the more active isoform PKM1. Furthermore, in comparison to PKM1 rescued cells, reintroducing PKM2 into knockdown cells rescued the deficiency of cell proliferation under hypoxic conditions.

This investigation also suggests that PKM2 may contribute to the adaptive response (hypoxia response) of cells to hypoxia, which is specifically relevant to tumorigenesis as solid cancers consistently face hypoxia intratumorally. It is thus a typical characteristic that cancers consistently execute hypoxia response. In the heart of this response lies the master transcription factor, hypoxia-inducible factor 1 (HIF-1) [47]. HIF-1 is a heterodimeric transcription factor, consisting of HIF-1α and HIF-1β. The β subunit is constitutively expressed, while the α subunit is directly regulated by oxygen (O$_2$) levels [48, 49]. Under normoxic conditions, HIF-1α is hydroxylated at prolines (P) 402 and 564 by three prolyl hydroxylase domain proteins (PHD1-3) in the presence of oxygen, α-ketoglutarate, iron, and ascorbate [50]. This results in the ubiquitination of prolyl-hydroxylated HIF-1α

by the von Hippel-Lindau (VHL) tumor suppressor and the subsequent degradation of HIF-1α [51, 52]. Under hypoxic conditions, HIF-1α is stabilized as a result of inhibiting prolyl hydroxylation, allowing HIF-1α to dimerize with HIF-1β in the nucleus. This leads to transcription of a set of genes to cope with reduced O_2 availability [53–55]. These target genes include those responsible for promoting glycolysis [56]. HIF-1 transactivates the glucose transporters GLUT1 and GLUT3, hexokinase (the first kinase in the glycolysis pathway), lactate dehydrogenase A (LDHA), and pyruvate dehydrogenase kinase 1 which phosphorylates and inhibits pyruvate dehydrogenase (PDH) [57] (Figure 1). Consistent with the Warburg effect's association with synthesis of cellular building blocks, HIF-1 also transactivates glucose-6-phosphate dehydrogenase (G6PD) to channel glucose-6-P into the pentose phosphate shunt for nucleotide and amino acid synthesis (Figure 1) [56]. Therefore, the collective actions of HIF-1 transcription activity seem to shift cells from oxidative metabolism to glycolysis (Figure 1). In line with these observations, PKM2 shares an intimate connection with HIF-1. The first intron of the *PKM2* gene contains the functional hypoxia-response element (HRE), thus also making it a target of HIF-1 [21].

PKM2 also possesses a positive feedback regulation towards HIF-1. PKM2 interacts with HIF-1α, a process that requires the prolyl hydroxylase 3 (PHD3). PHD3 binds to and causes hydroxylation of PKM2 at P303/408. This association and hydroxylation induces PKM2 to interact with HIF-1α, which plays a role in HIF-1-mediated transactivation of target genes including the *LDHA*, *PDK1*, and *VEGFA* (encoding the vascular endothelial growth factor) genes [21]. Additionally, PKM2 binds to p300 and enhances its recruitment to the HRE sites of HIF-1 target genes. Taken together, PKM2 functions as a HIF-1 coactivator by enhancing the Warburg effect in cancers [21, 22].

The regulation between HIF-1 and PKM2 also occurs under normoxic conditions, by changes in other signalling events which act to stabilize HIF-1α in cancer cells. HIF-1 is stabilized by mTOR and induced for degradation by VHL. Activation of mTOR is inhibited by tumor suppressors TSC1/TSC2 and facilitated by the PI3 K-AKT pathway [57]. Consistent with this knowledge, abnormal activation of the PI3 K-AKT-mTOR pathway and loss of function of tumor suppressors VHL, TSC1/2, and PTEN have been demonstrated to stabilize HIF-1α [57, 58]. Activation of mTOR by downregulation of TSC1/2 and PTEN induced PKM2 expression via stabilization of HIF-1α [59]. PKM2 makes essential contributions to mTOR-mediated aerobic glycolysis, as knockdown of PKM2 reduced glucose consumption and lactate production in cells with elevated mTOR activation. Furthermore, downregulation of PKM2 also suppressed mTOR-mediated tumorigenesis [59].

3.2. Positive Feedback Regulation between PKM2 and c-Myc. The *PKM2* gene produces both M1 and M2 isoforms through alternative splicing. The difference between these is the inclusion of exon 9 and exclusion of exon 10 for PKM1 and vice versa for PKM2 (Figure 2) [5, 15]. This mutually exclusive pattern of splicing is mediated by members of the

heterogeneous nuclear ribonucleoprotein (hnRNP) family, hnRNPA1, hnRNPA2, and hnRNP1/PTB (polypyrimidine track binding protein) [23, 60]. Binding of these proteins to the DNA sequence flanking exon 9 prevents its inclusion, resulting in the inclusion of exon 10 [23, 60, 61]. In order to achieve predominant expression of the M2 isoform, cancer cells have a strategy to preferentially splice the M2 isoform over M1 through c-Myc-mediated upregulation of hnRNPA1, hnRNPA2, and PTB (Figure 2) [23, 61]. This finding is supported by the discovery that cells with high levels of c-Myc activity also demonstrated high PKM2/PKM1 ratios [23, 62]. These observations are well in line with a large body of evidence indicating that c-Myc stimulates glycolysis and is required to coordinate with HIF-1 to regulate the cellular response to hypoxia [24, 46]. Thus, evidence suggests that PKM2 plays a role in c-Myc-mediated cancer metabolism and in c-Myc's communication with HIF-1. Adding to this attractive possibility is a recent demonstration that PKM2 also upregulates c-Myc transcription [63, 64], suggesting another positive feedback loop involving PKM2 in regulating the Warburg effect. Taken together, PKM2 is an integrated piece in the network of glycolysis regulation together with HIF-1 and c-Myc. The importance of hnRNPA1, hnRNPA2, and PTB in splicing PKM2 has also been explored by tumour suppression activity. The microRNAs mir-124, mir-137, and mir-340 inhibit colorectal cancer growth by repressing the expression of these hnRNAs favouring PKM1 splicing, thereby inhibiting aerobic glycolysis or the Warburg effect [65].

4. Regulation of PKM2 in the Warburg Effect during Tumorigenesis

Cancers have developed a complex regulation of PKM2 to meet the needs for energy and synthesis of nucleotides, amino acids, and lipids. These mechanisms center on regulating PKM2's expression, allosteric regulation, and modifications. The latter two mechanisms directly or indirectly affect PKM2 activity through physical interaction and by regulating the PKM2 dimer-tetramer dynamic.

4.1. Transcription Regulation. In addition to the above discussion of HIF-1 and c-Myc-mediated transcription and splicing of PKM2, transcription of the *PKM2* gene is also regulated by the SP1 and SP3 transcription factors [5, 22, 66]. The network of PI3 K-AKT-mTOR (mammalian target of rapamycin) plays a critical role in cell metabolism, proliferation, and survival and is one of the most frequently activated pathways in cancer owing to the activation of kinases and the inactivation of tumor suppressors, TSC1/2 (tuberous sclerosis 1/2) and PTEN [67]. Nutrient status is well known to modulate mTOR activation [68]. Under normoxic conditions, mTOR activity induces PKM2 expression through the combination of HIF-1α and c-Myc [59, 69]. Inhibition of mTOR has been found to reduce glycolysis and PKM2 expression [70]. Elevation in PTEN function reduces glucose uptake and the Warburg effect and inhibits PKM2 expression [25]. In a feedback

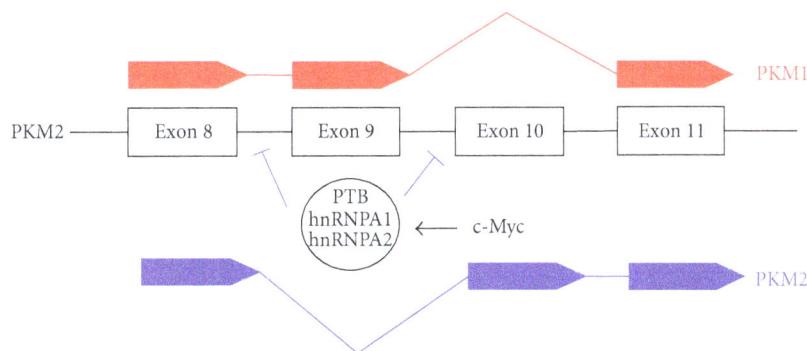

FIGURE 2: Schematic illustration of alternative splicing of PKM1 and PKM2. The proportion of the *PKM2* gene is shown. c-Myc upregulates the indicated complex which inhibits the splicing for exon 9, resulting in its exclusion in PKM2. PTB: polypyrimidine track binding protein; hnRNPA1 and hnRNPA2: heterogeneous nuclear ribonucleoprotein 1 and 2.

manner, PKM2 is able to sustain mTOR activation in serine-depleted medium by enhancing endogenous serine synthesis [71]. Taken together, evidence supports that the upregulation of PKM2 plays an important role in the mTOR-mediated Warburg effect in tumors.

4.2. Regulation of the Dimer-Tetramer Dynamics. Tumor cells express high levels of dimer PKM2 [14, 32]. Among the four PK isoforms, PKM2 is the only one to be allosterically regulated between a less active dimer and an active tetramer [18, 19]. These different forms of PKM2 regulate glucose metabolism through either the TCA cycle or glycolysis. Accumulating evidence supports the concept that the less active PKM2 dimer drives aerobic glycolysis, while the active PKM2 tetramer produces pyruvate for oxidative phosphorylation (Figure 1) [12, 72–74]. PKM2 is regulated by fructose-1,6-biphosphate (FBP), an upstream intermediate of glycolysis which when bound to PKM2 activates tetramerization through high affinity association [75–77]. Binding of tyrosine-phosphorylated peptides dissociates FBP from the PKM2 tetramer, resulting in conversion to the PKM2 dimer [72]. The less active PKM2 dimer is critical in mediating aerobic glycolysis in tumor cells based on high levels of lactate production and lower oxygen consumption [72]. Disrupting the binding of the phosphotyrosine peptide in a PKM2 mutant (M2KE) increased PKM2 kinase activity, which was associated with reduction in lactate production and elevation of oxygen consumption [72]. In supporting the low levels of cellular pyruvate kinase activity being critical for aerobic glycolysis, replacing PKM2 with PKM1 led to an increase in cellular pyruvate kinase activity, decreasing lactate production and elevating oxygen consumption [12, 74].

Collectively, evidence supports that the PKM2 dimer is critical in mediating aerobic glycolysis. In addition to the above mechanism regulating PKM2 activity, PKM2 was also controlled by tyrosine phosphorylation [73]. It was observed in 1988 that PKM2 was tyrosine-phosphorylated in v-Src-transformed chicken embryo cells. This phosphorylation reduced the affinity of PKM2 towards its substrate phosphoenolpyruvate (PEP) [78]. In vitro, v-Src was able to directly phosphorylate PKM2 [78]. Although this investigation

suggested that v-Src phosphorylated PKM2, the sites of phosphorylation remain unknown. Recent development demonstrated that PKM2 was phosphorylated at several tyrosine residues, including Y105, by fibroblast growth factor receptor type 1 (FGFR1) [73]. Phosphorylation at Y105 causes FBP to dissociate from the PKM2 tetramer, which results in PKM2 dimers and promotes the Warburg effect based on the production of lactate [73]. Conversely, abolishing Y105 phosphorylation by substitution with phenylalanine (Y105F) elevated the kinase activity, resulting in decreased lactate production and increased oxygen consumption [73]. Taken together, evidence demonstrates that phosphorylation at Y105 plays a role in the conversion of PKM2 tetramers to dimers.

More importantly, regulation of PKM2 dimer and tetramer conversion is critical for tumorigenesis. While the less active PKM2 dimer enhances xenograft tumor formation, enforced formation of active PKM2 (KE and Y105F mutations) and replacing PKM2 with PKM1 inhibited the formation of xenograft tumors [72, 73, 79]. In line with this concept, the conversion between dimer and tetramer PKM2 is also used in tumour suppression to inhibit tumorigenesis. The death-associated protein kinase (DAPK) tumor suppressor activates PKM2 by stabilizing the PKM2 tetramer via a direct association. This reduces cancer metabolism or the Warburg effect, which may be one aspect of DAPK-mediated tumor suppression [80, 81].

In line with these observations, several small molecule PKM2 activators have been identified. Among them, DASA-58 (the substituted N, N′-diarylsulfonamide NCGC00185916) and TEPP-46 (the thieno-[3,2-b]pyrrole [3,2-d]pyridazinone NCGC00186528) activate PKM2 by inducing PKM2 tetramerization. Unlike FBP-induced activation, the tetramer induced by these compounds is resistant to tyrosine-phosphorylated peptide-mediated conversion to the PKM2 dimer. This suggests that FBP and these small molecule activators bind PKM2 at distinct sites, but, importantly, all inhibit tumorigenesis [74, 82, 83]. Additionally, a new set of chemical platform bases, the quinolone sulfonamide-based PKM2 activators, have recently been reported. Similar to DASA-58 and TEPP-46, these activators also stabilize the PKM2 tetramer via binding to a pocket distinct from FBP

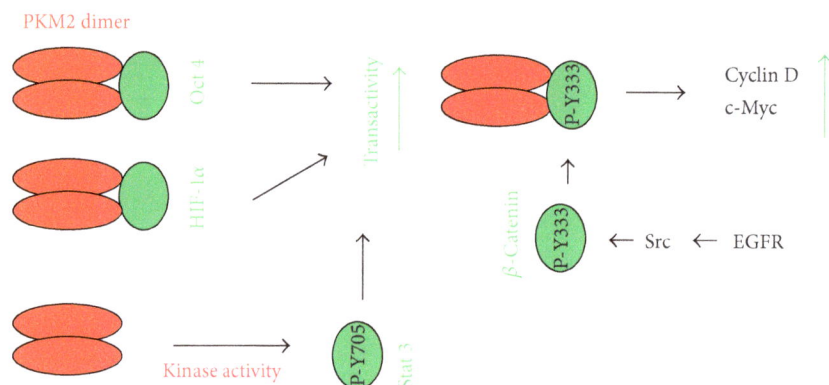

FIGURE 3: Diagram showing the nuclear function of PKM2. PKM2 dimers in the nucleus bind to Oct 4 and HIF-1α and enhance their transcription activity; EGFR signal activates Src tyrosine kinase, which phosphorylates β-catenin at tyrosine (Y) 333 (P-Y333). PKM2 binds Y333-phosphorylated β-catenin, contributing to β-catenin-mediated transcription of cyclin D and c-Myc; PKM2 dimer possesses kinase activity that phosphorylates Stat 3 at Y705, which enhances Stat 3's transcriptional activity.

binding and thus prevent the PKM2 tetramer from tyrosine-phosphorylated peptide-mediated disruption. Quinolone sulfonamide-based PKM2 activators reduce carbon flow towards the serine biosynthetic pathway, rendering cells to serine auxotrophy [84].

4.3. Factors Affecting PKM2 Activity via Physical Association. In addition to the above two small molecule PKM2 activators, a third activator was recently reported by the same research group based on modifications to one of their previous compounds [85]. The mechanism underlying this activation remains to be defined. A series of PKM2 activators (1-(sulfonyl)-5-(arylsulfonyl)indoline) were also reported very recently [86]. In contrast to these, potent small molecule PKM2 inhibitors which may in part induce cell death by inhibiting PKM2 activity have also been developed [87]. Furthermore, the pyruvate kinase activity of PKM2 can be inhibited by association with several distinct proteins. While the nuclear promyelocytic leukemia (PML) protein functions as a tumor suppressor, cytosolic PML was reported to specifically inhibit tetrameric but not dimeric PKM2 activity, thereby contributing to the Warburg effect [88]. Prolactin signal promotes cell proliferation by inducing its receptor to associate with PKM2, leading to PKM2 activity reduction [89]. The MUC1-C oncoprotein was reported to promote breast cancer tumorigenesis in part via inhibiting PKM2 activity. Although interaction of MUC1-C Cys3 with PKM2 C-domain Cys474 results in activation of PKM2, oncogenic signals from EGFR (epidermal growth factor receptor) can alter the association of MUC1-C and PKM2, thereby leading to inhibition of PKM2 activity [90]. EGFR phosphorylates MUC1-C at tyrosine 46, causing MUC1-C to interact with PKM2 at Lys433. This association inhibits tetrameric PKM2 activity and thereby increases aerobic glycolysis along with glucose uptake [90]. PKM2 was also found to interact with human papillomavirus 16 (HPV16) protein E7, which may contribute to HPV16-induced cervical cancer [91]. A potential therapeutic protein TEM8-Fc, consisting of a portion of the tumor endothelial marker 8 (TEM8)

and the Fc domain of human IgG1, was found to associate with PKM2 [92]. Whether this interaction contributed to TEM8-Fc-associated tumor suppression was not clear [92]. Consistent with the knowledge that PKM2 plays a critical role in regulating aerobic glycolysis and biosynthesis for cellular building blocks, PKM2 is activated by serine but inhibited by alanine and phenylalanine when bound to these amino acids [93].

4.4. Posttranslational Modifications of PKM2. A reduction in activity was reported by acetylation of PKM2 at lysine (K) 305 in response to high levels of glucose. This modification reduces PKM2 activity and its affinity towards the PEP substrate, resulting in PKM2 degradation via chaperone-mediated autophagy [94]. As a result, acetylation enhances cell proliferation by increasing the availability of glycolytic intermediates for anabolic synthesis [94, 95].

PKM2 also plays a role in cell survival to oxidative stress. Acute increases in intracellular levels of ROS (reactive oxygen species) induce oxidation of PKM2 at Cys358. This reduces PKM2 activity, which allows the accumulation of glucose-6-phosphate and thus shifts glucose flux through the pentose phosphate pathway (PPP) to generate reduced NADPH (Figure 1). As PPP is the major pathway of generating reduced NADPH, oxidation-mediated inhibition of PKM2 is therefore a mechanism of detoxification during oxidative stress. Consistent with this notion, substitution of C358 with S358 to produce oxidation-resistant mutants sensitized cells to oxidative stress and inhibited xenograft tumor formation [96]. A similar antioxidative stress function of PKM2 is also mediated through binding to CD44, a major cell adhesion molecule. Cancer stem cells are known to be CD44 positive, so this interaction is consistent with CD44 promoting cancer progression, metastasis, and chemoresistance [97, 98]. CD44's tumorigenic function is in part also attributable to its association with EGFR [99]. Consistent with these observations, PKM2 was reported to bind CD44, resulting in receptor tyrosine kinase-mediated phosphorylation of PKM2 and inhibition of PKM2 activity. This enhanced glucose flux through the PPP pathway to generate reduced NADPH and counteract oxidative stresses through detoxification [61, 100].

5. The Nuclear Function of PKM2

PKM2 displays intriguing nonglycolytic functions in the nucleus. In addition to its cytoplasmic presence to regulate aerobic glycolysis, PKM2 was also detected in the nucleus in response to interleukin-3 and apoptotic signals [101, 102]. Nuclear PKM2 binds Oct 4 through its C-terminal region (residues 307–531), enhancing Oct-4-mediated transcription [103] (Figure 3). Nuclear PKM2 was also reported to be a coactivator of HIF-1 [21] (Figure 3). EGFR signaling was reported to activate Src tyrosine kinase, which in turn phosphorylates β-catenin at Y333. PKM2 binds to tyrosine-phosphorylated β-catenin in the nucleus and contributes to β-catenin-mediated transactivation of cyclin D and c-Myc, thereby promoting both cell proliferation and tumor progression (Figure 3). This process requires the kinase activity of PKM2 [63, 104]. Since the binding of tyrosine-phosphorylated peptides maintains PKM2 in its dimer status [72], these observations suggest that dimerized PKM2 binds and enhances β-catenin function, in which a new kinase activity rather than pyruvate kinase activity might be involved. Indeed, it was very recently reported that the PKM2 dimer contributes to its nuclear function and possesses protein tyrosine kinase activity. Surprisingly, instead of using high-energy ATP, PKM2 uses the high-energy phosphate from PEP as a phosphate donor to phosphorylate its protein substrates [26]. The PKM2 dimer phosphorylates Stat 3 at Y705 in the nucleus and thus enhances Stat 3 transcription activity [26, 105] (Figure 3). Taken together, while tetramer PKM2 is a pyruvate kinase, dimer PKM2 can also act as a protein tyrosine kinase [26].

6. Concluding Remarks and Future Perspectives

The last decade has seen a high reemergence of interest in the Warburg effect, the typical cancer cell metabolism that was reported almost 90 years ago. The detailed molecular and genetic knowledge accumulated in the last few decades of extensive cancer research has rapidly advanced our understanding of cancer metabolism. Mutations in several enzymes of the TCA-cycle were discovered, including isocitrate dehydrogenases 1 and 2 (IDH1 and IDH2), succinate dehydrogenase (SDH), and fumarate hydratase (FH) [106–109]. These mutations collectively reduce TCA-cycle-mediated oxidative phosphorylation, resulting in an accumulation of metabolites for the biosynthesis of amino acids, nucleotides, and lipids as well as increases in glucose uptake [110]. The increases in glucose uptake together with aerobic glycolysis yield a robust elevation of lactate production. Although recent development suggests that the by-product of aerobic glycolysis (lactate) contributes to overall tumorigenesis [111, 112], it is also critical for cancer cells to efficiently export lactate to maintain the flux of glycolysis and to prevent cellular acidification [112]. Cancer cells accomplish this task in part by upregulation

of the monocarboxylate transporters (MCTs) [112]. Another strategy to reduce the cellular burden of lactate accumulation during aerobic glycolysis may be the prevention of a complete conversion of glucose to lactate (1 glucose for every 2 lactate molecules) by reducing the conversion of PEP to pyruvate. This would allow the glycolytic intermediates to be used for macromolecular synthesis. Therefore, predominantly using the less active PKM2 dimer fits this logic.

Accumulating evidence obtained in the last 10 years demonstrates that PKM2's glycolytic enzyme activity is regulated by oncogenes and tumor suppressors [21–25]. These regulations center on modulation of aerobic glycolysis. Favoring a shift of the dimer-tetramer dynamic towards dimerization is critical for PKM2 to promote the Warburg effect, leading to cell proliferation and tumorigenesis.

Surprisingly, in addition to its glycolytic pyruvate kinase activity in the cytosol, the PKM2 dimer also displays protein tyrosine kinase activity in the nucleus and nuclear PKM2 promotes the transcriptional activities of HIF, β-catenin, STAT 3, and Oct 4 [21, 26, 63, 103–105]. This all indirectly contributes to cancer metabolism and other aspects of tumorigenesis. In light of this new development, future research should determine the contributions of the cytosolic versus nuclear PKM2 dimer to aerobic glycolysis.

Effort is currently underway to target PKM2 for cancer therapy, which is part of the current attempt in targeting cancer metabolism. Several small molecule PKM2 inhibitors and activators have been developed [61]. As nearly complete knockdown of PKM2 does not completely inhibit cancer cell proliferation, the utility of PKM2 inhibition in targeting cancer should be cautious [61]. On the other hand, small molecule activators might be an attractive approach. However, several factors call for precautions in targeting PKM2. (1) PKM2 is also expressed in normal tissue [16, 17] and the function of PKM2 in normal tissues has not yet been determined; (2) genetic changes in PKM2 have not been reported in primary cancers; (3) despite modulation of PKM2 which affects formation of xenograft tumors, whether tissue-specific manipulation of PKM2 impacts tumorigenesis is still on the waiting list; (4) as PKM2 was detected in cancer stroma [113, 114], whether it plays a role in tumorigenesis by affecting cancer-associated fibroblasts is not clear; (5) while aerobic glycolysis has been a hot topic in the last decade, its impact on cancer stem cells (CSCs) has not been addressed. As it is becoming increasingly clear that CSCs play a critical role in tumorigenesis, especially in tumor progression and metastasis [115], it would appear critical to understand whether targeting cancer metabolism in general and PKM2 in particular will have an inhibitory effect on CSCs. This knowledge became important as it was suggested that glioma CSCs (GSCs) may not use aerobic glycolysis to the same degree as differentiated cancer cells. Thus, targeting PKM2 or cancer metabolism may still spare GSCs [116].

Acknowledgments

This work was in part supported by a Grant from Kidney Foundation of Canada (KFOC11001 7, 2011–2013) to D.

Tang and by St. Joseph's Healthcare Hamilton, ON, Canada, through financial support to the Hamilton Centre for Kidney Research (HCKR).

References

[1] O. Warburg, F. Wind, and E. Negelein, "The metabolism of tumors in the body," *The Journal of General Physiology*, vol. 8, no. 6, pp. 519–530, 1927.

[2] O. Warburg, "On the origin of cancer cells," *Science*, vol. 123, no. 3191, pp. 309–314, 1956.

[3] R. Curi, P. Newsholme, and E. A. Newsholme, "Metabolism of pyruvate by isolated rat mesenteric lymphocytes, lymphocyte mitochondria and isolated mouse macrophages," *Biochemical Journal*, vol. 250, no. 2, pp. 383–388, 1988.

[4] T. Pfeiffer, S. Schuster, and S. Bonhoeffer, "Cooperation and competition in the evolution of ATP-producing pathways," *Science*, vol. 292, no. 5516, pp. 504–507, 2001.

[5] B. Chaneton and E. Gottlieb, "Rocking cell metabolism: revised functions of the key glycolytic regulator PKM2 in cancer," *Trends in Biochemical Sciences*, vol. 37, no. 8, pp. 309–316, 2012.

[6] P. E. Porporato, S. Dhup, R. K. Dadhich, T. Copetti, and P. Sonveaux, "Anticancer targets in the glycolytic metabolism of tumors: a comprehensive review," *Frontiers in Pharmacology*, vol. 2, p. 49, 2011.

[7] P. P. Hsu and D. M. Sabatini, "Cancer cell metabolism: Warburg and beyond," *Cell*, vol. 134, no. 5, pp. 703–707, 2008.

[8] M. G. V. Heiden, L. C. Cantley, and C. B. Thompson, "Understanding the warburg effect: the metabolic requirements of cell proliferation," *Science*, vol. 324, no. 5930, pp. 1029–1033, 2009.

[9] L. Stryer, *Glycolysis in Biochemistry*, W.H. Freemen, New York, NY, USA, 1995.

[10] M. E. Muñoz and E. Ponce, "Pyruvate kinase: current status of regulatory and functional properties," *Comparative Biochemistry and Physiology B*, vol. 135, no. 2, pp. 197–218, 2003.

[11] G. Valentini, L. Chiarelli, R. Fortini, M. L. Speranza, A. Galizzi, and A. Mattevi, "The allosteric regulation of pyruvate kinase: a site-directed mutagenesis study," *Journal of Biological Chemistry*, vol. 275, no. 24, pp. 18145–18152, 2000.

[12] H. R. Christofk, M. G. Vander Heiden, M. H. Harris et al., "The M2 splice isoform of pyruvate kinase is important for cancer metabolism and tumour growth," *Nature*, vol. 452, no. 7184, pp. 230–233, 2008.

[13] T. Noguchi, K. Yamada, H. Inoue, T. Matsuda, and T. Tanaka, "The L- and R-type isozymes of rat pyruvate kinase are produced from a single gene by use of different promoters," *Journal of Biological Chemistry*, vol. 262, no. 29, pp. 14366–14371, 1987.

[14] K. Yamada and T. Noguchi, "Nutrient and hormonal regulation of pyruvate kinase gene expression," *Biochemical Journal*, vol. 337, no. 1, pp. 1–11, 1999.

[15] T. Noguchi, H. Inoue, and T. Tanaka, "The M1- and M2-type isozymes of rat pyruvate kinase are produced from the same gene by alternative RNA splicing," *Journal of Biological Chemistry*, vol. 261, no. 29, pp. 13807–13812, 1986.

[16] K. Bluemlein, N. M. Grüning, R. G. Feichtinger, H. Lehrach, B. Kofler, and M. Ralser, "No evidence for a shift in pyruvate kinase PKM1 to PKM2 expression during tumorigenesis," *Oncotarget*, vol. 2, no. 5, pp. 393–400, 2011.

[17] K. Imamura and T. Tanaka, "Multimolecular forms of pyruvate kinase from rat and other mammalian tissues: I. electrophoretic studies," *Journal of Biochemistry*, vol. 71, no. 6, pp. 1043–1051, 1972.

[18] S. Mazurek, C. B. Boschek, F. Hugo, and E. Eigenbrodt, "Pyruvate kinase type M2 and its role in tumor growth and spreading," *Seminars in Cancer Biology*, vol. 15, no. 4, pp. 300–308, 2005.

[19] S. Mazurek, "Pyruvate kinase type M2: a key regulator of the metabolic budget system in tumor cells," *International Journal of Biochemistry and Cell Biology*, vol. 43, no. 7, pp. 969–980, 2011.

[20] W. Luo and G. L. Semenza, "Pyruvate kinase M2 regulates glucose metabolism by functioning as a coactivator for hypoxia-inducible factor 1 in cancer cells," *Oncotarget*, vol. 2, no. 7, pp. 551–556, 2011.

[21] W. Luo, H. Hu, R. Chang et al., "Pyruvate kinase M2 is a PHD3-stimulated coactivator for hypoxia-inducible factor 1," *Cell*, vol. 145, no. 5, pp. 732–744, 2011.

[22] W. Luo and G. L. Semenza, "Emerging roles of PKM2 in cell metabolism and cancer progression," *Trends in Endocrinology and Metabolism*, vol. 23, no. 11, pp. 560–566, 2012.

[23] C. J. David, M. Chen, M. Assanah, P. Canoll, and J. L. Manley, "HnRNP proteins controlled by c-Myc deregulate pyruvate kinase mRNA splicing in cancer," *Nature*, vol. 463, no. 7279, pp. 364–368, 2010.

[24] D. M. Miller, S. D. Thomas, A. Islam, D. Muench, and K. Sedoris, "c-Myc and cancer metabolism," *Clinical Cancer Research*, vol. 18, no. 20, pp. 5546–5553, 2012.

[25] I. Garcia-Cao, M. S. Song, R. M. Hobbs et al., "Systemic elevation of PTEN induces a tumor-suppressive metabolic state," *Cell*, vol. 149, no. 1, pp. 49–62, 2012.

[26] X. Gao, H. Wang, J. J. Yang, X. Liu, and Z. R. Liu, "Pyruvate kinase M2 regulates gene transcription by acting as a protein kinase," *Molecular Cell*, vol. 45, no. 5, pp. 598–609, 2012.

[27] U. Brinck, E. Eigenbrodt, M. Oehmke, S. Mazurek, and G. Fischer, "L- and M2- pyruvate kinase expression in renal cell carcinomas and their metastases," *Virchows Archiv*, vol. 424, no. 2, pp. 177–185, 1994.

[28] J. Schneider, K. Neu, H. Grimm, H. G. Velcovsky, G. Weisse, and E. Eigenbrodt, "Tumor M2-pyruvate kinase in lung cancer patients: immunohistochemical detection and disease monitoring," *Anticancer Research*, vol. 22, no. 1A, pp. 311–318, 2002.

[29] H. W. Wechsel, E. Petri, K. H. Bichler, and G. Feil, "Marker for renal cell carcinoma (RCC): the dimeric form of pyruvate kinase type M2 (Tu M2-PK)," *Anticancer Research*, vol. 19, no. 4A, pp. 2583–2590, 1999.

[30] G. M. Oremek, S. Teigelkamp, W. Kramer, E. Eigenbrodt, and K. H. Usadel, "The pyruvate kinase isoenzyme tumor M2 (Tu M2-PK) as a tumor marker for renal carcinoma," *Anticancer Research*, vol. 19, no. 4A, pp. 2599–2601, 1999.

[31] T. Pottek, M. Müller, T. Blum, and M. Hartmann, "Tu-M2-PK in the blood of testicular and cubital veins in men with testicular cancer," *Anticancer Research*, vol. 20, no. 6D, pp. 5029–5033, 2000.

[32] E. Eigenbrodt, D. Basenau, S. Holthusen, S. Mazurek, and G. Fischer, "Quantification of tumor type M2 pyruvate kinase (Tu M2-PK) in human carcinomas," *Anticancer Research*, vol. 17, no. 4B, pp. 3153–3156, 1997.

[33] D. Lüftner, J. Mesterharm, C. Akrivakis et al., "Tumor type M2 pyruvate kinase expression in advanced breast cancer," *Anticancer Research*, vol. 20, no. 6D, pp. 5077–5082, 2000.

[34] J. Roigas, G. Schulze, S. Raytarowski, K. Jung, D. Schnorr, and S. A. Loening, "Tumor M2 pyruvate kinase in plasma of patients with urological tumors," *Tumor Biology*, vol. 22, no. 5, pp. 282–285, 2001.

[35] P. D. Hardt, M. Toepler, B. Ngoumou, J. Rupp, and H. U. Kloer, "Measurement of fecal pyruvate kinase type M2 (Tumor M2-PK) concentrations in patients with gastric cancer, colorectal cancer, colorectal adenomas and controls," *Anticancer Research*, vol. 23, no. 2A, pp. 851–853, 2003.

[36] J. Y. Lim, S. O. Yoon, S. Y. Seol et al., "Overexpression of the M2 isoform of pyruvate kinase is an adverse prognostic factor for signet ring cell gastric cancer," *World Journal of Gastroenterology*, vol. 18, no. 30, pp. 4037–4043, 2012.

[37] G. J. Gordon, L. Dong, B. Y. Yeap et al., "Four-gene expression ratio test for survival in patients undergoing surgery for mesothelioma," *Journal of the National Cancer Institute*, vol. 101, no. 9, pp. 678–686, 2009.

[38] J. Chen, Z. Jiang, B. Wang, Y. Wang, and X. Hu, "Vitamin K(3) and K(5) are inhibitors of tumor pyruvate kinase M2," *Cancer Letters*, vol. 316, no. 2, pp. 204–210, 2012.

[39] F. Jahns, A. Wilhelm, K. O. Greulich et al., "Impact of butyrate on PKM2 and HSP90β expression in human colon tissues of different transformation stages: a comparison of gene and protein data," *Genes and Nutrition*, vol. 7, no. 2, pp. 235–246, 2012.

[40] J. Chen, J. Xie, Z. Jiang, B. Wang, Y. Wang, and X. Hu, "Shikonin and its analogs inhibit cancer cell glycolysis by targeting tumor pyruvate kinase-M2," *Oncogene*, vol. 30, no. 42, pp. 4297–4306, 2012.

[41] B. Kefas, L. Comeau, N. Erdle, E. Montgomery, S. Amos, and B. Purow, "Pyruvate kinase M2 is a target of the tumorsuppressive microRNA-326 and regulates the survival of glioma cells," *Neuro-Oncology*, vol. 12, no. 11, pp. 1102–1112, 2010.

[42] C. Wang, S. Delogu, C. Ho et al., "Inactivation of Spry2 accelerates AKT-driven hepatocarcinogenesis via activation of MAPK and PKM2 pathways," *Journal of Hepatology*, vol. 57, no. 3, pp. 577–583, 2012.

[43] O. H. Kwon, T. W. Kang, J. H. Kim et al., "Pyruvate kinase M2 promotes the growth of gastric cancer cells via regulation of Bcl-xL expression at transcriptional level," *Biochemical and Biophysical Research Communications*, vol. 423, no. 1, pp. 38–44, 2012.

[44] C. F. Zhou, X. B. Li, H. Sun et al., "Pyruvate kinase type M2 is upregulated in colorectal cancer and promotes proliferation and migration of colon cancer cells," *International Union of Biochemistry and Molecular Biology Life*, vol. 64, no. 9, pp. 775–782, 2012.

[45] M. S. Goldberg and P. A. Sharp, "Pyruvate kinase M2-specific siRNA induces apoptosis and tumor regression," *Journal of Experimental Medicine*, vol. 209, no. 2, pp. 217–224, 2012.

[46] J. D. Gordan, C. B. Thompson, and M. C. Simon, "HIF and c-Myc: sibling rivals for control of cancer cell metabolism and proliferation," *Cancer Cell*, vol. 12, no. 2, pp. 108–113, 2007.

[47] W. Liu, S. M. Shen, X. Y. Zhao, and G. Q. Chen, "Targeted genes and interacting proteins of hypoxia inducible factor-1," *International Journal of Biochemistry and Cell Biology*, vol. 3, no. 2, pp. 165–178, 2012.

[48] G. L. Wang, B. H. Jiang, E. A. Rue, and G. L. Semenza, "Hypoxia-inducible factor 1 is a basic-helix-loop-helix-PAS heterodimer regulated by cellular O_2 tension," *Proceedings of the National Academy of Sciences of the United States of America*, vol. 92, no. 12, pp. 5510–5514, 1995.

[49] G. L. Semenza, "Hypoxia-inducible factor 1: master regulator of O_2 homeostasis," *Current Opinion in Genetics and Development*, vol. 8, no. 5, pp. 588–594, 1998.

[50] A. C. R. Epstein, J. M. Gleadle, L. A. McNeill et al., "C. elegans EGL-9 and mammalian homologs define a family of dioxygenases that regulate HIF by prolyl hydroxylation," *Cell*, vol. 107, no. 1, pp. 43–54, 2001.

[51] P. Jaakkola, D. R. Mole, Y. M. Tian et al., "Targeting of HIF-α to the von Hippel-Lindau ubiquitylation complex by O_2-regulated prolyl hydroxylation," *Science*, vol. 292, no. 5516, pp. 468–472, 2001.

[52] M. Ivan, K. Kondo, H. Yang et al., "HIFα targeted for VHL-mediated destruction by proline hydroxylation: implications for O_2 sensing," *Science*, vol. 292, no. 5516, pp. 464–468, 2001.

[53] W. G. Kaelin and P. J. Ratcliffe, "Oxygen sensing by metazoans: the central role of the HIF hydroxylase pathway," *Molecular Cell*, vol. 30, no. 4, pp. 393–402, 2008.

[54] L. P. Song, J. Zhang, S. F. Wu et al., "Hypoxia-inducible factor-1α-induced differentiation of myeloid leukemic cells is its transcriptional activity independent," *Oncogene*, vol. 27, no. 4, pp. 519–527, 2008.

[55] R. H. Wenger and M. Gassmann, "Oxygen(es) and the hypoxia-inducible factor-1," *Biological Chemistry*, vol. 378, no. 7, pp. 609–616, 1997.

[56] T. W. Meijer, J. H. Kaanders, P. N. Span, and J. Bussink, "Targeting hypoxia, HIF-1, and tumor glucose metabolism to improve radiotherapy efficacy," *Clinical Cancer Research*, vol. 18, no. 20, pp. 5585–5594, 2012.

[57] G. L. Semenza, "HIF-1: upstream and downstream of cancer metabolism," *Current Opinion in Genetics and Development*, vol. 20, no. 1, pp. 51–56, 2010.

[58] E. Laughner, P. Taghavi, K. Chiles, P. C. Mahon, and G. L. Semenza, "HER2 (neu) signaling increases the rate of hypoxia-inducible factor 1α (HIF-1α) synthesis: novel mechanism for HIF-1-mediated vascular endothelial growth factor expression," *Molecular and Cellular Biology*, vol. 21, no. 12, pp. 3995–4004, 2001.

[59] Q. Sun, X. Chen, J. Ma et al., "Mammalian target of rapamycin up-regulation of pyruvate kinase isoenzyme type M2 is critical for aerobic glycolysis and tumor growth," *Proceedings of the National Academy of Sciences of the United States of America*, vol. 108, no. 10, pp. 4129–4134, 2011.

[60] M. Chen, J. Zhang, and J. L. Manley, "Turning on a fuel switch of cancer: hnRNP proteins regulate alternative splicing of pyruvate kinase mRNA," *Cancer Research*, vol. 70, no. 22, pp. 8977–8980, 2010.

[61] M. Tamada, M. Suematsu, and H. Saya, "Pyruvate kinase m2: multiple faces for conferring benefits on cancer cells," *Clinical Cancer Research*, vol. 18, no. 20, pp. 5554–5561, 2012.

[62] C. V. Clower, D. Chatterjee, Z. Wang, L. C. Cantley, M. G. V. Heidena, and A. R. Krainer, "The alternative splicing repressors hnRNP A1/A2 and PTB influence pyruvate kinase isoform expression and cell metabolism," *Proceedings of the National Academy of Sciences of the United States of America*, vol. 107, no. 5, pp. 1894–1899, 2010.

[63] W. Yang, Y. Xia, H. Ji et al., "Nuclear PKM2 regulates β-catenin transactivation upon EGFR activation," *Nature*, vol. 480, no. 7375, pp. 118–122, 2011.

[64] I. Harris, S. McCracken, and T. W. Mak, "PKM2: a gatekeeper between growth and survival," *Cell Research*, vol. 22, no. 3, pp. 447–449, 2012.

[65] Y. Sun, X. Zhao, Y. Zhou, and Y. Hu, "miR-124, miR-137 and miR-340 regulate colorectal cancer growth via inhibition of the Warburg effect," *Oncology Reports*, vol. 28, no. 4, pp. 1346–1352, 2012.

[66] D. J. Discher, N. H. Bishopric, X. Wu, C. A. Peterson, and K. A. Webster, "Hypoxia regulates β-enolase and pyruvate kinase-M promoters by modulating Sp1/Sp3 binding to a conserved GC

element," *Journal of Biological Chemistry*, vol. 273, no. 40, pp. 26087–26093, 1998.

[67] J. A. McCubrey, L. S. Steelman, W. H. Chappell et al., "Mutations and deregulation of Ras/Raf/MEK/ERK and PI3K/PTEN/Akt/mTOR cascades which alter therapy response," *Oncotarget*, vol. 3, no. 9, pp. 954–987, 2012.

[68] A. Efeyan and D. M. Sabatini, "MTOR and cancer: many loops in one pathway," *Current Opinion in Cell Biology*, vol. 22, no. 2, pp. 169–176, 2010.

[69] J. P. Bayley and P. Devilee, "The Warburg effect in 2012," *Current Opinion in Oncology*, vol. 24, no. 1, pp. 62–67, 2012.

[70] M. A. Iqbal and R. N. Bamezai, "Resveratrol inhibits cancer cell metabolism by down regulating pyruvate kinase M2 via inhibition of mammalian target of rapamycin," *PLoS ONE*, vol. 7, no. 5, Article ID e36764, 2012.

[71] J. Ye, A. Mancuso, X. Tong et al., "Pyruvate kinase M2 promotes de novo serine synthesis to sustain mTORC1 activity and cell proliferation," *Proceedings of the National Academy of Sciences of the United States of America*, vol. 109, no. 18, pp. 6904–6909, 2012.

[72] H. R. Christofk, M. G. Vander Heiden, N. Wu, J. M. Asara, and L. C. Cantley, "Pyruvate kinase M2 is a phosphotyrosine-binding protein," *Nature*, vol. 452, no. 7184, pp. 181–186, 2008.

[73] T. Hitosugi, S. Kang, M. G. Vander Heiden et al., "Tyrosine phosphorylation inhibits PKM2 to promote the warburg effect and tumor growth," *Science Signaling*, vol. 2, no. 97, p. ra73, 2009.

[74] D. Anastasiou, Y. Yu, W. J. Israelsen et al., "Pyruvate kinase M2 activators promote tetramer formation and suppress tumorigenesis," *Nature Chemical Biology*, vol. 8, no. 10, pp. 839–847, 2012.

[75] K. Ashizawa, P. McPhie, K. H. Lin, and S. Y. Cheng, "An in vitro novel mechanism of regulating the activity of pyruvate kinase M2 by thyroid hormone and fructose 1,6-bisphosphate," *Biochemistry*, vol. 30, no. 29, pp. 7105–7111, 1991.

[76] K. Ashizawa, M. C. Willingham, C. M. Liang, and S. Y. Cheng, "In vivo regulation of monomer-tetramer conversion of pyruvate kinase subtype M2 by glucose is mediated via fructose 1,6-bisphosphate," *Journal of Biological Chemistry*, vol. 266, no. 25, pp. 16842–16846, 1991.

[77] J. D. Dombrauckas, B. D. Santarsiero, and A. D. Mesecar, "Structural basis for tumor pyruvate kinase M2 allosteric regulation and catalysis," *Biochemistry*, vol. 44, no. 27, pp. 9417–9429, 2005.

[78] P. Presek, M. Reinacher, and E. Eigenbrodt, "Pyruvate kinase type M2 is phosphorylated at tyrosine residues in cells transformed by Rous sarcoma virus," *FEBS Letters*, vol. 242, no. 1, pp. 194–198, 1988.

[79] C. V. Dang, "PKM2 tyrosine phosphorylation and glutamine metabolism signal a different view of the warburg effect," *Science Signaling*, vol. 2, no. 97, p. pe75, 2009.

[80] A. Erol, "Death-associated proliferation kinetic in normal and transformed cells," *Cell Cycle*, vol. 11, no. 8, pp. 1512–1516, 2012.

[81] I. Mor, R. Carlessi, T. Ast, E. Feinstein, and A. Kimchi, "Death-associated protein kinase increases glycolytic rate through binding and activation of pyruvate kinase," *Oncogene*, vol. 31, no. 6, pp. 683–693, 2012.

[82] M. B. Boxer, J. K. Jiang, M. G. Vander Heiden et al., "Evaluation of substituted N,N'-diarylsulfonamides as activators of the tumor cell specific M2 isoform of pyruvate kinase," *Journal of Medicinal Chemistry*, vol. 53, no. 3, pp. 1048–1055, 2010.

[83] J. K. Jiang, M. B. Boxer, M. G. Vander Heiden et al., "Evaluation of thieno[3,2-b]pyrrole[3,2-d]pyridazinones as activators of the tumor cell specific M2 isoform of pyruvate kinase," *Bioorganic & Medicinal Chemistry Letters*, vol. 20, no. 11, pp. 3387–3393, 2010.

[84] C. Kung, J. Hixon, S. Choe et al., "Small molecule activation of PKM2 in cancer cells induces serine auxotrophy," *Chemical Biology*, vol. 19, no. 9, pp. 1187–1198, 2012.

[85] M. J. Walsh, K. R. Brimacombe, H. Veith et al., "2-Oxo-N-aryl-1,2,3,4-tetrahydroquinoline-6-sulfonamides as activators of the tumor cell specific M2 isoform of pyruvate kinase," *Bioorganic & Medicinal Chemistry Letters*, vol. 21, no. 21, pp. 6322–6327, 2011.

[86] A. Yacovan, R. Ozeri, T. Kehat et al., "1-(sulfonyl)-5-(arylsulfonyl)indoline as activators of the tumor cell specific M2 isoform of pyruvate kinase," *Bioorganic & Medicinal Chemistry Letters*, vol. 22, no. 20, pp. 6460–6468, 2012.

[87] M. G. Vander Heiden, H. R. Christofk, E. Schuman et al., "Identification of small molecule inhibitors of pyruvate kinase M2," *Biochemical Pharmacology*, vol. 79, no. 8, pp. 1118–1124, 2010.

[88] N. Shimada, T. Shinagawa, and S. Ishii, "Modulation of M2-type pyruvate kinase activity by the cytoplasmic PML tumor suppressor protein," *Genes to Cells*, vol. 13, no. 3, pp. 245–254, 2008.

[89] B. Varghese, G. Swaminathan, A. Plotnikov et al., "Prolactin inhibits activity of pyruvate kinase M2 to stimulate cell proliferation," *Molecular Endocrinology*, vol. 24, no. 12, pp. 2356–2365, 2010.

[90] M. Kosugi, R. Ahmad, M. Alam, Y. Uchida, and D. Kufe, "MUC1-C oncoprotein regulates glycolysis and pyruvate kinase M2 activity in cancer cells," *PLoS ONE*, vol. 6, no. 11, Article ID e28234, 2011.

[91] S. Mazurek, "Pyruvate kinase type M2: a key regulator within the tumour metabolome and a tool for metabolic profiling of tumours," *Ernst Schering Foundation symposium proceedings*, no. 4, pp. 99–124, 2007.

[92] H. F. Duan, X. W. Hu, J. L. Chen et al., "Antitumor activities of TEM8-Fc: an engineered antibody-like molecule targeting tumor endothelial marker 8," *Journal of the National Cancer Institute*, vol. 99, no. 20, pp. 1551–1555, 2007.

[93] K. H. Ibsen and S. W. Marles, "Inhibition of chicken pyruvate kinases by amino acids," *Biochemistry*, vol. 15, no. 5, pp. 1073–1079, 1976.

[94] L. Lv, D. Li, D. Zhao et al., "Acetylation targets the M2 isoform of pyruvate kinase for degradation through chaperone-mediated autophagy and promotes tumor growth," *Molecular Cell*, vol. 42, no. 6, pp. 719–730, 2011.

[95] A. N. Macintyre and J. C. Rathmell, "PKM2 and the tricky balance of growth and energy in cancer," *Molecular Cell*, vol. 42, no. 6, pp. 713–714, 2011.

[96] D. Anastasiou, G. Poulogiannis, J. M. Asara et al., "Inhibition of pyruvate kinase M2 by reactive oxygen species contributes to cellular antioxidant responses," *Science*, vol. 334, no. 6060, pp. 1278–1283, 2011.

[97] M. Zöller, "CD44: can a cancer-initiating cell profit from an abundantly expressed molecule?" *Nature Reviews Cancer*, vol. 11, no. 4, pp. 254–267, 2011.

[98] O. Nagano and H. Saya, "Mechanism and biological significance of CD44 cleavage," *Cancer Science*, vol. 95, no. 12, pp. 930–935, 2004.

[99] S. J. Wang and L. Y. W. Bourguignon, "Role of hyaluronan-mediated CD44 signaling in head and neck squamous cell carcinoma progression and chemoresistance," *The American Journal of Pathology*, vol. 178, no. 3, pp. 956–963, 2011.

[100] M. Tamada, O. Nagano, S. Tateyama et al., "Modulation of glucose metabolism by CD44 contributes to antioxidant status and drug resistance in cancer cells," *Cancer Research*, vol. 72, no. 6, pp. 1438–1448, 2012.

[101] A. Hoshino, J. A. Hirst, and H. Fujii, "Regulation of cell proliferation by interleukin-3-induced nuclear translocation of pyruvate kinase," *Journal of Biological Chemistry*, vol. 282, no. 24, pp. 17706–17711, 2007.

[102] A. Steták, R. Veress, J. Ovádi, P. Csermely, G. Kéri, and A. Ullrich, "Nuclear translocation of the tumor marker pyruvate kinase M2 induces programmed cell death," *Cancer Research*, vol. 67, no. 4, pp. 1602–1608, 2007.

[103] J. Lee, H. K. Kim, Y. M. Han, and J. Kim, "Pyruvate kinase isozyme type M2 (PKM2) interacts and cooperates with Oct-4 in regulating transcription," *International Journal of Biochemistry and Cell Biology*, vol. 40, no. 5, pp. 1043–1054, 2008.

[104] F. Canal and C. Perret, "PKM2: a new player in the β-catenin game," *Future Oncology*, vol. 8, no. 4, pp. 395–398, 2012.

[105] G. Semenova and J. Chernoff, "PKM2 enters the morpheein academy," *Molecular Cell*, vol. 45, no. 5, pp. 583–584, 2012.

[106] D. W. Parsons, S. Jones, X. Zhang et al., "An integrated genomic analysis of human glioblastoma multiforme," *Science*, vol. 321, no. 5897, pp. 1807–1812, 2008.

[107] H. Yan, D. W. Parsons, G. Jin et al., "IDH1 and IDH2 mutations in gliomas," *The New England Journal of Medicine*, vol. 360, no. 8, pp. 765–773, 2009.

[108] T. Bourgeron, P. Rustin, D. Chretien et al., "Mutation of a nuclear succinate dehydrogenase gene results in mitochondrial respiratory chain deficiency," *Nature Genetics*, vol. 11, no. 2, pp. 144–149, 1995.

[109] I. P. M. Tomlinson, N. A. Alam, A. J. Rowan et al., "Germline mutations in FH predispose to dominantly inherited uterine fibroids, skin leiomyomata and papillary renal cell cancer the multiple leiomyoma consortium," *Nature Genetics*, vol. 30, no. 4, pp. 406–410, 2002.

[110] B. A. Teicher, W. M. Linehan, and L. J. Helman, "Targeting cancer metabolism," *Clinical Cancer Research*, vol. 18, no. 20, pp. 5537–5545, 2012.

[111] P. Sonveaux, F. Végran, T. Schroeder et al., "Targeting lactate-fueled respiration selectively kills hypoxic tumor cells in mice," *Journal of Clinical Investigation*, vol. 118, no. 12, pp. 3930–3942, 2008.

[112] S. Dhup, R. K. Dadhich, P. E. Porporato, and P. Sonveaux, "Multiple biological activities of lactic acid in cancer: influences on tumor growth, angiogenesis, and metastasis," *Current Pharmaceutical Design*, vol. 18, no. 10, pp. 1319–1330, 2012.

[113] G. Bonuccelli, D. Whitaker-Menezes, R. Castello-Cros et al., "The reverse Warburg effect: glycolysis inhibitors prevent the tumor promoting effects of caveolin-1 deficient cancer associated fibroblasts," *Cell Cycle*, vol. 9, no. 10, pp. 1960–1971, 2010.

[114] B. Chiavarina, D. Whitaker-Menezes, U. E. Martinez-Outschoorn et al., "Pyruvate kinase expression (PKM1 and PKM2) in cancer-associated fibroblasts drives stromal nutrient production and tumor growth," *Cancer Biology and Therapy*, vol. 12, no. 12, pp. 1101–1113, 2011.

[115] I. Baccelli and A. Trumpp, "The evolving concept of cancer and metastasis stem cells," *Journal of Cell Biology*, vol. 198, no. 3, pp. 281–293, 2012.

[116] E. Vlashi, C. Lagadec, L. Vergnes et al., "Metabolic state of glioma stem cells and nontumorigenic cells," *Proceedings of the National Academy of Sciences of the United States of America*, vol. 108, no. 38, pp. 16062–16067, 2011.

Permissions

The contributors of this book come from diverse backgrounds, making this book a truly international effort. This book will bring forth new frontiers with its revolutionizing research information and detailed analysis of the nascent developments around the world.

We would like to thank all the contributing authors for lending their expertise to make the book truly unique. They have played a crucial role in the development of this book. Without their invaluable contributions this book wouldn't have been possible. They have made vital efforts to compile up to date information on the varied aspects of this subject to make this book a valuable addition to the collection of many professionals and students.

This book was conceptualized with the vision of imparting up-to-date information and advanced data in this field. To ensure the same, a matchless editorial board was set up. Every individual on the board went through rigorous rounds of assessment to prove their worth. After which they invested a large part of their time researching and compiling the most relevant data for our readers. Conferences and sessions were held from time to time between the editorial board and the contributing authors to present the data in the most comprehensible form. The editorial team has worked tirelessly to provide valuable and valid information to help people across the globe.

Every chapter published in this book has been scrutinized by our experts. Their significance has been extensively debated. The topics covered herein carry significant findings which will fuel the growth of the discipline. They may even be implemented as practical applications or may be referred to as a beginning point for another development. Chapters in this book were first published by Hindawi Publishing Corporation; hereby published with permission under the Creative Commons Attribution License or equivalent.

The editorial board has been involved in producing this book since its inception. They have spent rigorous hours researching and exploring the diverse topics which have resulted in the successful publishing of this book. They have passed on their knowledge of decades through this book. To expedite this challenging task, the publisher supported the team at every step. A small team of assistant editors was also appointed to further simplify the editing procedure and attain best results for the readers.

Our editorial team has been hand-picked from every corner of the world. Their multi-ethnicity adds dynamic inputs to the discussions which result in innovative outcomes. These outcomes are then further discussed with the researchers and contributors who give their valuable feedback and opinion regarding the same. The feedback is then collaborated with the researches and they are edited in a comprehensive manner to aid the understanding of the subject.

Apart from the editorial board, the designing team has also invested a significant amount of their time in understanding the subject and creating the most relevant covers. They scrutinized every image to scout for the most suitable representation of the subject and create an appropriate cover for the book.

The publishing team has been involved in this book since its early stages. They were actively engaged in every process, be it collecting the data, connecting with the contributors or procuring relevant information. The team has been an ardent support to the editorial, designing and production team. Their endless efforts to recruit the best for this project, has resulted in the accomplishment of this book. They are a veteran in the field of academics and their pool of knowledge is as vast as their experience in printing. Their expertise and guidance has proved useful at every step. Their uncompromising quality standards have made this book an exceptional effort. Their encouragement from time to time has been an inspiration for everyone.

The publisher and the editorial board hope that this book will prove to be a valuable piece of knowledge for researchers, students, practitioners and scholars across the globe.

List of Contributors

Silvia Chifflet
Departamento de Bioquimica, Facultad de Medicina, Universidad de la Republica, Gral. Flores 2125, 11800 Montevideo, Uruguay

Julio A. Hernandez
Seccion Biofisica, Facultad de Ciencias, Universidad de la Republica, Igua 4225 esq. Mataojo, 11400 Montevideo, Uruguay

Nic E. Savaskan, Michael Buchfelder and Ilker Y. Eyupoglu
Department of Neurosurgery, University of Erlangen-Nuremberg, Schwabachanlage 6, 91054 Erlangen, Germany

Gunter Fingerle-Rowson
Clinic I for Internal Medicine, University Hospital Cologne, Kerpener Straße 62, 50924 Cologne, Germany

Midori Umekawa and Daniel J. Klionsky
Life Sciences Institute, University of Michigan, Ann Arbor, MI 48109-2216, USA

Ioannis P. Nezis
Department of Biochemistry, Institute for Cancer Research, Oslo University Hospital, Montebello, 0310 Oslo, Norway
Centre for Cancer Biomedicine, Faculty of Medicine, University of Oslo, Montebello, 0310 Oslo, Norway
Laboratory of Cell Biology, Department of Biological Applications and Technologies, University of Ioannina, 45110 Ioannina, Greece

David A. Zacharias
Whitney Laboratory, Department of Neuroscience, University of Florida, St. Augustine, FL 32080, USA

Matthew Mullen and Sonia Lobo Planey
Department of Basic Sciences, The Commonwealth Medical College, Scranton, PA 18509, USA

Simone Cardaci
Department of Biology, University of Rome "Tor Vergata", Via della Ricerca Scientifica, 00133 Rome, Italy

Maria Rosa Ciriolo
IRCCS San Raffaele Pisana, Via di Val Cannuta, 00166 Rome, Italy
Department of Biology, University of Rome "Tor Vergata", Via della Ricerca Scientifica, 00133 Rome, Italy

Simon C. M. Kwok
ORTD, Albert Einstein Medical Center, 5501 Old York Road, Korman 214, Philadelphia, PA 19141-3098, USA

Chandra Somasundaram
Cardiovascular Disease Research Program, JLC-Biomedical/Biotechnology Research Institute, North Carolina Central University, Durham, NC 27707, USA
Research Division, Texas Nerve and Paralysis Institute, Houston, TX 77030, USA
Intron Pharmaceuticals, Houston, TX 77005, USA

Rahul K. Nath
Research Division, Texas Nerve and Paralysis Institute, Houston, TX 77030, USA
Intron Pharmaceuticals, Houston, TX 77005, USA

Richard D. Bukoski
Cardiovascular Disease Research Program, JLC-Biomedical/Biotechnology Research Institute, North Carolina Central University, Durham, NC 27707, USA

Debra I. Diz
Hypertension & Vascular Research Center, Wake Forest University School of Medicine, Winston-Salem, NC 27157, USA

Andrew M. Fukuda
Departments of Physiology, Loma Linda University School of Medicine, Loma Linda, CA 92354, USA

Romeo Cecchelli and Vincent Berezowski
Universite Lille Nord de France, 59000 Lille, France
UArtois, LBHE, EA 2465, 62300 Lens, France
IMPRT-IFR114, 59000 Lille, France

Jerome Badaut
Departments of Physiology, Loma Linda University School of Medicine, Loma Linda, CA 92354, USA
Departments of Pediatrics, Loma Linda University School of Medicine, Loma Linda, CA 92354, USA

Mario Mauthe and Tassula Proikas-Cezanne
Autophagy Laboratory, Interfaculty Institute for Cell Biology, Eberhard Karls University Tubingen, Auf der Morgenstelle 15, 72076 Tubingen, Germany

Wenqi Yu and Friedrich Gotz
Microbial Genetics, Interfaculty Institute for Microbiology and Infectious Medicine, University of Tubingen, 72076 Tubingen, Germany

Oleg Krut and Martin Kronke
Institute for Medical Microbiology, Immunology and Hygiene, University of Cologne, 50935 Cologne, Germany

Horst Robenek
Leibniz Institute for Arteriosclerosis Research, University of Munster, 48149 Munster, Germany

Eduardo Cebollero
Department of Cell Biology and Institute of Biomembranes, University Medical Centre Utrecht, 3584 CX Utrecht, The Netherlands
Department of Biochemistry and Cell Biology and Institute of Biomembranes, Utrecht University, 3508 TD Utrecht, The Netherlands

Fulvio Reggiori
Department of Cell Biology and Institute of Biomembranes, University Medical Centre Utrecht, 3584 CX Utrecht, The Netherlands

Claudine Kraft
Max F. Perutz Laboratories, University of Vienna, 1030 Vienna, Austria

Zackie Aktary and Manijeh Pasdar
Department of Cell Biology, University of Alberta, Edmonton, AB, Canada T6G 2H7

Katrin Juenemann and Eric A. Reits
Department of Cell Biology and Histology, Academic Medical Center, Meibergdreef 9, 1105 AZ Amsterdam, The Netherlands

Katie Marchbank, Roland G. Roberts, Ellen Solomon and Caroline A. Whitehouse
Department of Medical and Molecular Genetics, Kings College London, London SE1 9RT, UK

Sarah Waters
The Randall Division for Cell and Molecular Biophysics and Cardiovascular Division, British Heart Foundation Centre of Research Excellence, King's College London, London SE1 1UL, UK

Hiroto Shinomiya
Department of Immunology and Host Defenses, Ehime University Graduate School of Medicine, Toon, Ehime 791-0295, Japan

Rajeswari Jinka
Department of Biochemistry, Acharya Nagarjuna University, Guntur 522510, India

Renu Kapoor, Pavana Goury Sistla, T. Avinash Raj and Gopal Pande
Centre for Cellular and Molecular Biology, Council of Scientific and Industrial Research (CSIR), Uppal Road, Hyderabad 500 007, India

Fulvio Reggiori
Department of Cell Biology and Institute of Biomembranes, University Medical Center Utrecht, 3584 CX Utrecht, The Netherlands

Masaaki Komatsu
Protein Metabolism Project, Tokyo Metropolitan Institute of Medical Science, Tokyo 156-8506, Japan

Anne Simonsen
Department of Biochemistry, Institute of Basic Medical Sciences, Faculty of Medicine, University of Oslo, 0317 Oslo, Norway

Kim Finley
Bio Science Center, San Diego State University, San Diego, CA 921825, USA

Xiaodong Feng
Department of Clinical and Administrative Sciences, California Northstate University College of Pharmacy, Rancho Cordova, CA 95670, USA

Marcia G. Tonnesen
Department of Dermatology, School of Medicine, State University of New York at Stony Brook, Stony Brook, NY 11794, USA
Dermatology Section, Veterans Affairs Medical Center, Northport, NY 11768, USA

Shaker A. Mousa
Pharmaceutical Research Institute, Albany College of Pharmacy and Health Sciences, Albany, NY 12208, USA

Richard A. F. Clark
Center for Tissue Engineering, State University of New York at Stony Brook, Stony Brook, NY 11794, USA

Nicholas Wong, Jason De Melo and Damu Tang
Division of Nephrology, Department of Medicine, McMaster University, Hamilton, ON, Canada L8S 4L8
Division of Urology, Department of Surgery, McMaster University, Hamilton, ON, Canada L8S 4L8
Father Sean O'Sullivan Research Centre, St. Joseph's Healthcare Hamilton, Hamilton, ON, Canada L8N 4A6
The Hamilton Center for Kidney Research, St. Joseph's Healthcare Hamilton, Hamilton, ON, Canada L8N 4A6